Lecture Notes in Mathematics

Edited by A. Dold and B. Eckmann

1256

Pseudo-Differential Operators

Proceedings of a Conference held in Oberwolfach, February 2–8, 1986

Edited by H. O. Cordes, B. Gramsch and H. Widom

Springer-Verlag
Berlin Heidelberg New York London Paris Tokyo

Editors

Heinz O. Cordes
Department of Mathematics, University of California
Berkeley, CA 94720, USA

Bernhard Gramsch
Fachbereich 17 – Mathematik, Johannes Gutenberg-Universität
Postfach 3880, 6500 Mainz
Federal Republic of Germany

Harold Widom
Department of Mathematics, University of California
Santa Cruz, CA 95064, USA

Mathematics Subject Classification (1980): 22E65, 35L05, 47A55, 47G05, 58G15, 58G16

ISBN 3-540-17856-2 Springer-Verlag Berlin Heidelberg New York
ISBN 0-387-17856-2 Springer-Verlag New York Berlin Heidelberg

Printed in Germany

Printing and binding: Druckhaus Beltz, Hemsbach/Bergstr.
2146/3140-543210

PREFACE

Pseudodifferential operators, which originated as a powerful tool in the study of partial differential equations, have now become also a field of independent interest. This Oberwolfach conference presented the opportunity of bringing together many of the world's leading workers in pseudodifferential operators and their applications so that all could learn of and discuss the latest developments in the field. Thirty-five lectures were given on a wide variety of topics, including the following:

Nonlinear hyperbolic equations;

Pseudodifferential operator calculus and asymptotic expansions;

Operators on manifolds with singularities;

Boundary value problems;

Fourier integral operators, Toeplitz operators and index theory;

Fréchet algebras;

L^p boundedness.

The papers in this volume are, with a few exceptions, expanded versions of lectures given during the conference.

We take this opportunity to express our appreciation to all the participants in this exciting conference and to the Mathematisches Forschungsinstitut Oberwolfach who made available to us their outstanding facilities and generous hospitality.

We also thank E. Schrohe and F. Ali Mehmeti and the secretary U. Schack for their help.

Our thanks also go to Springer Verlag for accepting this volume for publication and for their kind cooperation.

H.O. Cordes

B. Gramsch

H. Widom

CONTRIBUTIONS

LIST OF PARTICIPANTS

M.R. ADAMS	Athens, USA
E. ALBRECHT	Saarbrücken, W. Germany
F. ALI MEHMETI	Mainz, W. Germany
R.M. BEALS	New Brunswick, USA
G. BENGEL	Münster, W. Germany
J. BRÜNING	Augsburg, W. Germany
L. COBURN	Buffalo, USA
H.O. CORDES	Berkeley, USA
M. COSTABEL	Darmstadt, W. Germany
M. DAUGE	Nantes, France
R.V. DUDUCHAVA	Tbilisi, USSR
K. ERKIP	Ankara, Turkey
D. FUJIWARA	Tokyo, Japan
D. GELLER	Stony Brook, USA
P. GODIN	Brussels, Belgium
B. GRAMSCH	Mainz, W. Germany
A. GRIGIS	Palaiseau, France
G. GRUBB	Copenhagen, Denmark
V. GUILLEMIN	Cambridge, USA
S. HANSEN	Paderborn, W. Germany
L. HÖRMANDER	Djursholm, Sweden
C. IWASAKI	Osaka, Japan
N. IWASAKI	Kyoto, Japan
K. KALB	Mainz, W. Germany
S. KIRO	Rehovot, Israel
M. LANGENBRUCH	Münster, W. Germany
P. LAUBIN	Liège, Belgium
O. LIESS	Bonn, W. Germany
K. LORENTZ	Mainz, W. Germany
E. MEISTER	Darmstadt, W. Germany
R. MELROSE	Cambridge, USA
R. MENNICKEN	Regensburg, W. Germany
H. MOSCOVICI	Columbus, USA
T. MURAMATU	Ibaraki, Japan
M. NAGASE	Osaka, Japan
H.J. PETZSCHE	Dortmund, W. Germany

D. ROBERT	Nantes, France
H. SCHRÖDER	Augsburg, W. Germany
E. SCHROHE	Mainz, W. Germany
B.-W. SCHULZE	Berlin, DDR
J. SJÖSTRAND	Lund, Sweden
H. SOHRAB	Towson, USA
F.O. SPECK	Darmstadt, W. Germany
H.G. TILLMANN	Münster, W. Germany
G. UHLMANN	Seattle, USA
D. VOGT	Wuppertal, W. Germany
W. WENDLAND	Darmstadt, W. Germany
H. WIDOM	Santa Cruz, USA
M. YAMAZAKI	Tokyo, Japan
S. ZELDITCH	Baltimore, USA

LECTURES GIVEN AT THE CONFERENCE

M.R. ADAMS	A Lie group structure for Fourier integral operators.
F. ALI MEHMETI	A characterization of a generalized C^∞-notion on nets.
M. BEALS	Reflection of transversal progressing waves in nonlinear strictly hyperbolic mixed problems.
J. BRÜNING	L^2 index theorems for regular singular problems.
L.A. COBURN	Toeplitz operators on the Segal-Bargman space.
H.O. Cordes	On Fréchet *-algebras of pseudodifferential operators.
A.K. ERKIP	Elliptic boundary value problems in the half space.
D. FUJIWARA	A remark on the Taniguchi-KumanoGo theorem of Fourier Integral operators.
D. GELLER	An analytic Weyl calculus and analysis on the Heisenberg group.
P. GODIN	Analytic regularity of uniformly stable shock fronts with analytic data.
B. GRAMSCH	On the Oka principle for some classes of pseudo-differential operators.
A. GRIGIS	On the asymptotics of gaps in Hill's equation.
G. GRUBB	Functional calculus of pseudo-differential boundary problems.
V. GUILLEMIN	The trace formula for vector bundles.
S. HANSEN	An Airy operator calculus.
L. HÖRMANDER	The lifespan of classical solutions of non-linear hyperbolic equations of second order.
C. IWASAKI	Pseudo-differential operators on Gevrey classes.
N. IWASAKI	Examples of effectively hyperbolic equations.
S. KIRO	On the global existence of real analytic solutions of linear partial differential equations.
P. LAUBIN	Second microlocalization and operators with involutive double characteristics.
R.B. MELROSE	Rings of pseudodifferential operators associated to boundaries, cones and cusps.
H. MOSCOVICI	Higher indices of elliptic operators.

T. MURAMATU	Estimates for the norm of pseudo-differential operators by means of Besov spaces.
M. NAGASE	Sufficient conditions for pseudo-differential operators to be L^p-bounded.
D. ROBERT	Weyl formula for pseudo-differential operators of negative order.
E. SCHROHE	Complex powers of elliptic pseudodifferential operators.
B.-W. SCHULZE	Pseudo-differential operators and Mellin expansions with conormal asymptotics.
J. SJÖSTRAND	Semiclassical resonances.
H.H. SOHRAB	A class of pseudo-differential C*-algebras.
F.-O. SPECK	Diffraction problems with impedance conditions.
G. UHLMANN	On an inverse boundary value problem.
W. WENDLAND	On the numerical solution of certain pseudodifferential equations.
H. WIDOM	Asymptotic expansions for pseudodifferential operators on bounded domains.
M. YAMAZAKI	Propagation of singularities of solutions to nonlinear partial differential equations.
S. ZELDITCH	Selberg trace formulae, pseudodifferential operators and equidistribution of closed geodesics.

INTERACTION OF RADIALLY SMOOTH NONLINEAR WAVES

Michael Beals*
Department of Mathematics
Rutgers University
New Brunswick, N.J. 08903

0. Introduction

We consider solutions to nonlinear wave equations of the form

$$\square u \equiv (\partial_n^2 - \sum_{i=1}^{n-1} \partial_i^2)u = f(x,u,Du) \tag{0.1}$$

on an open set $\mathcal{O} \subset \mathbb{R}^n$. Here f is assumed to be a smooth function of its arguments, Du stands for $(\partial_1 u,\cdots,\partial_n u)$, and u is assumed to be sufficiently smooth that the right hand side makes sense; for example, $u \in H^s(\mathcal{O})$, $s > \frac{n}{2} + 1$. If f is linear in (u,Du), singularities in the wave front set of u are well known to propagate along the null bicharacteristics for $\square$ (Hörmander [8]). In the nonlinear case, Bony [5] introduced microlocal techniques in the nonlinear setting to show that the propagation statement remains true for singularities of strength H^s, but that additional nonlinear singularities (of strength roughly H^{2s}) could in general appear. Rauch and Reed [11] demonstrated the absence of such anomalous singularities when $n = 2$. But in higher dimensions such singularities (in fact of strength roughly H^{3s}) do appear, as shown in Beals [1], [2].

Natural conditions on the solution u in the past or on the Cauchy data on $\{x_n = 0\}$ are known to prevent or to limit sharply

---

* American Mathematical Society Research Fellow
Alfred P. Sloan Research Fellow
Supported by NSF Grant #DMS-8603158

the appearance of these nonlinear singularities. A solution to (0.1) "conormal" in the past with respect to a smooth characteristic hypersurface (that is, infinitely differentiable in directions tangential to the hypersurface without loss of H^s smoothness) remains conormal in the future. The same holds for a pair of smooth characteristic hypersurfaces intersecting transversally in the future (Bony [6]). Weaker hypotheses than conormality, which essentially allow a reduction to the proof in the case $n = 2$, will also suffice to control the spreading of singularities; see Rauch-Reed [14], [15], Beals [3]. These results again apply to singularities which are essentially interacting pairwise. In the case of the interaction of three progressing waves in $\mathbb{R}^3$, singular along three transversally intersecting characteristic hyperplanes, Rauch and Reed [13] gave an example that demonstrated the appearance of a single nonlinear singularity on the surface of the light cone over the point of triple intersection. In Melrose-Ritter [9] and Bony [7] it is proved that this new singularity is the only one which occurs; solutions to (0.1) conormal in the past with respect to three smooth characteristic hypersurfaces intersecting transversally at 0 remain conormal in the future with respect to the family consisting of the hypersurfaces and the surface of the light cone with vertex 0. The techniques needed are considerably more intricate than the commutator argument used to handle a pair of hypersurfaces as in [6]. Essentially, the singular point must be blown up and a new linear propagation theorem in the new coordinates must be proved.

In this paper we give a simpler approach to the C^∞ regularity of a solution in the "flat" case, that is, with singularities in the past (or on the initial surface) on hyperplanes intersecting at the origin. Although full conormality with respect to the light cone over the origin is not obtained, it is shown under a natural weak hypothesis that a solution will be C^∞ inside that light cone

(Theorems 1.1 and 1.2). The hypothesis is on smoothness in the radial direction $\sum_{i=1}^{n} x_i \partial_i$, or the spatial radial direction $\sum_{i=1}^{n-1} x_i \partial_i$ in the case of the initial value problem. The proof involves a simple microlocal technique - reduction to a microlocally elliptic problem on the interior of the light cone. The conclusion applies in any number of dimensions.

In section 2, the main regularity theorems, finite propagation speed, and the known results for pairwise interactions are shown to yield the result that in $\mathbb{R}^3$ the only nonlinear singularity appearing after the interaction of three (or more) hyperplanes of conormal singularities is on the surface of the generated light cone (Theorem 2.1). More complicated (non-conormal) singularities are also allowed, along the lines of [14] and [3]. In particular, initial data "radially smooth" and possibly singular on a continuum of angles are shown to yield solutions to (0.1) with singularities propagating as in the linear case (Theorem 2.2).

Examples in $\mathbb{R}^4$ are considered in section 3. The geometry of the singular supports is complicated considerably by the lower dimensional interactions, for instance triple intersections, while the true 4-dimensional interaction is simple. It is shown that even the simplest genuine 4-dimensional problem, involving the interaction of singularities on four characteristic hyperplanes, will in general be expected to produce singularities on a dense subset of the complement of the light cone over the origin, while the solution of (0.1) remains C^∞ on the interior of that cone (Theorem 3.3).

The author would like to thank the Mittag-Leffler Institute for its hospitality during the time that part of the research for this paper was performed. Thanks also go to the American Mathematical Society and the Alfred P. Sloan Foundation for fellowship support.

Notation. On $\mathbb{R}^n$, $n \geq 3$, $\square$ will denote the wave operator $\partial_n^2 - \sum_{i=1}^{n-1} \partial_i^2$. Char $\square$ is the characteristic set $\{\xi \neq 0: \xi_1^2+\cdots+\xi_{n-1}^2 = \xi_n^2\}$. $\mathcal{O} \subset R^n$ will be an open set, divided into two pieces $\mathcal{O}^-$ and $\mathcal{O}^+$ by the initial surface $\{x_n = 0\}$. It is assumed that all backward characteristics for $\square$ from points in $\mathcal{O}^+$ pass through $\mathcal{O}^-$ (in other words, the "past" determines the "future"). $\tilde{\mathcal{O}} = \mathcal{O} \cap \{x_n = 0\}$, and $\tilde{\mathcal{O}}^+$ is an open set with compact closure contained in $\tilde{\mathcal{O}}$. $H^s(\mathcal{O})$ stands for the Sobolev space of functions locally in H^s on $\mathcal{O}$. If (x^0,ξ^0) $T^*(\mathbb{R}^n)/0$, $u \in H^s_{\text{microloc}}(x^0,\xi^0)$ means that $Pu \in H^s_{\text{loc}}(x^0)$ for all zero-order pseudodifferential operators P with conic support sufficiently near ξ^0. $WF(u)$ is the wave front set, as in [8].

§1. General Regularity Results

We wish to treat the "flat" case of interaction of progressing waves for the semilinear wave equation

(1.1) $\square u = f(x,u)$ on $\mathcal{O}$, or

(1.2) $\square u = f(x,u,Du)$ on $\mathcal{O}$.

Here f is assumed to be a smooth function of its arguments, but otherwise arbitrary. Thus we consider a solution conormal in the past with respect to a family of characteristic hyperplanes which simultaneously intersect, say at the origin. The equations for these surfaces are then given by $x_n = \omega \cdot x'$, $\omega \in S^{n-2}$. Since these surfaces are homogeneous, in particular the radial vector field

(1.3) $M = \sum_{i=1}^{n} x_i \partial_i$

is tangential to all of them. Because this vector field interacts with $\square$ in a particularly nice fashion, it seems natural to consider solutions of (1.1) or (1.2) where only differentiability in the direction of M is postulated. Note that this vector field is also tangential to the surface of the light cone $\{x_1^2+\cdots+x_{n-1}^2 = x_n^2\}$, which is the only location of additional nonlinear singularities in the case of conormal interaction of three or more hyperplanes as above.

If $s > \frac{n}{2}$ in the case of (1.1), or $s > \frac{n}{2} + 1$ for (1.2), the usual contraction mapping argument yields the local existence of solutions u to these equations with $u \in H^s(\mathcal{O})$ on appropriate sets $\mathcal{O}$.

Theorem 1.1. Let $u \in H^s(\mathcal{O})$, $s > \frac{n}{2}$ (respectively, $s > \frac{n}{2} + 1$) satisfy (1.1) (respectively (1.2)). Let M be the radial vector field and suppose that $M^j u \in H^s(\mathcal{O}^-)$ for all j. Then $u \in C^\infty(\{x_1^2+\cdots+x_{n-1}^2 < x_n^2\} \cap \mathcal{O}^+)$.

Proof. The commutator of $\square$ with M is $[\square, M] = 2\square$. Assume first that $\square u = f(x,u)$. Then by induction, $\square(M^j u) = f_j(x,u,\ldots M^j u)$ for smooth functions f_j, as long as applications of the chain rule are justified. To see that they are, suppose by induction that $M^{j-1} u \in H^s(\mathcal{O})$. Then

$$\square(M^j u) = M\square(M^{j-1}u) + 2\square(M^{j-1}u)$$

$$= Mf_{j-1}(x,u,\ldots,M^{j-1}u) + 2f_j(x,u,\ldots,M^{j-1}u).$$

The right-hand side is in $H^{s-1}(\mathcal{O})$ by Schauder's Lemma and the inductive hypothesis. Since $M^j u \in H^s(\mathcal{O}^-)$, the usual linear energy estimates (e.g., Taylor [17]) yield that $M^j u \in H^s(\mathcal{O})$.

We have thus established that $M^j u \in H^s(\mathcal{O})$ for all j. In particular, if $(x^0,\xi^0) \in WF(u)$, it follows that ξ^0 is in the hyperplane P_{x^0} perpendicular to the vector x^0, as long as $x^0 \neq 0$. Indeed, on a sufficiently small conic neighborhood of (x^0,ξ^0) with $\xi^0 \notin P_{x^0}$, M is a microlocally elliptic operator. Hence $M^j u \in H^s(\mathcal{O})$ implies that $u \in H^{s+j}_{\text{microloc}}(x^0,\xi^0)$.

Let $x^o \in \mathcal{O}$ satisfy $(x_1^o)^2+\cdots+(x_{n-1}^o)^2 < (x_n^o)^2$. Then $(\mathrm{Char}\,\square) \cap P_{x^o} = \emptyset$. Suppose inductively that $u \in H^r_{loc}(x^o)$, $r \geq s$. Then by Schauder's Lemma $f(x,u) \in H^r_{loc}(x^o)$. If $b(x,\xi)$ is a symbol of order 0 microlocally elliptic at (x^o,ξ^o) and with conic ξ-support sufficiently near to P_{x^o}, then $b(x,D)\square$ is microlocally elliptic at (x^o,ξ^o). Thus by elliptic regularity, $b(x,D)\square u = b(x,D)f(x,u) \in H^r_{loc}(x^o)$ implies that $b(x,D)u \in H^{r+2}_{loc}(x^o)$. On the other hand, if $\tilde{b}(x,\xi)$ is a symbol of order 0 with conic ξ-support away from P_{x^o}, then the statement above about $WF(u)$ yields that $\tilde{b}(x,D)u \in H^\infty_{loc}(x^o)$. Therefore by a microlocal partition of unity, $u \in H^{r+2}_{loc}(x^o)$. By repeating this bootstrap argument, $u \in C^\infty$ near x^o.

In the case $\square u = f(x,u,Du)$, as above we now have $\square(M^j u) = g_j(x,u,\cdots,DM^{j-1}u) + h_j(x,u,\cdots,DM^{j-1}u)\cdot DM^j u$ for smooth functions g_j, h_j. If by induction $M^{j-1}u \in H^s(\mathcal{O})$, then $\square(M^j u) - h_j \cdot D(M^j u) = g_j$, with $g_j, h_j \in H^{s-1}(\mathcal{O})$ by Schauder's Lemma, using $s > \frac{n}{2} + 1$. By linear energy estimates for semilinear operators with H^s coefficients (see e.g. Beals-Reed [4]), it follows that $M^j u \in H^s(\mathcal{O})$ since $M^j u \in H^s(\mathcal{O}^-)$. The rest of the proof is as above, except that now $u \in H^s_{loc}(x^o)$ implies that $f(x,u,Du) \in H^{s-1}_{loc}(x^o)$ so with $b(x,\xi)$ as above, $b(x,D)u \in H^{s+1}_{loc}(x^o)$. This improvement in smoothness allows the bootstrap argument to go through. //

An analogous conclusion applies to solutions of the initial value problem when the data is smooth in the direction of the spatial radial vector field

$$(1.4) \qquad \tilde{M} = \sum_{i=1}^{n-1} x_i \partial_i .$$

Note that if the initial displacement and velocity (u_0, u_1) are in $H^s(\tilde{\theta}) \times H^{s-1}(\tilde{\theta})$, and T is sufficiently small, then a solution u to the problem (1.1) (respectively (1.2)) with the given Cauchy data exists on $\tilde{\theta}^+ \times (0,T)$ as long as $s > \frac{n-1}{2}$ (respectively $s > \frac{n+1}{2}$). This solution satisfies

$$u \in C((0,T); H^s(\tilde{\theta}^+)) \cap C^1((0,T); H^{s-1}(\tilde{\theta}^+)).$$

<u>Theorem 1.2</u>. Let $(u_0, u_1) \in H^s(\tilde{\theta}) \times H^{s-1}(\tilde{\theta})$, $s > \frac{n+1}{2}$ (respectively $s > \frac{n+1}{2}$) and let u satisfy (1.1) (respectively (1.2)), with $u(x',0) = u_0$, $\partial_n u(x',0) = u_1$. Let $\tilde{M}$ be the spatial radial vector field and suppose that $(\tilde{M}^j u_0, \tilde{M}^j u_1) \in H^s(\tilde{\theta}) \times H^{s-1}(\tilde{\theta})$ for all j. Then $u \in C^\infty(\{x_1^2 + \ldots + x_{n-1}^2 < x_n^2\} \cap \{\tilde{\theta}^+ \times (0,T)\})$.

<u>Proof</u>. First we show by induction that

$(1.5)_j$ $\quad M^j u(x',0) = \tilde{M}^j u(x',0)$ (M as in (1.3)),

$(1.6)_j$ $\quad \partial_n (M^j u)(x',0) = p_j(\tilde{M}) \partial_n u(x',0)$, with p_j a polynomial in one variable of degree j, and

$(1.7)_j$ $\quad M^j u \in C((0,T); H^s(\tilde{\theta}^+)) \cap C^1((0,T); H^{s-1}(\tilde{\theta}^+))$.

These properties hold for $j = 0$ by the remarks above. If true for $j-1$, then

$$\begin{aligned} M^j u(x',0) &= ((\tilde{M} + x_n \partial_n) M^{j-1} u)(x',0) \\ &= (\tilde{M} M^{j-1} u)(x',0) \quad \text{since } M^{j-1} \partial_n u \in H^{s-1} \text{ by } (1.7)_{j-1} \\ &= \tilde{M}^j u(x',0) \quad \text{by } (1.5)_{j-1}. \end{aligned}$$

And $\quad (\partial_n M^j u)(x',0) = ((M+1) \partial_n M^{j-1} u)(x',0)$

$$= ((\tilde{M}+1)\partial_n M^{j-1}u)(x',0) \quad \text{from } (1.7)_{j-1} \text{ and the equation (1.1) or (1.2)}$$

$$= p_j(\tilde{M})\partial_n u(x',0) \quad \text{by } (1.6)_{j-1}.$$

Finally, $(1.7)_j$ follows from $(1.5)_j$, $(1.6)_j$, $(1.7)_{j-1}$, and the usual energy estimates, since $M^j u$ satisfies

$$\Box M^j u = f_j(x,u,\cdots,M^j u) \quad \text{or}$$

$$\Box M^j u = g_j(x,u,\cdots,DM^{j-1}u) + h_j(x,u,\cdots,DM^{j-1}u)\cdot DM^j u.$$

The remainder of the proof now proceeds exactly as that given for Theorem 1.1. //

It should be noted that the same method of proof yields analogous results when differentiability with respect to M (or $\tilde{M}$ in the case of the initial value problem) up to only finite order is postulated. If for example $M^j u \in H^s(\mathcal{O}^-)$ for $j \leq J$, then $M^j u \in H^s(\mathcal{O})$ for $j \leq J$ and it follows that $u \in H^{s+J}_{\text{microloc}}(x^0,\xi^0)$ for any (x^0,ξ^0) where M is microlocally elliptic. As above, one then concludes that $u \in H^{s+J}(\{x_1^2+\cdots+x_{n-1}^2 < x_n^2\} \cap \mathcal{O}^+)$. In this case one can also relax the hypotheses on the forcing functions $f(x,u)$ and $f(x,u,Du)$ to require only appropriate finite differentiability in their arguments.

§2. The interaction of plane waves in two space dimensions

The regularity results of the previous section, when combined with finite propagation speed and known results for simpler interactions, lead immediately to characterizations of the locations of singularities for solutions with special types of regularity. We first consider the theorem of Melrose-Ritter and Bony on the interaction of conormal waves. Recall that a distribution $u \in H^s_{\text{loc}}$ is called conormal with respect to the nonsingular hypersurface Σ if $M_1\cdots M_j u \in H^s_{\text{loc}}$ for all smooth vector fields $M_1,\ldots,M_j$

tangential to Σ. (If Σ is locally given by $x_1 = 0$, these vector fields are generated over C^∞ by $x_1\partial_1, \partial_2, \cdots, \partial_n$.) In [9] and [7] are introduced appropriate notions of conormality with respect to the variety generated by three transversally intersecting characteristic hypersurfaces and the surface of the light cone over the point of triple intersection. It is then proved that a solution conormal in the past remains conormal after the triple intersection; in particular it is smooth away from the union of the characteristic hypersurfaces and the surface of the light cone. We here recapture that smoothness result, though not full conormality at the light cone. Note that the result actually applies to any number of hypersurfaces greater than or equal to three which simultaneously intersect at a single point.

Theorem 2.1. Let $u \in H^s(\mathcal{O})$, $\mathcal{O} \subset \mathbb{R}^3$, $s > \frac{3}{2}$ (respectively $s > \frac{5}{2}$) be a solution to

(2.1) $\square u = f(x,u)$, respectively

(2.2) $\square u = f(x,u,Du)$.

Suppose that u is conormal with respect to $\Sigma = \{\Sigma_1, \Sigma_2, \Sigma_3\}$ in $\mathcal{O}-$, where Σ is a family of characteristic hyperplanes intersecting transversally at 0. Then

$$u \in C^\infty(\mathcal{O} \setminus \Sigma_1 \cup \Sigma_2 \cup \Sigma_3 \cup \{\sqrt{x_1^2+x_2^2} = x_3\}).$$

Proof. By finite propagation speed, the values of u on $\mathcal{O} \setminus \{\sqrt{x_1^2+x_2^2} < x_3\}$ are determined by those of u on a set where there are at most pairwise intersections of the hyperplanes in Σ. By the commutator argument of Bony [6], u is conormal with respect to Σ there, so in particular $u \in C^\infty(\mathcal{O} \setminus \Sigma_1 \cup \Sigma_2 \cup \Sigma_3 \cup \{\sqrt{x_1^2+x_2^2} < x_3\})$. On the other hand, $M = x_1\partial_1 + x_2\partial_2 + x_3\partial_3$ is simultaneously tangential to Σ_1, Σ_2 and Σ_3, so the hypotheses of Theorem 1.1 are satisfied and $u \in C^\infty(\{\sqrt{x_1^2+x_2^2} < x_3\} \cap \mathcal{O})$. //

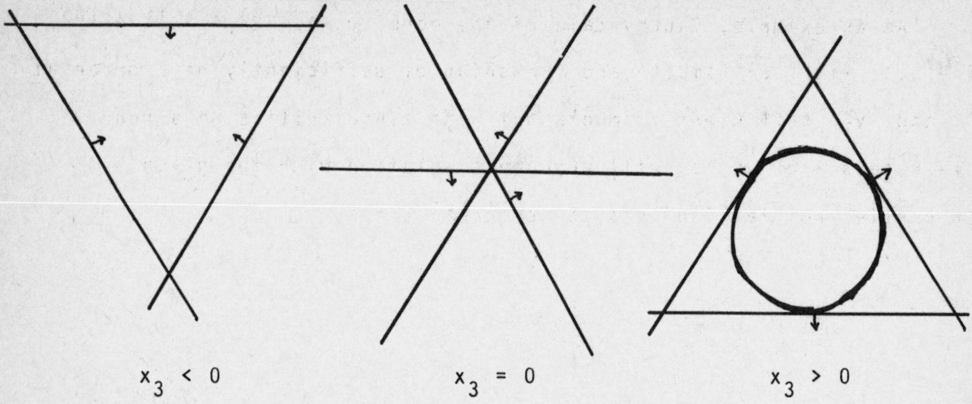

Figure 1

Time slices of the location of singularities for a solution to (2.1) or (2.2) under conormal hypotheses in the past.

Weaker assumptions than conormality still allow the conclusion that, with the exception of the surface of the light cone over the origin, the only singularities of the solution to the nonlinear problem are in the location of those for the corresponding linear problem. In particular, the singularities need not be located on a discrete set of hyperplanes. This result is the analogue of the striated case treated by Rauch-Reed [14], where appeared no degenerate point such as the origin here, and of the angularly smooth solutions in [3]. We state it in the context of the initial value problem, for which it seems most natural.

Theorem 2.2. Let $(u_0, u_1) \in H^s(\tilde{\theta}) \times H^{s-1}(\tilde{\theta})$, $\tilde{\theta} \subset \mathbb{R}^2$, $s > 1$ (respectively $s > 2$), and let u satisfy (2.1) (respectively (2.2)). Suppose that the data are radially smooth, in the sense that $(x_1\partial_1 + x_2\partial_2)^j u_0 \in H^s(\tilde{\theta})$ and $(x_1\partial_1 + x_2\partial_2)^j u_1 \in H^{s-1}(\tilde{\theta})$ for all j. Then on $\tilde{\theta}^+ \times (0,T)$,

$$\text{sing supp } u \subset \{x_1^2 + x_2^2 = x_3^2\} \cup \{(r\cos\theta \mp t\sin\theta, r\sin\theta \pm t\cos\theta, t) : (r\cos\theta, r\sin\theta) \in \text{sing supp}(u_0, u_1)\}.$$

As an example, Cauchy data of the form $u_i(x_1,x_2) = v_i(\theta)\phi_i(r)$, $i = 0,1$, with ϕ_i smooth and vanishing of sufficiently high order at 0 and v_i sufficiently smooth but with singularities on a non-discrete set of angles will produce a solution with the given properties not falling into the conormal class.

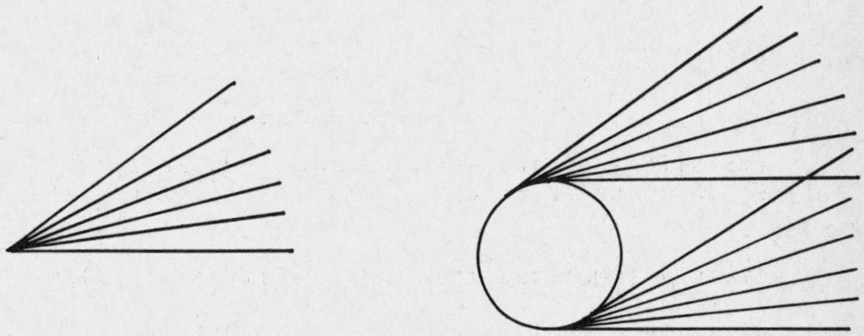

$x_3 = 0$ $\qquad\qquad$ $x_3 > 0$

Figure 2

Time slices of the location of singularities for radially smooth solutions to (2.1) or (2.2).

Proof. Again by Theorem 1.2, u is smooth on $\{\sqrt{x_1^2+x_2^2} < x_3\}$. Outside of this set, by finite propagation speed, the values of u are determined by those of (u_0,u_1) away from the origin. Such data satisfy the "striated" hypothesis of Rauch-Reed [14], and outside of the light cone the two families of characteristic planes generated by this stratification give nondegenerate foliations of space-time. The results of [14] then apply to yield the desired conclusion. But to see this fact directly in the present situation, and to illustrate the philosophy of treating the radial variable as a parameter, we instead follow the argument of [3] with the roles of θ and r reversed. On $\{\sqrt{x_1^2+x_2^2} > x_3 > 0\}$ we look for a change of coordinates after which the d'Alembertian can be written as the sum of an operator involving two variables and an operating involving M. Thus let $x_1 = rs$, $x_2 = rt$,

$x_3 = r\,c(s,t)$, with c to be determined. This form is chosen so that $\partial_r = s\partial_1 + t\partial_2 + c\partial_3 = \frac{1}{r}M$ is a multiple of the radial vector field. Since $\partial_s = r\partial_1 + rc_s\partial_2$ and $\partial_t = r\partial_2 + sc_t\,\partial_3$, we have

$$(\partial_1 r, \partial_2 r, \partial_3 r) = \frac{r}{D}(-rc_s, -sc_t, s),$$

$$(\partial_1 s, \partial_2 s, \partial_3 s) = \frac{r}{D}(c - tc_t, sc_t, -s),$$

$$(\partial_1 t, \partial_2 t, \partial_3 t) = \frac{r}{D}(tc_s, c - sc_s, -t).$$

Here D is the Jacobian, $D = r^2(c - sc_s - tc_t)$.

In the expression for $\square$ in the new variables, the coefficients of $\partial_r\partial_s$ and $\partial_r\partial_t$, which we wish to vanish, are $2(\partial_3 r\partial_3 s - \partial_1 r\partial_1 s - \partial_2 r\partial_2 s)$ and $2(\partial_3 r\partial_3 t - \partial_1 r\partial_1 t - \partial_2 r\partial_2 t)$. It follows that we require

(2.3) $\quad s = c_s(c - tc_t) + sc_t^2, \quad t = tc_s^2 + c_t(c - sc_s).$

Hence $\quad t(cc_s - tc_sc_t + sc_t^2) = s(cc_t - sc_sc_t + tc_s^2),$

$$tc_s(c - tc_t - sc_s) = sc_t(c - sc_s - tc_t).$$

Thus (for non-vanishing Jacobian) we require $tc_s = sc_t$, and hence from (2.3), $s = cc_s$, $t = cc_t$. In other words, $c^2 = s^2+t^2+k$ for some constant k. In this case, $c - sc_s - tc_t = \dfrac{k}{\sqrt{s^2+t^2+k}}$, so we choose $k \neq 0$. Since we are interested in $x_1^2 + x_2^2 > x_3^2$, we choose without loss of generality $c = \sqrt{s^2+t^2-1}$.

After this change of variables, it is easily verified that modulo lower order terms, $\square = -\partial_r^2 + \frac{1}{r^2}\{(s^2-1)\partial_s^2 + (t^2-1)\partial_t^2 + 2st\partial_s\partial_t\}$. Thus the equation $\square u = f(x,u,Du)$ may be written on $r > 0$ as

$$\{(s^2-1)\partial_s^2 + (t^2-1)\partial_t^2 + 2st\,\partial_s\partial_t\}u = g(x,u,Du,u_{rr}).$$

Given our known smoothness of u_{rr} (from the Proof of Theorem 1.2), it follows as in [3] that singularities propagate as in the case of two dimensions, with r as a parameter. Since by Rauch-Reed [11] there are no new nonlinear singularities for a second order equation in one space dimension, this means that singularities propagate along the characteristics for

$$(2.4)\quad (s^2-1)\partial_s^2 + (t^2-1)\partial_t^2 + 2st\,\partial_s\partial_t .$$

To find these curves explicitly, first eliminate the cross term by changing to polar coordinates: $s = a\cos b$, $t = a\sin b$. Then modulo lower order terms, (2.4) becomes $(a^2-1)\partial_a^2 - \frac{1}{a^2}\partial_b^2$, which we replace with $a^2(a^2-1)\partial_a^2 - \partial_b^2$ on $a > 0$. If $\tilde{a} = \operatorname{arcsec} a$, then the principal part of this operator in the coordinates $(\tilde{a},b)$ is $\partial_{\tilde{a}}^2 - \partial_b^2$, and thus characteristics (parametrized by α) are lines of the form $\tilde{a} = \pm b + \alpha$. In the original coordinates, these curves are

$$(2.5)\quad \frac{\sqrt{x_1^2+x_2^2}}{\sqrt{x_1^2+x_2^2+x_3^2}} = \sec(\pm \arctan \frac{x_2}{x_1} + \alpha).$$

At $x_3 = 0$, (2.5) becomes $\alpha = \mp \arctan \frac{x_2}{x_1}$. Thus for the family of characteristics through $(r\cos\theta, r\sin\theta, 0)$, we have $\alpha = \mp\theta$. We are looking at $\sqrt{x_1^2+x_2^2-x_3^2} = r$ constant. Set $t = x_3 = \sqrt{x_1^2+x_2^2-r^2}$. Then

$$\operatorname{arcsec}\left(\frac{\sqrt{t^2+r^2}}{r}\right) = \pm(\arctan\frac{x_2}{x_1} - \theta), \text{ so } \arctan\frac{t}{r} = \pm(\arctan\frac{x_2}{x_1} - \theta).$$

Finally, we have

$$\frac{x_2}{x_1} = \frac{\pm\frac{t}{r} + \tan\theta}{1 \mp \frac{t}{r}\tan\theta} = \frac{\pm t\cos\theta + r\sin\theta}{r\cos\theta \mp t\sin\theta}.$$

It follows that the characteristics through $(r\cos\theta, r\sin\theta, 0)$ are $(r\cos\theta \mp t\sin\theta,\ r\sin\theta \pm t\cos\theta,\ t)$.

An example to which Theorem 2.2 applies would be the problem with initial data of the form $x_1^\alpha x_2^\beta$ with α,β non-integer and sufficiently large. This is the simplest singular structure for initial conditions that are genuinely two dimensional. Singularities for a solution of $\Box u = f(x,u,Du)$ with this data are then contained in the sets as shown in Figure 3.

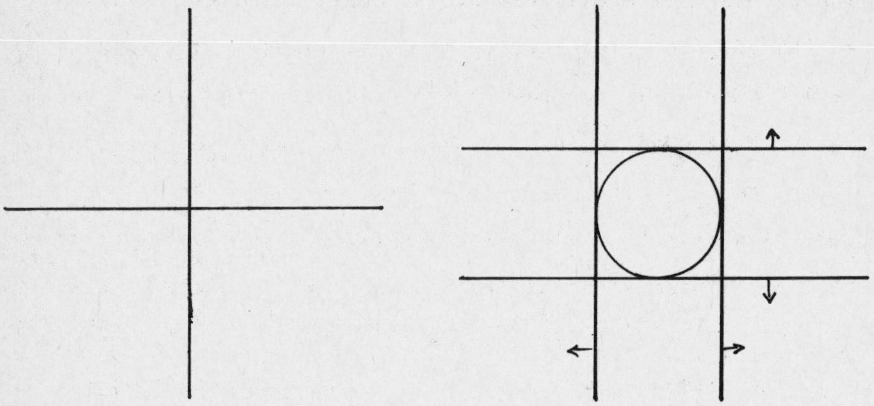

$x_3 = 0$ $x_3 > 0$

Figure 3

Singularities of the solution to a natural initial value problem.

§3. Interaction in three space dimensions

The results of section 1 apply equally well to higher dimensions, for example to the interaction in $\mathbb{R}^n$ of solutions conormal in the past with respect to a family of n hyperplanes intersecting transversally at the origin. The only true n-dimensional interaction occurs at the origin, producing (possible) singularities on the surface of the forward light cone over the origin, but leaving the solution smooth inside that cone. Outside, by finite propagation speed, the simultaneous n-interaction is not observed. As in section 2, after the nonsingular change of variables $x' = rs'$, $s' \in \mathbb{R}^{n-1}$, $|s'| > 1$, $x_n = r\sqrt{|s'|^2-1}$ on this set, the problem is reduced to an $(n-1)$-dimensional one, with r appearing as a parameter and u_{rr} incorporated into $f(x,u,Du)$.

On the other hand, even for the simplest case of four planes in three space dimensions, these lower dimensional interactions themselves produce a much more complicated picture than in lower dimensions. For example, before the quadruple interaction at the origin, four planes in $\mathbb{R}^4$ in general position will intersect in threes along four distinct lines. Consider the variable in the direction of one of these lines as a parameter; then the triple intersection of three singularity-carrying planes in the remaining three variables will in general produce singularities along the surface of a half-cone. Thus in $\mathbb{R}^4$ the singular set will in general include half-conoids of the form (surface of half-cone in $\mathbb{R}^3$) × $\mathbb{R}$. Singularities along these sets will interact with the others and with those along the hyperplanes; when triple intersections occur, later generation singularities will appear. (For such phenomena in one-space dimension for a higher order equation see Rauch-Reed [12]; in two-space dimensions with more complicated initial singularities see Melrose-Ritter [10].) Such interactions, even of more than three surfaces simultaneously, remain a lower dimensional phenomenon. Geometrically this property is demonstrated by the fact that, by the homogeneity of all the surfaces involved, all multiple intersections occur along lines, except for the bang at the origin taken care of by Theorem 1.1 or 1.2.

We examine two examples in $\mathbb{R}^4$ in detail. For one we give the natural conormal hypotheses in the past; the interactions are minimal in the past, but very complicated after time zero. In the other we consider the simplest truly 4-dimensional initial value problem - data yielding a solution in the linear case conormal with respect to four characteristic hyperplanes - and show that the smallest set of singularities given by the algorithm for triple interactions is dense in the exterior of the light cone over the origin.

We consider a solution on $\mathcal{O} \subset \mathbb{R}^4$ to

(3.1) $\Box u = f(x,u,Du)$, $u \in H^s(\mathcal{O})$, $s > \frac{5}{2}$, satisfying

(3.2) u is conormal in $\mathcal{O}^-$ with respect to the family $\Sigma = \{\Sigma_1, \Sigma_2,$

$\Sigma_3, \Sigma_4\}$ of characteristic hyperplanes $\Sigma_i = \{x_i - x_4 = 0\}$, $i = 1,2,3$, $\Sigma_4 = \{x_1 + x_2 + x_3 + \sqrt{3}\, x_4 = 0\}$. Due to the triple interactions occurring in $\{x_4 < 0\}$, this family is not the natural one for a solution to the nonlinear problem. It should be enlarged to include the four half-conoids H_1, H_2, H_3, H_4 from the lines of triple intersection. Thus we replace (3.2) with

(3.3) u is conormal in Θ^- with respect to the family $\Sigma = \{\Sigma_1, \Sigma_2, \Sigma_3, \Sigma_4, H_1, H_2, H_3, H_4\}$. For (varying) definitions of conormality with respect to such intersecting planes and cones, see Melrose-Ritter [9], [10], Bony [7]. The equations of the H_i will be computed below; given the time orientation they will occur in the octants $\{x_1 < x_4, x_2 < x_4, x_1 + x_2 + x_3 < -\sqrt{3}\, x_4\}$, the other two with (x_1, x_2, x_3) permuted, and $\{x_1 < x_4, x_2 < x_4, x_3 < x_4\}$. For $x_4 < 0$ these sets are nonintersecting, so there the only nonlinear singularities (by [9], [7], or Theorem 1.2 above, with an extra parameter) will be contained in the four half-conoids. See Figure 4 for a time slice; by homogeneity the picture remains unchanged for $x_4 < 0$.

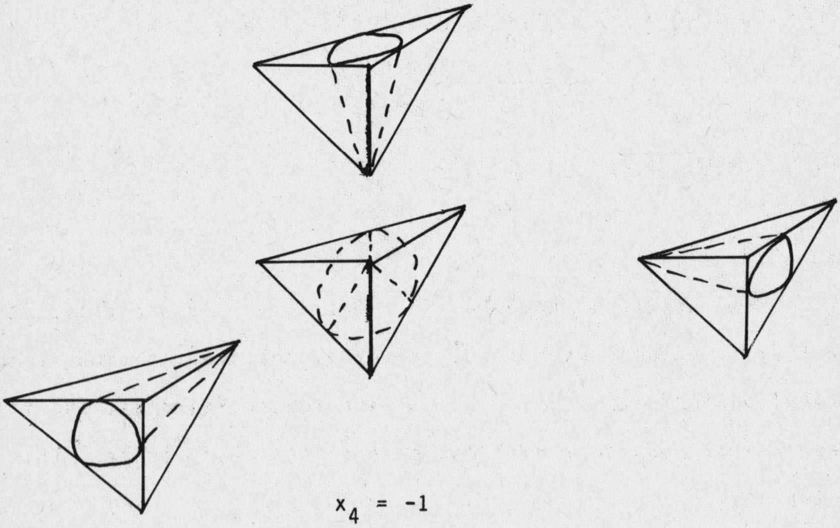

$x_4 = -1$

Figure 4

The set of singularities in $x_4 < 0$ for a solution satisfying (3.1) and (3.3).

Note that, given a line in $\mathbb{R}^4$ which is the intersection of three characteristic hypersurfaces for $\Box$ in general position, the conoid resulting from the triple intersection is uniquely determined (independent of the hypersurfaces). Indeed, without loss of generality the line is $\{x_1 = 0,\ x_2 = 0,\ x_3 + x_4 = 0\}$ and the hypersurfaces are the hyperplanes $\sum_{j=1}^{3} A_{ij}x_j + x_4 = 0$, $i = 1,2,3$. If the line is the intersection, then $A_{i3} = \alpha$, $i = 1,2,3$. If the planes are characteristic, then $\sum_j A_{ij}^2 = 1$, so we must have $|\alpha| < 1$. After an orthogonal transformation in (x_1, x_2) we can take $A_{31} = \sqrt{1-\alpha^2}$, $A_{32} = 0$, and $A_{22} = \sqrt{1-A_{12}^2 - \alpha^2}$; without loss of generality $A_{12} = \sqrt{1-a_{11}^2-\alpha^2}$. To find the generated conoid we transform to coordinates in which one of the variables appears as a parameter. In this case it is simplest to choose

$$a = \gamma_1\ (A_{11}x_1 + \sqrt{1 - A_{11}^2 - \alpha^2}\ x_2 + \alpha x_3 + x_4)$$

$$b = \gamma_2\ (A_{21},x_1 + \sqrt{1 - A_{21}^2 - \alpha^2}\ x_2 + \alpha x_3 + x_4)$$

$$c = \gamma_3\ (\sqrt{1 - \alpha^2}\ x_1 + \alpha x_3 + x_4)$$

$$d = x_4.$$

The γ_i will be chosen momentarily. Since ∂_d is tangential to the three hyperplanes $a = 0$, $b = 0$, $c = 0$, it suffices to compute the principal part of $\Box$ modulo ∂_d in the new coordinates, and find the corresponding light cone over the origin in (a,b,c) space. This operator is

$$\Box \sim 2\{\gamma_1\gamma_2(1 - A_{11}A_{21} - \sqrt{(1-A_{11}^2 - \alpha^2)(1-A_{21}^2 - \alpha^2)} - \alpha^2)\partial_a\partial_b$$

$$+ \gamma_1\gamma_3(1-A_{11}\sqrt{1-\alpha^2} - \alpha^2(\partial_a\partial_c + \gamma_2\gamma_3(1-A_{21}\sqrt{1-\alpha 2} - \alpha^2)\partial_\alpha\partial_c\}.$$

Thus the choice $\gamma_3 = 1 - A_{11}A_{21} - \sqrt{(1-A_{11}^2 - \alpha^2)(1-A_{21}^2 - \alpha^2)} - \alpha^2$,

$\gamma_2 = 1 - A_{11}\sqrt{1-\alpha^2} - \alpha^2$, $\gamma_1 = 1-A_{21}\sqrt{1-\alpha^2} - \alpha^2$ yields

$\square \sim 2\gamma_1\gamma_2\gamma_3(\partial_a\partial_b + \partial_a\partial_c + \partial_b\partial_c)$. The corresponding dual cone is

$\{0 = a^2 + b^2 + c^2 - 2ab - 2ac - 2bc\}$. After a great deal of algebraic simplification, it follows that the conoid in the original coordinates is

(3.4) $\{0 = (1-\alpha^2)(x_1^2 + x_2^2) - (\alpha x_3 + x_4)^2\}$, independent of A_{11} and A_{21} as expected.

More generally, given the line $\{x_1 = Ax_4, x_2 = Bx_4, x_3 = Cx_4\}$, we can compute the equation of the generated conoid H from (3.4), after a rotation (in $\mathbb{R}^3$) into the line $\tilde{x}_1 = 0$, $\tilde{x}_2 = 0$, $\alpha\tilde{x}_3 + x_4 = 0$. Since this change requires $\alpha = \frac{-1}{\sqrt{A^2+B^2+C^2}}$, the condition $|\alpha| < 1$ becomes $A^2 + B^2 + C^2 > 1$. The transformation

$$\tilde{x}_1 = -\frac{b}{\sqrt{a^2+b^2}}x_1 + \frac{a}{\sqrt{a^2+b^2}}x_2,$$

$$\tilde{x}_2 = \frac{ac}{\sqrt{a^2+b^2}}x_1 + \frac{bc}{\sqrt{a^2+b^2}}x_2 - \frac{a^2+b^2}{\sqrt{a^2+b^2}}x_3$$

$$\tilde{x}_3 = ax_1 + bx_2 + cx_3,$$

where $a = \frac{A}{\sqrt{A^2+B^2+C^2}}$ and similarly for b, c, satisfies this requirement. From (3.4) and algebraic simplification comes the following characterization.

Lemma 3.1. The conoid generated by the triple interaction of three characteristic hypersurfaces for $\square$ along the line $\{x_1 = Ax_4, x_2 = Bx_4, x_3 = Cx_4\}$, $A^2 + B^2 + C^2 > 1$, has the form

$$(3.5)\quad H = \{0 = (B^2 + C^2 - 1)x_1^2 + (A^2 + C^2 - 1)x_2^2 + (A^2 + B^2 - 1)x_3^2$$

$$- (A^2 + B^2 + C^2)x_4^2 - 2(ABx_1x_2 + ACx_1x_3 + BCx_2x_3)$$

$$+ 2x_4(Ax_1 + Bx_2 + Cx_3)\}.$$

A moment's thought about the unit speed of propagation of light indicates that such a conoid should lie in the exterior of the light cone over the origin in $\mathbb{R}^4$, except at places where it grazes that cone. In order to verify this property, set $R = x_1^2 + x_2^2 + x_3^2$, $S = Ax_1 + Bx_2 + Cx_3$, and $D = A^2 + B^2 + C^2 \geqslant 1$. Then from (3.5),

$$(3.6)\quad x_4 = \frac{1}{D}(S \pm \sqrt{(D-1)(DR^2 - S^2)}) \quad \text{on } H.$$

We are interested in the relationship with the light cone $\{R^2 - x_4^2 = 0\}$. On H, from (3.6),

$$D^2(R^2 - x_4^2) = [(D-1)S^2 + (DR^2 - S^2)] \mp 2S\sqrt{(D-1)(DR^2 - S^2)}$$

$$= (S\sqrt{D-1} \mp \sqrt{DR^2 - S^2})^2.$$

Since $S^2 \leqslant (A^2 + B^2 + C^2)(x_1^2 + x_2^2 + x_3^2) = DR^2$, it follows that the left hand side is nonnegative. Hence $x_1^2 + x_2^2 + x_3^2 - x_4^2 \geqslant 0$ on H, and the intersection of H with $x_1^2 + x_2^2 + x_3^2 = x_4^2$ is $\{S = R\}$, that is, $x_1^2 + x_2^2 + x_3^2 = (Ax_1 + Bx_2 + Cx_3)^2$. Since $A^2 + B^2 + C^2 \geqslant 1$, the intersection does occur. See Figure 5: on a time-slice $x_4 = C$, it is the latitude on the sphere $x_1^2 + x_2^2 + x_3^2 = c^2$ given by the intersection with the plane $Ax_1 + Bx_j + Cx_3 = c$.

For the hyperplanes as in (3.2), the half-conoids in (3.3) are given by (3.5) as follows:

$$H_1 = \{0 = (7+4\sqrt{3})(x_1^2+x_2^2) + x_3^2 - (9+4\sqrt{3})x_4^2$$

(3.6) $$-2(x_1x_3 - (2+\sqrt{3})(x_1x_3 + x_2x_3)) + 2x_4(x_1+ x_2 - (2+\sqrt{3})x_3),$$

$$x_1 < x_4,\ x_2 < x_4\};$$

H_2 and H_3 as in H_1 with (x_1,x_2,x_3) permuted;

$$H_4 = \{0 = x_1^2 + x_2^2 + x_3^2 - 3x_4^2 - 2(x_1x_2 + x_1x_3 + x_2x_3)$$

$$+ 2x_4(x_1 + x_2 + x_3),\ x_1 < x_4,\ x_2 < x_4,\ x_3 < x_4\}.$$

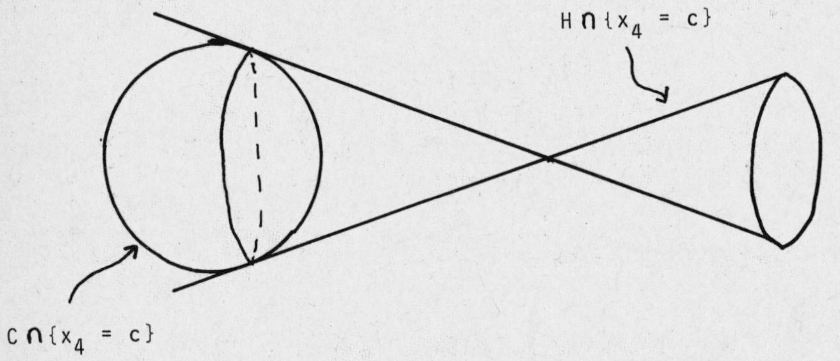

Figure 5

For fixed x_4, the projection on $\mathbb{R}^3$ of the surface of the light cone C and one of the conoids H.

In $x_4 > 0$ there are additional triple intersections among the sets Σ_i, H_j, i,j = 1,2,3,4. The simplest example is $H_1 \cap H_2 \cap H_3$, which by symmetry includes intersections in the set $\{x_1 = x_2 = x_3\}$. From (3.6), these intersections are the half-lines $\{x_1 = x_2 = x_3 = (-8+5\sqrt{3})x_4, x_4 > 0\}$. In general, new half-conoids of singularities determined by these (and many other) lines of triple interaction will appear in solutions to (3.1), (3.3). On the other hand, it is expected that these and later generation singularities are the only ones arising in the conormal case (see Melrose-Ritter [10]; the only point to be proved concerns the nature of the conormal singularity at points of tangency of the generated cone and its generating hypersurfaces). In any case, the later generation singularities are weaker than the original H^s. Since these interactions take place outside the light cone, and Theorem 1.1 handles the region $x_1^2 + x_2^2 + x_3^2 < x_4^2$, the overall smoothness picture is in principle fairly well understood.

Rather than analyze subsequent interactions in the example above, we turn to the symmetric picture in $\mathbb{R}^4$ in order to demonstrate the order of complication outside the light cone. After the linear transformation of $\mathbb{R}^4$ given by

$$w = x_1 + x_4,\ x = -\frac{1}{3}x_1 + \frac{\sqrt{8}}{3}x_2 + x_4,$$

$$y = -\frac{1}{3}x_1 - \frac{\sqrt{2}}{3}x_2 + \frac{\sqrt{6}}{3}x_3 + x_4,\ z = -\frac{1}{3}x_1 - \frac{\sqrt{2}}{3}x_2 - \frac{\sqrt{6}}{3}x_3 + x_4,$$

$\Box$ is transformed to a multiple of

$$(3.7)\quad \tilde{\Box} = \partial_w\partial_x + \partial_w\partial_y + \partial_w\partial_z + \partial_x\partial_y + \partial_x\partial_z + \partial_y\partial_z .$$

The surface of the light cone over the origin determined by $\tilde{\Box}$ is then given by

$$(3.8)\quad \tilde{C} = \{0 = w^2 + x^2 + y^2 + z^2 - (wx + wy + wz + xy + xz + yz)\}.$$

Since (1,1,1,1) is a time-like vector, the interior of the light cone is $\{w^2 + x^2 + y^2 + z^2 - (wx + wy + wz + xy + xz + yz) < 0\}$.

The four hyperplanes $w = 0$, $x = 0$, $y = 0$, and $z = 0$ are characteristic for $\tilde{\Box}$. Consider an initial value problem with data given along some space-like hyperplane such that the resulting solution of the linearized problem $\tilde{\Box} u = 0$ is conormal with respect to each of these four hyperplanes and singular on each. Since forward and backward time are not distinguished, triple interactions in the nonlinear problem will in general generate singularities on full conoids as in Lemma 3.1, rather than half-conoids as in the example above. Using the inverse of the transformation given above, and after a tedious amount of algebra, we find the following description in symmetric coordinates.

Lemma 3.2. The conoid generated by the triple interaction of three characteristic hypersurfaces for $\tilde{\Box}$ along the line $\{w = at, x = bt, y = ct, z = dt\}$, $a^2 + b^2 + c^2 + d^2 - (ab + ac + ad + bc + bd + cd) \geq 0$, has following form:

$$\begin{aligned}\tilde{H} = \{0 = {} & [b^2 + c^2 + d^2 - 2(bc+bd+cd)]w^2 + [a^2 + c^2 + d^2 - 2(ac+ad+cd)]x^2 \\ & + [a^2 + b^2 + d^2 - 2(ab+ad+bd)]y^2 + [a^2 + b^2 + c^2 - 2(ab+ac+ad)]z^2 \\ & - 2[c^2 + d^2 + ab - (a+b)(c+d)]wx - 2[b^2 + d^2 + ac - (a+c)(b+d)]wy \\ & - 2[b^2 + c^2 + ad - (a+d)(b+c)]wz - 2[a^2 + d^2 + bc - (a+d)(b+c)]xy \\ & - 2[a^2 + d^2 + bc - (a+c)(b+d)]xz - 2[a^2 + b^2 + cd - (a+b)(c+d)]yz\}.\end{aligned}$$

The four conoids generated by the four sets of triple interactions among the hyperplanes $w = 0$, $x = 0$, $y = 0$, $z = 0$, are thus the sets $H_1 = \{w^2 + x^2 + y^2 - 2wx - 2wy - 2xy = 0\}$, and H_2, H_3, H_4 given by permutations of (w,x,y,z). Up to permutations, the following lines are all of those obtained in the next generation of interactions (of the form (plane, plane, conoid), (plane, conoid, conoid), and (conoid, conoid, conoid)):

$$(3.9) \quad \{x = 0, y = 0, z = w\}, \{w = 0, y = x, z = x\}, \{w = x, y = 0, z = 4x\}, \{w = 4x, y = x, z = x\}, \{w = 4y, x = 9y, z = y\}.$$

Clearly the picture quickly becomes very complicated. We consider perhaps the simplest (because of symmetry) sequence of successive interactions.

Index the line $\{w = at,\ x = bt,\ y = ct,\ z = dt\}$ by the ordered quadruple (a,b,c,d), and note that $(1,\alpha,1,1)$ with $\alpha = 4$ is one of the lines obtained above. From Lemma 3.2, the corresponding conoid is

$$(3.10)\quad \tilde{H}(1,\alpha,1,1) = \{0 = (\alpha^2-4\alpha)(w^2+y^2+z^2) - 3x^2 \\ - 2(\alpha^2-2\alpha)(wy+wz+yz) + 2\alpha(wx+xy+xz)\}.$$

Permute the position of the α, and examine the four corresponding interactions of triples of conoids. For example, $\tilde{H}(1,\alpha,1,1) \cap \tilde{H}(1,1,\alpha,1) \cap \tilde{H}(1,1,1,\alpha)$ meets the set $\{x = y = z\}$ by symmetry, and from (3.10), $0 = (\alpha^2-4\alpha)w^2-2(2\alpha^2-5\alpha)wx - 3x^2$ on that intersection. This yields $\{y = x,\ z = x,\ w = [\frac{2\alpha^2-5\alpha\pm2(\alpha-1)\sqrt{\alpha(\alpha-3)}}{\alpha^2 - 4\alpha}]x\}$ as one of the lines of triple intersection. In other words, from the conoids generated by the lines $\{(\alpha,1,1,1),\ (1,\alpha,1,1),\ (1,1,\alpha,1),\ (1,1,1,\alpha)\}$ are obtained the new set of lines $\{(\tilde{\alpha},1,1,1),\ (1,\tilde{\alpha},1,1),\ (1,1,\tilde{\alpha},1),\ (1,1,1,\tilde{\alpha})\}$ with

$$(3.11)\quad \tilde{\alpha} = \frac{2\alpha^2-5\alpha\pm2(\alpha-1)\sqrt{\alpha(\alpha-3)}}{\alpha^2 - 4\alpha}.$$

(In the degenerate case $\alpha = 4$, we obtain $\tilde{\alpha} = -\frac{1}{8}$.)

From above, we can start with $\alpha_{-1} = 4$, obtain $\alpha_0 = -\frac{1}{8}$, and examine the next few generations. (We stop when an earlier α is repeated.)

$$\alpha_{1+} = -\frac{16}{11},\ \alpha_{1-} = 4,\ \alpha_{2+} = -\frac{1}{8},$$

$$\alpha_{2-} = \frac{121}{40},\ \alpha_{3+} = -\frac{16}{11},\ \alpha_{3-} = -\frac{100}{143}.$$

Since rational answers are obtained at each stage, despite the square root in (3.11), and since one solution leads to repetitions, it is reasonable to hope for a more tractable description of the map. Noting the perfect squares in the numerators, we try writing $\alpha = \frac{a^2}{c}$; then for $\alpha(\alpha-3)$ to be a perfect square we obtain $c = \frac{a^2-b^2}{3}$ for integers (a,b). Then $\alpha = \frac{3}{1-(\frac{b}{a})^2} \equiv \frac{3}{1-r^2}$, and from (3.11),

$$\tilde{\alpha} = \frac{1+5r^2\pm 2r(2+r^2)}{-1+4r^2} = \frac{3}{1-(\frac{4\pm 4r-7r^2\pm 2r^3}{1\pm 4r+5r^2\pm 2r^3})} = \frac{3}{1-(\frac{(\pm r-2)^2(\pm 2r+1)}{(\pm r+1)^2(\pm 2r+1)})} = \frac{3}{1-(\frac{\pm r-2}{\pm r+1})^2}.$$

Thus the mapping in terms of r is given by $T_\pm r = \frac{\pm r-2}{\pm r+1}$. Note that $T_-(T_\pm r) = \mp r$, resulting in the same α as previously. Thus it is enough to consider the iteration of the map T_+, which we will denote by L:

(3.12) $$Lr = \frac{r-2}{r+1}.$$

Write $r = \frac{b}{a}$, so that $Lr = \frac{b-2a}{a+2b}$, and consider the map on pairs given by

(3.13) $$\tilde{L}\binom{a}{b} = \begin{pmatrix} 1 & 1 \\ -2 & 1 \end{pmatrix}\binom{a}{b}, \quad \tilde{L}^{-1}\binom{\tilde{a}}{\tilde{b}} = \frac{1}{3}\begin{pmatrix} 1 & -1 \\ 2 & 1 \end{pmatrix}\binom{\tilde{a}}{\tilde{b}}.$$

Note that iterations of $\tilde{L}$ starting with relatively prime integers (a,b) do not produce any common prime factors $p \neq 3$, since $p \neq 3$, $p|\tilde{a}$, $p|\tilde{b}$ implies $p|\tilde{L}^{-1}\tilde{a}$, $p|\tilde{L}^{-1}\tilde{b}$. Moreover, if the initial (a_0, b_0) are equal to (2,2) mod 3, then neither factor is divisible by 3 after iteration of $\tilde{L}$, since, mod 3, $\tilde{L}\binom{2}{2} = \binom{1}{1}$, $\tilde{L}\binom{1}{1} = \binom{2}{2}$. We are interested in starting at $\alpha_0 = -\frac{1}{8} = \frac{3}{1-(\frac{b_0}{a_0})^2}$; that is, without loss of generality, $a_0 = -1$, $b_0 = 5$. Then if $\binom{a_n}{b_n} = \tilde{L}^n\binom{a_0}{b_0}$, a_n and b_n have no common prime factors. It follows from (3.12) that $L^n r = \pm r_0$ if and only if $\tilde{L}^n\binom{a_0}{b_0} = \pm\binom{a_0}{b_0}$. Equivalently, $\tilde{L}$ has an integer root of unity as an eigenvalue. But the eigenvalues of $\tilde{L}$ are $\lambda_\pm = 1 \pm \sqrt{2}\, i$, $|\lambda_\pm| = \sqrt{3}$. It follows that

(3.14) if $r_n = L^n r_0$, L as in (3.12), $r_0 = 5$, then $\{r_n\}$ is an infinite set.

The map $Lr = \frac{r-2}{r+1}$ is a linear fractional transformation sending $\mathbb{R} \cup \{\infty\}$ to itself. Its fixed points in $\mathbb{C}$ are $r = \pm\sqrt{2}\, i$. Conjugate by linear fractional transformations to bring this map to a rotation on S^1: set $Tz = \frac{z-\sqrt{2}\,i}{z+\sqrt{2}\,i}$, $T^{-1}z = \sqrt{2}\, i(\frac{z+1}{-z+1})$. Then $TLT^{-1}z = \frac{1+2\sqrt{2}\,i}{-3}z = e^{i\lambda}z$, $\lambda = \pi + \arctan(2\sqrt{2})$. Two possibilities occur (Sinai [16]): either TLT^{-1} is ergodic on S^1 and the orbit of each each point is dense in S^1 (equivalently, $\arctan(2\sqrt{2})$ is not a rational multiple of π), or TLT^{-1} is periodic on S^1. But in the latter case, L would be periodic on $\mathbb{R} \cup \{\infty\}$, contradicting (3.14). It follows that the orbit of each point of S^1 under TLT^{-1} is dense, and hence the orbit of each point of $\mathbb{R}$ under the map L given by (3.12) is dense in $\mathbb{R}$. Hence the iterated map given by (3.11) with the positive square root, i.e., $\alpha_n = \frac{3}{1-r_n^2}$ with r_n given by (3.14), yields a dense subset of $(-\infty,0) \cup (3,\infty)$.

It follows from Lemma 3.2, the remarks following Lemma 3.1 (after the linear transformation to (w,x,y,z)), and (3.9), that among the hypersurfaces generated by successive interactions along lines of triple intersection is a family of conoids sweeping out a dense subset of the complement of the light cone over the origin. These results are summarized in the following.

Theorem 3.3. Let $u \in H^s(\mathcal{O})$, $\mathcal{O} \subset \mathbb{R}^4$, $s > 2$ (respectively $s > 3$) be a solution to $\tilde{\Box}\, u = f(w,x,y,z,u)$, respectively $\tilde{\Box}\, u = f(w,x,y,z,u,Du)$ on $\mathcal{O}$. Suppose that on some space-like hyperplane Σ, u and its normal derivative are conormal with respect to the family of hyperplanes given by $\Sigma \cap \{w = 0\}$, $\Sigma \cap \{x = 0\}$, $\Sigma \cap \{y = 0\}$, $\Sigma \cap \{z = 0\}$. Then

$$u \in C^\infty(\{w^2+x^2+y^2+z^2 - (wx+wy+wz+xy+xz+yz) < 0\} \cap \mathcal{O}).$$

On the other hand, the collection of characteristic hypersurfaces generated by successive interactions along lines of triple intersection starting with the family $\{w = 0\}$, $\{x = 0\}$, $\{y = 0\}$, $\{z = 0\}$ includes a dense subset of $\mathcal{O} \cap \{w^2+x^2+y^2+z^2 - (wx+wy+wz+xy+xz+yz) < 0\}$.

In particular, it is expected that a solution to the problem given in Theorem 3.3 should, for generic choice of f and initial data subject to these given conditions, exhibit only finite regularity on the complement of the light cone over the origin. Note that Theorem 1.2 allows the conclusion about smoothness inside the light cone, regardless of the complication of defining conormality with respect to an infinite family of hypersurfaces whose envelope is the surface of the light cone.

References

[1] M. Beals, Spreading of singularities for a semilinear wave equation, Duke Math. J. 49 (1982), 275-286.

[2] ________, Self-spreading and strength of singularities for solutions to semilinear wave equations, Ann. of Math. 118 (1983), 187-214.

[3] ________, Nonlinear wave equations with data singular at one point, Contemp. Math. 27 (1984), 83-95.

[4] M. Beals and M. Reed, Propagation of singularities for hyperbolic pseudodifferential operators with nonsmooth coefficients, Comm. Pure Appl. Math. 35 (1982), 169-184.

[5] J.-M. Bony, Calcul symbolique et propagation des singularités pour les équations aux deriveés partielles nonlineaires, Ann. Scien. de l'Ecole Norm. Sup. 14 (1981), 209-246.

[6] ________, Interaction des singularités pour les équations aux deriveés partielles nonlineaires, Sem. Goulaouic-Meyer-Schwartz, exp. no. 22 (1979-80).

[7] ________, Interaction des singularités pour les équations de Klein-Gordon non lineaires, Sem. Goulaouic-Meyer-Schwartz, exp. no. 10 (1983-84).

[8] L. Hörmander, Fourier integral operators I, Acta Math. 127 (1971), 79-183.

[9] R. Melrose and N. Ritter, Interaction of nonlinear progressing waves for semilinear wave equations, Ann. of Math. 121 (1985), 187-213.

[10] ________, Interaction of progressing waves for semilinear wave equations II, preprint.

[11] J. Rauch and M. Reed, Propagation of singularities for semilinear hyperbolic equations in one space variable, Ann. of Math. 111 (1980), 531-552.

[12] ________, Nonlinear microlocal analysis of semilinear hyperbolic systems in one space dimension, Duke Math. J. 49 (1982), 379-475.

[13] ________, Singularities produced by the nonlinear interaction of three progressing waves: examples, Comm. P.D.E. 7 (1982), 1117-1133.

[14] ________, Striated solutions of semilinear, two speed wave equations, Indiana U. Math. J. 34 (1985), 337-353.

[15] ________, Bounded, stratified, and striated solutions of hyperbolic systems, preprint.

[16] Ya. G. Sinai, Introduction to Ergodic Theory, Princeton University Press, Princeton, N.J., 1976.

[17] M. Taylor, Pseudodifferential Operators, Princeton University Press, Princeton, N.J., 1981.

TOEPLITZ OPERATORS AND FUNCTION THEORY IN n-DIMENSIONS

C.A. Berger, L.A. Coburn and K.H. Zhu

1. Introduction. For Toeplitz operators on the Bergman space of any domain in C^n, there is a maximal conjugation-closed algebra Q of functions in L^∞ with the property that $T_f T_g - T_{fg}$ is a compact operator for all f, g in Q. For the (Bergman) Segal-Bargmann space of Gaussian square-integrable entire functions on C^n, Q has been characterized in terms of functions which oscillate slowly at infinity and properties of certain solutions of the heat equation [1, 2, 3]. For the Bergman space of area square-integrable holomorphic functions on the open unit disc in C, [4] characterized Q as those functions which are of "vanishing mean oscillation near the boundary." In this note, we return to the Segal-Bargmann space and show that Q can be characterized as those functions which are of "vanishing mean oscillation at infinity" with respect to Lebesgue volume measure on balls of constant radius.

The organization of this note is as follows: in §2, we provide the necessary definitions and some preliminary results; in §3, we prove the main result; in §4 we discuss some properties of the algebra VMO_∞.

2. Preliminary Analysis. Let $B(z,R)$ denote the closed ball of radius R centered at z in C^n. Let $|B(z,R)|$ be the usual Euclidean volume of $B(z,R)$ (of course, $|B(z,R)| = |B(0,R)|$). In this note, all functions will be essentially bounded, measurable complex-valued functions defined on C^n, i.e. elements of $L^\infty(C^n)$. We denote the algebra of bounded continuous functions by BC and the algebra of continuous functions which vanish at infinity by C_o. For f in L^∞, we consider the heat transform

Research supported by grants of the NSF.

$$\tilde{f}(a) = (2\pi)^{-n} \int f(z) e^{-|z-a|^2/2}\, dv(z) .$$

It is easy to check that $\tilde{f}(a)$ is the solution to the initial-value problem for the heat equation at $t = 1/2$ with initial values $f(z)$ at $t = 0$.

We recall from [1, 3] that the algebra of "eventually slowly varying" functions is given by

$$\mathrm{BCESV} = \{f \in \mathrm{BC}: \lim_{|z|\to\infty} \{\sup_{|w-z|\le 1} |f(z) - f(w)|\} = 0\}$$

$$= \{f: f - \tilde{f} \in C_o\}$$

while there is a natural ideal $\mathcal{I}$ in Q given by

$$\mathcal{I} = \{f \in L^\infty: (|f|^2)^\sim \in C_o\}$$

$$= \{f \in L^\infty: (|f|)^\sim \in C_o\}$$

and $Q = \mathrm{BCESV} + \mathcal{I}$. Note that BCESV is a closed subalgebra of Q (in the supremum norm). It is easy to check that C_o is contained in $\mathcal{I}$ properly [1, 3] and

$$C_o = \mathrm{BCESV} \cap \mathcal{I} .$$

For f in L^∞, we consider the averaging transforms

$$\hat{f}(z,R) = |B(z,R)|^{-1} \int_{B(z,R)} f(w)\, dv(w) .$$

Note that, for fixed $R > 0$, $\hat{f}(z,R)$ is <u>always</u> in BC.

We can now <u>define</u> the space $\mathrm{VMO}_\infty(R)$ by

$$\mathrm{VMO}_\infty(R) = \{f: \lim_{|a|\to\infty} \int_{B(a,R)} |f(w) - \hat{f}(a,R)|\, dv(w) = 0\} .$$

In the next section, we will prove that $Q = \mathrm{VMO}_\infty(R)$ <u>for any</u> <u>$R \geq 0$</u>. Thus, as a Corollary, we will see that $\mathrm{VMO}_\infty(R)$ is independent of the

the choice of $R > 0$.

We conclude this section by defining the auxilliary space $\tilde{Q}$ by

$$\tilde{Q} = \{f \in L^\infty : \ (|f|^2)^\sim(z) - |\tilde{f}(z)|^2 \in C_o\} .$$

and proving several preliminary results.

Lemma 1. Q is contained in $\tilde{Q}$.

Proof. We first check that $\tilde{Q}$ is a linear subspace of L^∞. To do this, consider the bilinear functional on L^∞ given by

$$\langle f,g\rangle_a = (f\bar{g})^\sim(a) - \tilde{f}(a)\tilde{\bar{g}}(a) .$$

Several applications of the Cauchy-Schwarz inequality show that $\langle f,f\rangle_a \geq 0$ and that $\langle f,g\rangle_a$ is in C_o for f and g in $\tilde{Q}$. Linearity of $\tilde{Q}$ follows by direct calculation.

Using linearity of $\tilde{Q}$, we need only check that BCESV and $\mathcal{I}$ are separately contained in $\tilde{Q}$. But

$$0 \leq (|f|^2)^\sim(z) - |\tilde{f}(z)|^2 \leq (|f|^2)^\sim(z)$$

so that f in $\mathcal{I}$, implying $(|f|^2)^\sim \in C_o$, also implies that f is in $\tilde{Q}$. For f in BCESV, we see that $|f|^2$ is also in BCESV so that

$$|f|^2 - (|f|^2)^\sim , \quad f - \tilde{f}$$

are both in C_o. It follows easily that $(|f|^2)^\sim - |\tilde{f}|^2$ is in C_o so that f is in $\tilde{Q}$.

Lemma 2. We have

$$(|f|^2)^{\sim}(z) - |\tilde{f}(z)|^2 \geq$$

$$(1/2)e^{-R^2}(2\pi)^{-2n} \int_{B(z,R)} \int_{B(z,R)} |f(w) - f(u)|^2 \, dv(w)dv(u) \; .$$

<u>Proof</u>. This is a direct calculation using the equality

$$(|f|^2)^{\sim}(z) - |\tilde{f}(z)|^2 =$$

$$(1/2)(2\pi)^{-2n} \int \int |f(w) - f(u)|^2 \, e^{-|w-z|^2/2-|u-z|^2/2} \, dv(w)dv(u) \; .$$

<u>Lemma</u> 3. $\displaystyle\int_{B(z,R)} \int_{B(z,R)} |f(w) - f(u)|^2 \, dv(w)dv(u)$

$$= 2|B(z,R)| \int_{B(z,R)} |f(w) - \hat{f}(z,R)|^2 \, dv(w)$$

<u>Proof</u>. Direct calculation.

<u>Theorem</u> 4. For f in L^∞, the following are equivalent:

a) $f \in VMO_\infty(R)$

b) $\displaystyle\lim_{|z|\to\infty} \int_{B(z,R)} \int_{B(z,R)} |f(w) - f(u)|^2 \, dv(w)dv(u) = 0$.

<u>Proof</u>. We note, by standard inequalities

$$|B(0,R)|^{-1} \left\{ \int_{B(z,R)} |f(w) - \hat{f}(z,R)| dv(w) \right\}^2 \leq \int_{B(z,R)} |f(w) - \hat{f}(z,R)|^2 \, dv(w)$$

$$\leq 2\|f\|_\infty \int_{B(z,R)} |f(w) - \hat{f}(z,R)| \, dv(w) \; .$$

The desired equivalence now follows from Lemma 3.

<u>Corollary</u>. $VMO_\infty(R) \subset VMO_\infty(R')$ if $R \geq R'$.

3. Main Result. We can prove

Theorem 5. We have $Q = \tilde{Q} = VMO_\infty(R)$.

Proof. By Lemma 1, Q is contained in $\tilde{Q}$. Using Lemma 2, we see that, if f is in $\tilde{Q}$, then

$$\lim_{|z|\to\infty} \int_{B(z,R)} \int_{B(z,R)} |f(w) - f(u)|^2 \, dv(w)dv(u) = 0 .$$

It follows from Theorem 4 that $f \in VMO_\infty(R)$ and so $\tilde{Q}$ is contained in $VMO_\infty(R)$.

To establish the equalities, it suffices to show that $VMO_\infty(R)$ is contained in $Q = BCESV + \mathcal{I}$. For f in $VMO_\infty(R)$, we check first that $\hat{f}(z,R/2)$ is in BCESV. This is straightforward. For $|z - w| \leq R$,

$$\hat{f}(z,R/2) - \hat{f}(w,R/2)$$

$$= |B(z,R/2)|^{-1} |B(w,R/2)|^{-1} \int_{B(z,R/2)} \int_{B(w,R/2)} (f(u) - f(s)) \, dv(u)dv(s)$$

and so

$$|\hat{f}(z,R/2) - \hat{f}(w,R/2)|$$

$$\leq \left\{ |B(0,R/2)|^{-2} \int_{B(z,R/2)} \int_{B(w,R/2)} |f(u) - f(s)|^2 \, dv(u)dv(s) \right\}^{1/2} .$$

It follows, for $|z - w| \leq R$, that

$$|\hat{f}(z,R/2) - \hat{f}(w,R/2)|$$

$$\leq \left\{ |B(0,R/2)|^{-2} \int_{B(\frac{z+w}{2},R)} \int_{B(\frac{z+w}{2},R)} |f(u)-f(s)|^2 \, dv(u)dv(s) \right\}^{1/2} .$$

Hence, for $|z|$ large and $|z - w| \leq R$, Theorem 4 implies that

$$\lim_{|z|\to\infty} |\hat{f}(z,R/2) - \hat{f}(w,R/2)| = 0 .$$

An elementary covering argument now shows that the above limit is also zero for $|z - w| \leq 1$. Hence, $\hat{f}(z,R/2)$ is in BCESV.

Next, we want to show that $f(z) - \hat{f}(z,R/2)$ is in $\mathcal{I}$ for f in $VMO_\infty(R)$. We note that by the Corollary to Theorem 4, f is in $VMO_\infty(R/2)$. Let $g(z) = f(z) - \hat{f}(z,R/2)$. We need to check that

$$\int |g(w)|\, e^{-|w-a|^2/2}\, dv(w)$$

is in C_0 . Given $\epsilon > 0$ there is a $D(\epsilon) > 0$, independent of a so that

$$\int |g(w)|\, e^{-|w-a|^2/2}\, dv(w) \leq \epsilon + \int_{B(a,D)} |g(w)|\, dv(w) .$$

An elementary covering argument shows that for R given, there are m points a_j in $B(a,D)$ with

$$B(a,D) \subset \bigcup_{j=1}^{m} B(a_j,R/2) .$$

Here, m is independent of a . Thus,

$$\int_{B(a,D)} |g(w)|\, dv(w) \leq \sum_{j=1}^{m} \int_{B(a_j,R/2)} |g(w)|\, dv(w)$$

and it suffices to consider

$$\int_{B(a_j,R/2)} |g(w)|\, dv(w) \leq \int_{B(a_j,R/2)} |f(w) - \hat{f}(a_j,R/2)|\, dv(w)$$

$$+ \int_{B(a_j,R/2)} |\hat{f}(a_j,R/2) - \hat{f}(w,R/2)|\, dv(w) .$$

Now as $|a| \to \infty$, the first term on the right-hand side tends to zero since f is in $VMO_\infty(R/2)$. The second term also tends to zero as $|a| \to \infty$ since $\hat{f}(z,R/2)$ is in BCESV. It follows that $f(z) - \hat{f}(z,R/2)$ is in $\mathcal{I}$ and so f is in Q.

<u>Corollary</u> 1. For any R,R' greater than zero, $VMO_\infty(R) = VMO_\infty(R')$.

<u>Proof</u>. $VMO_\infty(R) = Q = VMO_\infty(R')$.

Henceforth, we write $VMO_\infty = VMO_\infty(R)$. We also have

<u>Corollary</u> 2. For f in Q, $\hat{f}(z,R) - \tilde{f}(z)$ is in C_o.

<u>Proof</u>. Recall that, by [3] and Theorem 5,

$$f = \tilde{f} + (f - \tilde{f})$$

$$f = \hat{f}(.,R/2) + (f - \hat{f}(.,R/2))$$

give decompositions of Q into BCESV + $\mathcal{I}$. Since BCESV $\cap\, \mathcal{I} = C_o$, it follows at once that $\tilde{f} - \hat{f}(.,R/2)$ is in C_o for any $R > 0$.

4. <u>The structure of VMO_∞</u>. For $z = (z_1,\ldots,z_n)$ in C^n and λ in C, we write $\lambda z = (\lambda z_1,\ldots,\lambda z_n)$ and $f_\lambda(z) = f(\lambda z)$. We note here some additional consequences of Theorems 4 and 5.

<u>Theorem</u> 6. The algebra $Q = \tilde{Q} = VMO_\infty$ is translation invariant and invariant under the map $f \to f_\lambda$.

<u>Proof</u>. Translation invariance is clear from Theorem 4. Note, for f in $VMO_\infty(|\lambda|)$ and $\lambda \neq 0$, that

$$\int_{B(z,1)} \int_{B(z,1)} |f_\lambda(w) - f_\lambda(u)|^2 \, dv(w)dv(u)$$

$$= |\lambda|^{-4n} \int_{B(\lambda z, |\lambda|)} \int_{B(\lambda z, |\lambda|)} |f(w) - f(u)|^2 \, dv(w)dv(u) \; .$$

Thus, f_λ is in $VMO_\infty(1)$ and Corollary 1 of Theorem 5 completes the proof.

REFERENCES

[1] Berger, C. A. and Coburn, L. A., A symbol calculus for Toeplitz operators, Proc. Nat Acad Sci. USA 83 (1986) 3072-3073.

[2] Berger, C. A. and Coburn, L. A., Toeplitz operators and quantum mechanics, to appear in J. Functional Analysis.

[3] Berger, C. A. and Coburn, L. A., Toeplitz operators on the Segal-Bargmann space, to appear in Transactions AMS.

[4] Zhu, K. H., VMO, ESV and Toeplitz operators on the Bergman space, preprint.

Department of Mathematics and Computer Science
Lehman College of CUNY
Bronx, New York 10468

Department of Mathematics
State University of New York at Buffalo
Buffalo, New York 14214

Department of Mathematics
University of Washington
Seattle, Washington 98195

INDEX THEORY FOR REGULAR SINGULAR OPERATORS AND APPLICATIONS

J. Brüning
Institut für Mathematik
Universität Augsburg
Memminger Str. 6
D - 8900 Augsburg

1. This is a report on joint work with R. Seeley. In dealing with singular elliptic problems which admit separation of variables one frequently encounters regular singularities in the classical sense i.e. one has to solve ordinary differential equations of the type

$$(\partial_x + x^{-1}a(x))u(x) = f(x)$$

or

$$(-\partial_x^2 + x^{-2}a(x))u(x) = f(x)\,,\; x > 0\,,$$

where a is smooth in $x \geq 0$. Cheeger [Ch] used this approach systematically to study the geometric operators on manifolds with cone-like singularities. In a series of papers [B+S1,2,3] we have developed the notion of first and second order regular singular operators abstractly, derived the asymptotic expansion of the trace of the resolvent in the second order case, and applied this to prove an index theorem for first order regular singular operators. In the following we will describe how these techniques can be used to calculate the L^2 index of the geometric operators on complete manifolds with finitely many ends all of which are warped products; the full details will appear elsewhere. The resulting index theorem will then be applied to the Gauß–Bonnet operator.

Let us recall first the notion of a regular singular first order differential operator on a Riemannian manifold M (cf. [B+S,3] §1) which we present here in a slightly more general form. So let $D : C^\infty(E) \to C^\infty(F)$ be a first order elliptic differential operator between the smooth sections of two hermitian vector bundles E and F over M. We think of M as a singular Riemannian manifold with singularities in an open subset U such that $M \setminus U$ is a smooth compact manifold with boundary. The nature of the singularities of course influences the structure of the geometric operators on U. From this fact we abstract certain axioms concerning D; it will be called a regular singular differential operator if the following is true.

(RS 1) There is a compact Riemannian manifold N, with $\dim N + 1 = \dim M$, and a hermitian vector bundle G over N such that there are bijective linear maps

$$\Phi_E : C_0^\infty(E \mid U) \to C_0^\infty(I, C^\infty(G)),$$
$$\Phi_F : C_0^\infty(F \mid U) \to C_0^\infty(I, C^\infty(G)),$$

where $I := (0, \varepsilon)$ for some ε, $0 < \varepsilon \leq 1$.

(RS 2) Φ_E and Φ_F extend to unitary maps $L^2(E \mid U) \to L^2(I; L^2(G))$ and $L^2(F \mid U) \to L^2(I, L^2(G))$, respectively.

(RS 3) For $\varphi \in C^\infty(I)$ with φ constant near 0 and ε let M_φ be the multiplication operator on $L^2(I, L^2(G))$. Then $\Phi_E^* M_\varphi \Phi_E = \Phi_F^* M_\varphi \Phi_F = M_{\overline{\varphi}}$ for some $\overline{\varphi} \in C^\infty(M)$, and $\overline{\varphi} \in C_0^\infty(M)$ if φ vanishes in a neighborhood of 0.

(RS 4) On $C_0^\infty(E \mid U)$ we have

$$\Phi_F D \Phi_E^* = \partial_x + x^{-1}(S_0 + S_1(x))$$

where

a) S_0 is a self-adjoint first order elliptic differential operator on $C^\infty(G)$, and spec $S_0 \cap \{-1/2, 1/2\} = \emptyset$;

b) $S_1(x)$ is a first order differential operator depending smoothly on $x \in (0, \varepsilon)$;

c) $\|S_1(x)(|S_0| + 1)^{-1}\| + \|(|S_0| + 1)^{-1} S_1(x)\| = o(1)$ as $x \to 0$.

The main example for this situation is a manifold with asymptotically cone-like singularitites. In this case we assume that U above is isometric to $(0, \varepsilon) \times N$ with metric $dx^2 + x^2 ds_N(x)^2$ where $\varepsilon > 0$, x is the standard coordinate on $(0, \varepsilon)$, N is a compact (not necessarily connected) Riemannian manifold, and $ds_N(x)^2$ is a family of metrics on N variing smoothly in $[0, \varepsilon)$. It is then readily verified that the geometric operators on M are regular singular in the above sense.

2. Now let M be a complete Riemannian manifold with finitely many ends. We assume that there is an open $U \subset M$ such that $M \setminus U$ is a smooth compact manifold with boundary and $U = \bigcup_{i=1}^{k} U_i$ where each U_i is isometric to a warped product $(y_{0i}, \infty) \times_{f_i} N_i$, $1 \leq i \leq k$. To simplify the exposition we will assume that $k = 1$ so U is $(y_0, \infty) \times N$, for some $y_0 > 0$ and some compact Riemannian manifold N, equipped with the metric $dy^2 + f(y)^2 ds_N^2$, where ds_N^2 is the metric on N and f is some positive function in $C^\infty[y_0, \infty)$. A lengthy but straightforward calculation shows that the geometric operators on U are unitarily equivalent to

$$\partial_y + \frac{1}{f(y)} S_0 + \frac{f'(y)}{f(y)} S_1 \tag{1}$$

in the sense of RS4) where S_0 is a suitable self-adjoint first order differential operator on $C^\infty(G)$, G a bundle over N, and S_1 is a zero order differential operator on $C^\infty(G)$ (cf. Section 5 below for the example of the Gauß–Bonnet operator).

We therefore consider a first order elliptic differential operator $D : C^\infty(E) \to C^\infty(F)$ between the smooth sections of two hermitian bundles E, F over M which are unitarily equivalent to an operator of the form (1) over U in the above sense. It is natural to investigate the L^2-index of D i.e. the quantity

$$L^2\text{-ind } D := \dim \ker D \cap L^2(E) - \dim \ker D' \cap L^2(F) \tag{2}$$

where $D' : C^\infty(F) \to C^\infty(E)$ is the formal adjoint of D, defined by $(Du, v) = (u, D'v)$ for all $u \in C_0^\infty(E)$, $v \in C_0^\infty(F)$. Note that D' has automatically similar properties as D, in particular

$$D' \simeq -\partial_y + \frac{1}{f(y)} S_0 + \frac{f'(y)}{f(y)} S_1' \,. \tag{3}$$

There are various L^2-index theorems applying to this situation, dealing e.g. with cylinders [A+P+S], asymptotically Euclidean spaces ([S] Theorem 1), or Riemannian manifolds with cusps ([S] Theorem 2). We will present an L^2-index theorem unifying and extending these results; the main point is to link the L^2-index with the index of a regular singular operator in the sense of RS1) - RS4). To do so we need of course a condition on f since in general the L^2-index will not be finite; a counterexample is provided by the Gauß–Bonnet operator on $\mathbb{R}^n = [0,\infty) \times S^{n-1}$ with a rotationally invariant metric $dy^2 + f(y)^2 ds^2_{S^{n-1}}$ such that $\int_1^\infty \frac{dy}{f(y)} < \infty$ (cf. [D]). The condition we impose is

$$\lim_{y\to\infty} f'(y) = 0 \,. \tag{4}$$

implying

$$(5) \qquad f(y) = o(y)\,,\ y \to \infty\,.$$

It is well known that all warped products are conformally equivalent to Riemannian products i.e. cylinders; elaborating on this idea we show that a weighted version of D, i.e. gDg for a suitable positive function $g \in C^\infty(M)$, is regular singular if (4) holds. To do so, define

$$(6) \qquad F(y) := \int_{y_0}^{y} \frac{du}{f(u)}$$

such that $F \in C^\infty(y_0, \infty)$; in view of (5) we have the estimate

$$(7) \qquad F(y) \geq \log y^N - c_N$$

for all $N > 0$. Next pick a positive function $g \in C^\infty(M)$ such that $g \mid U$ depends on y only and

$$(8) \qquad g^2(y) = f(y)e^{F(y)} \quad \text{for } y \text{ sufficiently large.}$$

Then the function

$$(9) \qquad s(y) := \int_{y_0}^{y} \frac{du}{g^2(u)}$$

equals $e^{-F(y)}$ for large y and defines a diffeomorphism from (y_0, ∞) to $(0, x_1)$ for some $x_1 > 0$. Thus we obtain a linear transformation $\Phi : C_0^\infty((0, x_1), C^\infty(G)) \to C_0^\infty((y_0, \infty), C^\infty(G))$ given by

$$(10) \qquad \Phi u(y) := \frac{1}{g(y)} u(s(y))\,.$$

Clearly, Φ extends to a unitary map $L^2((0, x_1), L^2(G)) \to L^2((x_0, \infty), L^2(G))$, and it is easily calculated that $D_g := gDg$ transforms as

$$\Phi^* D_g \Phi = -\partial_x + \frac{g^2}{f} \circ s^{-1}(x)(S_0 + f' \circ s^{-1}(x) S_1) . \tag{11}$$

The definition of g and s and (4) then imply

LEMMA 1 *D_g is a regular singular differential operator.*

The discussion of the closed extensions of D_g and their Fredholm properties can now be carried out essentially along the lines of [B+S,3] §§2 and 3. The only difference lies in the fact that we have relaxed condition RS4,c) above where in [B+S,3] we required instead

$$\begin{aligned} &\|S_1(x)(|S_0|+1)^{-1}\| + \|(|S_0|+1)^{-1} S_1(x)\| = O(x^\alpha) \\ &\text{as } x \to 0 \text{ for some } \alpha > 1/2 , \end{aligned} \tag{12}$$

whereas the elimination of the $\pm 1/2$ eigenvalues in RS4,a) was not necessary. In the case at hand we may assume that the restriction on spec S_0 is satisfied; otherwise we replace S_0 by μS_0 and f by μf for a suitable $\mu > 0$ which will not affect condition (4). Then we obtain the following result.

THEOREM 1 *The closed extensions of D_g in $L^2(E)$ are classified by the subspaces of the finite dimensional space $W_0 := \mathcal{D}(D_{g,\max}) / \mathcal{D}(D_{g,\min})$. All closed extensions are Fredholm operators, and if $D_{g,W}$ denotes the closed extension corresponding to $W \subset W_0$ we have*

$$\text{ind } D_{g,W} = \text{ind } D_{g,\min} + \dim W .$$

3. The next task is to compare ind $D_{g,W}$ with L^2-ind D for a suitably chosen W. If $u \in \ker D \cap L^2(E)$ then clearly $0 = Du = \frac{1}{g} D_g \frac{1}{g} u$. It is easy to see from (8) and (4) that $\frac{1}{g} \in L^\infty(M)$ so we obtain an injection

$$\ker D \cap L^2(E) \ni u \mapsto \frac{1}{g} u \in \ker D_{g,\max} . \tag{13}$$

This map is bijective onto $\ker D_{g,\max} \cap \frac{1}{g} L^2(E)$ so we would like to define

$$\mathcal{D}(D_{g,W}) := \mathcal{D}(D_{g,\max}) \cap \frac{1}{g} L^2(E)\,,$$
$$D_{g,W} := D_{g,\max} \mid \mathcal{D}(D_{g,W})\,.$$

With the modifications of [B+S,3] §2 mentioned above and the crucial condition (4) it then follows that

$$\mathcal{D}(D_{g,\min}) \subset \mathcal{D}(D_{g,\max}) \cap \frac{1}{g} L^2(E)\,. \tag{14}$$

This implies that $D_{g,W}$ is a closed extension of D_g hence a Fredholm operator in view of Theorem 1. It is also not difficult to see that under the map analogous to (13) we obtain an injection

$$\ker D' \cap L^2(F) \hookrightarrow \ker D^*_{g,W}\,. \tag{15}$$

We define

$$h_0 := \dim W\,,\ h_1 := \dim \ker D^*_{g,W} - \dim \ker D' \cap L^2(F)\,. \tag{16}$$

Using Theorem 1 we arrive at the following L^2-index theorem.

THEOREM 2

$$\begin{aligned} L^2\text{-ind } D &= \text{ind } D_{g,W} + h_1 \\ &= \text{ind } D_{g,\min} + h_0 + h_1\,. \end{aligned} \tag{17}$$

It is now necessary to describe the terms on the right hand side of (17) more explicitly. The calculation of ind $D_{g,\min}$ is largely parallel to the index calculation in [B+S,3] and will be carried out in the next section. To clarify the role of h_0 and h_1 we need an additional assumption which is also satisfied by the geometric operators (cf. Section 5), namely: if Q denotes the orthogonal projection in $H := L^2(G)$ onto ker S_0 we have

$$S_1 \text{ is symmetric in } H \text{ and } (I - Q)S_1 Q = 0\,. \tag{18}$$

If $u \in C^1((y_0,\infty),H)$ solves $Du = 0$ we obtain from (1) and (18)

$$(Qu)'(y) + \frac{f'(y)}{f(y)} QS_1Qu(y) = 0\,,\ y \in (y_0,\infty)\,. \tag{19}$$

We now write the spectral decomposition of QS_1Q in the form

$$QS_1Q = \bigoplus_{t\in\mathbb{R}} tQ_t \tag{20}$$

where of course only finitely many Q_t are nonzero. Then the general solution of (19) is

$$Qu(y) = \sum_{t\in\mathbb{R}} \left(\frac{f(y)}{f(y_0)}\right)^{-t} Q_t u(y_0)\,, \tag{21}$$

and since we are only interested in L^2-solutions of D and D' it is natural to decompose further

$$Q = Q_0 \oplus Q_0' \oplus Q_1$$

where

$$\begin{aligned} Q_0 &:= \bigoplus_{f^{-t}\in L^2} Q_t\,,\\ Q_0' &:= \bigoplus_{f^{t}\in L^2} Q_t\,,\\ Q_1 &:= \bigoplus_{f^{-t},f^{t}\notin L^2} \,. \end{aligned} \tag{22}$$

The analysis of h_0 requires a good description of $\mathcal{D}(D_{g,\min})$ which is provided by a result analogous to [B+S,3] Lemma 3.2 namely

$$\mathcal{D}(D_{g,\min}) = \{u \in \mathcal{D}(D_{g,\max}) \mid \|\Phi^* u(x)\| = O(x^{1/2}) \ \text{ as } x \to 0\}\,.$$

Analyzing the solutions of the transformed equation along the lines of [B+S,3] Lemma 3.2 then proves

LEMMA 2

$$h_0 = \dim Q_0 .$$

In dealing with h_1 it seems advantageous to study the original equation directly. In fact, under the isomorphism $v \mapsto \tilde{v} := gv$ we have

$$\begin{aligned}\ker D^*_{g,W} &= \{\tilde{v} \in C^\infty(F) \mid D'\tilde{v} = 0, \frac{1}{g}\tilde{v} \in L^2(F), \\ &\quad (D\tilde{u}, \tilde{v}) = 0 \text{ for all } \tilde{u} \in L^2(E) \text{ with } gD\tilde{u} \in L^2(F)\} \\ &=: \mathcal{H}^*_W .\end{aligned}$$

The homogeneous equation $D'\tilde{v}(y) = 0$ is conveniently transformed by the change of variables

$$y(z) := F^{-1}(z), \quad \tilde{w}(z) := \tilde{v}(y(z)), \quad z \in (0, \infty),$$

leading to

$$[\partial_z - S_0 - f'(F^{-1}(z))S_1]\tilde{w}(z) = 0 . \tag{23}$$

The L^2-solutions of this equation can be studied by standard methods. Then it follows that

$$L^2\text{-ker } D' = \{\tilde{v} \in \mathcal{H}^*_W \mid Q_1\tilde{v}(y) = 0 , \ y \geq y_0\} .$$

Introducing the map

$$\tau_y : \mathcal{H}^*_W \ni \tilde{v} \mapsto Q_1\tilde{v}(y) \in Q_1 H ,$$

defined for $y \geq y_0$, we therefore find

LEMMA 3 *For all $y \geq y_0$*

$$\begin{aligned}h_1 &= \dim \text{im } \tau_y = \dim \{Q_1\tilde{v}(y) \mid \tilde{v} \in \mathcal{H}^*_W\} \\ &\leq \dim Q_1 .\end{aligned}$$

In particular, $h_1 = 0$ if $Q_1 = 0$ which is the case e.g. if $f(y) = e^{-y}$, that is if M is a manifold with a cusp. It seems, however, very difficult to compute h_1 in general. We will give an example below with $h_1 > 0$, cf. Theorem 5.

4. It remains to compute ind $D_{g,\min}$. This is parallel to the work in [B+S,3] §4 though now the manifold may have infinite volume. The above discussion shows that ind $D_{g,\min}$ is the same for all g satisfying (8) for x sufficiently large. Thus it is natural that we define g to be constant on the part of M where $y \leq R$ for some large R in order to obtain the regularized interior contribution to the index independent of g. Taking the limit $R \to \infty$ in this approach is, however, technically somewhat delicate, and we are lead to impose a further condition on the growth of f, namely

$$\text{(24)}\qquad \begin{aligned} &\text{if } Q_\alpha f := f^{\alpha_0}(f')^{\alpha_1}\cdots(f^{(k)})^{\alpha_k},\ \alpha_i \geq 0, \text{ is any} \\ &\text{monomial such that } \alpha_0 \leq \sum_{j\geq 2}(j-1)\alpha_j \text{ then} \\ &\lim_{y\to\infty} Q_\alpha f(y) = 0\,. \end{aligned}$$

Note that this condition contains (4) and that it is satisfied if $f(y) = e^{-y}$ or $f(y) = y^\beta$, $\beta < 1$, for large y. Also, (24) can be viewed as the analogue of condition (4.31) in [B+S,3] for the case under consideration. Then g_R will be a positive function in $C^\infty(M)$ satisfying

$$\text{(25)}\qquad \begin{aligned} &g_R^2(y) = f(R) \quad \text{if } y \leq R + \frac{1}{2}f(R)\,, \\ &g_R^2(y) = f(y)e^{F(y)} \quad \text{if } y \text{ is sufficiently large}, \\ &\int_R^\infty \frac{du}{g_R(u)^2} = 1\,, \end{aligned}$$

and we define

$$s_R(y) := \int_y^\infty \frac{du}{g_R(u)^2}\,.$$

Then an isometry Φ_R is defined as in (11) which transforms $D_{g_R} \mid C_0^\infty((R,\infty),H)$ to

$$-\partial_x + a_R(x)S_0 + b_R(x)S_1$$

on $C_0^\infty((0,1),H)$ where $a_R(x) = 1/x$ near $x = 0$ and $\lim_{x\to 0} b_R(x) = 0$. The condition (24) ensures that uniformly on $[1/2,1]$

$$\begin{aligned} &\lim_{R\to\infty} a_R(x) = 1\,, \\ &\lim_{R\to\infty} a_R^{(j+1)}(x) = \lim_{R\to\infty} b_R^{(j)}(x) = 0\,,\ j \geq 0\,. \end{aligned} \tag{26}$$

We can then modify $D_{g_R,\min}$ to an operator $D_R : \mathcal{D}(D_{g_R,\min}) \to L^2(F)$ in such a way that $b_R(x) = 0$ if $x \in [0,1/2]$ and $a_R(x) = 1$ near $x = 1/2$, and ind $D_R =$ ind $D_{g_R,\min}$. Using suitable cut-off functions and computing separately the contributions to the constant term in the asymptotic expansion of

$$\mathrm{tr}(e^{-tD_R^*D_R} - e^{-tD_RD_R^*})$$

coming from $y \leq R$, $R \leq y \leq R + \frac{1}{2}f(R)$, and $y \geq R + \frac{1}{2}f(R)$, we obtain three terms. Since the sum gives the index of $D_{g_R,\min}$ and hence does not depend on R we can take the limit $R \to \infty$. The first contribution involves only the "index form" ω_D of D and equals

$$\lim_{R\to\infty} \int_{y\leq R} \omega_D \tag{27}$$

proving in particular the existence of the limit. The index form is obtained as follows: the operators e^{-tD^*D} and e^{-tDD^*} have kernels with respect to the given Riemannian measure which when restricted to the diagonal in $M \times M$ yield smooth sections of the bundles $\mathrm{Hom}(E,E)$ and $\mathrm{Hom}(F,F)$, respectively. These kernels have pointwise asymptotic expansions as $t \to 0$ and ω_D denotes the difference of the fiber traces of the constant terms in this expansion. The second contribution turns out to be $o(1)$ as $R \to \infty$ in view of (26) since $(-\partial_x + S_0)(\partial_x + S_0) = (\partial_x + S_0)(-\partial_x + S_0)$. The third contribution is computed in [B+S,4]; it is independent of R and equals

$$\frac{1}{2}(\eta(S_0) - \dim\ker S_0) \tag{28}$$

where

$$\eta_{S_0}(z) := \sum_{s\in\mathrm{spec}\, S_0\setminus\{0\}} \mathrm{sgn}\ s\ |s|^{-z}$$

is the η-function introduced in [A+P+S]. For general elliptic operators η_{S_0} is known to be meromorphic in $\mathbb{C}$ with only simple poles, and 0 is a regular value. Then

$$\eta(S_0) := \eta_{S_0}(0) .$$

Combining Theorem 2 with (27) and (28) we obtain

THEOREM 3 (L^2-index theorem) *Let M be a complete Riemannian manifold as in Section 2, such that the warping factor f satisfies condition (24). Let $D : C^\infty(E) \to C^\infty(F)$ be a first order elliptic differential operator on M satisfying (1) and (18). Then*

$$L^2\text{-ind } D = \int_M \omega_D + \frac{1}{2}(\eta(S_0) - \dim \ker S_0) + h_0 + h_1 . \tag{29}$$

Theorem 3 generalizes in a straightforward way to manifolds with k ends $U_i = (x_{0i}, \infty) \times_{f_i} N_i$ all of whose warping factors satify (24). Then on each end we have the representation

$$D \simeq \partial_y + \frac{1}{f_i} S_{0i} + \frac{f_i'}{f_i} S_{1i}$$

where (18) is required now for all i. Then we put

$$S_j := \bigoplus_{i=1}^{k} S_{ji} \, , \; j = 0, 1 \, ,$$

and

$$h_0 = \sum_{i=1}^{k} h_{0i} = \sum_{i=1}^{k} \dim Q_{0i} . \tag{30}$$

h_1 is again defined by (16) and satisfies the estimate analogous to Lemma 3,

$$h_1 \leq \sum_{i=1}^{k} \dim Q_{1i} . \tag{31}$$

5. We want to explain the various ingredients of Theorem 3 in the case of the Gauß–Bonnet operator D_{GB}. So we assume again that M is a complete Riemannian manifold with finitely many ends $U_i = (y_{0i}, \infty) \times_{f_i} N_i$ such that all warping factors satisfy (24). Denoting by $\Omega(M) = \bigoplus_{j \geq 0} \Omega^j(M)$ the smooth forms on M, by $\Omega^{\mathrm{ev}}(M)$ and $\Omega^{\mathrm{odd}}(M)$ those of even and odd degree, respectively, and by d and d^* the exterior derivative and its adjoint with respect to the natural L^2-structure on $\Omega(M)$, D_{GB} is defined by

$$D_{GB} := d + d^* : \Omega^{\mathrm{ev}}(M) \to \Omega^{\mathrm{odd}}(M)\,.$$

It is well known that D_{GB} is a first order elliptic differential operator. If M is compact then it is easily seen that with $H^j(M) := \{\omega \in \Omega^j(M) \mid (d^*d + dd^*)\omega = 0\}$, the space of harmonic j-forms,

$$L^2\text{-ind}\ D_{GB} = \sum_{j \geq 0} (-1)^j \dim\ H^j(M)\,.$$

By de Rham's theorem $H^j(M)$ is isomorphic to the jth singular cohomology group of M with real coefficients so

$$L^2\text{-ind}\ D_{GB} = \chi(M)\,,$$

the Euler characteristic of M. In the noncompact complete case the harmonic forms have to be replaced by the L^2-harmonic forms i.e. we introduce

$$\mathcal{H}^j(M) := \{\omega \in \Omega^j(M) \mid (dd^* + d^*d)\omega = 0\,,\quad \int_M \omega \wedge *\omega < \infty\}\,.$$

It follows from a well known theorem of Andreotti and Vesentini that $\omega \in \mathcal{H}^j(M)$ iff $d\omega = d^*\omega = 0$. Hence we obtain

$$L^2\text{-ind}\ D_{GB} = \sum_{j \geq 0} (-1)^j \dim\ \mathcal{H}^j(M) =: \chi_{(2)}(M)\,,$$

the L^2-Euler characteristic of M. It is natural to ask whether $\chi_{(2)}(M)$ is a topological invariant. That this is not the case can be seen already from the fact that the finiteness of $\chi_{(2)}(M)$ depends on the metric and not on the topology alone, cf. [D]. The L^2-index theorem above will give a formula for $\chi_{(2)}(M)$ if we can show that D_{GB} satisfies our assumptions. For this purpose we note that any $\omega \in \Omega^j(U_i)$ can be written as

$$\omega = \omega_j(y) + \omega_{j-1}(y) \wedge dy$$

where $\omega_\ell \in C^\infty((y_{0i}, \infty), \Omega^\ell(N_i))$, $\ell = j-1, j$.

A lengthy but straightforward calculation then gives the following result.

LEMMA 4 *On $\Omega^{\mathrm{ev}}(U_i)$ we have*

$$D_{GB} \simeq \partial_y + \frac{1}{f_i(y)} S_{0i} + \frac{f_i'(y)}{f_i(y)} S_{1i}$$

acting on $C^\infty((y_{0i}, \infty), \Omega(N_i))$. Here

$$(32) \qquad S_{0i} \begin{pmatrix} \omega_0 \\ \vdots \\ \omega_{n_i} \end{pmatrix} = \begin{pmatrix} 0 & d_{N_i}^* & & \\ d_{N_i} & \ddots & \ddots & \\ & \ddots & \ddots & d_{N_i}^* \\ & & d_{N_i} & 0 \end{pmatrix} \begin{pmatrix} \omega_0 \\ \vdots \\ \omega_{n_i} \end{pmatrix}$$

where ω_j denotes the component in $\Omega^j(N_i)$ and d_{N_i}, $d_{N_i}^*$ denote the intrinsic operations on N_i. Moreover,

$$(33) \qquad S_{1i} \begin{pmatrix} \omega_0 \\ \vdots \\ \omega_{n_i} \end{pmatrix} = \begin{pmatrix} c_0 & & 0 \\ & \ddots & \\ 0 & & c_{n_i} \end{pmatrix} \begin{pmatrix} \omega_0 \\ \vdots \\ \omega_{n_i} \end{pmatrix}$$

where $c_j = (-1)^j (j - \frac{n_i}{2})$.

Note that $n_i =: n = \dim M - 1$ for all i; as in the compact case we assume from now on that $\dim M$ is even i.e.

$$\dim M = 2k = n + 1, \; k \geq 1.$$

So D_{GB} satisfies condition (1). Now it is easily checked that

$$(34) \qquad \ker S_{0i} = \bigoplus_{j \geq 0} H^j(N_i)$$

and consequently D_{GB} also satisfies (18) for all i. Hence the L^2-index theorem applies and we obtain

$$\chi_{(2)}(M) = \int_M \omega_{GB} + \frac{1}{2}(\eta(S_0) - \dim \ker S_0) + h_0 + h_1 , \tag{35}$$

where S_0, h_0, h_1 are defined at the end of §4. We have to investigate the terms on the right hand side of (35) more closely. Clearly, (34) implies that

$$\dim \ker S_0 = \sum_{i,j} \dim H^j(N_i) = \sum_{i,j} b_j(N_i) \tag{36}$$

where b_j is the jth Betti number. Next, the calculations in [B+S,3] Lemma 5.1 prove

LEMMA 5 *If S_0 denotes the operator in (32) on an arbitrary compact Riemannian manifold N then*

$$\eta(S_0) = 0 .$$

Using (30) and (33) we also arrive at

$$h_0 = \sum_{\substack{i,j \\ f_i^{-c_j} \in L^2}} b_j(N_i) . \tag{37}$$

Now consider $\int_M \omega_{GB}$, the integral of the Gauß–Bonnet integrand. If M is compact then the Chern–Gauß–Bonnet Theorem asserts that

$$\int_M \omega_{GB} = \chi(M) . \tag{38}$$

For a general complete manifold M with finitely many ends we say that the Chern–Gauß–Bonnet theorem holds if (38) is true. This is not true in general as the example $M = \mathbb{R}^n$ shows. On the other hand, the surface case has been studied thoroughly in a classical paper by Cohn–Vossen [CV]; he gives various sufficient conditions for (38) and shows that in great generality the inequality

$$\int_M \omega_{GB} \leq \chi(M)$$

is true. Further work concerns the case of locally symmetric spaces [H] and the case of bounded geometry [Ch-G]. In our situation there seems to apply only the result of Rosenberg [R] Theorem 1.9 which says that (38) holds if

$$\lim_{x\to\infty} f_i(x) = \lim_{x\to\infty} f_i'(x) = 0 \quad \text{for all } i\,.$$

We will show that the Chern-Gauß–Bonnet theorem also holds under our assumptions.

LEMMA 6 *If all warping factors satisfy the condition (24) then the Chern–Gauß–Bonnet theorem holds for M.*

PROOF The function $f_0(y) \equiv 1$ satisfies (24). Then we pick $\varphi \in C_0^\infty(\mathbb{R})$ with $\varphi = 1$ in a sufficiently large neigborhood of 0 and try to deform all warping factors f_i to f_0 near infinity i.e. we put

$$f_{i,\theta} := \varphi f_i + (1-\varphi)(\theta f_0 + (1-\theta) f_i)\,,\; \theta \in [0,1]\,.$$

It is easily checked that $f_{i,\theta}$ satisfies (24), too. In this way we obtain a family $D_{GB,\theta}$ of elliptic first order differential operators. Now we construct a smooth family of functions g_θ satisfying

$$\begin{aligned} &g_\theta^2(y) = 1\,,\; y_0 \leq y \leq y_0 + 1/3\,,\\ &g_\theta^2(y) = f_\theta(y) e^{F_\theta(y)} \quad \text{if } e^{-F_\theta(y)} \leq 1/3\,,\\ &\int_{y_1}^\infty \frac{du}{g_\theta(u)^2} = 1\,, \end{aligned}$$

where $F_\theta(y) = \int_{y_0}^y \frac{du}{f_\theta(u)}$, and we define $s_\theta(y) := \int_y^\infty \frac{du}{g_\theta^2(u)}$. Transforming the square integrable forms on the Riemannian manifold with warping function f_θ using the transformation (10) generated by g_θ and s_θ maps the closure D_θ of the operators $g_\theta D_{GB} g_\theta$ to a family of Fredholm operators with domain independent of θ, variing continuously with θ. Using (27) and (28) we thus conclude that

$$\int_M \omega_D^0 = \int_M \omega_D^1 \tag{37}$$

where ω_D^j is the index form of $D_{GB,j}$, $j = 0, 1$. Moreover, it follows easily from the Gauß–Bonnet theorem for manifolds with boundary that

$$\int_M \omega_D^1 = \chi(M)\,. \tag{38}$$

The Lemma follows from (37) and (38).

□

Since (24) holds e.g. for $f(y) = y^\beta$ with $\beta < 1$, Lemma 6 applies to warping factors which are not covered in [R]. As pointed out in this paper it is not necessary to control the derivatives of f of order greater than 1; thus it seems likely that the Chern–Gauß–Bonnet theorem will hold if only (4) is satisfied for all i.

Summing up we have proved

THEOREM 4 *Let M be a complete connected Riemannian manifold with finitely many ends U_i, $1 \le i \le k$, and assume that each end is a warped product with warping factor f_i satisfying (24). Then*

$$\chi_{(2)}(M) = \chi(M) + \frac{1}{2}\left(\sum_{f_i^{-c_j} \in L^2} b_j(N_i) - \sum_{f_i^{-c_j} \notin L^2} b_j(N_i) \right) + h_1$$

where h_1 is an integer satisfying

$$0 \le h_1 \le \sum_{f_i^{-c_j}, f_i^{c_j} \notin L^2} b_j(N_i)\,.$$

We conclude this section with the surface case which allows the explicit calculation of h_1 under much weaker conditions than stated in Theorem 4. In particular, it shows that $h_1 > 0$ in general.

THEOREM 5 *Let M be a complete connected surface with finitely many ends U_i, $1 \le i \le k$, and assume that each end is a warped product with warping factor f_i satisfying*

$$\int_{y_{0i}}^{\infty} \frac{du}{f_i(u)} = \infty\,. \tag{39}$$

Then

$$\chi_{(2)}(M) = \begin{cases} \chi(M) + k & \textit{if } \mathrm{vol}\, M < \infty\,, \\ \chi(M) + k - 2 & \textit{if } \mathrm{vol}\, M = \infty\,. \end{cases} \tag{40}$$

This implies that

$$h_1 = \begin{cases} 0 & \textit{if } \mathrm{vol}\, M < \infty\,, \\ 2[\sharp\{i \mid f_i \notin L^1\} - 1] & \textit{if } \mathrm{vol}\, M = \infty\,. \end{cases} \tag{41}$$

PROOF Assume first that vol $M < \infty$ which is equivalent to $f_i \in L^1$ for $1 \leq i \leq k$; in view of Theorem 4 this yields $h_1 = 0$. By (33) we have $c_j = -1/2$ for $j = 0, 1$, hence we see from (30) that $h_0 = 2k$. Also, dim ker $S_0 = 2k$. On each U_i the circles $y = \mathrm{const}$ have constant geodesic curvature equal to $\frac{f_i'(y)}{f_i(y)}$ so it follows from (4) and the Gauß–Bonnet theorem for surfaces with boundary that

$$\int_M \omega_{GB} = \chi(M)\,.$$

Plugging this into (35) and observing Lemma 5 we obtain

$$\chi_{(2)}(M) = \chi(M) - k + 2k = \chi(M) + k\,. \tag{42}$$

Next, if vol $M = \infty$ h_1 may be nonzero since

$$\int_{y_{0i}}^{\infty} \frac{du}{f_i(u)} = \infty$$

for all i, by $f_i(y) = o(y)$ as a consequence of (4). Now

$$\chi_{(2)}(M) = \dim\, \mathcal{H}^0(M) - \dim\, \mathcal{H}^1(M) + \dim\, \mathcal{H}^2(M)\,,$$

and we have

$$\dim\, \mathcal{H}^0(M) = \dim\, \mathcal{H}^2(M) = \begin{cases} 1 & \text{if vol } M < \infty\,, \\ 0 & \text{if vol } M = \infty\,, \end{cases} \tag{43}$$

since M is connected and the Hodge $*$ operator induces an isomorphism $\mathcal{H}^0(M) \to \mathcal{H}^2(M)$. It is also easily checked that dim $\mathcal{H}^1(M)$ is a conformal invariant of M (cf. [D] for these facts). So (40) follows from (42) and (43) if we can show that under our assumptions M is conformally equivalent to a finite volume surface $\tilde{M}$ with all ends warped products with warping factors f_i satisfying (4). To achieve this we first choose a positive C^∞ function $\overline{f}$ on M such that on U_i $\overline{f}(y) = f_i(y)^{-2}$ if y is sufficiently large. Next we construct a diffeomorphism $\psi : M \to M$ such that $\psi =$ id on $y \leq R$ for R sufficiently large and $\psi(y,n) = (\int_{y_0 i}^{y} \frac{du}{f_i(u)}, n)$ if y is sufficiently large and $n \in N_i$. Denoting by g the original metric on M we obtain a conformally equivalent metric setting $\tilde{g} := (\psi^{-1})^* \overline{f} g$. Clearly, this construction can always be carried out if we have (39), and it gives a conformal equivalence to a manifold with cylindrical ends. But then we can also obtain a conformal equivalence to a manifold all of whose warping factors equal e^{-y} for y sufficiently large which completes the argument.

Finally, (41) follows from (40) and (35) by comparison.

□

REFERENCES

[A+P+S] M.F. Atiyah, V.K. Patodi, and I.M. Singer: *Spectral asymmetry and Riemannian geometry.* I. Math. Proc. Camb. Philos. Soc. 77 (1975), 43 - 69.

[B+S,1] J. Brüning and R. Seeley: *Regular singular asymptotics.* Adv. Math. 58 (1985), 133 - 148.

[B+S,2] J. Brüning and R. Seeley: *The resolvent expansion for second order regular singular operators.* Preprint Augsburg 1985.

[B+S,3] J. Brüning and R. Seeley: *An index theorem for first order regular singular operators.* Preprint Augsburg 1986.

[B+S,4] J. Brüning and R. Seeley: *On the regularity of η-functions.* In preparation.

[Ch] J. Cheeger: *Spectral geometry of singular Riemannian spaces.* J. Differ. Geom. 18 (1983), 575 - 657.

[Ch-G] J. Cheeger and M. Gromov: *On the characteristic numbers of complete manifolds of bounded curvature and finite volume.* Preprint.

[CV] S. Cohn-Vossen: *Kürzeste Wege und Totalkrümmung auf Flächen.* Comp. Math. 2 (1935), 69 - 133.

[D] J. Dodziuk: *L^2-harmonic forms on rotationally symmetric Riemannian manifolds.* Proc. AMS 77 (1979), 395 – 400.

[H] G. Harder: *A Gauß–Bonnet formula for discrete arithmetically defined groups.* Ann. Sci. École Norm. Sup. 4 (1971), 409 – 455.

[R] S. Rosenberg: *On the Gauß–Bonnet theorem for complete manifolds.* Trans. AMS 287 (1985), 745 – 753.

[S] M. Stern: *L^2-index theorems on warped products.* Thesis, Princeton University 1984.

The Laplace comparison algebra of spaces with conical and cylindrical ends

H.O.Cordes and S.H.Doong
Dept. Math., University of California
Berkeley, Calif. 94707, USA

In recent times the discussion of singular elliptic boundary problems over non-compact manifolds has been revived by a spectrum of results in analysis, differential equations and geometry (cf. Bruening-Seeley [$BS_{1,2,3}$] , Cheeger [Che], Cheeger-Gromov-Taylor [CGT], Choquet-Bruhat-Christadolou [CBC], Cordes [C_2], Lockhart-McOwen [LM_1], Melrose-Mendoza [MM_1], Mazzeo-Melrose [MaM], Schulze [$Schu_{1,2}$].) For an earlier approach, pertaining to estimates and asymptotic expansions on cylinders cf. Agmon-Nirenberg [AN_1]; for further references cf. the bibliography in [$Schu_2$] .

Much of the work mentionned is concerned with asymptotic estimates of solutions, involving the Mellin transform, and with the Fredholm index of operators on cones or cylinders. Analytical tools are those of common theory of elliptic differential operators, and of microlocal analysis.

Our present paper is based on an approach of the first author and his associates, using theory of C^*-algebras, in particular the representation of commutative C^*-algebras as algebras of continuous functions, to explore algebras of singular integral operators over a noncompact space Ω with differentiable structure (cf. [C_1], [C_2],V, for definitions and a general discussion).

Such C^*-algebras of singular integral operators are called <u>comparison algebras</u> (allowing comparison of differential operators to a realization H of an elliptic Schroedinger type differential expression, called comparison operator). Usually a comparison algebra $\mathcal{C}$ is generated by the multiplications a of a given function class $A^{\#}$, and by $D\Lambda$, with first order differential expressions D of a given class $\mathcal{D}^{\#}$, and the inverse square root $\Lambda = H^{-1/2}$, where the $D\Lambda$ play the role of Riesz-operators often used for $\mathbb{R}^n$. For a Riemannian space Ω a natural choice is $H=1-\Delta$, with the Laplace operator Δ. Then we speak of a <u>Laplace comparison algebra</u>.

The comparison algebras we study either have compact commutators - then the Fredholm operators of the comparison algebra C are characterized as operators $A \in C$ with nonvanishing symbol σ_A, where the symbol function is defined over a well described compact space $\mathbb{M}$ - or else possess a two-link chain

$$(0.1) \qquad C \supset E \supset K \quad ,$$

with the commutator ideal E of C and the ideal K of compact operators of the Hilbert space used. In the latter case both quotients C/E and E/K turn out to be function algebras, so that the process of solving a linear equation $Au=f$, for $A \in C$, involves two inversions, one mod E, and another one, mod K. A corresponding necessary and sufficient criterion for A to be Fredholm then is more complicated.

Moreover, in two such examples (cf.[CPo] and [CMe_1]) it was seen that the second symbol, defined on E, could be extended to the entire algebra C again, with the extended symbol taking singular integral operators as values. This implied a simple necessary and sufficient criterion for the Fredholm property of operators in C, not requiring two inversions.

In the present paper we explore Laplace comparison algebras for spaces with the following type of 'ends': Either (i) Ω is an asymptotic outgoing cone at infinity - just like $\mathbb{R}^n$, or a hyperboloid- , or else (ii) Ω is a product of a cone and a compact space at infinity. Case (ii) includes (i), and also includes the case of a cylindrical end (where the cone has dimension 1). In case (i) the end is a cone $[\varepsilon,\infty)\times\Theta$, with compact Θ, and metric

$$(0.2) \qquad ds^2 = dr^2 + r^2 d\theta^2 \quad , \quad 0<\varepsilon<r<\infty \quad ,$$

with a metric $d\theta^2$ on Θ. In this case C has compact commutator, and at infinity the symbol space looks like that of the Laplace comparison algebra of the Euclidean $\mathbb{R}^n$, (cf.[C_1],III,IV). This case is discussed in sec's 1,2,3. Our methods are discussed in detail there.

In sec.4 we discuss the global pseudo-differential operator structure of the generators of C: Under slightly stronger assumptions they are global 'ψdo-s' in the sense of Schrohe [$Schr_1$]. This has the implication that, in case (i), a subalgebra S of C is accessible to the global ψdo-methods - i.e.,K-parametrix construction for md-elliptic operators - , as practiced in [CP] for $\mathbb{R}^n$, and [$Schr_{1,2}$] for manifolds with an SG-structure.

In sec.5 we discuss case (ii): Ω , at each end, is a product $(\varepsilon,\infty)\times\Theta\times B$ of a different kind: The first two factors alone make an (outside) cone, while the first and third factor alone make a cylinder with 'crossection' B. In other words, our metric is

$$(0.3) \qquad ds^2 = dr^2 + r^2 d\theta^2 + ds'^2 \ , \ 0 < \varepsilon < r < \infty \ .$$

with $d\theta^2$ and ds'^2 on the compact spaces Θ and B, respectively. If dim B>0, then C no longer has compact commutators, and we get a multiple symbol as in [CPo]: In a compactification of $T^*\Omega$ the symbol is complex-valued, but over a certain space at $r=\infty$ the symbol assumes values in the algebra of singular integral operators over the compact manifold B .

While we are not concerned with index formulas, it should be pointed out that the Fredholm index still is invariant under homotopies of the symbol function. On compact manifolds as well as on spaces with outside-cone-shape the symbol function is entirely complex-valued. Then the index will be a homotopy invariant of a simple compact space associated with Ω and its metric. On the other hand, for the algebra of sec.5 the symbol in part will be operator-valued, and a symbol formula may have to involve intrinsically certain spectral properties of operators on an infinite dimensional space. In that respect we point to the fact that formulas involving the eta-function of an operator derived from the symbol at infinity already have been derived by Atiyah-Patodi-Singer [APS] .

As in [C_2] and [CPo] our operators act on complex-valued functions, and we omit vector bundles. Again this is justified by [C_2],X,3, an abstract procedure allowing to translate all our Fredholm results to the case of operators on bundle crossections.

Although only L^2-algebras are discussed, we point to the abstract machinery of [C_2], which at once allows generalization of results to Sobolev spaces H_s, weighted as well as unweighted, including $s=\infty$. For more details, cf. the ends of sec.4 and sec.5.

1. A class of asymptotically flat metrics on $\mathbb{R}^n$.

First consider $\mathbb{R}^n$ under the general positive definite metric

$$(1.1) \qquad ds^2 = g_{jk}(x)dx^j dx^k \ , \ g_{jk} \in C^\infty(\mathbb{R}^n) \ ,$$

with summation convention (sum from 1 to n over upper-lower pairs).

Assume (1.1) equivalent to the Euclidean metric: With constants $c,C>0$, independent of x, and $\delta_{jk}=1$, $j=k$, $=0$ else, we have

(1.2) $$c\,\delta_{jk}\xi_j\xi_k \le g_{jk}\xi_j\xi_k \le C\,\delta_{jk}\xi_j\xi_k\ , \text{ for all } \xi\in\mathbb{R}^n\ .$$

Then g^{jk}, defined by $((g^{jk}))=((g_{jk}))^{-1}$, also is bounded, and a bounded tensor field $b^{j_1..j_N}_{k_1..k_M}$ over $\mathbb{R}^n$ is a tensor field with bounded coefficients, and the field has limit 0 at infinity iff all coefficients tend to zero, as $|x|\to\infty$. We use Landau symbols $O(1)$ and $o(1)$ to indicate that a tensor field is bounded or has limit 0 at infinity.

We assume a condition of 'zero oscillation at infinity': Let

(1.3) $$\partial^\alpha_x g_{jk}=\partial^{\alpha_1+..+\alpha_n} g_{jk}/\partial x^{1\alpha^1}..\partial x^{n\alpha^n}\to 0,\ \text{as } |x|\to\infty,\ |\alpha|=\alpha_1+..+\alpha_n\le 3$$

The Laplace operator Δ and surface measure dS are given by

(1.4) $$\Delta=\kappa^{-1}\partial_{x^j}\kappa g^{jk}\partial_{x^k}\ ,\ dS=\kappa dx\ ,\ \kappa=\sqrt{g}\ ,\ g=\det((g_{jk}))\ ,$$

where g^{jk} and κ are bounded again. Also, the Christoffel symbols

(1.5) $$\Gamma^j_{kl}=g^{jm}\Gamma_{klm}\ ,\ \Gamma_{klm}=1/2(g_{km|x^l}+g_{lm|x^k}-g_{kl|x^m})$$

vanish at infinity (i.e., have limit zero, as $|x|\to\infty$), together with their partial derivatives up to order 2 (including). In particular we thus conclude that the Riemannian covariant derivative

(1.6) $$b^j{}_{|k}=\partial_{x^k}b^j-\Gamma^j_{mk}b^m$$

of a bounded vector field $X=b^j\partial_{x^j}$ coincides with the ordinary derivative at ∞, in the sense that $\lim_{|x|\to\infty}(b^j{}_{|k}-\partial_{x^k}b^j)=0$.

Specifically $b^j{}_{|k}=o(1)$ holds if and only if $\partial_{x^k}b^j\to 0$, as $x\to\infty$. Also the Riemannian space $\mathbb{R}^n$, with the metric (1.1), is asymptotically flat, insofar as the Riemannian curvature tensor

(1.7) $$R^l_{ijk}=\Gamma^l_{ik|x^j}-\Gamma^l_{ij|x^k}+\Gamma^m_{ik}\Gamma^l_{mj}-\Gamma^m_{ij}\Gamma^l_{mk}$$

vanishes at infinity. In fact, even the first covariant derivatives of the tensor (1.7) vanish at ∞.

First we aim at showing that the metric (1.1) generates the same Laplace comparison algebra, in the sense of [C_2], as the Euclidean metric. Here we use the generating classes

(1.8) $$A_0^{\#}=\{a\in C^\infty(\mathbb{R}^n): a=O(1)\ ,\ \partial^\alpha a=o(1)\ ,\ \alpha\neq 0\}\ ,$$

and

(1.9) $$\mathcal{D}_0^{\#} = \{ b^j \partial_{x^j} + p : b^j , p \in A_0^{\#} \} .$$

In [C_2] we defined the Laplace comparison algebra $C = C(A_0^{\#}, \mathcal{D}_0^{\#})$ as the C^*-subalgebra of $L(H)$, $H = L^2(\mathbb{R}^n, dS)$, generated by

(1.10) $$a , D\Lambda , a \in A_0^{\#} , D \in \mathcal{D}_0^{\#} ,$$

with the multiplication operators a , taking u(x) into a(x)u(x) , and the inverse square root Λ of the unbounded self-adjoint operator $H=1-\Delta$, Δ as in (1.4).

More precisely, H denotes the unique self-adjoint realization of $1-\Delta$, the closure of the minimal operator H_0 (with domain dom $H_0 = C_0^\infty(\mathbb{R}^n)$) of the differential expression $H=1-\Delta$. Note that $\mathbb{R}^n$, under the metric (1.1), is a complete Riemannian space, so that the operator H_0 has a unique self-adjoint extension (cf.[Ga_1] [Ro_1], or [C_2],IV). Then the maps $D\Lambda : \Lambda^{-1} C_0^\infty \to C_0^\infty$ extend to continuous operators $A: H \to H$, since $\Lambda^{-1} C_0^\infty$ is dense in H. Also,

(1.11) $$(Du,Du) \leq c_0(u,Hu) = c_0(\Lambda^{-1}u, \Lambda^{-1}u) \text{ for all } u \in C_0^\infty ,$$

with the inner product (.,.) of H, and a constant c_0 (cf.[C_2],V,1) Clearly the multiplication operators $u \to au$ are bounded too, and the operator norm $\|a\| = \sup\{\|au\| : \|u\| \leq 1\}$ coincides with the sup norm of a The C^*-subalgebra C of $L(H)$ with generators (1.10) then is defined as the smallest C^*-subalgebra of $L(H)$ containing (1.10).

We now preassume some of the discussions of [C_2]. By [C_2],V, thm.4.1 the algebra C is the smallest norm closed subalgebra of $L(H)$ containing (1.10), since adjoints are automatically included. Moreover, C contains the entire ideal $K(H)$ of all compact operators of H , and it has compact commutators. Indeed, our $A_0^{\#}$, $\mathcal{D}_0^{\#}$ are contained in the classes of thm.4.1, but contain the minimal classes of V,1 so that V,lemma 1.1. applies, all referring to [C_2]

Note that the Hilbert spaces $L^2(\mathbb{R}^n) = L^2(\mathbb{R}^n, dx)$ and $L^2(\mathbb{R}^n, dS)$, $dS = \kappa dx$, coincide as sets: Since κ and κ^{-1} are bounded, we get

(1.12) $$\|u\|_\kappa^2 = \int |u|^2 \kappa dx \leq \|\kappa\|_{L^\infty} \|u\|^2 , \quad \|u\|^2 = \int |u|^2 dx \leq \|\kappa^{-1}\|_{L^\infty} \|u\|_\kappa^2 ,$$

i.e., the norms $\|.\|$ and $\|.\|_\kappa$ are equivalent. Hence also the two Banach space topologies of $\|.\|$ and $\|.\|_\kappa$ are identical, and we have $L(L^2(\mathbb{R}^n)) = L(L^2(\mathbb{R}^n, dS))$ as sets.

On the other hand, the inner products of dx and dS in general are different, and we get two different adjoints on $L(H)$.

Accordingly, the algebra C, for any choice of the metric (1.1), is a subalgebra of $L(H)$, $H=L^2(\mathbb{R}^n)$, but is a C^*-algebra under its own involution, that of its corresponding inner product.

Notice that the algebra $C=C_0$ arising from the Euclidean metric (i.e., $\kappa\equiv 1$, $dS=dx$, $\Delta=\sum\partial_{x^j}^2$) has been studied in [$C_1$]. In particular we found $C_0/K(H)$ isometrically isomorphic to a function algebra $C(\partial\mathbb{Q})$ with the set $\partial\mathbb{Q}$ of infinite points of a certain compactification of the cotangent space of $\mathbb{R}^n$ (i.e.,of $\mathbb{R}^{2n}$).

In the present section we discuss the following result.

Theorem 1.1. For every choice of the metric (1.1) meeting (1.2) and (1.3) we have $C=C_0$, hence $C/K(H)=C(\partial\mathbb{Q})$, with the compactification $\mathbb{Q}$ of the Euclidean metric. With reference to the explicit form $\mathbb{Q}=\mathbb{P}^n\times\mathbb{B}^n$ of [C_1], the symbols of the generators (1.10) are

$$(1.13)\qquad \sigma_a(x,\xi)=a(x),\ \ \sigma_{D\Lambda}(x,\xi)=(ib^j(x)\xi_j+p(x))/(g^{jk}(x)\xi_j\xi_k+1)^{1/2},$$

to be interpreted as follows: The functions at right of (1.13) are bounded continuous over the cotangent space $T^*\mathbb{R}^n=\mathbb{R}^{2n}$ and (by definition of $\mathbb{Q}$) extend continuously onto $\mathbb{Q}$. Then (1.13) states that the restrictions to $\partial\mathbb{Q}$ of these extensions give the symbols of the generators. Also, with Δ_0 and Λ_0 denoting Δ and Λ of the Euclidean metric, we have $\Lambda_0^{-1}\Lambda$, $\Lambda\Lambda_0^{-1}\in C$, and, with the interpretation described for (1.13),

$$(1.14)\quad \sigma_{\Lambda_0^{-1}\Lambda}(x,\xi)=\sigma_{\Lambda\Lambda_0^{-1}}(x,\xi)=((\delta^{lk}\xi_l\xi_k+1)/(g^{lk}\xi_l\xi_k+1))^{1/2},$$

Proof. First we invoke the transformation of dependent variable

$$(1.15)\qquad u=\kappa^{-1/2}v\ ,\ \int|u|^2\kappa dx=\int|v|^2dx\ ,$$

which equates the inner product of the metric (1.1) with that of the Euclidean metric. In other words, we conjugate $A\in L(L^2(\mathbb{R}^n,dS))$ with the unitary map $L^2(\mathbb{R}^n,dS)\to L^2(\mathbb{R}^n,dx)$ defined by $u\to\kappa^{1/2}u$.

Proposition 1.2. We have

$$(1.16)\quad \kappa^{-1/2}D\Lambda\kappa^{1/2}-D\Lambda\in K(H)\ ,\ \kappa^{-1/2}\Lambda D\kappa^{1/2}-\Lambda D\in K(H)\ ,\ D\in\mathcal{D}_0^{\#}\ ,$$

so that the comparison algebra C of the metric (1.1) remains invariant under conjugation with $\kappa^{1/2}$.

Proof. This is a matter of using a resolvent integral for the operator Λ. Let $R(\lambda)=(H+\lambda)^{-1}$, $-\lambda\in\mathbb{C}\setminus sp(H)$, referring to the unique self-adjoint realization H of H_0 again. We have $H \geq 1$, so that $Sp(H) \subset [1,\infty) \subset \mathbb{R} \subset \mathbb{C}$. Therefore $R(\lambda) \in L(H)$ is defined for $0 \leq \lambda < \infty$, norm continuous in λ. We have $\|R(\lambda)\| \leq (1+\lambda)^{-1}$, and

$$(1.17) \qquad \Lambda^{s} = \pi^{-1} \sin \pi s/2 \int_0^\infty R(\lambda)\lambda^{-s/2} d\lambda \ , \quad 0 < s < 2 \ ,$$

where the right hand side integral exists as an improper Riemann integral, in norm convergence. For details cf. $[C_2]$,VI,(1.6).

The functions $\psi(x) = \kappa^{1/2}(x)$, $\chi(x) = \kappa^{-1/2}(x)$ are bounded and $C^\infty(\mathbb{R}^n)$, and $\partial_x^\alpha \psi$, $\partial_x^\alpha \chi = o(1)$, for $|\alpha| \leq 2$. The commutators $L_1 = [\Delta,\psi]$ and $L_1' = [\Delta,\chi]$ are first order expressions of the form (1.9), with coefficients b^j, $p = o(1)$.

For (1.16) we may show that $[\psi, D\Lambda]$, $[\chi, D\Lambda] \in K(H)$, since ψ,χ are bounded, and $\kappa^{-1/2} D\Lambda \kappa^{1/2} - D\Lambda = \chi[D\Lambda,\psi]$, for example. The compactness of commutators follows from $[C_2]$,V,thm.3.1. We give an outline of the proof, since (1.17) is used anyway: Write $[D\Lambda,\psi] = D\Lambda(\Lambda^{-1}[\Lambda,\psi]) + [D,\psi]\Lambda$, where the last term is compact, since the function $[D,\psi]$ vanishes at ∞, by $[C_1]$,III,1.8.1, for example. Hence we only must show compactness of $\Lambda^{-1}[\Lambda,\psi]$. Using (1.17) get

$$(1.18) \qquad \Lambda^{-1}[\Lambda,\psi] = \pi^{-1} \int_0^\infty \lambda^{-1/2} (\Lambda^{-1} R(\lambda)) L_1 R(\lambda) \, d\lambda \ ,$$

where the integrand is compact and $O(\lambda^{-1/2}(1+\lambda)^{-1})$, and also norm continuous in λ, by a calculation. Hence the improper Riemann integral in (1.18) exists in norm convergence and defines a compact operator. The handling of the domains of the unbounded operators in (1.18) requires some precautions which are discussed in detail in $[C_2]$,V,3. The other commutator above may be treated similarly, q.e.d.

Under (1.15) the expression $H = 1-\Delta$, with Δ of (1.4), goes into

$$(1.19) \quad H^\sim = \kappa^{1/2} H \kappa^{-1/2} = q^\sim - \Delta^\sim = -\partial_{x^j} g^{jk} \partial_{x^k} + q^\sim, \quad q^\sim = \kappa^{1/2} H(\kappa^{-1/2}) \ ,$$

as shown by a calculation (cf. $[C_2]$,III,(3.2), for a guidance). Here (1.2),(1.3) imply $q^\sim = 1 - \kappa^{1/2}\Delta(\kappa^{-1/2})$, with second term $= o(1)$.

To complete the proof of thm.1.1 we now must show:

<u>Proposition 1.3</u>. Let $\Lambda^\sim = H^{\sim -1/2}$, with the unique realization of $H^\sim$, self-adjoint with respect to dx. Then $\Lambda_0^{-1}\Lambda^\sim \in C_0$, and the C_0-symbol of this operator is given by the function of (1.14).

Indeed, we clearly get $\chi^{-1}\Lambda\chi = \Lambda^\sim$. Also, $D^\sim = \chi^{-1} D \chi$ is given by

$$(1.20) \qquad D^\sim = b^j \partial_{x^j} + p - d_* \ , \quad d_* = 1/2 \, b^j \kappa_{|x^j}/\kappa \ ,$$

where $d_* = o(1)$, hence $d_* \Lambda^\sim$ is compact and surely in C_0. Thus

$$(1.21) \qquad \chi^{-1} D \Lambda \chi = D^\sim \Lambda^\sim = (D\Lambda_0)(\Lambda_0^{-1}\Lambda^\sim) + K \ , \ K \in K(H) \ .$$

Under prop.1.3 it follows that the operators (1.21) are in C_0. Similarly for the adjoints $(D\Lambda)^* = (\Lambda D^*)^{**}$. Since the multiplication generators are the same we conclude that $C = \chi^{-1} C \chi \subset C_0$. Then, by prop.1.3 again, we may calculate the C_0-symbols of the generators of C and find that they strongly separate points of $\partial \mathbb{Q}$. Thus the Stone-Weierstrass theorem gives $C/K = C_0/K$, implying $C = C_0$, and (1.13), (1.14) follow, completing the proof of thm.1.1.

The proof of prop.1.3 is similar to that of $[C_2]$,VI,lemma 1.7. We use (1.17) for commutator relations needed. For an efficient notation let, for a moment, K and H denote $1 - \sum \partial_{x^j}^2$ and $H^\sim$, respectively, so that $\Lambda_0 = K^{-1/2}$, $\Lambda^\sim = H^{-1/2}$. Clearly all powers H^s, K^s, $s \in \mathbb{R}$, are meaningfull, and (1.17) applies.

<u>Proposition 1.4</u>. For $0 \leq s \leq 1/2$ we have

$$(1.22) \qquad H^s [H^{-s}, K^{-s}] H^s \in K(H) \ , \text{ and } H^{2s}[H^{-s}, K^{-s}] \in K(H) \ .$$

<u>Proof</u>. With the resolvents $R(\lambda) = (H+\lambda)^{-1}$, $S(\lambda) = (K+\lambda)^{-1}$, and integrals as in (1.17) (and either $(\sigma,\tau) = (s,s)$, or $(\sigma,\tau) = (2s,0)$) write

$$(1.23) \quad H^\sigma [H^{-s}, K^{-s}] H^\tau = (\sin^2 \pi s)/\pi^2 \int_0^\infty \int_0^\infty \lambda^{-s} d\lambda \mu^{-s} d\mu H^\sigma [R(\lambda), S(\mu)] H^\tau ,$$

(cf. $[C_2]$,VI,(1.6) for detail) where, with dom H_0=dom $K_0 = C_0^\infty$,

$$(1.24) \quad [R(\lambda), S(\mu)] v = R(\lambda) S(\mu) [K,H] S(\mu) R(\lambda) v \ , \ v \in \operatorname{im}(H_0+\lambda)(K_0+\mu),$$

by a simple calculation. To continue we need the following:
<u>Proposition 1.5</u>. We have $\operatorname{im}(H_0+\lambda)(K_0+\mu)$ dense in H for all $\lambda, \mu > 0$.
<u>Proof</u>. For $\lambda = \mu$, and H=K this is a direct consequence of the fact that H_0^2 is essentially self-adjoint (cf.$[C_2]$,IV,V). For the general case we need the same result, and the fact that the operators HK^{-1} and KH^{-1} are bounded and inverses of each other. Indeed,write

$$(1.25) \qquad KH^{-1} = \Lambda^{\sim 2} - (\partial_{x^j} \Lambda^\sim)(\partial_{x^j} \Lambda^\sim) - (\partial_{x^j} \Lambda^\sim)(\Lambda^{\sim -1} [\partial_{x^j} \Lambda^\sim, \Lambda^\sim]) \ .$$

Use $[C_2]$,V,lemma 3.6, for $r=s=1, \rho=\varepsilon=0$ to show that the last term is compact. One finds the right hand side of (1.25) bounded, and in C. Similarly for HK^{-1}, and $(HK^{-1})(KH^{-1}) = (KH^{-1})(HK^{-1}) = 1$. The corresponding statements hold for $(K+\mu)(H+\lambda)^{-1}$ and $(H+\lambda)(K+\mu)^{-1}$ as well, since $K \geq 1$ and $K+\mu \geq K$ imply boundedness of $K(K+\mu)^{-1}$.

Similarly for $H(H+\lambda)^{-1}$. Thus one may show prop.1.5 for $\lambda=\mu=0$ only.

To show that im H_0K_0 is dense let $f \in H$ satisfy $(f,HKu)=0$ for all $u\in C_0^\infty(\mathbb{R}^n)$. Write this as $((HK^{-1})^*f,K^2u)=0$, $u\in C_0^\infty$. Since K_0^2 is essentially self-adjoint and ≥ 1, we have im K_0^2 dense, hence $(HK^{-1})^*f=0$. Since HK^{-1} was seen invertible, its adjoint is invertible as well. Thus $f=0$, q.e.d.

Continuing with prop.1.4, note that $J=[K,H]$ has order ≤ 3, and

$$(1.26)\qquad J = \partial_{x^j}[\Delta,g^{jk}]\partial_{x^k} - [\Delta,q^\sim] \ , \ \Delta = \sum\partial_{x^l}{}^2 \ .$$

where

$$(1.27)\qquad [\Delta,\phi] = \sum(\phi_{|x^l}\partial_{x^l} + \partial_{x^l}\phi_{|x^l}) \ .$$

Hence (1.24) is a finite sum of terms $R(\lambda)S(\mu)ZS(\mu)R(\lambda)$ with

$$(1.28)\qquad Z = D\phi \ , \ \phi D \ , \ D^2\phi D \ , \ D\phi D^2 \ ,$$

where D is any first derivative ∂_{x^j} , and $\phi \in C^\infty(\mathbb{R}^n)$ is o(1). Note that D commutes with $K=1-\Delta$ and $S(\mu)$. Thus we may write

$$(1.29)\qquad R(\lambda)S(\mu)ZS(\mu)R(\lambda) = R(\lambda)ES(\mu)FS(\mu)GR(\lambda) \ ,$$

with first order expressions E,F,G , either =1 or with coefficients vanishing at infinity, always at least one expression $\neq 1$.

Next we note the estimates

$$(1.30)\qquad \|H^sR(\lambda)\|\leq(1+\lambda)^{s-1},\ \|K^sS(\mu)\|\leq(1+\mu)^{s-1},\ \lambda,\mu\in(0,\infty),\ 0\leq s\leq 1,$$

as an easy consequence of the spectral theorem. Also, trivially,

$$(1.31)\qquad (u,Hu) \leq c(u,Ku) \leq c'(u,Hu) \ , \ u \in C_0^\infty(\mathbb{R}^n) \ .$$

This implies boundedness of H^sK^{-s}, $-1/2\leq s\leq 1/2$, by a result of Heinz and Loewner (cf.[C_2],I,5). Since $E\Lambda^\sim,\ldots,G\Lambda^\sim \in L(H)$, the operator at right of (1.24) is bounded. By continuous extension the commutator $[R(\lambda),S(\mu)]$ is a finite sum of terms (1.29), where the unbounded terms E,F,G may be compensated either by the resolvent to the left or to the right, at our choice.

The remainder of the proof consists of just balancing all the various effects occurring in the integrand

$$(1.32)\qquad \lambda^{-s}\mu^{-s}H^\sigma R(\lambda)ES(\mu)FS(\mu)GR(\lambda)H^\tau \quad .$$

Convergence of the integral needs a power $(1+\lambda)^{s-1-\delta}(1+\mu)^{s-1-\delta}$, with some $\delta>0$. Accordingly, in view of (1.29) we can absorb a total power (left and right) $H^{s-\delta+1}$, using the two factors $R(\lambda)$. For $0\le s<1/2$ this means that both $\sigma=\tau=s$ and $\sigma=2s$, $\tau=0$ give a converging integral of compact operators, as long as two of the first order expressions E,F,G can be absorbed by the factors $S(\mu)$. This can be done, because it still leaves a factor $(1+\mu)^{-1}$, hence a total power $\mu^{-s}(1+\mu)^{-1}$ to generate a convergent integral. For $s=1/2$, on the other hand, one may split $K^{-1/4}$ from the second $S(\mu)$ and $H^{-1/4}$ from the last $R(\lambda)$, for a term $K^{-1/4}GH^{-1/4}=(K^{-1/4}H^{1/4})(GH^{-1/2}+[H^{-1/4},G]H^{-1/4}) \in L(H)$, by $[C_2]$,V,lemma 3.6, while the integrand still is $O(\lambda^{-1/2}(1+\lambda)^{-3/4}\mu^{-1/2}(1+\mu)^{-3/4})$.

This completes the proof of prop.1.4.

To complete the proof of prop.1.3 we first observe that $A_1=HK^{-1} \in C_0$ was seen in the proof of prop.1.5. We have a formula like (1.25) where the last term is compact. Thus we calculate the symbol as $\sigma_{A_1}(x,\xi)=(g^{jk}\xi_j\xi_k+1)/(1+|\xi|^2)$. We already noticed above that $A_s = H^sK^{-s} \in L(H)$, $0\le s\le 1/2$. Let $B_s=A_s+K(H)$ be the coset (mod K) of A_s. We just show that $B_{1/2} \in L(H)/K(H)$ is the unique positive self-adjoint square root of $B_1 \in C_0/K(H) \subset L/K$. Since both algebras involved are C^*-algebras it follows that $B_{1/2} \in C_0/K$, and that it also is the positive root of B_1 in the smaller algebra. Thus $A_{1/2} \in C_0$ and its symbol is given by (1.14).

Indeed, the first relation (1.22) gives $(K^{-s}H^s)^{**}- H^sK^{-s}\in K$, so that B_s is hermitian, $0\le s\le 1/2$. From the second relation (1.22) we get $(H^sK^{-s})^2 = H^{2s}K^{-2s}+H^{2s}[H^{-s},K^{-s}](H^sK^{-s}) = H^{2s}K^{-2s}+C$, $C\in K$, so that $B_s{}^2=B_{2s}$ for $0\le s\le 1/2$. For $s=1/2$ we find that $B_{1/2}$ is a hermitian square root of B_1. We also get $B_{1/2} = (B_{1/4})^2\ge 0$, $B_{1/2}\in C_0/K$, $A_{1/2} \in C_0$, $\sigma_{A_{1/2}}\ne 0$, $A^{-1}_{1/2} =\Lambda_0^{-1}\Lambda^{\sim}\in C_0$, proving prop.1.3, and thm.1.1.

Note that the above proof of thm.1.1 will have to be repeated, below, to accomodate a variety of modified assumptions. While we were giving fairly general details, above, we later on will leave details to the reader.

2. Riemann spaces with pseudo-conical ends.

In this section we consider a compact (connected) manifold Ω_0 with boundary $\partial\Omega_0$. Assume that Ω_0 and its boundary are C^∞ , and let $\dim \Omega_0 = n$. For convenience we introduce a positive definite Riemannian metric on Ω_0 and denote the distance of $x,y \in \Omega_0$ by $\delta = \delta(x,y)$. In particular let $\delta(x) = \inf\{\delta(x,y):y\in\partial\Omega_0\}$ be the distance from x to the boundary $\partial\Omega_0$, and note that $\delta(x)$ is a C^∞-function for x sufficiently close to the boundary $\partial\Omega_0$.

On the other hand we introduce another Riemannian metric ds on the interior Ω of Ω_0 , and assume that ds becomes singular at the boundary $\partial\Omega_0$, in the following sense:

If $\delta(x),\theta(x)$, with the local coordinates $\theta(x)=(\theta^1,\ldots,\theta^{n-1})$ of the foot point of the lot from x to $\partial\Omega_0$, are used as local coordinates in a compact neighbourhood N_{x_0} of the boundary point x_0 then ds^2 is equivalent to

$$(2.1) \qquad ds^{\sim 2} = \delta^{-3}(d\delta^2 + \delta^2 d\theta^2) \text{ for } x \in N_{x_0} .$$

More precisely by equivalence we mean that

$$(2.2) \qquad ds^2 = \delta^{-3}(\gamma d\delta^2 + \delta^2 \sum \gamma_{jk} d\theta^j d\theta^k) \text{ , for } x \in N_{x_0} ,$$

with functions $\gamma(x)$, $\gamma_{jk}(x) \in C^\infty(N_{x_0})$, such that γ and the symmetric matrix $((\gamma_{jk}))_{j,k=1,\ldots,n-1}$ are bounded and bounded below by by positive constants in N_{x_0} . We then have c' , c" > 0 such that

$$(2.3) \qquad c' ds^{\sim 2} \le ds^2 \le c'' ds^{\sim 2} \text{ , for all } x \in N_{x_0} .$$

Moreover, we require that near each component $\partial\Omega_0'$ of $\partial\Omega_0$, for sufficiently small constant δ, the expression $ds''^2 = ds_\delta''^2 = \gamma_{jk} d\theta^j d\theta^k$ defines a metric on $\partial\Omega_0'$, depending smoothly on δ , also at $\delta=0$.

A Riemannian space Ω with metric ds^2 as above will be called <u>pseudo-conical</u>. Evidently such a space Ω is noncompact, unless $\partial\Omega_0 = \emptyset$. The motivation for this notation is the fact that, in suitable coordinates near $\partial\Omega$, the metric ds will be equivalent to the Euclidean metric on the chart, while the chart will have the general shape of a cone

$$(2.4) \qquad V = V(\varepsilon,Q) = \{x \in \mathbb{R}^n: |x| \ge \varepsilon , x/|x| \in Q\} ,$$

with $\varepsilon>0$, and an open set Q on the unit sphere $|x|^2 = \sum {x^j}^2 = 1$.

To verify this, let (δ,θ) be the above local coordinates in N_{x_0} . Then introduce new coordinates (r,θ) by setting

$$(2.5) \qquad r = \delta^{-1/2} , \theta = \theta ,$$

leaving the θ-variables unchanged. In these coordinates we get

(2.6) $$ds^2 = 4\gamma dr^2 + r^2\gamma_{jk}d\theta^j d\theta^k \ .$$

Assuming $N_{x_0}=(0,\delta_0)\times P$, with a neighbourhood P of x_0 on $\partial\Omega_0$, we get the infinite cylinder $(\delta_0^{-1/2},\infty)\times P$ in the coordinates (r,θ). Interpreting these as spherical coordinates in $\mathbb{R}^n$ (with radius r and $\theta = (\theta_1,\ldots,\theta_{n-1})$ shifted well away from the singularities of the coordinate system), the above cylinder will go into a cone (2.4), while indeed ds^2 is bounded above and below by cdx^2.

Moreover, the space Ω, near $\partial\Omega_0'$, has a product structure $(0,\varepsilon)\times\partial\Omega_0'$ with metric globally given by (2.6), i.e.

(2.6') $$ds^2 = 4\gamma dr^2+r^2ds_\delta''^2 \ , \quad \delta=r^{-2} \ ,$$

where γ and ds'' also are smooth at $\delta=0$, so that near the end $r=\infty$ the space approximately has the shape of an outside cone.

Moreover, since γ and γ_{jk} are smooth even at $\partial\Omega_0'$, the metric tensor, in these coordinates, satisfies (1.2),(1.3). We even get

(2.7) $$\partial_{x^j} = \kappa_j(\theta)\delta^{3/2}\partial_\delta + \delta^{1/2}\ \kappa_j^l(\theta)\partial_{\theta^l} \ ,$$

with bounded C^∞-functions κ_j, κ_j^l, defined for $\theta \in P$, implying that <u>all</u> partial derivatives vanish at $|x|=\infty$, and we even have

(2.8) $$\partial_x^\alpha g_{jk}(x) = O(|x|^{-|\alpha|}) \ , \quad \text{for every } \alpha \ .$$

We can extend the coefficients $g_{jk}(x)$ from the cone (2.4) into the entire $\mathbb{R}^n$. For example, assume $Q \subset\subset Q'$, with another open set Q' of the sphere. Then use

(2.9) $$(1-\chi_1\chi_2)\delta_{jk} + \chi_1\chi_2 g_{jk}$$

as such an extension, with cut-off functions $\chi_1(\theta)$, $\chi_2(r)$, $0\le\chi_j\le1$, $\chi_1=1$ near Q, $\chi_1\in C_0^\infty(Q')$, $\chi_2=1$ in (ε,∞), $\chi_2\in C_0^\infty((0,\infty))$. Then (1.1), (1.2), (1.3) hold in all of $\mathbb{R}^n$.

We have shown that, on a space with pseudo-conical ends, there exists a special type of 'boundary coordinates'. Clearly the manifold Ω_0 with boundary appears as a special compactification of Ω with set $\partial\Omega_0$ of infinite points. For every infinite point x_0 and sufficiently small neighbourhood N_{x_0} the intersection $N_{x_0}\cap\Omega$ is a chart with coordinates $x=(x^1,\ldots,x^n)$ as described, containing a conical set (2.4), on which the metric tensor extends to all of $\mathbb{R}^n$ in such a way that (1.1), (1.2), (1.3) hold. We will refer to the coordinates described as <u>distinguished coordinates</u>. Actually, a chart $U \subset \Omega$ with $\phi: U \to V(\varepsilon,\mathcal{Q})$, $\phi(U) = V$, V as in (2.4), arising from a product neighbourhood $N = (0,\delta)\times\mathcal{Q}$ in the manner described, will be called a <u>distinguished boundary chart</u>.

For distinguished boundary charts $\phi: U \to V$, $\phi': U' \to V'$ we introduce the order relation $U' \ll U$, to mean that (i) $U' \subset U$, (ii) $\phi' = \phi|U'$, (iii) $V = V(\varepsilon,\mathcal{Q})$, $V' = V(\varepsilon',\mathcal{Q}')$, where $\varepsilon' > \varepsilon$, $\mathcal{Q}' \subset\subset \mathcal{Q}$.

We now ask the question about the Laplace comparison algebra of a space Ω with above properties, using generating sets compatible with those of sec.1 in distinguished coordinates. To explore the ideal structure of such an algebra C it is natural to use the technique of 'algebra surgery' of $[C_2]$,VIII,3. We will find that commutators are compact. The general surgery techniques apply for a characterization of the symbol space. However a direct application of the results is impossible since cuts will be noncompact.

The generating sets $A^\#$ and $\mathcal{D}^\#$ are given as follows:

$A^\#$ and $\mathcal{D}^\#$ consist of all $a \in C^\infty(\Omega)$, and all first order differential expressions D with C^∞-coefficients, such that for each $x_0 \in \partial\Omega_0$ there exists a distinguished chart $V_{x_0} = N_{x_0} \cap \Omega$ such that, in the distinguished coordinates, we have $a|V_{x_0} = a_{x_0}|V_{x_0}$, $D|V_{x_0} = D_{x_0}|V_{x_0}$, with some $a_{x_0} \in A_0^\#$, $D_{x_0} \in \mathcal{D}_0^\#$, respectively.

Again C denotes the Banach-subalgebra of $L(H)$, $H = L^2(\Omega, dS)$, $dS = \sqrt{g}dx$ = surface measure of the metric ds^2, generated by the continuous linear operators a, $D(1-\Delta)^{-1/2}$: $a \in A^\#$, $D \in \mathcal{D}^\#$.

<u>Theorem 2.1</u>. The algebra C contains $K(H)$, and has compact commutators. Its symbol space $\mathbb{M}$ coincides with the set $\partial\mathbb{P}^*\Omega$ of infinite points in the compactification $\mathbb{P}^*\Omega$ of the cotangent bundle $T^*\Omega$ of Ω generated by the bounded continuous functions

(2.10) $a(x)$, $d(x,\xi) = (b^j(x)\xi_j + p(x))/(g^{jk}(x)\xi_j\xi_k + 1)^{1/2}$, $(x,\xi) \in T^*\Omega$,

where $a \in A^\#$, $D = -ib^j\partial_{x^j} + p \in \mathcal{D}^\#$. Moreover, we have

(2.11) $\sigma_a(x,\xi) = a(x)$, $\sigma_{D\Lambda}(x,\xi) = d(x,\xi)$, $(x,\xi) \in \partial\mathbb{P}^*\Omega = \mathbb{P}^*\Omega \setminus T^*\Omega$.

<u>Proof</u>. First of all the algebra C contains K and has compact com-

mutators, and is a C^*-algebra, by $[C_2]$,V,lemma 1.1 and V,thm.4.1. The latter must be applied with $q=q\hat{}=1$, $\Omega=\Omega\hat{}$, $\partial\Omega=\emptyset$, $\varepsilon=0$. It is found that $A^\#_m \subset A^\# \subset A^\#_c$, $A^\#_m \subset \mathcal{D}^\# \subset \mathcal{D}^\#_c$, with $A^\#_c$, $\mathcal{D}^\#_c$, $A^\#_m$, $\mathcal{D}^\#_m$ of $[C_2]$,V,4.

In V,(4.4),(4.5) we write $D=\sum\omega_j D+\sum\chi_j D$, with an atlas of finitely many interior charts U_j and distinguished boundary charts $V_j=V_{x_j}$, and partition $1=\sum\omega_j+\sum\chi_j$. Each $\omega_j D$ has compact support, and $\chi_j D$ has support in V_j . We may choose ω_j , χ_j as a partition of unity for the space Ω_0 , using the N_{x_j} instead of V_j as boundary charts. By definition of $A^\#$, $\mathcal{D}^\#$ one confirms ω_j, $\chi_j \in A^\#$, and $\omega_j D$, $\chi_j D \in \mathcal{D}^\#$. For later convenience we assume the set $\mathcal{Q}$ of (2.4) as a connected open subset of the sphere.

Accordingly the commutator ideal of $\mathcal{C}$ will be $K(H)$, and we focus on the quotient $\mathcal{C}/K$ only. Here we attempt an extension of $[C_2]$,VIII,thm.3.1, for the noncompact spaces $\Omega_1=\Omega$, $\Omega_2=\mathbb{R}^n$, with common subset $U=V_{x_0}$, a distinguished chart, with the chart homeomorphism used as identification. We may assume $U=V_{x_0}$ of the precise conical shape of (2.4). Clearly then thm.3.1 is not applicable, since we do not have ∂U compact.

4 On the other hand it is easy to repeat all the arguments of the proof under the special assumptions of the present case again. Let the dual map of the injection $\mathcal{C}_{A^\#} \to \mathcal{C}$ (with the C^*-function algebra $\mathcal{C}_{A^\#}$ generated by $A^\#$) be called ι again. The maximal ideal space $M_{A^\#}$ of $\mathcal{C}_{A^\#}$ is another larger compactification of Ω similar to the Stone-Cech compactification. Clearly $\iota:\mathbb{M}\to M_{A^\#}$ is surjective.

<u>Proposition 2.2</u>. Let $\mathcal{Z}$ denote the closed 2-sided ideal of $\mathcal{C}$ consisting of all operators with symbol vanishing outside of $\iota^{-1}U^{clos}$ (we refer to the closure U^{clos} of U in $M_{A^\#}$). Then $\mathcal{Z}$ is generated by $K(H)$ and the operators

$$\text{(2.12)} \qquad a\ ,\ D\Lambda\ ,\ a\in A^\#\ ,\ D\in\mathcal{D}^\#\ ,\ \operatorname{supp} a\ ,\ \operatorname{supp} D\subset U^\sim\subset U\ ,$$

where $U^\sim = V(\varepsilon^\sim,\mathcal{Q}^\sim) \ll U = V(\varepsilon,\mathcal{Q})$.

Proof. This follows similarly to $[C_2]$,VIII,prop.3.2: The operators (2.12) are clearly contained in Z . For an operator $A \in Z$ and a sequence χ_j of cut-off-functions with $\chi_j \in C^\infty(U)$, $0 \le \chi_j \le 1$, $\lim \chi_j = 1$ at each $x \in U$, supp $\chi_j \subset U^\sim$ for some $U^\sim \subset U$, we get $\sigma_A - \chi_j \sigma_A \to 0$ in $C(\mathbb{M})$, hence $\chi_j A \to A$ in C/K . Therefore there exists $C_j \in K(H)$ with $\|A - \chi_j A - C_j\| \to 0$, $j \to \infty$. Also A is approximated by a finite sum of finite products of generators. For such a product $P = G_1 \dots G_n$ write $\chi_j P = (\chi_{j,1} G_1)(\chi_{j,2} G_2) \dots\dots (\chi_{j,N} G_N) + C$, where $C \in K$ and $\chi_{j,1}$ have the same properties as χ_j with supports in larger cones of the same kind, supp $\chi_{j,1} \subset U^\wedge \subset\subset U$. (One constructs $\chi_{j,1}$ on Ω_0 as C^∞-cut off functions near x_0 with supp $\chi_{j,1} \subset N_{x_0}$, $\chi_j = \chi_{j,1} \dots \chi_{j,N}$).

All $\chi_{j1} G_1$ are of the form (2.12). Thus indeed A is approximated mod K by sums of products of (2.12) , q.e.d.

The set $U = V_{x_0}$ also appears as a subset of $\mathbb{R}^n$,and we may just as well introduce the ideal $Z_0 \subset C_0 \subset L(H_0)$, $H_0 = L^2(\mathbb{R}^n)$ of operators with symbol vanishing outside $\iota_0^{-1} U^{\mathrm{clos}}$, with ι_0 just as ι, with respect to $\mathbb{R}^n$ and C_0. Hence prop.2.2 also applies to Z_0. Note that $\mathbb{R}^n$ itself is a pseudo-conical space and that U is a distinguished boundary chart. The functions and first order differential expressions of (2.12) just as well are functions or expressions over Ω, or over $\mathbb{R}^n$.

Proposition 2.3. The commutative C^*-algebras $Z/K(H)$ and $Z_0/K(H_0)$ are isometrically isomorphic, (hence their corresponding structure spaces are homeomorphic). Moreover, the isomorphism is such that the cosets of a and $D\Lambda$ of the generators (2.12) are carried into the cosets of a and $D\Lambda_0$, respectively, with the corresponding interpretation of a and D as function (folpde) either over Ω or over $\mathbb{R}^n$, and with $\Lambda_0 = (1-\Delta_0)^{-1/2}$, Δ_0 denoting the Laplace operator (1.4) of the tensor g_{jk} of ds^2 in U , extended to $\mathbb{R}^n$ in the manner described.

Note that prop.2.3 will complete the proof of thm.2.1: The symbol space $\mathbb{M}$ of C contains the wave front space $\mathbb{W}$ of Ω, introduced in $[C_2]$,VI as the symbol space of the minimal algebra J_0 of Ω. $\mathbb{W}$ was seen homeomorphic to the bundle of spheres of radius infinity, compactifying the cotangent spaces. Also, (2.11) was already verified over $\mathbb{W}$, cf.$[C_2]$,VI,thm.2.2. Thus we only look at the set $\mathbb{M} \setminus \mathbb{W} = M_s^c$. We know that $\mathbb{M}_s^c \subset \iota^{-1}(\partial M_{A^\#})$, with $\partial M_{A^\#} = M_{A^\#} \setminus \Omega$. In turn, the restriction π to $\partial M_{A^\#}$ of the dual of $C(\Omega_0) \to C_{A^\#}$

provides a surjective map $\partial M_{A^\#} \to \partial\Omega_0$.

The isomorphism of prop.2.3 clarifies the structure of $\mathbb{M}$ near $\iota^{-1}\pi^{-1}x_0$, $x_0 \in \partial\Omega_0$. Observe that $\mathbb{Q}$ of sec.1 is determined by the symbols of C_0, as a class of bounded continuous functions over $\mathbb{R}^{2n}=T^*\mathbb{R}^n$. By prop.2.3 the set of such functions with support in $\iota^{-1}(U^{clos})$ coincides with the corresponding set of symbols of C_0. Thus the homeomorphism extends from $\mathbb{W}$ to $\mathbb{M}=\mathbb{W}\cup\mathbb{M}_s^c$, and (2.11) holds.

For the proof of prop.2.3 we introduce yet another comparison algebra. Consider U as a noncompact manifold of its own, a joint submanifold of Ω and $\mathbb{R}^n$, with the induced Riemannian metric, and with Hilbert space $H_U=L^2(U,dS)$. Generate a Banach-subalgebra C_U of $L(H_U)$ from the linear operators

(2.13) $\qquad a\ ,\ D\Lambda_U$: with a and D as in (2.12) .

Here we define

(2.14) $$\Lambda_U = H_U^{-1/2}\ ,\ H_U = q - \Delta\ ,\ q = 1+c_0\chi/\zeta^2\ ,$$

where $\chi\in C^\infty$, $\chi\geq 0$, denotes a cut-off function, equal to 1 in $U\setminus U^\wedge$, vanishing in $U^\sim$, with $U\rangle\rangle\ U^\wedge\ \rangle\rangle U^\sim$, all of the form (2.4). Also, $\zeta \in C^\infty(U)$, $\zeta>0$, is equivalent to the (Euclidean) distance from ∂U, and the constant $c_0>0$ is large enough to insure that the minimal operator $H_{U,0}$ is essentially self-adjoint (cf.[C_2],IV, thm.1.1. Actually we need a slight generalization of thm.1.1, combining the theorem with a separation of variables in the radial direction). Then H_U also denotes the unique selfadjoint extension of $H_{U,0}$ and the square root Λ_U of (2.14) is well defined.

The Hilbert space H_U may be interpreted as a closed subspace of the space H by extending an $L^2(U)$-function zero in $\Omega\setminus U$. Then one also obtains $L(H_U)$ as a closed subalgebra of $L(H)$, reduced by the orthogonal direct decomposition $H = H_U \oplus H_{\Omega\setminus U}$. If χ_U denotes the characteristic function of the subset $U\subset\Omega$ (i.e., $\chi_U=1$ on U, $=0$ on $\Omega\setminus U$) then the above identification is defined by the map $A \to A\chi_U$, with $A\chi_U: H \to H$, for $A \in L(H_U)$, in the evident sense.

<u>Theorem 2.4</u>. The algebra $C_U\subset L(H_U)$ is adjoint invariant, and independent of χ in (2.14). Moreover, the above identification $L(H_U)\to L(H)$ takes C_U into a subalgebra of the ideal $Z \subset C$. In fact,

(2.15) $$Z = C_U + K(H)\ ,\ C_U = \chi_U Z \chi_U\ .$$

For D of (2.12) we have

(2.16) $\quad D\Lambda - D\Lambda_U\chi_U \in K(H)$ whenever $\chi D = 0$, for χ of (2.14).

Again thm.2.4 implies prop.2.3, hence thm.2.1, since the same conclusion works on the ideal Z_0 and the subset U of $\mathbb{R}^n$. We trivially have $K(H_U) = L(H_U) \cap K(H)$. Therefore (2.15) implies

(2.17) $$Z/K(H) \cong C_U/K(H_U) \ .$$

Since the same holds for $U \subset \mathbb{R}^n$ we also get (2.17) for $Z_0/K(H_0)$, so that both must be *-isomorphic. The representation of symbols then is a consequence of (2.16).

Theorem 2.4 will be proven in sec.3 below.

3. Proof of theorem 2.4.

Regarding the proof of thm.2.4 we first notice that

(3.1) $\quad D\Lambda_U - (\Lambda_U D)^{**} \in K(H_U)$, for all D of (2.12).

All comparison algebras contain the compact ideal. Thus we get $(D\Lambda_U)^* = (\Lambda_U D^*)^{**} \in C_U$, so that C_U is adjoint invariant, hence a C^*-subalgebra of $L(H_U)$.

Next we focus on the stated independence of the comparison algebra C_U of the cutoff function χ in (2.14). This is a task very similar to that in thm.1.1 (or also of [C_2],VI,thm.1.1). In the previously treated two cases the assumptions were slightly different, forcing a somewhat different approach. However, the general method remains applicable. We only indicate the changes.

For a moment write again H and K for two expressions H_U with different cut-off functions $\chi=\chi'$ and $\chi = \chi''$, but with the same U and all other coefficients. It then follows that H-K = q' , where $q' \in C^\infty(\Omega_0)$, and supp $q' \subset U^\wedge \setminus U^\sim$, with $U^\sim \subset\subset U^\wedge \subset\subset U$, $U^\wedge$, $U^\sim$ of the form (2.4). The proof is a matter of showing that $H^{1/2}K^{-1/2} - 1 \in C_{U,H}$ again, where, for a moment, $C_{U,H}$ denotes C_U, with respect to H. This implies $C_{U,K} \subset C_{U,H}$ and "=" for reason of symmetry.

Attempting to repeat the argument of prop.1.3 we get $1-HK^{-1} = -q'K^{-1}$ bounded and in $C_{U,K}$, and the operators are mutual quasi-inverses. Hence both of them are in $C_{U,H} \cap C_{U,K}$.

The proof of prop.1.5 will remain intact, since we were insisting on a choice of c_0 in (2.14) such that $H^2_{U,0}$ is essentially self-adjoint, and since HK^{-1}, KH^{-1} are bounded. Then we need an analogue to prop.1.4, for the new operators H , K , and H_U. Using

(1.23) we focus on the new commutator [K,H] of (1.24).

We have H=K+q', hence [K,H]=[K,q'], an expression of the form (2.12), with coefficients o(1). In stead of the product of up to 3 first order expressions in (1.29) we have E=G=1 here, while the new form of F insures compactness of the integrand. Therefore prop.1.4 holds. Thus, again, let $B_s = H^s K^{-s} + K(H_U)$ be the coset of $A_s \in L(H_U)$, $0 \le s \le 1$. It again follows that $B_{1/2}$ is self-adjoint and positive, and that $(B_{1/2})^2 - 1 = B_1 - 1 \in C_{U,H}/K(H_U)$. Thus we have $B_{1/2}$ the positive square root of B_1 , in this case. Writing it as a resolvent integral, we find $B_{1/2} - 1 \in C_{U,H}/K(H_U)$, $A_{1/2} - 1 \in C_{U,H}$. Thus C_U is independent of the choice of χ .

Next we verify (2.16). For D of (2.12) let D=0 in $U \setminus U^\sim$, $U,U^\sim$ of (2.4), $U^\sim \subset\subset U$. For $\psi \in C^\infty(\Omega_0)$ with support in U, and =1 near $U^\sim$, we have $D = D\psi$, and $\psi\chi_U = \psi$. Using $[C_2]$,V,lemma 3.3, we get

$$(3.2) \qquad Iu = D\Lambda u - D\Lambda_U \chi_U u = D(\Lambda - \Lambda_U)\psi u + Cu, \text{ for all } u \in C_0^\infty(\Omega),$$

with some $C \in K(H)$ (We also used that $K\chi_U \in K(H)$, for $K \in K(H_U)$) . Here we express Λ and Λ_U by (1.17), using resolvents $R(\lambda) = (H-\lambda)^{-1}$, and $R_U(\lambda) = (H_U - \lambda)^{-1}$. It follows that

$$(3.3) \qquad (I-C)u = 1/\pi \int_0^\infty \lambda^{-1/2} d\lambda D\psi(R(\lambda) - R_U(\lambda))\psi u \ .$$

Here we write

$$(3.4) \qquad \psi(R-R_U)\psi = [\psi,R]\psi - (\psi R_U - R\psi)\psi \ .$$

For $v \in \mathrm{im}(H_{U,0}+\lambda)$, using $H=H_U$ in supp ψ, we get

$$(3.5) \qquad \psi R_U v - R\psi v = R(H+\lambda)\psi R_U v - R\psi(H_U+\lambda)R_U v = R[H,\psi]R_U v \ .$$

By essential self-adjointness of $H_{U,0}$ we may extend to H_U, and get

$$(3.6) \qquad \psi(R-R_U)\psi = R[H,\psi](R-R_U)\psi \ .$$

Here one uses that $[\psi,R] = R[H,\psi]R$, and that $[H,\psi]$ is of the form (2.12) again, as easily verified. Using (3.6) in (3.3) we get

$$(3.7) \qquad I = C + 1/\pi \int_0^\infty \lambda^{-1/2} d\lambda DR(\lambda)[H,\psi](R(\lambda) - R_U(\lambda))\psi \ .$$

In fact, $[H,\psi]$ not only is of the form (2.12), but its coefficients are o(1). Thus $[H,\psi]\Lambda^{1+\varepsilon} \in K(H)$ and $[H,\psi]\Lambda_U^{1+\varepsilon} \in K(H_U)$, $\varepsilon > 0$, (cf. $[C_2]$,V,3). The integrand is in K, and an estimate like

(1.30) implies that it is $O(\lambda^{-1/2}(1+\lambda)^{-1/2-\varepsilon})$. Hence the integral converges in norm and is a compact operator, confirming (2.16).

Now the remainder of the proof follows easily: Every finite sum of finite products of generators (2.12) differs from the corresponding sum of finite products of generators (2.13) (i.e., only Λ is always replaced by Λ_U) only by an additive term in $K(H)$. Thus we have $C_U^0 \subset C$. In fact, the symbol of an $A \in C_U^0$ vanishes outside $\iota^{-1}U^{clos}$, so that we even get $A \in Z$.

Note that $L(H_U)$ is a closed subalgebra of $L(H)$, with the same (operator-) norm. Taking closure in C_U^0 gives $C_U \subset Z$.

Vice versa, every $A \in Z$ may be approximated (in $L(H)$) by $A_j = A_{U,j} + K_j$, with $K_j \in K(H)$, $A_{U,j} \in C_U^0$. Then also $\chi_U(A_{U,j}+K_j)\chi_U$ converges in $L(H_U)$ to $A_U \in C_U$ (since evidently $\chi_U K_j \chi_U \in K(H_U) \subset C_U$). However, $A_j - \chi_U A_j \chi_U \in K(H)$, hence its limit also is in $K(H)$. We get $Z \subset C_U + K(H)$. Equality follows since $Z \supset K(H)$ and $Z \supset C_U$ was seen above.

The same conclusion above shows that $\chi_U Z \chi_U \subset C_U$, while we saw that $C_U^0 = \chi_U C_U^0 \chi_U \subset \chi_U Z \chi_U$, confirming the second relation (2.15).

This completes the proof of thm.2.4.

4. Global pseudodifferential operators in the algebra C.

In this section we will focus on a finitely generated subalgebra S^0 of our algebra C, of a space Ω with pseudo-conical ends. We will show that every $A \in S^0$ is a pseudo-differential operator of type $SG_1^0(\Omega)$ in the sense of Schrohe [$Schr_1$],[$Schr_2$]. Actually, we will show what seems to be a slightly stronger statement: The operators of S^0 are finite sums of local ψdo's in the algebra $Op\psi c_0$ of [CGO],[C_3]or [CP], in the distinguished boundary charts, and Hoermander type ψdo's in interior charts, and a 'regularizing operator'. Here 'regularizing' refers to an operator taking temperate distributions into rapidly decreasing functions, where it must be noted that both latter concepts make sense on our pseudo-cones.

We define $S^0 \subset C^0$ as the algebra finitely generated by a and $D\Lambda$, with $a \in A_S^{\#}$, $D \in \mathcal{D}_S^{\#}$, with the following classes $A_S^{\#}$, $\mathcal{D}_S^{\#}$:

$$A_S^{\#} = \{a \in A^{\#} : a_{x_0} \in A_{0,S}^{\#} \text{ for every } x_0 \in \partial\Omega_0\},$$

$$\mathcal{D}_S^{\#} = \{D \in \mathcal{D}^{\#} : D_{x_0} \in \mathcal{D}_{0,S}^{\#} \text{ for every } x_0 \in \partial\Omega_0\},$$

Here we refer to the definition of $A^{\#}$ and $\mathcal{D}^{\#}$ in sec.2, and to the following classes over $\mathbb{R}^n$:

$$A^{\#}_{0,S} = \{a\in A^{\#}_0 : a^{(\alpha)}(x)=\partial_x^\alpha a(x)=O(\langle x\rangle^{-|\alpha|}), \text{for all } \alpha\}.$$
(4.1)
$$\mathcal{D}^{\#}_{0,S} = \{D\in\mathcal{D}^{\#}_0 : D=b^j\partial_{x^j}+p\ ,\ b_j,\ p\in A^{\#}_{0,S}\}$$

Clearly the closure S of S^0 is a C^*-subalgebra of C, containing $K(H)$, with compact commutators. The dual of the injection $S/K\to C/K$ is a surjection $\pi_S:\mathbb{M}\to\mathbb{M}_S$, with the symbol space $\mathbb{M}_S$ of S. The duals of the injections $C(\Omega_0)\to C_{A_S^\#}\to C_{A^\#}$ define surjections

(4.2) $$M_{A^\#}\to M_{A_S^\#}\to\Omega_0\ ,\quad \partial M_{A^\#}\to\partial M_{A_S^\#}\to\partial\Omega_0\ .$$

From the principles of $[C_1]$,IV it follows that π_S is the lifting of the first map (4.2) onto $\partial\mathbb{P}^*\Omega=\mathbb{M}$. In other words, the symbol space $\mathbb{M}_S$ is obtained by identifying all points $(x,\xi)\in\mathbb{M}$ with $x\in\pi^{-1}(\partial\Omega_0)$, mapping into a given $y\in\partial M_{A_S^\#}$, under $(4.2)_2$.

Proposition 4.1. Every finite atlas consisting of interior charts (with compact closure) and distinguished boundary charts, in the sense of sec.2, induces an SG_1-compatible structure on Ω.
Proof. This is a straight-forward calculation, based on the fact that each of our boundary charts is diffeomorphic to a conical set of type (2.4). In particular, the transition between two types of boundary coordinates satifies cdn.'comp.1' of $[\mathrm{Schr}_2]$,4.1,we have the 'shrinking condition' met for every chart, and have already used a partition of unity, as required, q.e.d.

From now on we assume that Ω is equipped with the SG_1-structure of prop.4.1. For a ψdo A with symbol $a\in SG_1^m(\Omega)$, in the sense of $[\mathrm{Schr}_{1,2}]$ we will write $A\in OpSG_1^m(\Omega)$.

Theorem 4.2. All operators a , $D\Lambda$, ΛD , for $a\in A_S^\#$, $D\in\mathcal{D}_S^\#$, are pseudodifferential operators in $OpSG_1^{(0,0)}$. Accordingly, the entire algebra S^0 consists of ψdo's in $OpSG_1^0$, where $0=(0,0)$.

For the proof let $1=\sum\omega_j$ be the partition of the proof of thm.2.1, where we now do not distinguish in notation between interior charts and boundary charts. The corresponding (interior or boundary) charts will be called U_j. They are identified with corresponding subsets of $\mathbb{R}^n$, (as in (2.4) for boundary charts). Let $\psi_j\in C^\infty(U_j)$, $0\le\psi_j\le 1$, supp $\omega_j\subset U_j^\wedge\langle\langle U_j^\sim\subset$supp $\psi_j\subset U_j$, $\psi_j=1$ in $U_j^\wedge$, where $U_j^\sim\langle\langle U_j^\wedge\langle\langle U_j$, where "$\langle\langle$" means "$\subset\subset$" for interior charts.

For $A = a$, $D\Lambda$, ΛD , one of the generators of $\mathcal{S}^0$, write

$$A = \sum_j \omega_j A \omega_j + \sum_{j\neq 1} \omega_j A \omega_1 \psi_j + \sum_{j\neq 1} \omega_j A \omega_1 (1-\psi_j) . \tag{4.3}$$

Notice that the terms of the first two sums in (4.3) contain multipliers left and right of A both having support in $U_j^{\vee} \langle\langle U_j$, while the multipliers of the last sum have disjoint supports, i.e. $\text{supp}\ \omega_j \Subset U_j^{\wedge} \langle\langle U_j^{\blacktriangle}$, $\text{supp}\ \omega_1(1-\psi_j) \subset (\Omega \backslash U_j^{\blacktriangle})^{\text{clos}}$. The operators of the first two sums thus may be thought to map $L^2(\mathbb{R}^n) \to L^2(\mathbb{R}^n)$.

Since SG_1^0 is an algebra, thm.4.2 follows from prop.4.3,below:

Proposition 4.3. For χ with $\text{supp}\ \chi \Subset U_j^{\wedge}\langle\langle U_j$, and A as above we have $\chi A \chi \in Op\psi c_0 = OpSG_1^0(\mathbb{R}^n)$. For ϕ, $\chi \in C^\infty(\Omega_0)$ with disjoint supports we have $\phi A \chi$ (infinitely) regularizing, in the sense that

$$\gamma^N \Lambda^{-N} \phi A \chi \Lambda^{-N} \gamma^N \in L(H) \ , \ N=1,2,\dots \ , \tag{4.4}$$

with the function $\gamma(x) = \sum(\omega_j(x)\langle x\rangle)$, where in the j-th term of this sum $\langle x\rangle$ stands for $\langle x\rangle=(1+|x|^2)^{1/2}$, in the coordinates of U_j.

Proof. Regarding the second statement we refer to [C_2],IX,prop.5.7 which needs the same modification we have been practicing througout this paper, extending statements for 'compactly supported surgery' to 'conical surgery'. We shall not discuss the changes here.

The first statement is obvious for A=a, in view of (4.1). For the other two cases it follows trivially from prop.4.4,below.

Proposition 4.4. For χ of prop.4.3 we have

$$\chi \Lambda \chi \in Op\psi c_{(-1,0)} = OpSG_1^{(-1,0)}(\mathbb{R}^n) . \tag{4.5}$$

Proof. In view of the results of [CLG] it suffices to show that

$$\langle x\rangle^{|\alpha|}((\text{ad}\ \partial)^\alpha (\text{ad}\ x)^\beta A)\Lambda_0^{-|\beta|-1} \in L(H_0) , \tag{4.6}$$

for all multi-indices α,β , where $(\text{ad}\ \partial)^\alpha = \Pi(\text{ad}\ \partial_{x^j})^{\alpha_j}$, $\text{ad}\ \partial_{x^j} B = [\partial_{x^j}, B]$, and similarly for $(\text{ad}\ x)^\beta$. In (4.6) we refer to the distinguished coordinates of $U \subset \mathbb{R}^n$, of course, and let $H_0=L^2(\mathbb{R}^n)$, $\Lambda_0 = (1-\sum \partial_{x^j}^2)^{-1/2}$.

Indeed, by [CLG], thm.1.2, A is a ψdo in $Op\psi t_0$, (4.6) implying $(\text{ad}\ \partial)^\alpha(\text{ad}\ x)^\beta A \in L(H_0)$, i.e., e_A' to be C^∞ over the Heisenberg group. Moreover, (4.6) gives $A=a(x,D)$ with $(\langle x\rangle^{|\alpha|} a(x,\xi)\langle \xi\rangle^{|\beta|+1})_{(\alpha)}^{(\beta)}$

the symbol of an $A_{\alpha\beta}\in L(H)$, for all α,β. Thus $\langle x\rangle^{|\alpha|}a(x,D)\langle D\rangle^{|\beta|+1}$ is a ψdo in $Op\psi t_0$, giving the desired estimate

$$a^{(\beta)}_{(\alpha)}(x,\xi) = O(\langle x\rangle^{-|\alpha|}\langle\xi\rangle^{-1-|\beta|}) \text{ , for all } \alpha \text{ , } \beta \text{ .} \tag{4.7}$$

For the proof of (4.6) we introduce a short-hand notation: $L_{j,k}$ denotes the class of all expressions in $SG_1^{(j,k)}$ which are finite sums of products $\gamma^k D_1\ldots D_r$, $r\leq j$, with $D_\mu\in \mathcal{D}_S$. By P_{lkr} we mean the collection of all finite sums of products

$$L_0R(\lambda)L_1R(\lambda)\ldots\ldots L_{s-1}R(\lambda)L_s \text{ , } s\geq r \text{ ,} \tag{4.8}$$

with s factors $R(\lambda) = (H+\lambda)^{-1}$, and $L_\mu\in L_{j_\mu k_\mu}$, $\mu=0,1,\ldots,s$, and

$$j_0+j_1+\ldots+j_s-2s=l \text{ , } k_0+k_1+\ldots+k_s=k \text{ .} \tag{4.9}$$

We assume $s,j_\mu,k_\mu\in\mathbb{Z}$, $j_\mu\geq 0$, $s\geq 0$, otherwise arbitrary.

We focus on a distinguished boundary chart $U_j=U$. The case of an interior chart is similarly, not to be discussed. We no longer distinguish in notation between different functions χ, as described, and note thate $\chi=\chi\cdot\chi$ (where one of the factors χ is =1 in 'supp' of the other one). Also, $\partial_j=\partial$ and $x_j=x$ denote one of the local partial derivatives (coordinates). Note however, that, in the commutator $[\partial,A]$, $A=\chi\Lambda\chi$, we always may substitute ∂ by $\chi\partial$. Similarly for commutators $[x,A]$, and higher order commutators.

<u>Proposition 4.5</u>. For $L\in L_{jk}$, and the function γ of (4.4) we have

$$L(\gamma)^s = O(\gamma^{s+k}) \text{ , } s\in\mathbb{R} \text{ , } j,k\in\mathbb{Z} \text{ .} \tag{4.10}$$

The proof is a calculation.

<u>Proposition 4.6</u>. For $A\in P_{lkr}$, $l<0$, $0\leq\delta\leq|l|$ we have

$$\|\gamma^{-k}A\Lambda^{-\delta}\|_{L^2(\Omega)} = O((1+|\lambda|)^{-(|l|-\delta)/2}),\ 0\leq\lambda<\infty \text{ .} \tag{4.11}$$

<u>Proof</u>. The key is relation (4.12), derived similarly as (1.24):

$$[R(\lambda),\gamma^{-1}] = \gamma^{-1}R(\lambda)[H,\gamma]R(\lambda)\gamma^{-1},\ 0\leq\lambda<\infty \text{ .} \tag{4.12}$$

Again $H|C_0^\infty$ is e.s.a., and $\gamma C_0^\infty=C_0^\infty$, so that $(H_0+\lambda)\gamma C_0^\infty$ is dense. By (4.12) we may assume $k_1=k_2=\ldots=k_r=0$, so that all powers of γ are occurring in L_0. Excessive powers of γ occur while commuting, but enough are at the left hand side; the others may be ignored. Also,

$$[R(\lambda),\gamma] = R(\lambda)[\gamma,H]R(\lambda) \text{ ,} \tag{4.13}$$

where $[\gamma,H] \in L_{1,0}$, while $\gamma^{-1} \in L_{0,0}$. Furthermore, while commuting $D \in \mathcal{D}_S^{\#}$ or $\gamma^{\pm 1}$ with $R(\lambda)$ we will increase s by 1 .

Also, since $l<0$, we can redistribute the j_μ such that

(4.14) $\quad j_0=\dots j_{\nu-1}=2$, $\nu=[s+1/2]$, $j_\nu=0$ or 1 , $j_{\nu+1}=\dots=j_r=0$,

where the commutator relations for $[D,R(\lambda)], D\in\mathcal{D}_S^{\#}$ have to be used .

<u>Proposition 4.7</u>. We have $(\Lambda^m\chi\Lambda_0^{-m})^{**} \in L(H_0,H)$.
<u>Proof</u>: It suffices to show that $\quad A_m=\Lambda_0^{-m}\chi\Lambda^m=(\Lambda^m\chi\Lambda_0^{-m})^* \in L(H,H_0)$.

For even $m=2k$ write $A_m= (1-\Delta_0)^k\chi\Lambda^{2k}$. Note that $(1-\Delta_0)^k\chi\in L_{2k,0}$ it may be written as a finite sum of products of at most m expressions in $\mathcal{D}_S^{\#}$. For odd $m=2k+1$ write $\Lambda_0^{-m}\chi=(\Lambda_0-\sum(\partial_j\Lambda_0)\partial_j)(1-\Delta_0)^k\chi$,

showing that now $A_m = \sum_{j=0}^n S_jL_j\Lambda^m$, $S_j\in L(H_0)$, $L_j\in L_{m,0}$. Thus, in any case, we must show that $D_1\dots D_r\Lambda^m \in L(H)$, $r\leq m$. (Note $L(H,H_0)$ and $L(H)$ mean the same, due to the factor χ at left.) This again amounts to well known commutator rules (cf.[C_2],IX,thm.7.5.),q.e.d

After the above, for (4.6), it suffices to show that

(4.15) $$\gamma^{|\alpha|}((\mathrm{ad}(\chi\partial))^\alpha(\mathrm{ad}(\chi x))^\beta A)\Lambda^{-1-|\beta|} \in L(H) \ , \ A = \chi\Lambda\chi \ .$$

For a first step write $[\chi\partial,\chi\Lambda\chi]=[\chi\partial,\chi]\Lambda\chi+\chi[\chi\partial,\Lambda]\chi+\chi\Lambda[\chi\partial,\chi]$, confirm that $[\chi\partial,\chi] \in L_{0,-1}$, and use the resolvent formula (1.17) for

(4.16) $$[\chi\partial,\chi\Lambda\chi]=L\Lambda+M\Lambda^3+\int_0^\infty\lambda^{-1/2}P(\lambda)d\lambda,\ P(\lambda)\in P_{-3,-2,2}\ ,\ L\in L_{0,-1},$$

$M\in L_{2,-1}$. Prop.4.6 implies $P(\lambda)=O((1+|\lambda|)^{-3/2})$, yielding convergence of the integral of $\gamma[\chi\partial,A]\Lambda^{-1}$, for (4.15) with $|\alpha|=1$.

For successive commutators we now look at the commutators of $\chi\partial$ or χx with the integrand of (4.15). In particular,

(4.17) $$[\chi\partial,L] \in L_{j,k-1}\ ,\ [\chi x,L] \in L_{j-1,k}\ ,\ \text{for } L \in L_{jk}\ ,$$

and

(4.18) $$[\chi\partial,R(\lambda)]=R(\lambda)[H,\chi\partial]R(\lambda) \in P_{-2,-1,2}\ ,\ [\chi x,R(\lambda)] \in P_{-3,0,2},$$

due to $[H,\chi\partial]\in L_{2,-1}$, $[H,\chi x]\in L_{1,0}$. Also, the formula ([C_2],VI(1.6)),

(4.19) $$\int_0^\infty\lambda^{-1/2}R^m(\lambda)d\lambda = (-1)^{m+1}/\pi\binom{m-1/2}{m-1}\Lambda^{2m-1}\ ,$$

and the evident algebraic properties of the classes L_{jk} , P_{lkr} will have to be used. While we have listed all ingredients of an induction proof, we leave details to the reader.

Corollary 4.8. For every chart $U=U_j$, and a function χ as above, the ψdo-symbol $a\in\psi c_{-1,0}$ of $A=\chi\Lambda\chi$, $\Lambda=(1-\Delta)^{-1/2}=H^{-1/2}$, is given by

$$(4.20)\qquad a=\chi^2(x)(g^{jk}(x)\xi_j\xi_k+1)^{-1/2} \text{ , mod } \psi c_{(-2,-1)} \text{ .}$$

Moreover, for a general $A\in S^0$ the algebra symbol σ_B of $B=\chi A\chi$ equals the restriction to $\partial\mathbb{Q}$ of the extension to $\mathbb{Q}$ of the ψdo-symbol of B

Proof. Use that $HA^2=\chi^4+H[\chi,\Lambda]\chi^2\Lambda\chi+H\Lambda[\chi^3,\Lambda]\chi$ where the commutator terms are in ψc_{-e} . Here H is md-elliptic. Using calculus of ψdo's one concludes that $a^2=\chi^4(x)(g^{jk}(x)\xi_j\xi_k+1)^{-1}$ mod $\psi c_{(-3,-1)}$. In the set where $\chi=1$ we then get $a=(g^{jk}\xi_j\xi_k+1)^{-1/2}(1+b)^{1/2}$, with b of order -e, and then $(1+b)^{1/2}=1+c$, $c\in\psi c_{-e}$. Using $c\cdot c=c$ again, one then confirms (4.20). Then calculus of ψdo's may be used to obtain all symbols of $A\in S^0$, modulo SG_1^{-e} , and cor.4.8 follows.

Thm.4.2 and cor.4.8 show the relation between $OpSG_1^0$ and S :

(1) The algebras $OpSG_1^0(\Omega)$ and C (of sec.2) contain the common subalgebra S^0 , above. The Fredholm operators in S^0 are precisely the md-elliptic operators of order 0, or also the operators with non-vanishing symbol σ , defined over $\mathbb{M}$. As a consequence of cor.4.8 both conditions mean the same. For a proof one either may apply thm.2.1 or else construct a K-parametrix in the sense of [CP],[Schr$_j$] . Each method applies for more general A ; the ψdo-method for $A\in OpSG_1^0(\Omega)$, and the C^*-algebra method for $A\in C$.

(2) Even for a general SG-compatible structure the metric

$$(4.21)\qquad ds^2=\sum\omega_j dx^2 \text{ ,}$$

with dx^2 in the j-th term denoting the Euclidean metric of the j-th chart, will make the space essentially pseudo-conical; If the boundary charts have the shape of (2.4) then one will get a precisely pseudo-conical space, in the sense of sec.2. In general conditions 'comp 1,2' still will insure a conical shape.

(3) Starting from our general algebra C^0 a chain of 'Sobolev algebras' C_s , $s=(s_1,s_2)\in\mathbb{R}^2$, may be formed, by norm closing the finitely generated algebra C^0 in the L^2-Sobolev spaces H_s, with norms

$$(4.22)\qquad \|u\|_s=\|\gamma^{s_2}\Lambda^{-s_1}u\| \text{ , } \|.\| \text{ the norm of } H \text{ .}$$

Basic in this process is the proposition, below.

Proposition 4.9 For every $A\in C^0$ we have

$$(4.23) \qquad (\gamma^{s_2}\Lambda^{-s_1})^{-1}A(\gamma^{s_2}\Lambda^{-s_1}) - A \in K(H) \ , \ s \in \mathbb{R}^2 \ .$$

Proof. Apply the resolvent expansion of $[C_2]$,V,thm. 6.8, and a corresponding expansion, substituting Λ^{-1} with γ. The proof then runs parallel to that of $[C_1]$,IV,lemma 3.1 (or $[C_2]$,IX,(4.4), with $E = K(H)$).

With prop.4.9 one can repeat the argument of $[C_1]$,IV,sec's 3 and 4 , leading to an algebra

$$(4.24) \qquad \Psi\mathcal{D} = \cup\{\Psi\mathcal{D}_s : s\in R^2\} \ , \text{ with } \Psi\mathcal{D}_s=\{A\gamma^{s_2}\Lambda^{-s_1} : A \in C_\infty\},$$

where C_∞ again denotes the Frechet-*-closure of C^0 in the locally convex topology generated by all norms (4.22). In particular $C_\infty \subset L(H)$ is a ψ^*-algebra again, in the sense of Gramsch $[G_1]$.

The (doubly) graded algebra $\Psi\mathcal{D}$ then contains global ψdo's of all (multiplication and differentiation) orders, which may be md-elliptic, supplying a theory of ψdo's with Green inverses.

(4) While there is a strong parallel of the above two theories in the case of pseudo-conical spaces, it appears that our Fredholm results on spaces with cylindrical ends or with ends being products of cones and cylinders (as discussed in [CPo] and in sec.5 below) are not at all similar to those obtainable by the 'K-parametrix-method', since there we constructed a second singular-integral-operator-valued 'γ-symbol' over infinity, to be invertible, in addition to the requirement $\sigma\neq 0$ for the complex-valued 'σ-symbol'.

Of course, the existence of an SG-compatible structure on a space Ω with cylindrical ends is trivial, since the metric may be conically expanded at each end. Schrohe $[Schr_1]$,thm.4.24, constructs an explicit such structure by the coordinate transform $(\theta,t) \to (u,t)=(t\theta,t)$, in local cylindrical coordinates, for t large. He mentions the example of $[C_2]$,VIII,4 ,in this connection.

In the interest of avoiding confusion, perhaps we should mention here, that the md-elliptic operators of Schrohes SG-structure on a cylindrical end do not match our (or Lockhart-McOwen's) uniformly elliptic operators on cylindrical ends. For example, the (uniformly elliptic) Laplace operator of the cylinder's metric $dt^2+d\theta^2$ transforms into a second order elliptic operator with principal part matrix degenerating at $t=\infty$, so not elliptic or md-elliptic of any multiplication order, in the (u,t)-coordinates, as a calculation shows. This explains the very different type of result, in thm.5.3, below.

(5) It is known, on the other hand, that cylindrical ends, conical ends and conical tips all are conformally equivalent. In 'spherical coordinates' we have $ds^2=dr^2+r^2d\theta^2$, as metric of the conical end, ($r>>1$, $\theta\in B$). The metric of the tip is $r^{-4}(dr^2+r^2d\theta^2)$ (using $r\to 1/r$) . For the cylinder we get $r^{-2}(dr^2+r^2d\theta^2)$ (using $r \to e^r$) . Since these are conformal transformations, the principal part of Δ changes only by the factor of conformity. so that $\Delta_{tip} \equiv r^4\Delta_{end}$, $\Delta_{cyl} \equiv r^2\Delta_{end}$ (mod lower order terms). These operators are elliptic of order 4 and 2, resp., but have no zero order terms hence are not md-elliptic. Only the operators $r^4-\Delta_{tip}$, and $r^2-\Delta_{cyl}$ (with singular potentials) are md-elliptic. This again attests to the dissimilarity of results.

5. More general spaces.

In this section we will look at a class of noncompact spaces which are not pseudo-conical in the sense of sec.2, but with ends being products of a cone and a compact space, leading to a poly-cylindrical shape of the type discussed in [CPo].

Consider a compact manifold Ω_0 with boundary $\partial\Omega_0$, with an auxiliary C^∞-Riemannian metric $d\hat{s}^2$ defined on all of Ω_0 . Our manifold Ω again will be the interior of Ω_0 . We assume each component $\partial\Omega_0'$ of $\partial\Omega_0$ to be a product $\Theta\times B$ of n'-1- (and n"-)dimensional compact manifolds Θ (and B), respectively, with n'+n"=n. There exists a neighbourhood N of $\partial\Omega_0'$ in Ω_0 of the form $I\times\Theta\times B$, with $I=[0,1]$. Just define $N=\{x\in\Omega:\delta=\mathrm{dist}(x,\partial\Omega_0')/\varepsilon\le 1\}$. For small $\varepsilon>0$ we use the pair (δ,y) with $0\le\delta\le 1$ and with the foot point y of the lot from x to $\partial\Omega_0'$ as coordinates. Then N is a product as desired.

Assume for each connected component of $\partial\Omega_0$ a product structure as explained, and even a product neighbourhood $I\times\Theta\times B$, where where Θ and B need not to be the same for different components. The metric $d\hat{s}^2$ and its relation to the product structure of N are inessential, while, on the other hand, the product structure of N is to be kept invariant under permissible changes of coordinates.

Write $y=(\theta,z)$, $\theta\in\Theta$, $z\in B$. Consider a metric ds^2 on Ω, singular near $\partial\Omega_0'$ of the form

$$(5.1)\qquad d\tilde{s}^2 = \delta^{-3}(d\delta^2 + \delta^2 d\theta^2) + dz^2 \ , \ (\delta,\theta,z) \in N \ .$$

More precisely, similarly as in (2.2), we assume that

$$(5.2)\qquad ds^2 = \delta^{-3}(\gamma d\delta^2 + \delta^2\gamma_{jk}d\theta^j d\theta^k) + \beta_{jk}dz^j dz^k \ ,$$

with real-valued C^∞-functions γ , γ_{jk} , β_{jk} defined in local coordinates of the strip N , such that γ , $((\gamma_{jk}))$, $((\beta_{jk}))$ all are bounded and bounded below by positive constants. In particular, $ds''^2 = \beta_{jk}dz^j dz^k$ defines a C^∞- Riemannian metric on B , independent of δ,θ , and we also assume γ , γ_{jk} independent of z , and $\gamma \in C^\infty(I\times\Theta)$, $ds'^2 = \gamma_{jk}d\theta^j d\theta^k$ a C^∞-metric on Θ, for constant δ, depending smoothly on δ , just as in sec.2.

With the change $r=\delta^{-1/2}$ of coordinates as in sec.2, regarding (δ,θ), with z unchanged, we get local coordinates of the form (t,z) , where $t\in \mathbb{R}^{n'}$ varies over a conical set U as in (2.4), with n replaced by n', while z varies over B . Moreover, we have

$$(5.3)\qquad ds^2=g_{jk}dt^j dt^k+d\sigma''^2 \ , \ g_{jk}=g_{jk}(t) \ , \ ds''^2=\beta_{jk}(z)dz^j dz^k \ ,$$

where again the metric tensor g_{jk} may be considered extended to $\mathbb{R}^{n'}$ to satisfy (1.2),(1.3) (for n'). We again get (2.8). Moreover, g_{jk}, as a function of (δ,θ) is $C^\infty(I\times\Theta)$, by a calculation.

We now again introduce distinguished charts and distinguished coordinates: If Θ' and B' are charts on Θ and B, respectively corresponding to the subsets T' and B' of $\theta \in \mathbb{R}^{n'-1}$ and $z \in \mathbb{R}^{n''}$, then a distinguished chart is given by $V'=(0,1)\times\Theta'\times B' \subset N$, above. Also, distinguished coordinates are given by (r,θ,z) (in spherical form, with $r=\delta^{-1/2}$), and by (t,z), where $t \in \mathbb{R}^{n'}$ is the cartesian coordinate of the point (r,θ), with respect to any regular change from rectangular to spherical coordinates. Moreover, for a chart Θ' of Θ the set $W = (0,1)\times\Theta'\times B \subset N$ needs not to be a chart, of course. However, we may introduce our spherical coordinates (r,θ) and cartesian coordinates t in all of W , i.e. the entire W is homeomorphic to a set U×B of the poly-cylinder $\mathbb{R}^{n'}\times B$, with a conical subset U of $\mathbb{R}^{n'}$.
Such a set W will be called a distinguished poly-chart.

Let $A^\#$ denote the class of all restrictions to Ω of functions in $C^\infty(\Omega_0)$, and let $\mathcal{D}^\#$ denote the set of all first order C^∞-expressions on Ω which in every distinguished boundary chart U×B are linear combinations of 1, ∂_{t_j} and ∂_{z_l} with coefficients extending to $C^\infty(\Omega_0)$-functions. Let C again be the Laplace comparison algebra on the space Ω with metric ds^2 of (5.2), (5.3), with these $A^\#$, $\mathcal{D}^\#$. These classes are more restricted than those of sec.2.

In case of n'=n, i.e. B is a (0-dimensional) point, we get a C^*-sub-algebra T of S of sec.4. The symbol space of the (compact commutator algebra) T is the boundary $\partial T^*\Omega_0$ of the directional compactification of the cotangent space of Ω_0, i.e., the disjoint union of the wave front space $\mathbb{W}=\infty S^*\Omega$, and the parts over $\partial\Omega_0$ of the entire directionally compactified cotangent space $T^*\Omega_0 \cup \infty S^*\Omega_0$.

On the other hand, for $n'\geq 1$ and $\Omega_0 = B^{n'}\times B$, with the unit ball $B^{n'} = \{|s|\leq 1\}$ of $\mathbb{R}^{n'}$ and a compact manifold B of dimension n" we shall find that the algebra C precisely coincides with the poly-cylinder algebra C of [CPo], called C_0 here, for distinction (thm.5.1, below). We write $\Omega^\Delta = \mathbb{R}^{n'}\times B$, $\Omega^\Delta{}_0 = B^{n'}\times B$. The product neighbourhood N of $\partial\Omega^\Delta$ is defined as $F_\varepsilon\times B$, with the spherical shell $F_\varepsilon = \{s\in\mathbb{R}^{n'}: 1-\varepsilon\leq|s|\leq 1\}=[1-\varepsilon,1]\times S^{n'-1}\times B$, $\Theta=S^{n'-1}$, induced by spherical coordinates in $\mathbb{R}^{n'}$. We write $\delta=1-|s|$, and then relate the shell F_ε to $T = \{t\in\mathbb{R}^{n'}: |t|\geq\varepsilon^{-1/2}\}$, of the form (2.4), the outside of a ball in $\mathbb{R}^{n'}$, using the transformation $r=\delta^{-1/2}$ of sec. 2 again, leaving $z\in B$ and angular coordinates $\theta\in S^{n'-1}$ unchanged. This defines distinguished boundary coordinates in $U=T\times B\subset\mathbb{R}^{n'}\times B$. Our metric ds over $\Omega^\Delta=\mathbb{R}^{n'}\times B$ is defined by (5.2) or (5.3) in U, and is C^∞, positive definite, otherwise arbitrary, in $\Omega^\Delta\setminus U$. Thus, over Ω^Δ we have two Riemannian metrics - the above metric ds^2 ,and the poly-cylinder metric of [CPo], (1.1), called ds_0^2 here .

It is essential also, that the <u>same</u> metric ds''^2 over B must be used for both ds^2 of (5.3) and ds_0^2 of [CPo],(1.1). The generating classes of [CPo],(1.2) are smaller than our above classes, written for this special case, but will generate the same C^*-algebra C_0, by the Stone-Weierstrass theorem, if used with the Laplace operator Δ_0 of ds_0^2 . On the other hand, with the larger classes above and the Laplace operator Δ of ds^2 another C^*-algebra C^Δ is generated. We then first must prove that both algebras coincide.

In the general case $n'\geq 1$ we again want to perform algebra surgery to compare pieces of the symbol spaces of C related to the distinguished poly-charts with those of the corresponding spaces for the algebra in [CPo] .

It is clear that commutators no longer will be compact. Also an interpretation of the generators of the algebra C as pseudodifferential operators related to an SG_1-structure no longer seems to be useful, since the corresponding charts have a cylindrical, not a conical form in the z-direction.

We get ready for algebra surgery by identifying a distinguished poly-chart $W = (0,\varepsilon)\times\Theta'\times B$ with the corresponding subset $U\times B$ of the poly-cylinder $\mathbb{R}^{n'}\times B$. Here we assume that U has the precise conical form (2.4). Again the results of $[C_2]$,VIII,3 will not be applicable directly, since the cut ∂W is noncompact.

Theorem 5.1. The algebra $\mathcal{C}^{\blacktriangle}$ coincides with the algebra C_0 of the polycylinder $\Omega^{\blacktriangle}$ discussed in [CPo].

For the proof we again only indicate the changes to be made in the proof of thm.1.1, discussing the special case n'=0. The essential point is that the permitted changes of the metric still only account to compact changes of the generators, although the commutator ideal no longer is $K(H)$.

First, the change of dependent variable to equate the volume elements of the metrics ds and ds_0 as in prop.1.2 remains intact. For ds of (5.3) we get $dS=\sqrt{g}dtdS''$, with the volume element dS" of ds", while ds_0 has volume element $dS_0=dtdS''$. Here $\kappa=\sqrt{g}$ is independent of z, since the g_{jk} were independent of z. In particular, κ has all derivatives vanishing at ∞. The classes $A^{\#}$ and $\mathcal{D}^{\#}$ remain 'effectively invariant'.- Only the expressions of $\mathcal{D}^{\#}$ suffer a perturbation by an additive 0-order term vanishing at ∞, yielding a compact term, if multiplied by Λ. Therefore formulas (1.16) remain true. Since every comparison algebra contains $K(H)$, we indeed may change the dependent variable, just as in sec.1, and have the comparison algebra unchanged.

Then it again suffices to show that $K^{1/2}H^{-1/2}\in C_0$, where now

$$K=1-\Delta_t-\Delta_z,\quad \Delta_t=\sum\partial_{t^j}^2,\quad \Delta_z=\beta^{-1/2}\partial_{z^j}\beta^{1/2}\beta^{jk}\partial_{z^k},\quad \beta=\det((\beta_{jk})), \tag{5.4}$$

and

$$H=\kappa^{1/2}(1-\Delta)\kappa^{-1/2}=q-L_t-\Delta_z\,,$$
$$q=1-g^{-1/4}\partial_{t^j}\sqrt{g}\,g^{jk}\partial_{t^k}(g^{-1/4})\,,\quad L_t=\partial_{t^j}g^{jk}\partial_{t^k}\,, \tag{5.5}$$

with the Laplace operator Δ of ds. Clearly q depends on t only, and

$$q-1=O(\langle t\rangle^{-1})\,,\quad q^{(\alpha)}=O(\langle t\rangle^{-1-|\alpha|})\,. \tag{5.6}$$

The commutator $J=[K,H]$ of (1.24) still is a 3-rd order expression as in (1.26),(1.27), with Δ and ∂_{x^j} replaced by Δ_t and ∂_{t^j}. J is independent of z, and contains no ∂_{z^j}. It again is a sum of terms (1.28), where $D=\partial_{t^k}$ commutes with $S(\mu)=(K+\mu)^{-1}$. Thus the right hand side of (1.23) is compact again, and (1.22) holds. Then the argument of prop.1.3 shows that $T=K^{1/2}H^{-1/2}\in C_0$.

Now we will have two symbols σ_T and γ_T, and find that

$$(5.7)\qquad \sigma_T=((|\tau|^2+\beta^{jk}(z)\zeta_j\zeta_k)/(g^{jk}(t)\tau_j\tau_k+\beta^{jk}(z)\zeta_j\zeta_k))^{1/2},$$

as $(t,z)\in \mathbb{B}^{n'}\times B=\Omega^{\Delta}{}_0$: $\infty(\tau,\zeta)\in \partial\mathbb{B}^n$, i.e., $((t,z),\ \infty(\tau,\zeta))\in S^*_\infty\Omega^{\Delta}{}_0$. (Note that, for $n''\neq 0$, the space $\mathbb{M}$ only uses the infinite cospheres over $\partial\Omega^{\Delta}{}_0$, while for $n''=0$ the entire cotangent fiber over a point of $\partial\Omega^{\Delta}{}_0$ is contained in $\mathbb{M}$.) For the other symbol γ_T we note that the coset mod K of T is the inverse positive square root of the coset of $R = HK^{-1} = (q-L_t-\Delta_z)(1-\Delta_t-\Delta_z)^{-1}$ again. Since $K\subset\ker\gamma$,

it follows that $\gamma_T(e) = (\gamma_R(e))^{-1/2}$, for all $e\in\mathbb{E}=\partial\mathbb{B}^{n'}\times\mathbb{R}^{n'}$. Using the rules of [CPo], sec.3, we get

$$(5.8)\qquad \gamma_R(\infty t,\tau) = (1+g^{jk}(\infty t)\tau_j\tau_k-\Delta_z)(\langle\tau\rangle^2-\Delta_z))^{-1} .$$

This indeed is a positive definite self-adjoint operator. We get

$$(5.9)\quad \gamma_T(\infty t,\tau) = (1+g^{jk}(\infty t)\tau_j\tau_k-\Delta_z)^{-1/2}(\langle\tau\rangle^2-\Delta_z)^{1/2},\ (\infty t,\tau)\in\mathbb{E} .$$

Then we use Stone-Weierstrass to confirm $C^{\Delta} = C_0$.

For algebra surgery we follow sec.2, looking at $H_W= q-\Delta$, with q of (2.14) (but now q depends on $t\in\mathbb{R}^{n'}$ only, rather than on $x\in\mathbb{R}^n$, and is defined in U×B, with a conical set $U\subset\mathbb{R}^{n'}$). Using that $\Delta = M_t + \Delta_z$ over $\partial\Omega_0'$, with an elliptic $M_t=L_t-1$ over U, independent of z, we use separation of variables to show that $H_{W,0}{}^2$ is e.s.a. Then indeed we get an analogue of thm.2.4:

<u>Theorem 5.2</u>. Let $H_W=L^2(W,dS)$, and $C_W\subset L(H_W)$ be generated by

$$(5.10)\qquad a\ ,\ D\Lambda_W\ ,\ a\in A^{\#}_W\ ,\ D\in\mathcal{D}^{\#}_W\ ,$$

with the generating sets $A^{\#}_W$, $\mathcal{D}^{\#}_W$ of all $a\in A^{\#}$, $D\in\mathcal{D}^{\#}$ with support in a set $U^{\sim}\times B = W^{\sim}$, $U^{\sim}\subset\subset U$, and where $\Lambda_W = H_W^{-1/2}$. Let χ_W denote the characteristic function of $W\subset\Omega$. Then C_W is independent of the cut-off function χ used in the potential q of H_W, and C_W is adjoint invariant. Moreover, we have

(5.11) $D\Lambda-D\Lambda_W\chi_W\in K(H)$ whenever $\chi D=0$ for the function χ of (2.14)

The C^*-subalgebra Z of $C\subset L(H)$ generated by

$$K(H)\ ,\ a\ ,\ D\Lambda\ ,\ a\in A^{\#}_W\ ,\ D\in\mathcal{D}^{\#}_W\ ,\ \Lambda = (1-\Delta)^{-1/2}\ ,$$

satisfies the relation

$$(5.12)\qquad Z = C_W\chi_W + K(H)\ ,\ C_W = \chi_W Z\chi_W .$$

For any function $\chi \in A_W^{\#}$, independent of z, and $A \in C$, we have χA , $A\chi \in Z$, and the commutator $[\chi,A]$ is compact ($\in K(H)$) .

The proof of thm.5.2 uses the same ideas as that of thm.2.4, but must accomodate the fact that C now has noncompact commutators Note that only $K(H)$, not E, appears in both (5.11) and (5.12), since $[(q-q^{\sim}),H_W]$, for potentials q and $q^{\sim}$ involving different cut off functions, and $[H,\psi]$, with ψ of (3.2) (in the t-variable, independent of z) are first order differential expressions with coefficients vanishing at $|t|=\infty$. Also, commutation of ψ with Λ or Λ_U , in the sense of [C_2], lemma 3.3 yields a compact perturbation term in (3.2), in spite of our cutting through the secondary symbol space.

Theorem 5.2 again implies a formula like (2.17). Again, since the poly-cylinder of [CPo] satisfies all assumptions, the same conclusion applies for the ideal $Z_0 \subset C_0$ of all operators A on the poly-cylinder $\mathbb{R}^{n'} \times B$ with symbol vanishing over the complement of W . Therefore, since the algebra C_W/K_W is the same, in both cases, we again find that $Z/K(H)$ and $Z_0/K(H_0)$ are isometrically isomorphic.

On the other hand, in [CPo] we were giving a precise description of the algebra $C_0^{\vee} = C_0/K(H_0)$, using two homomorphisms

(5.13) $\sigma_0^{\vee}: C_0^{\vee} \to C(\mathbb{M}_0)$, and $\gamma_0^{\vee}: C_0^{\vee} \to CB(\mathbb{E}_0, C_B)$,

with the algebra C_B of singular integral operators over B, (i.e. the unique comparison algebra of the compact space B), such that

(5.14) $\ker \sigma_0^{\vee} = E_0^{\vee}$, $\ker \gamma_0^{\vee} = CO(\mathbb{W}_0)$,

with $E_0^{\vee} = E_0/K(H_0)$, E_0 the commutator ideal of C_0 .
The product $\sigma_0^{\vee} \times \gamma_0^{\vee}$ is a $*$-isomorphism $C_0^{\vee} \to C(\mathbb{M}_0) \times CB(\mathbb{E}_0, C_B)$, hence also an isometry, by a well known theorem on C^*-algebras.

For the general C of this section we get the following:

Theorem 5.3. The algebra C admits two symbol homomorphisms

(5.15) $\sigma: C \to C(\mathbb{M})$, $\mathbb{M} = S_{\infty}^{*}\Omega_0$,

and

(5.16) $\gamma: C \to CB(\mathbb{E}, C_B)$, $\mathbb{E} = \cup(\pi^{-1}\partial M))$,

where the union in (5.16) is a disjoint union, taken over the ends of Ω, and where π is the bundle projection in the cotangent bundle of the manifold $M=[0,1)\times\Theta$ with boundary $\{0\}\times\Theta$, referring to the product decomposition of the neighbourhood $N=I\times\Theta\times B$ at the given end. We have

(5.17) $$\ker\sigma = E\ ,\ \ker\gamma = J_0\ ,\ \ker\sigma\cap\ker\gamma = K(H)\ ,$$

with the commutator ideal E of C and the minimal comparison algebra J_0 of Ω .

For a generator A (=a or =DΛ) of C the two symbols are given as follows: σ_a , for $a\in A^{\#}$, is the function a itself, constant over each fiber of $\mathbb{M} = S^*_\infty\Omega_0$. For $D\in\mathcal{D}^{\#}$ we have (1.13) again, in any set of local coordinates x, (but may replace p(x) and 1 by 0, since we always have $|\xi|=\infty$). Referring to a distinguished polychart, γ_a is the multiplication by $a(t,.)\in C(B)$, for $(t,\tau)\in\pi^{-1}(\partial M)$, while DΛ, for $D = -ib^j\partial_{t^j}+p+D_z(t)$, $D_z(t)$ a homogeneous folpde on B, depending smoothly on t, also at $\delta=0$, i.e., $t=\infty$, yields

(5.18) $$\gamma_{D\Lambda} = (b^j(t,z)\tau_j+p(t,z)+D_z(t))(1+g^{jk}(t)\tau_j\tau_k+\Delta_z)^{-1/2}\ ,$$

as $(t,\tau)\in\pi^{-1}\partial M$ again. (Here '$(t,\tau)\in\pi^{-1}\partial M$' refers to an infinite coordinate τ , i.e., in the (δ,θ)-coordinates, a point with $\delta=0$.)

An operator $A\in C$ is Fredholm if and only if (i) $\sigma_A\neq 0$ on all of $\mathbb{M}$, (ii) $\gamma_A(e)$ is invertible in C_B , for every $e\in\mathbb{E}$, and (iii) $\gamma_A^{-1}(e)$ is bounded over $\mathbb{E}$.

<u>Proof</u>: We shall get restricted to the case of $\partial\Omega_0$ having only one connected component, i.e., Ω having only one end. The general case is easily reduced to this case, using the surgery of $[C_2]$,VIII, since the separation of an end from the rest of the manifold involves only a compact cut.

Choose a locally finite partition $1=\omega+\sum_1^N\chi_j$, where $\chi_j\in A^{\#}$, each χ_j has support in a distinguished boundary chart $W_j=U_j\times B$, and $\omega=\chi_0\in C_0^\infty(\Omega)$. We also assume that χ_j, $j\neq 0$, are independent of z. For $A\in C$ we get $A=\omega A+\sum_1^N\chi_jA=\sum_0^N\chi_jA$, where $\omega A\in J_0$, $\chi_jA\in Z_j$, with $Z_j=Z$ of thm.5.2, formed with W_j .

We then have $Z_j/K(H)$ isometrically isomorphic to $Z_{j0}/K(H_0)$, by thm.5.2, where $Z_{j0}\subset C_0\subset L(H_0)$, $H_0=L^2(\mathbb{R}^{n'}\times B)$. Thus $\sigma_0(A_j)$ and $\gamma_0(A_j)$, for $A_j\in Z_j$, may be defined as functions over $\mathbb{M}_0=S^*_\infty\Omega_0$, and $\mathbb{E}_0=\partial\mathbb{B}^{n'}\times\mathbb{R}^{n'}$, $\sigma_0=0$, $\gamma_0=0$ outside $\iota_0^{-1}(W)$ and $\pi_0^{-1}(\partial M\cap U_j^{clos})$, by just assigning the symbols of the coset $\check{A}_{j0}$ corresponding to $\check{A}_j$

One may use the coordinate map $W \to N$, and its t-component to transfer both symbols onto subsets $\mathbb{M}_j$ of $\mathbb{M} = S^*_\infty \Omega_0$ and $\mathbb{E}_j$ of $\pi^{-1}(\partial M \cap U_j^{clos})$ respectively.

In this way one obtains a pair of homomorphism

(5.19) $$\sigma_j : Z_j \to CO(\mathbb{M}_j) \ , \ \gamma_j : Z_j \to CO(\mathbb{E}_j, C_B), \ j=1,\dots,N,$$

where the $\mathbb{M}_j \subset \mathbb{M}$,and the $\mathbb{E}_j \subset \mathbb{E}$ form locally finite covers of $\mathbb{M}\backslash\mathbb{W}$, and $\mathbb{E}$, respectively. Moreover, let Z_0 be the ideal of J_0 $B \in J_0$ with $\sigma_B = 0$ outside $K \supset\supset \text{supp}\ \omega$, and $\mathbb{M}_0 = \iota^{-1}(K)$. then $\{\mathbb{M}_j : j=0,\dots,N\}$ forms a cover of $\mathbb{M}$, and the lifting of $1=\sum_0^N \chi_j$ to $\mathbb{M}$ under ι is a subordinate partition of unity.

Then we define

(5.20) $$\sigma_A = \sigma_{\omega A} + \sum_1^N \sigma_j(\chi_j A) \ , \ \gamma_A = \sum_1^N \gamma_j(\chi_j A) \ ,$$

where $\sigma_{\omega A} = \sigma_0(\chi_0 A)$ is the (complex-valued) J_0-symbol of $\omega A \in J_0$, extended zero outside $\iota^{-1}(K)$, and where σ_j and γ_j are extended zero outside $\mathbb{M}_j$ or $\mathbb{E}_j$. By examining the symbols of the generators, explicitly given by (5.7) and (5.9) , and the formulas of [CPo], sec.2, one finds that, for an $A \in Z' \cap Z''$, Z', Z'' corresponding to two distinguished poly-charts $W'=U'\times B$, $W'' = U''\times B$ with nonvoid intersection, one finds that the symbols $\sigma'(A)$ and $\sigma''(A)$, defined with Z' , Z'' , just like $\sigma_j(A)$ with Z_j above, are related to each by the corresponding coodinate transform. Similarly $\gamma'(A)$ and $\gamma''(A)$ are related by the transform relating the coordinates of the corresponding M' and M'' . Accordingly, for $A \in Z'$ we get

(5.21) $$\sigma_A = \sum_0^N \sigma_j(\chi_j A) = \sigma'(\sum_0^N \chi_j A) = \sigma'(A) \ , \ \gamma_A = \gamma'(A) \ .$$

For every $m \in \mathbb{M}\backslash\mathbb{W}$ we can find a distinguished poly-chart $W'=U'\times B$, containing m, and cut-off function $\chi=1$ near m , $\chi \in A^\#$, supp $\chi \subset W\hat{} = U\hat{}\times B$, $U\hat{} \subset\subset U'$, independent of z . Then $\sigma_A(m) = \sigma'(\chi A)(m)$. Since σ' is a homomorphism, and since one confirms that $\sigma_{AB}(m) = \sigma'(\chi A)\sigma'(\chi B)(m)$ (using the existence of χ' with similar properties and $\chi\chi' = \chi$), it follows that σ of (5.21) is a homomorphism at each m . Similarly for γ . Therefore (5.21) indeed defines a pair of homomorphisms, as stated, and one confirms by a calculation that the generators of C have symbols as described. Clearly ker σ contains the commutator ideal E, since $C/(\ker \sigma)$ is commutative. On the other hand the isometry $Z'^{\vee} \leftrightarrow Z_0'^{\vee}$ provides an isometry $E^{\vee} \cap Z'^{\vee} \leftrightarrow E_0^{\vee} \cap Z_0'^{\vee}$ as well. Then the precise knowledge of $E_0^{\vee}$ in [CPo] insures that we have ker $\sigma = E$. For A

$\in \ker \gamma$ and $e\in \mathbb{E}$ we get $\gamma'(\chi'A) = 0$, for a suitable cut-off χ' . But γ' is the homomorphism γ of [CPo] , with kernel $J_0(\Omega^{\Delta})$. Therefore we get $\sigma'(\chi'A) = 0$ on $\mathbb{M}\setminus\mathbb{W}$. This holds for all χ' , so that $A \in J_0$ follows . Then (5.17) follows, and the theorem is proven (Regarding the Fredholm properties of $A\in C$ we conclude as in [CPo])

Final remark: We again may introduce the weighted Sobolev norms (4.22), where now Λ is formed as always, while γ depends on t

only, not on z, i.e., $\gamma=\omega + \sum\langle t\rangle \chi_j$, as in (4.4), with our partition of (5.20). Note, γ is not related to the operator-valued symbol of this section, hence will be denoted by υ instead of γ.

It seems that even in this case the discussion of sec.4,(3) may be repeated, with some modifications pertaining to the more complicated symbol structure. Even the Frechet algebra C_∞ seems to be a ψ^*-algebra in the sense of Gramsch. In particular, an argument like that of $[C_2]$,X, prop.5.1 should imply that, for an $A\in C_\infty$, also the operator-valued γ-symbol has a kernel dimension independent of s. This should imply that the C_∞-spectrum and the C-spectrum of $A\in C_\infty$ coincide. We plan to discuss details in a later publication.

6. References.

[AN_1] S.Agmon and L.Nirenberg, Properties of solutions of ordinary differential equations in Banach space; Commun. Pure Appl. Math. 16 (1963) 121-239.

[APS] M.Atiyah, V.Patodi, and I.Singer, Spectral asymmetry and Riemannian Geometry; Math. Proc. Camb. Phil. Soc. 77 (1975) 43-69.

[BC_j] M.Breuer and H.O.Cordes,On Banach algebras with σ-symbol;I: J.Math.Mech. 13 (1964) 313-324; II: J.Math.Mech. 14 (1965) 299-314 .

[BS_1] J.Bruening and R.Seeley, Regular singular asymptotics; Adv. Math. 58 (1985) 133-148.

[BS_2] __________ The resolvent expansion for second order regular singular operators; Preprint Augsburg 1985.

[BS_3] __________ An index theorem for first order regular singular operators; Preprint Augsburg 1986.

[Che] J.Cheeger, Spectral geometry of spaces with cone-like singularities;Proc.Nat.Acad.Sci. USA 76 (1979) 2103.

[CGT] J.Cheeger, M.Gromov, and M.Taylor, Finite propagation speed, kernel estimates for functions of the Laplace operator, and the geometry of complete Riemannian manifolds; J. Diff. Geom. 17 (1982) 15-53.

[Ch_1] P.Chernoff, Essential selfadjointness of powers of generators of hyperbolic equations ; J.Functional Analysis 12 (1973) 402-414 .

[CBC] Y.Choquet-Bruhat and D.Christodolou, Elliptic systems in $H_{s,\delta}$-spaces on manifolds, Euclidean at infinity; Acta Math. 146 (1981) 129-150.

[CC_1] P.Colella and H.O.Cordes, The C^*-algebra of the elliptic boundary problem; Rocky Mtn.J.Math. 10 (1980) 217-238 .

[C_1] H.O.Cordes, Elliptic pseudo-differential operators, an abstract theory. Springer Lecture Notes in Math. Vol.756, Berlin, Heidelberg, New York 1979.

[C_2] __________, Spectral theory of linear differential operators and comparison algebras. LNM lecture notes, Cambridge Univ. Press, to appear.

[C_3] __________, Techniques of pseudo-differential operators; to appear.

[CGO] _________, On geometrical optics, lecture notes, Berkeley 1982 (available from the author).

[CP] __________, A global parametrix for pseudo-differential operators over $\mathbb{R}^n$; Preprint SFB 72 (U.of Bonn) (1976) (available as preprint from the author).

[CS] __________, The algebra of singular integral operators in $\mathbb{R}^n$; J.Math.Mech. 14 (1965) 1007-1032.

[CS2] __________, An algebra of singular integral operators with two symbol homomorphisms; Bulletin AMS ,75 (1969) 37-42 .

[CWS] _________, Selfadjointness of powers of elliptic operators on noncompact manifolds; Math.Ann. 195 (1972) , 257-272 .

[CPo] __________, On the two-fold symbol chain of a C^*-algebra of singular integral operators on the poly-cylinder To appear.

[CE] ____, and A.Erkip, The N-th order elliptic boundary problem for non-compact boundaries; Rocky Mtn.J.of Math. 10 (1980) 7-24 .

[CHe_1] ____, and E.Herman , Gelfand theory of pseudo-differential operators; American J.of Math. 90 (1968) 681-717.

[CMe_1] _____, and S.Melo, An algebra of singular integral operators with periodic coefficients; to appear.

[Du_1] R. Duduchava, On integrals of convolutions with discontinuous coefficients; Math. Nachr. 79 (1977) 51-69.

[Du_2] ________, On multi-dimensional singular integral operators; I: The half-space case; J. Operator Theory 11 (1984) 41-76 ; II: the case of a compact manifold J. Operator Theory 11 (1984) 199-214.

[Ga_1] M.P.Gaffney, A special Stoke's theorem for complete Riemannian manifolds; Annals of Math. 60 (1954) 140-145.

[G_1] B.Gramsch, Relative Inversion in der Stoerungstheorie von Operatoren und ψ-Algebren; Math.Ann. 269 (1984) 27-71

[He_1] E.Herman, The symbol of the algebra of singular integral operators; J.Math.Mech. 15 (1966) 147-156 .

[LM_1] R.Lockhart and R.McOwen, Elliptic differential operators on noncompact manifolds; Acta Math.151 (1984) 123-234.

[MaM] R.Mazeo and R.Melrose, Meromorphic extension of the resolvent on complete spaces with asymptotically constant negative curvature; to appear.

[MM_1] R.Melrose and G.Mendoza,Elliptic boundary problems on spaces with conic points; Journees des equations differentielles, St. Jean-de-Monts, 1981

[MM_2] ________, Elliptic operators of totally characteristic type; Preprint MSRI, Berkeley 1982.

[$Schr_1$] E.Schrohe, Potenzen elliptischer Pseudodifferentialoperatoren; Thesis, Mainz 1986.

[$Schu_1$] W.Schulze, Ellipticity and continuous conormal asymptotics on manifolds withconical singularities; to appear, Math. Nachrichten.

[$Schu_2$] ______, Mellin expansions of pseudo-differential operators and conormal asymptotics of solutions; Proc. Oberwolfach Conf. on Topics in Pseudo-differential Operators, 1986.

[Ro_1] W.Roelcke, Ueber Laplaceoperatoren auf Riemannschen Mannigfaltigkeiten mit diskontintuierlichen Gruppen; Math.Nachr.21 (1960) 132-149 .

[Se_1] R.T.Seeley,Integro-differential operators on vector bundles; Transactions AMS 117 (1965) 167-204.

WEYL'S FORMULA FOR A CLASS OF PSEUDODIFFERENTIAL OPERATORS WITH NEGATIVE ORDER ON $L^2(\mathbf{R}^n)$

Monique Dauge and Didier Robert
U.A. C.N.R.S. 758,
Département de Mathématiques et d'Informatique
2, Rue de la Houssinière
F. 44072 NANTES Cédex, FRANCE

Introduction

The starting point of this work is a paper by Birman and Solomjak [BI-SO 1] in which they study the eigenvalues asymptotics of a class of integral compact operators on $L^2(\mathbf{R}^n)$. These operators are pseudodifferential, with symbols that are homogeneous (or quasi-homogeneous) of negative order with respect to the phase variable ; there are very few regularity assumptions with respect to the space variable.

If such an operator A is self-adjoint, $(\lambda_j^+(A))$ denoting the positive eigenvalues decreasing sequence of A, [BI-SO 1] gives an equivalent to the $\lambda_j^+(A)$, by powers of j when $j \to +\infty$, and the corresponding result for negative eigenvalues (cf § 2.A). The method of [BI-SO 1] is based on Courant's mini-max principle and consists in reducing to a model problem on the torus $\mathbf{T}^n$.

In this work, we intend to extend and precise [BI-SO 1] results in two directions :

(1) relaxing the homogeneity assumption

(2) getting a remainder estimate.

As the symbol may vanish and change of sign, there are some difficulties we will overcome by making an hypoelliptic regularization of the operators and then using a Mellin functionnal pseudodifferential calculus.

Here is the structure of our paper

§1 : Assumptions and the main result.

§2 : Special cases and examples.

§3 : Further information about spectral theory of globally hypoelliptic pseudodifferential operators on $\mathbf{R}^n$.

§4 : Proof of the main result.

Annex A : Weyl-Ky Fan inequalities.

Annex B : Composition formula with precised remainder.

Annex C : Computation for an example.

The main results of this paper have been announced in the authors' note [DA-RO].

§1 : Assumptions and the main result

Our operators are described by the properties of their Weyl's symbols, according to L. Hörmander's formalization [HO 1]. For the sake of simplicity, we restrict ourselves to the diagonal metrics. Let us recall some definitions.

On $T^* \mathbf{R}^n$, identified to $\mathbf{R}^{2n} = \mathbf{R}^n_x \times \mathbf{R}^n_\xi$, let ϕ and φ be two weight functions, with their values in $]0,+\infty[$.

(1.1) Definition

Let m be a weight function, $m : \mathbf{R}^{2n} \to]0,+\infty[$.

(1) If there exists C_0, $C_1 > 0$ such that :

$$|y|\,\phi(x,\xi) + |\eta|\,\varphi(x,\xi) \leqslant C_0(\phi\varphi)(x,\xi) \Rightarrow C_1^{-1}\, m(x,\xi) \leqslant m(x+y,\xi+\eta) \leqslant C_1\, m(x,\xi)$$

then, m is called <u>(ϕ,φ)-continuous.</u>

(2) If, moreover, there exists C, M > 0 such that :

$$m(x+y,\xi+\eta) \leqslant C\, m(x,\xi)\,[1+|y|\,\phi(x,\xi) + |\eta|\,\varphi(x,\xi)]^M$$

then, m is called <u>(ϕ,φ)-temperate.</u>

In this whole work, we assume :

(H_1) ϕ^{-1},φ^{-1} are bounded on $\mathbf{R}^{2n}$ and they are (ϕ,φ)-continuous and (1,1)-temperate.

If m is a (ϕ,φ)-temperate weight, according to [BE] we denote by $S(m\,;\,\phi,\varphi)$ the class of symbols $a \in C^\infty(\mathbf{R}^{2n})$ such that for all $k \in \mathbf{N}$ the following quantities are finite :

(1.2) $$s_k(a\,;\,m) = \sup_{\mathbf{R}^n,\,|\alpha+\beta|=k} |m^{-1}\,\phi^{|\alpha|}\,\varphi^{|\beta|}\,\partial_\xi^\alpha\,\partial_x^\beta\, a| .$$

The (s_k) are a family of semi-norms for $S(m\,;\,\phi,\varphi)$.

To a symbol a in $S(m\,;\,\phi,\varphi)$, we associate the operator $A = \mathrm{Op}^W a$, according to the Weyl's quantization :

$$Au(x) = \int e^{i\langle x-y,\xi\rangle}\, a\left(\frac{x+y}{2},\xi\right) u(y)\, dy\, đ\xi$$

with $đ\xi = (2\pi)^{-n}\, d\xi$.

Here are complementary assumptions about the weights :

(H_2) $\exists c, \varepsilon_0 > 0$ such that $C(1+|x|+|\xi|)^{\varepsilon_0} \leqslant (\phi\varphi)(x,\xi)$ on $\mathbf{R}^{2n}$

(W) m is (ϕ,φ)-temperate and $m \in S(m\,;\,\phi,\varphi)$

(N) $\exists K, K' > 0$, $\exists \gamma, \gamma' > 0$ such that $K'\, m^{\gamma'} \leqslant (\phi\varphi)^{-1} \leqslant K\, m^{\gamma}$ on $\mathbf{R}^{2n}$.

With (H_2), (N) states that m has a negative order.

Let be $a \in S(m\,;\,\phi,\varphi)$ with real values. (H_1), (H_2), (W) and (N) imply that $\mathrm{Op}^W a$ is self-adjoint and compact, with the kernel :

$$K_a(x,y) = \int e^{i<x-y,\xi>}\, a(\frac{x+y}{2}, \xi)\, d\xi .$$

The spectrum of A is discrete with only one accumulation point in 0. Denote by $\lambda_j^+(A)$ (resp. $\lambda_j^-(A)$) the decreasing (resp. increasing) sequence of positive (resp. negative) eigenvalues of A, repeated according to their multiplicities. For $\lambda > 0$, we introduce the counting functions :

$$N_\pm(\lambda ; A) = \# \{j \in \mathbf{N} / \pm \lambda_j^\pm (A) > \lambda\}.$$

Like the classical Weyl's formula for the Dirichlet's problem on a bounded domain, we compare $N_\pm(\lambda ; A)$ when $\lambda \to +\infty$, with the volume functions $V_\pm(\lambda ; a)$ in the classical phase space :

$$V_\pm(\lambda ; a) = \int_{\pm a > \lambda} dx\, d\xi .$$

For the functions V, we need, as in [TU-SU], a control of the derivative : for a decreasing function $f :]0,+\infty[\to [0,+\infty[$, we say that f verifies (T) if there exists $\lambda_0 > 0$, $\gamma_1, \gamma_2 > 0$ such that :

$$(T) \begin{cases} f \text{ is derivable on }]0,\lambda_0] \\ \gamma_1 f(\lambda) < -\lambda f'(\lambda) < \gamma_2 f(\lambda) \text{ for all } \lambda \in]0,\lambda_0]. \end{cases}$$

By integrating (T), we obtain the useful inequality :

$$(T') \quad \left(\frac{\lambda_2}{\lambda_1}\right)^{\gamma_1} < \frac{f(\lambda_1)}{f(\lambda_2)} < \left(\frac{\lambda_2}{\lambda_1}\right)^{\gamma_2} \qquad \text{for } 0 < \lambda_1 < \lambda_2 < \lambda_0$$

Our main result is :

(1.3) Theorem

With the hypotheses (H_1), (H_2), (W) and (N), let $a \in S(m ; \phi,\varphi)$ be a real symbol and $A = Op^W a$. We assume that the volume functions $V_+(. ; m)$, $V_+(. ; a)$ and $V_-(. ; a)$ verify (T). Then, there is $\zeta > 0$ such that :

(1) $N_+(\lambda ; A) = V_+(\lambda ; a) + O(\lambda^\zeta V_+(\lambda ; m))$ when $\lambda \to 0_+$;

(2) if moreover, $V_+(\lambda ; m) = O(V_+(\lambda ; a))$, we have :

(1.4) $N_+(\lambda ; A) = V_+(\lambda ; a)\,(1+O(\lambda^\zeta))$ when $\lambda \to 0_+$;

(1.5) $\lambda_j^+ (A) = \Lambda^+(j ; a)\,(1+O(j^{-\zeta}))$ when $j \to +\infty$

where $\Lambda^+(t ; a)$ is the inverse function of $V_+(. ; a)$, which is well defined for t large enough.

We have corresponding statements linking N_- and V_-.

(1.6) Remarks

(1) The condition (T) about $V_-(. ; a)$ may be replaced by :

$$V_-(\lambda ; a) = O(\lambda^{\zeta_0} V(\lambda ; m)) \text{ with } \zeta_0 > 0.$$

(2) If $a = a_0 + a_1$ with $a_1 \in S(m(\phi\,\varphi)^{-\varepsilon} ; \phi,\varphi)$ and $\varepsilon > 0$, $V_\pm(\lambda ; a)$ may be replaced by $V_\pm(\lambda ; a_0)$

Asymptotics of (1.4) type are well known for wide classes of globally hypoelliptic pseudodifferential operators with positive order - cf §3 - Among the numerous works about this, let us quote : [TU-SU], [RO], [HO 2], [HE-RO 2], [MO], [FE]. However, in our case, the symbol need not to have a constant sign and we cannot reduce our operators to those of the above works : in particular, the separation between positive and negative parts of the spectrum is awkward.

About a connected problem, G. Grubb [GR] has noted that an hypoelliptic regularization can yield a result for positive operators. Our method of proof consists in sharpening this remark. We shall be led to use accurate results about functionnal calculus for a class of hypoelliptic operators (3.12) and we shall deduce from that a Weyl's formula of [HO 2] type (3.21). Those results will be stated in §3 and, moreover, we will show that they have other applications (3.16), (3.17), (3.20).

§2 Special cases and examples

2.A. Homogeneous symbols

1st case : $a = a_0 + a_1$ where a_0 is homogeneous of degree $\sigma < 0$ with respect to ξ variable for $|\xi| > 1/2$, and we suppose that there is some $s < \sigma$ such that :

(i) $\partial_x^\alpha a_0(x,\xi) \leqslant C_\alpha(1+|x|)^{s-|\alpha|}$ $\quad\forall\alpha \in \mathbf{N}^n,\ x \in \mathbf{R}^n,\ |\xi|=1$

(ii) $\partial_x^\alpha \partial_\xi^\beta a_1(x,\xi) \leqslant C_{\alpha,\beta}(1+|x|)^{s-|\alpha|}(1+|\xi|)^{s-|\beta|}$ $\quad\forall\alpha,\beta \in \mathbf{N}^n\ (x,\xi) \in \mathbf{R}^{2n}$

We can apply our result with the weigths :

$\phi(x,\xi) = 1+|\xi|$; $\varphi(x,\xi) = 1+|x|$; $m(x,\xi) = (1+|x|^2)^{s/2}(1+|\xi|^2)^{\sigma/2}$.

For $A = Op^W a$, we obtain :

(2.1) $N_\pm(\lambda ; A) = \lambda^{n/\sigma}(\gamma_\pm + O(\lambda^\zeta))$, $\lambda \to 0$, with $\zeta > 0$

where :

(2.2) $\gamma_\pm = \dfrac{(2\Pi)^{-n}}{n} \displaystyle\iint_{\mathbf{R}^n \times S^{n-1}} a_0^\pm(x,\xi)^{-n/\sigma}\, dx\, d\xi$.

Note that the condition $s < \sigma$ yields the integrability of $a_0^\pm(x,\xi)^{-n/\sigma}$ on $\mathbf{R}^n \times S^{n-1}$.

(2.1) is equivalent to :

$\lambda_j^\pm(A) = j^{\sigma/n}(C_\pm + 0(j^{\zeta\,\sigma/n}))$, $j \to +\infty$, where $C_\pm = (\gamma_\pm)^{-\sigma/n}$.

We can compare those results with [BI-SO 1] ones :

there are the following differences :

a) [BI-SO 1] symbols satisfy low regularity assumptions

b) they only get an equivalent

c) they require supplementary conditions about the behavior with respect to x variable :

. $s = -\infty$ if $\sigma \lesssim -n$

. $s < -n/2$ if $\sigma > -n/2$.

2^d **case** : we have obviously the same type of result if we permutate x and ξ roles. We are now interested with the case when the homogeneities with respect to x and ξ are equal to each other :

$$a(x,\xi) = b(x)\ c(\xi)$$

where b and c are homogeneous of degree $\sigma < 0$ outside the ball with radius 1/2. We will compute in annex C the volume functions of a. Thus, for $\zeta > 0$.

(2.1') $N_{\pm}(\lambda\ ;\ A) = \lambda^{n/\sigma} \operatorname{Log} \frac{1}{\lambda} \gamma_{\pm} + \lambda^{n/\sigma}\ \delta_{\pm} + 0(\lambda^{n/\sigma+\zeta}),\ \lambda \to 0_+$

where :

(2.2') $\gamma_{\pm} = \frac{(2\Pi)^{-n}}{n|\sigma|} \int_{S^{n-1}} b^{\pm}(\tilde{x})^{-n/\sigma}\, d\tilde{x}\ .\int_{S^{n-1}} c^{\pm}(\tilde{\xi})^{-n/\sigma}\, d\tilde{\xi}$

and $\delta_{\pm}$ may be computed as a function of b and c (annex C).

Note that a is not an homogeneous function of degree 2σ outside a neighborhood of 0 in $\mathbf{R}^{2n}$. Let us compare with the following case :

3rd **case** : let $\omega = (x,\xi)$. We suppose that a is C^{∞} and such that, for $|\omega| \geqslant 1/2$: $a(\omega) = |\omega|^{2\sigma} a(\frac{\omega}{|\omega|})$. Then :

(2.1") $N_{\pm}(\lambda\ ;\ A) = \lambda^{n/\sigma}\ (\gamma_{\pm} + 0(\lambda^{\zeta}))$ with $\varepsilon > 0$

(2.2") $\gamma_{\pm} = \frac{(2\Pi)^{-n}}{2n} \int_{S^{2n-1}} a^{\pm}(\tilde{\omega})^{-n/\sigma}\, d\tilde{\omega}\ .$

Note that the 2^d and 3rd cases are not in the scope of [BI-SO 1].

2.B : Condition (T) about the volume functions

The condition (T) allows us to get rid of the homogeneity hypothesis. Here are assumptions about a under which condition (T) about $V_{\pm}(\lambda\ ;\ a)$ is fulfiled.

(i) $a \in C^1(\mathbf{R}^{2n})$ and $a(x,\xi) \to 0$ when $|x| + |\xi| \to +\infty$

(ii) there is a vector field X on $\mathbf{R}^{2n}$ such that :

(ii_1) $\exists\ \eta_1,\eta_2 > 0 : \eta_1\ a \leqslant \langle X,\nabla a\rangle \leqslant \eta_2\ a$

(ii_2) $\exists\ \delta_1,\delta_2 > 0 : \delta_1 \leqslant -\operatorname{div} X \leqslant \delta_2$.

According to a classical result about the Leray's form on an hypersurface, we have, with $d\,S_{\lambda}$ denoting the euclidian measure on $\{a = \lambda\}$:

$$-\lambda V'_+(a\ ;\ \lambda) = \int_{a=\lambda} \lambda \frac{d\,S_{\lambda}}{|\nabla a|} = \int_{a=\lambda} a \frac{d\,S_{\lambda}}{|\nabla a|}\ .$$

With (ii_1) that gives :

(2.3) $\frac{1}{\eta_2} \int_{a=\lambda} \langle X,\nabla a\rangle \frac{d\,S_{\lambda}}{|\nabla a|} \leqslant -\lambda\, V'(a\ ;\ \lambda) \leqslant \frac{1}{\eta_1} \int_{a=\lambda} \langle X,\nabla a\rangle \frac{d\,S_{\lambda}}{|\nabla a|}\ .$

With the Green's formula, we have :

$$\int_{a=\lambda} \langle X,\nabla a\rangle \frac{d S_\lambda}{|\nabla a|} = - \int_{a \geqslant \lambda} \operatorname{div} X \; dx\, d\xi$$

with (ii_2) and (2.3), that yields condition (T) about $V_+(\lambda \,;\, a)$.

Note that if a is quasi-homogeneous, with weights k_i on x_i and h_i on ξ_j, we may take :

$$F(x,\xi) = -(k_1 x_1 , \dots , k_n x_n \,;\, h_1 \xi_1 , \dots , h_n \xi_n)$$

2.C : Schrödinger operators

Let us consider : $A = (-\Delta+E)^{\sigma/2} V(-\Delta+E)^{\sigma/2}$ with Δ the Laplace's operator on $\mathbf{R}^n$, $E > 0$, $\sigma < 0$, and the potential V such that there is $s < 0$:

(2.4) $|\partial^\alpha V(x)| \leqslant C_\alpha (1+|x|)^{2s-|\alpha|}$

If $s < \sigma$, this example is a particular case of 2.A,1. Thus, let us assume that $\sigma < s$.

According to remark (1.6/2) we may consider the " principal symbol " a_0 of A instead of the complete symbol.

$$a_0(x,\xi) = (|\xi|^2+E)^\sigma V(x)$$

According to 2.B, the condition (T) will be fulfiled for $V_\pm(\lambda \,;\, a_0)$ if, for $C,C',R > 0$, $k_1 , \dots , k_n > 0$, we have :

(2.5) $C\, V(x) \leqslant - \sum_{i=1}^{n} k_i x_i \partial_{x_i} V(x) \leqslant C'\, V(x)$, pour $|x| > R$

In that case, we have :

$$N_\pm(\lambda \,;\, A) = V_\pm(\lambda \,;\, a_0) + O(\lambda^\zeta f_{s,\sigma}(\lambda))$$

where $f_{s,\sigma}(\lambda) = \lambda^{n/2s}$ if $s < \sigma$ and $f_{s,\sigma}(\lambda) = \lambda^{n/2s} \operatorname{Log} \frac{1}{\lambda}$ if $s = \sigma$.

As an application, let us consider the stationnary problem for Schrödinger equation :

(2.6) $(\Delta+gV)\, \Psi = E\Psi$

where $E > 0$ is the energy and $g \in \mathbf{R}$ is a coupling constant. For a fixed E, we search values of g for which (2.6) has a non-null solution Ψ in $D(\Delta) \cap D(V)$. For V verifying (2.4) and $g \neq 0$, (2.6) is equivalent to

(2.7) $\Psi \in L^2(\mathbf{R}^n)\setminus\{0\}$, $(-\Delta+E)^{-1/2} V(-\Delta+E)^{-1/2}\, \Psi = \frac{1}{g}\, \Psi$

It is the above case, with $\sigma = -1$. Thus, (2.6) has non-trivial solutions for two sequences $(g_k^+(E))$ and $(g_k^-(E))$ where $(g_k^+(E))$ is positive, increasing and $(g_k^-(E))$ is negative, decreasing. If infinite, each of those sequences are not bounded. If V verifies (2.5), with :

$$N_\pm(g,E) = \# \{k / g_k^\pm (E) \geqslant \pm g\} \qquad (g > 0)$$

we have, when $g \to +\infty$:

(2.8) $$N_{\pm}(g,E) = \frac{(2\Pi)^{-n}}{n}\,\Gamma_n \int (g\,V(x)-E)_{\pm}^{n/2}\,dx + 0(g^{-\zeta}\int (g(1+|x|^2)^{\zeta}-E)^{n/2}\,dx$$

where Γ_n is the volume of the unit ball of $\mathbf{R}^n$.

Many works have given formulas of (2.8) type ([SI],[MA]).

2.D : Equation $A\Psi = \lambda B\Psi$.

Let $A = Op^W a$ and $B = Op^W b$ be two self-adjoint operators on $L^2(\mathbf{R}^n)$, possibly unbounded with domains D(A) and D(B). With (ϕ,φ) weight functions verifying (H_1), (H_2), we suppose that :

(i) $\begin{cases} A \text{ is positive and invertible,} \\ a = a_0 + a_1 \text{ where } a_0 \text{ is a temperate weight such that} \\ a_0 \in S(a_0\ ;\ \phi,\varphi) \text{ and } a_1 \in S(a_0(\phi\varphi)^{-\varepsilon}\ ;\ \phi,\varphi) \text{ with } \varepsilon > 0 \end{cases}$

(ii) $\begin{cases} b \in S(q\ ;\ \phi,\varphi) \text{ where } q \text{ is a temperate weight such that} \\ q \in S(q\ ;\ \phi,\varphi)\ ; \\ m := a_0^{-1} q \text{ verifies the hypotheses (W) and (N)}. \end{cases}$

And we consider the spectral problem :

(2.9) $A\Psi = \lambda B\Psi\ , \qquad \Psi \in D(A) \cap D(B) \setminus \{0\}$.

By mean of composition by the operator $C = A^{-1/2}$, (2.9) is equivalent to :

$\Psi = \lambda\, C\, B\, C\, \Psi\ , \qquad \Psi \in L^2(\mathbf{R}^n) \setminus \{0\}$

Thanks to (ii), theorem (1.3) can be applied to $C B C$, and if the volume functions satisfy the condition (T), we obtain, for a $\zeta > 0$, when $\lambda \to +\infty$:

(2.10) $$N_{\pm}(\lambda\ ;\ A,B) = (2\Pi)^{-n} \iint_{a(x,\xi) < \pm\lambda\, b(x,\xi)} dx\, d\xi + 0(\lambda^{-\zeta} \iint_{a(x,\xi) < \lambda\, q(x,\xi)} dx\, d\xi)$$

where $N_+(\lambda\ ;\ A,B)$ (resp. $N_-(\lambda\ ;\ A,B)$) is the number of eigenvalues of (2.9) belonging to $]0,\lambda]$ (resp. $]-\lambda,0[$).

Many papers are devoted to Weyl's formulas of (2.10) type, with various hypotheses (see [BI-SO 3] and [FL-LA], as well as their bibliographies). Most of them suppose that A is an elliptic differential operator and B is the multiplication by a function ρ. In particular, for Fleckinger and Lapidus in [FL-LA], A is of Schrödinger's type and ρ may be discontinuous (for instance, the caracteristic function of a compact set). However, those works generally only give an equivalent. Let us note that we find the same result as the one announced by Boitmatov and Kostyuchenko in [BO-KO] under assumptions that seem to us more general and more natural.

§3 Further information about spectral theory of globally hypoelliptic pseudodifferential operators on $\mathbf{R}^n$.

3.A : Introduction

Let (ϕ, φ) be weight functions satisfying (H_1). Let p be a symbol satisfying the condition (W), i.e :

$p \in S(p\ ;\ \phi,\varphi)$, and the further condition (S) which is inverse of the condition (N) :

(S) $\quad \exists C, C' > 0,\ \delta, \delta' > 0$ such that $C\, p^{\delta} \leqslant (\phi\varphi) \leqslant C'\, p^{\delta'}$ on $\mathbf{R}^{2n}$.

Let P be $Op^W\, p$. It is known that under conditions (W) and (S), P is essentially self-adjoint from $\mathcal{S}(\mathbf{R}^n)$, and semi-bounded from below. If, moreover, (ϕ,φ) satisfies the " ellipticity " condition (H_2), P has a compact resolvent ([TU-SU], [BE], [RO], [HO 1]).

In the following sections of this paragraph, we will study spectral properties of P, by the mean of a functionnal calculus developped according [HE-RO 1] scheme : parametrix, holomorphie functionnal calculus, Mellin transform. In this whole paragraph, we suppose that (H_1), (H_2), (W) and (S) are true. By possibly translating p, we suppose that P is positive and invertible :

$P \geqslant \gamma > 0$.

3.B : Functionnal calculus : parametrix.

For $z \in \mathbf{C} \setminus \mathbf{R}_+$, we have to study $(P-z)^{-1}\, P^K$ with K positive integer (the reason of the introduction of this parameter K will appear in section 3.C). Thus, we build a parametrix for the equation :

(3.1) $\quad Q_z^{(K)} \circ (P-z) = P^K$.

As a first approximation, we get the symbol :

(3.2) $\quad q_{z\,;\,0}^{(K)} = p^K\, (p-z)^{-1}$.

Then, by recurrence over j, we define the symbols :

(3.3) $\quad q_{z\,;\,j}^{(K)} = (p-z)^{-1} \left[p_j^{(K)} - \sum_{k=0}^{j-1} \sum_{|\alpha+\beta|=j-k} \Gamma(\alpha,\beta)(\partial_\xi^\alpha D_x^\beta\, p)(\partial_\xi^\beta D_x^\alpha\, q_{z\,;\,k}^{(K)}) \right]$

where $\Gamma(\alpha,\beta) = (\alpha\,!\ \beta\,!)^{-1}\, 2^{-|\alpha|}\, (-2)^{-|\beta|}$ and $p_j^{(K)}$ is the jth term in the asymptotics of the symbol of P^K. More precisely, $p_j^{(K)} \in S(p^K (\phi\varphi)^{-j}\ ;\ \phi,\varphi)$ and is a polynomial expression of the $\partial_\xi^\alpha\, \partial_x^\beta\, p$ for $|\alpha+\beta| \leqslant j$.

In particular, $q_{z\,;\,1}^{(K)} = 0$.

By recurrence on $j > 1$, we get :

(3.4) $\quad q_{z\,;\,j}^{(K)} = \sum_{k=0}^{2j-1} (-1)^k\, d_{jk}^{(K)}\, (p-z)^{-k-1}$

where the $d_{jk}^{(K)}$ are universal polynomial functions of the $\partial_\xi^\alpha\, \partial_x^\beta\, p$ for $|\alpha+\beta| \leqslant j$. $d_{jk}^{(K)}$ belongs to $S(p^{K+k}\,(\phi\varphi)^{-j}\,;\,\phi,\varphi)$. For $K = 0$, $d^{(0)}_{jk}$ is simply denoted by d_{jk} (cf [HE-RO 1]).

Weyl's quantization allows the following further details.

(3.5) Lemma

(1) If $j = 2\ell$, $\ell \in \mathbf{N}^*$, then $d^{(K)}_{jk} = 0$ for each $k \geqslant 3\ell$

(2) if $j = 2\ell + 1$, $\ell \in \mathbf{N}$, then $d^{(K)}_{jk} = 0$ for each $k \geqslant 3\ell + 1$

Proof

Let n_j be the greatest power of $(p-z)^{-1}$ in the expression of $q^{(k)}_{z\,;\,j}$. It is sufficient to show by recurrence on $j \in \mathbf{N}$ that :

$$n_j \leqslant m_j$$

where $m_{2\ell} = 3\ell + 1$ and $m_{2\ell+1} = 3\ell + 2$.

We have : $n_0 = 1$. On the other hand, $n_1 \leqslant 1$ because :

$$q^{(K)}_{z\,;\,1} = -(1/2)\,(p-z)^{-1}\,\{p,(p-z)^{-1}\} + (p-z)^{-1}\,p^{(K)}_1\,,$$

where $\{\ ,\ \}$ is the Poisson's bracket. We note that, in the sum (3.3), the contribution of the term corresponding to $k = j - 1$ (the highest value of k) is :

$$-\frac{1}{2}\,(p-z)^{-1}\,\{p,q^{(K)}_{z\,;\,j-1}\}\,.$$

Now, with (3.4), we are getting :

$$\{p,q^{(K)}_{z\,;\,j-1}\} = \sum_{0 \leqslant k \leqslant n_{j-1}-1} (-1)^k\,\{p,d^{(K)}_{jk}\,(p-z)^{-k-1}\}$$

As the Poisson's bracket of p with $(p-z)^{-k-1}$ is zero, we find that the greatest power of $(p-z)^{-1}$ in $\{p,q^{(K)}_{z\,;\,j-1}\}$ is equal to n_{j-1}. Then :

$$\textbf{(3.6)}\quad n_j \leqslant \sup\,\{n_{j-1} + 1,\ \max_{k=0}^{j-2}\,(n_k + j - k + 1)\}\,.$$

If it is supposed that $n_k \leqslant m_k$ for $k \leqslant j - 1$, then (3.6) implies that :

$$n_j \leqslant \sup\,\{m_{j-1} + 1,\ m_{j-2} + 3\}.$$

By looking at the cases when j is even, or odd, we obtain that $n_j \leqslant m_j$. ∎

Let us denote now by $r^{(K)}_{z\,;\,N}$ the symbol of the remainder of the parametrix of (3.1) at the order N, i.e :

$$\textbf{(3.7)}\quad Op^W(r^{(K)}_{z\,;\,N}) = Op^W\,\Big(\sum_{j=0}^{N} q^{(K)}_{z\,;\,N}\Big) \circ (P-z) - P^K$$

The composition formula with precised remainder (B.1) allows us to derive a sharp estimate of $r^{(K)}_{z\,;\,N}$.

Let us denote by $d(z)$ the distance in $\mathbf{C}$ between z and the half-axis $[\,\frac{\gamma}{2}\,,+\infty[$, where

γ is such that $P \geqslant \gamma > 0$ (see 3.A).

(3.8) Lemma

$\rho \geqslant 3/2$ is the number attached to (ϕ,φ) in (B.1). There exists $N_0 > 0$, depending only on p, such that :
for each integers $N \geqslant 1$ and $K \geqslant 0$, and each $\alpha, \beta \in \mathbf{N}^n$, there is a $C = C(p,N,K,\alpha,\beta)$ such that we have the estimate on $\mathbf{R}^{2n}$:

$$|\partial_\xi^\alpha \partial_x^\beta r_{z\,;\,N}^{(K)}| \leqslant C\, d(z)^{-1} \left(\frac{|z|}{d(z)}\right)^{\rho N+N_0+|\alpha|+|\beta|} p^{K+1} (\phi\varphi)^{-N-1} \phi^{-|\alpha|} \varphi^{-|\beta|}$$

for $z \in \mathbf{C} \setminus [\frac{\gamma}{2}, +\infty[$.

Moreover the constant C depends only on a finite number of semi-norms (1.2) of p in $S(p\,;\,\phi,\varphi)$.

Proof :

It follows those of [RO], [HE-RO 1], by using in addition the estimates in (B.1) about the remainder of order N-j in the composition :

$$Op^W (q_{z\,;\,j}^{(K)}) \circ (P-z)\,;$$

for $0 \leqslant j \leqslant N$. The number of derivatives on $q_{z\,;\,j}^{(K)}$ determines by (3.4) the number of powers of $(p-z)^{-1}$, which gives the power of $\frac{|z|}{d(z)}$ in (3.8). (B.1) yields that the number of derivatives on $q_{z\,;\,j}^{(K)}$ is smaller than $\rho N+N_0$, where N_0 is independant of N.

3.C : Functionnal calculus : complex powers.

Under supplementary hypothesis (H_2) on (ϕ,φ), we are now able to study P^s for $s \subset \mathbf{C}$, by the Cauchy's formula : for a suitable path Γ surrounding the half-axis $[\frac{\gamma}{2}, +\infty[$, and for an integer K such that $\mathrm{Re}\, s < K$, we have :

(3.9) $$P^s = \frac{i}{2\Pi} \int_\Gamma z^{s-K} P^K \circ (P-z)^{-1}\, dz\,.$$

Let :

(3.10) $$p_{s\,;\,j} = \frac{i}{2\Pi} \int_\Gamma z^{s-K} q_{z\,;\,j}^{(K)}\, dz\,.$$

The Cauchy's formula and an argument of analytic extension in the s variable, give us that $p_{s\,;\,j}$ does not depend on Γ or K. Moreover, we have the classical formulas (and thanks to (3.5)) :

$$p_{s\,;\,0} = p^s \qquad p_{s\,;\,1} = 0$$

$$p_{s\,;\,j} = \sum_{2 \leqslant k \leqslant [3j/2]} \frac{s(s-1)\ldots(s-k+1)}{k\,!}\, d_{jk}\, p^{s-k}\,.$$

Moreover, we have the following estimates on the remainders :

(3.11) Lemma

We assume that (H_2) is true and we recall that δ is such that $C\, p^{\delta} \leqslant \phi\varphi$ (condition (S)).

There exist positive integers N_1 and $N_2 > N_0$ - cf (3.8) - depending only of ϕ, φ, p such that for each $L, K : L < 0 < K$ and each $N \geqslant N_1 + K/\delta$ we have :

(i) $s \to P^s\text{-}Op^W\left(\sum_{j=0}^{N} p_{s\,;\,j}\right)$ is holomorphic in the strip $L \leqslant \mathrm{Re}\, s \leqslant K$, with values in the space of trace class operators on $L^2(\mathbf{R}^n)$.

(ii) There is $C = C(L,K,N)$ depending only on a finite number of semi-norms of p, such that :

$$\left\| P^s\text{-}Op^W\left(\sum_{j=0}^{N} p_{s\,;\,j}\right)\right\|_{Tr} \leqslant C(1+|\mathrm{Im}\, s|)^{\rho N+N_2}$$

in the strip $L \leqslant \mathrm{Re}\, s \leqslant K$, with ρ as in (B.1) and (3.8).

Proof

With (3.7), (3.9) and (3.10), we get :

$$P^s - Op^W \sum_{j=0}^{N} p_{s\,;\,j} = \frac{i}{2\Pi} \int_{\Gamma} z^{s-K}\, Op^W\, r^{(K)}_{z\,;\,N} \circ (P-z)^{-1}\, dz\ .$$

Thanks to (3.8), we derive that $Op^W\, r^{(K)}_{z\,;\,N}$ is of trace class when $p^K(\phi\varphi)^{-N}$ is bounded by $(\phi\varphi)^{-N_1}$, for N_1 large enough. With (S), we have :

$$p^K \leqslant (\phi\varphi)^{K/\delta}\ .$$

Then, we have (i). The point (ii), is deduced from the estimate (3.8) by a method like that of [HE-RO 1] the exponent of $(1+|\mathrm{Im}\, s|)$ derives from the exponent of $\frac{|z|}{d(z)}$ in (3.8).

3.D : Functionnal calculus : Mellin transform

For a function f in the Schwartz's space $\mathcal{S}(\mathbf{R}^n)$, we may define $f(P)$ as a pseudo-differential operator, the symbol of which is approximated by

$$\sum_{j=0}^{N} p_{f\,;\,j}\ , \text{ where :}$$

$$p_{f\,;\,0} = f(p)\ ;\ p_{f\,,\,1} = 0$$

$$p_{f\,;\,j} = \sum_{2 \leqslant k \leqslant [3j/2]} \frac{d_{jk}}{k!}\, f^{(k)}(p)$$

Here are estimates for the remainder.

(3.12) Proposition

Under the same assumptions than in (3.11), and with the same constants N_1 and

N_2, we have :

For each $N \geqslant N_1$, there exists $C = C(p,N)$ depending only on a finite number of semi-norms of p, such that, with $K(N) := [\delta(N-N_1)]$:

$$\text{(3.13)} \quad \| f(P) - Op^W (\sum_{j=0}^{N} p_{f\,;\,j}) \|_{Tr} \leqslant$$

$$C\{ \int_0^{+\infty} \lambda^{-K(N)-1/2} |(\lambda\, \partial_\lambda)^{\rho\, N+1+N_2} f_{K(N)}(\lambda)|\, d\lambda + \sum_{k=2}^{\delta N} |f^{(k)}(0)|\}$$

where $f_J(\lambda) = f(\lambda) - \sum_{k=0}^{J} f^{(k)}(0) \frac{\lambda^k}{k!}$ (Taylor's remainder).

Proof

We denote by $M[f] = \int_0^{+\infty} \lambda^{-s} f(\lambda) \frac{d\lambda}{\lambda}$, the Mellin transform of f, which is meromorphic on **C**, the poles of which coïncide with the positive integers. Thanks to the Mellin inversion formula, we get, for any $\sigma_1 < 0$:

$$\text{(3.14)} \quad f(P) - Op^W \sum_{j=0}^{N} p_{f\,;\,j} = \frac{1}{2i\Pi} \int_{Re\, s = \sigma_1} M[f](s)(P^s - Op^W \sum_{j=0}^{N} p_{s\,;\,j})\, ds$$

With $\sigma_2 = K(N) + 1/2$, and the Cauchy's formula, we get that (3.14) is equal to (3.14') :

$$\text{(3.14')} \quad \frac{1}{2i\Pi} \int_{Re\, s = \sigma_2} M[f](s)(P^s - Op^W \sum_{j=0}^{N} p_{s\,;\,j})\, ds + \sum_{k=0}^{K(N)} (P^k - Op^W \sum_{j=0}^{N} p_{k\,;\,j}) \frac{f^{(k)}(0)}{k!} .$$

Now, thanks to (3.11), we have :

$$\| \int_{Re\, s = \sigma_2} M[f]\,(s)\,(P^s - Op^W \sum_{j=0}^{N} p_{s\,;\,j})\, ds \|_{Tr} \leqslant$$

$$C \int_{Re\, s = \sigma_2} |M[f]\,(s)|\, (1+|Im\, s|)^{\rho\, N+N_2}\, d\, Im\, s$$

All that gives (3.12) - see also [HE-RO 1]. ∎

3.E : Applications

(3.15) Remark

In the statements (3.11), (3.12), the hypothesis (H_2) allows us to get trace norm estimates for the remainders. If we drop (H_2), we are still able to estimate those remainders in the scale of weighted Sobolev spaces $H^m_{\phi,\varphi}$ - see [BE].

For instance, if $\varphi = 1$ and $\phi(x,\xi) = (1+|\xi|)^2)^{1/2}$, then $H^m_{\phi,\varphi}$ is the usual Sobolev space $H^m(\mathbf{R}^n)$. If moreover, p is an elliptic symbol of order m, the method of proof of (3.12) shows that, for each $f \in \mathcal{S}(\mathbf{R})$:

$$f(p) \in \bigcap_{k \in \mathbf{N}} S(\phi^{-k}\,;\,\phi,\varphi) \qquad \text{(regularizer)}$$

We deduce from that, the theorem (1.1) of [KO-SU] namely, the kernel $K(x,y)$, of $f(P)$ is such that :

$$\forall N \in \mathbf{N},\ \forall \alpha,\beta \in \mathbf{N}^n,\ \exists C,\ |\partial_x^\alpha \partial_y^\beta K(x,y)| \leqslant C(1+|x-y|)^{-N}, \qquad \text{on } \mathbf{R}^{2n} :$$

As a consequence of (3.12), we get, under the same assumptions :

(3.16) Corollary

Let [a,b] be an interval such that $p^{-1}([a,b])$ is a compact set. Then, the spectrum of p contained in]a,b[is discrete.

Proof

As a consequence of the hypothesis and (3.12), f(P) is compact for each $f \in C_0^\infty(]a,b[)$. That implies (3.16).

(3.17) Corollary (asymptotic expansion)

We assume the same hypotheses than in (3.11) and (3.12) and moreover, that $f \in \mathcal{S}(\mathbf{R})$ and is constant in a neighborhood of 0. Then, we have the following asymptotic expansion of Tr f(tP) when $t \to 0_+$:

$$\operatorname{Tr} f(tP) \sim \iint f(tp)\, dx\, d\xi +$$

$$\sum_{2 \leqslant j} \sum_{2 \leqslant k \leqslant 3j/2} t^k \iint \frac{d_{jk}}{k\,!} f^{(k)}(tp)\, dx\, d\xi$$

Proof

Let f_t be the function defined by $f_t(\lambda) = f(t\lambda)$. It is sufficient to evaluate the behavior of

$$\| f_t(P) - \sum_{j=0}^{N} Op^W p_{f_t\,;\,j} \|_{Tr}$$

when $t \to 0_+$, thanks to (3.12) : in (3.13), the first term is a $O(t^{[\delta(N-N_1)]-1/2})$ and the second term is zero. ∎

(3.18) Remarks

(i) There exists $\eta > 0$ such that for all k,j :

$$t^k \iint d_{jk}\, f^{(k)}(tp)\, dx\, d\xi = O(t^{\delta j-\eta}), \qquad t \to 0_+$$

(ii) If f has some derivatives non-zero in $\lambda = 0$, we still have an asymptotic expansion with the extra-terms :

$$\sum_{k \geqslant 2} \frac{t^k}{k!} \operatorname{Tr}(P^k - \sum_{\ell=0}^{k} Op^W p_{k\,;\,\ell})\, f^{(k)}(0)$$

We have showed that the contribution for k=2 is zero and that the others are generally non zero.

We can also derive a trace formula for short range Schrödinger operators, which may be compared with the results of [CV] and [GU]. Let $V \in C^\infty(\mathbf{R}^n)$ be the potential, such that there is $s < -n$ and $\theta \in]0,1]$:

(3.19) $|\partial_x^\alpha V(x)| \leqslant C_\alpha (1+|x|)^{s-\theta|\alpha|}$.

Let us consider quantic hamiltonians $H_0 = -\frac{\Delta}{2}$ and $H = H_0 + V$. For $f \in \mathcal{S}(\mathbf{R})$, constant in a neighborhood of zero, we have :

(3.20) Corollary

For each $t \in \mathbf{R}$, $f(tH)-f(t\,H_0)$ is trace class and when $t \to 0$, we have the asymptotics :

$$\mathrm{Tr}(f(tH)-f(t\,H_0)) \sim \sum_{j \geqslant 1} t^{j-n/2}\, C_j(V,f)$$

where, in particular :

$$C_1(V,f) = (2\Pi)^{-n}\, \gamma_n \left(\int_{\mathbf{R}^n} V(x)\,dx\right) . \left(\int_{\mathbf{R}} (2\rho)^{n/2-1} . f'(\rho)\,d\rho\right)$$

$$C_2(V,f) = (2\Pi)^{-n}\, \gamma_n \frac{1}{2} \left(\int_{\mathbf{R}^n} V^2(x)\,dx\right) . \left(\int_{\mathbf{R}} (2\rho)^{n/2-1} . f''(\rho)\,d\rho\right)$$

$$C_3(V,f) = -(2\Pi)^{-n}\, \gamma_n \frac{1}{6} \left(\int_{\mathbf{R}^n} V^3(x) + \frac{1}{4}|\nabla V(x)|^2\,dx\right) \left(\int_{\mathbf{R}} (2\rho)^{n/2-1} f^{(3)}(\rho)\,d\rho\right) .$$

This statement could be derived from trace formulas in [CV], [GU], when V has a compact support. Howerer, this result seems new for potentials verifying (3.19) only - which ensures that $V \subset L^1(\mathbf{R}^n)$.

3 F Weyl's formula

We prove here a uniform estimate of the remainder of Weyl's formula for an hypoelliptic operator $P = \mathrm{Op}^W p$. We denote by $\tilde{N}(\lambda\,;P)$ and $\tilde{V}(\lambda\,;p)$ the corresponding counting and volume functions :

$$\tilde{N}(\lambda\,;P) = \#\{j \in \mathbf{N} / \lambda_j(P) \leqslant \lambda\}$$

$$\tilde{V}(\lambda\,;p) = (2\,\Pi)^{-n} \int_{p \leqslant \lambda} dx\,d\xi \;.$$

(3.21) Theorem

As in (3.12), we suppose that (H_1), (H_2), (W) et (S) are verified for ϕ,φ, and p respectively. Then, for each $\theta \in]0, \frac{\delta}{\rho}[$, there exist C_1 , C_2 depending only on a finite number of semi-norms of p, such that, for all $\lambda > 0$:

(3.22) $|\tilde{N}(\lambda\,;P)-\tilde{V}(\lambda\,;p)| \leqslant (2+C_1\,\lambda^{3\theta-2\delta})\,[\tilde{V}(\lambda+\lambda^{1-\theta}\,;p) - \tilde{V}(\lambda-\lambda^{1-\theta}\,;p)] + C_2$.

Proof

We have : $\tilde{N}(\lambda\,;P) = \mathrm{Tr}(\mathbf{1}_{[0,\lambda]}(P))$, and, as in [TU-SU] and [HO 2], we are going to regularize $\mathbf{1}_{[0,\lambda]}$. Let $I(\lambda,\theta)$ and $J(\lambda,\theta)$ be the intervals :

$$I(\lambda,\theta) = [\lambda-\lambda^{1-\theta}/2\,,\,\lambda+\lambda^{1-\theta}/2]$$

$$J(\lambda,\theta) = [\lambda-\lambda^{1-\theta}, \lambda+\lambda^{1-\theta}].$$

Let $f_{\lambda,\theta}$ and $g_{\lambda,\theta}$ be cut off functions such that :

$$f_{\mu,\theta}(\mu) = \begin{cases} 1 & \text{if } \mu \notin I(\lambda,\theta) \text{ and } \mu < \lambda \\ 0 & \text{if } \mu \notin I(\lambda,\theta) \text{ and } \mu > \lambda \end{cases} ;$$

$g_{\lambda,\theta} \in C_0^\infty(J(\lambda,\theta))$ and $g_{\lambda,\theta} \equiv 1$ on $I(\lambda,\theta)$;

(3.23) $\forall k \in \mathbf{N},\ |(\mu\, \partial_\mu)^k f_{\lambda,\theta}(\mu)|+|(\mu\, \partial_\mu)^k g_{\lambda,\theta}(\mu)| \leqslant C_k \lambda^{k\theta}$.

Then, we have :

(3.24) $|\tilde{N}(\lambda ; P) - \mathrm{Tr}\, f_{\lambda,\theta}(P)| \leqslant \mathrm{Tr}\, \mathbf{1}_{I(\lambda,\theta)}(P)$

(3.25) $\mathrm{Tr}\, \mathbf{1}_{I(\lambda,\theta)}(P) \leqslant \mathrm{Tr}\, g_{\lambda,\theta}(P)$

Now, the first term of $\mathrm{Tr}\, f_{\lambda,\theta}(P)$ is $\int f_{\lambda,\theta}(p)$; $|\int f_{\lambda,\theta}(p) - \tilde{V}(\lambda ; p)|$ is smaller than $\tilde{V}(\lambda+\lambda^{1-\theta} ; p) - \tilde{V}(\lambda-\lambda^{1-\theta} ; p)$ and so is the first term of $\mathrm{Tr}\, g_{\lambda,\theta}(P)$.

The jth term of $\mathrm{Tr}\, f_{\lambda,\theta}(P)$ is bounded by $\lambda^{(3\theta-2\delta)j/2}\, (\tilde{V}(\lambda+\lambda^{1-\theta} ; p) - \tilde{V}(\lambda-\lambda^{1-\theta} ; p))$. As the term corresponding to $j=1$ is zero, the highest power of λ is $3\theta-2\delta$ (for $3\theta-2\delta$ is negative !). Thanks to (3.12), we find that the remainder of order N for $\mathrm{Tr}\, f_{\lambda,\theta}(P)$ is a $O(\lambda^{N(\rho\theta-\delta)+\delta_0})$, and so it is for $\mathrm{Tr}\, g_{\lambda,\theta}(P)$.

All that, joined to (3.24), (3.25), give (3.22). ■

(3.26) Remarks

If $\phi = \varphi$, we can take $\rho = 3/2$ - see remark (B.2) -. Then, we have roughly the same assumptions that [TU-SU], and, thus, we improve their result which corresponds to $\rho = 2$, and we reach the same limit than [HO 2] : $\theta < 2\delta/3$.

However, under general hypothesis (H_1), we have but $\rho = 2$, what allows us to reach $\theta < \delta/2$. Nevertheless our estimate is more precise than that of [HO 2], which does not give the behavior in λ of the coefficient multiplicating $(\tilde{V}(\lambda+\lambda^{1-\theta} ; p)-\tilde{V}(\lambda-\lambda^{1-\theta} ; p))$.

§4 : Proof of the main theorem (1.3)

4 A Introduction

The matter consists in proving that :

(4.0) $N_\pm(\lambda ; A) = V_\pm(\lambda ; a) + O(\lambda^\zeta\, V_+(\lambda ; m))$

That will result from the two following propositions :

(4.1) Proposition

$N_\pm(\lambda ; A) \leqslant V_\pm(\lambda ; a) + O(\lambda^\zeta\, V_+(\lambda ; m))$

Let us denote :

$N(\lambda ; A) = (N_+ + N_-)(\lambda ; A)$ (counting function of the singular values of A)

$V(\lambda ; A) = (V_+ + V_-)(\lambda ; a)$ (volume function of $|a|$)

(4.2) Proposition

$$N(\lambda ; A) = V(\lambda ; a) + 0(\lambda^{\zeta} V(\lambda ; m))$$

Indeed (4.2) gives that :

$$V_+(\lambda ; a) = N_+(\lambda ; A) + 0(\lambda^{\zeta} V(\lambda ; m)) + N_-(\lambda ; A) - V_-(\lambda ; a)$$

With (4.1) for (-), we get :

$$V_+(\lambda ; a) \leqslant N_+(\lambda ; A) + 0(\lambda^{\zeta} V(\lambda ; m)),$$

which, with (4.1) for (+), gives (4.0) for (+). ■

In order to prove (4.1) and (4.2), we first need the two basic lemmas :

(4.3) Lemma

For two functions $a,b : \mathbf{R}^{2n} \to \mathbf{R}$, for $\eta \in [0,1]$, we have :

$$V_{\pm}(\lambda ; a+b) \leqslant V_{\pm}(\eta\lambda ; a) + V_{\pm}((1-\eta)\lambda ; b) .$$

Indeed, $a+b \geqslant \lambda$ implies that $a \geqslant \eta\lambda$ or $b \geqslant (1-\eta)\lambda$.

(4.4) Lemma

For two compact self-adjoint operators A,B, we have :

$$N_{\pm}(\lambda ; A+B) \leqslant N_{\pm}(\eta\lambda ; A) + N_{\pm}((1-\eta)\lambda ; B)$$

It is deduced from Weyl - Ky Fan inequalities for positive or negative eigenvalues, that we will prove in the annex A - because those inequalities are stated only for singular values in [GO-KR].

4 B Preliminary lemmas for weights in the class $S(m ; \phi,\varphi)$

We recall that m verifies the properties (W) , (N), and (T) for its volume function - see § 1.

(4.5) Notation

We denote by a capital letter $(M,B_0,...)$ the operator the Weyl symbol of which is the corresponding small letter (m,b,...) : $M = Op^W m$, $B_0 = Op^W b_0$,...

(4.6) Lemma

There is $m_1 \in S(m(\phi\varphi)^{-1} ; \phi,\varphi)$ such that, denoting $m_0 = m+m_1$, M_0 is positive and invertible.

Proof

As $m \in S(m ; \phi;\varphi)$, then $m^{-1} \in S(m^{-1} ; \phi,\varphi)$. Thanks to (N), we have $m^{-\gamma} \geqslant K\,\phi\varphi$, and then, $Op^W m^{-1}$ is semi-bounded. Thus, there is $\mu > 0$, such that $Op^W(m^{-1}+\mu)$ is strictly positive. Its inverse has its Weyl's symbol like m_0.

(4.7) Lemma

Let be $\delta \geqslant 0$ and $d \in S(m^{1+\delta} ; \phi,\varphi)$. Then, with the parameters λ_0, γ_1 appearing in condition (T) for m, we have :

$$\forall \lambda \in]0,\lambda_0], N(\lambda ; D) \leqslant C(d) \lambda^\tau V(\lambda ; m) \qquad \text{with } \tau = \gamma_1 \delta(1+\delta)^{-1},$$

where C(d) depends only on a finite number of semi-norms of d in $S(m^{1+\delta} ; \phi,\varphi)$.

(4.8) Convention

That means that C(d) is a polynomial combination of semi-norms $s_k(m^{1+\delta} ; \phi,\varphi)$ with coefficients depending only on m,ϕ,φ.

Proof

As in (4.6), there is $\mu > 0$ such that $Op^W(m^{-1-\delta}+\mu)$ is positive and invertible. Let L be its inverse. As $(m^{-1-\delta} d)$ is bounded like all its derivatives, thanks to composition and Calderòn-Vaillancourt's theorems, we get that $L^{-1} \circ D$ is a bounded operator of $L^2(\mathbf{R}^n)$ with a norm $C_1(d)$ depending only on a finite number of semi-norms of d. By mini-max principle, we have so :

$$\lambda_j(D^2) \leqslant C_1(d) \lambda_j(L^2) \qquad \forall j .$$

Therefore :

$$N(\lambda ; D) \leqslant C_1(d) N(\lambda ; L) .$$

Now, we deduce from theorem (3.21), the rough estimate :

$$N(\lambda ; L) \leqslant C V(\lambda ; m^{1+\delta}).$$

From consequence (T') of condition (T), we draw :

$$V(\lambda ; m^{1+\delta}) \leqslant C \lambda^\tau V(\lambda ; m).$$

The three last inequalities yield (4.7). ∎

Now, we are going to adapt Weyl's formula (3.21) to compact invertible operators.

(4.9) Proposition

Let b be a weight verifying (W) and (N) ; in particular, we have the estimate :

$$(\phi \varphi)^{-1} \leqslant K(b) b^\gamma$$

and there is $b_1 \in S(b(\phi \varphi)^{-1} ; \phi,\varphi)$ such that $B_0 = Op^W(b+b_1)$ is positive and invertible -see (4.6).

Then, with $\theta_0 = \inf(\frac{\gamma}{2}, \frac{\gamma}{\rho})$, we have for each $0 < \theta \leqslant \theta_0$ and $\lambda \leqslant 1/2$:

$$|N(\lambda\ ;\ B_0)-V(\lambda\ ;\ b)| \leqslant (2+C_1\ \lambda^{\theta})\ [V(\lambda-\lambda^{1+\theta}\ ;\ b)-V(\lambda+\lambda^{1+\theta}\ ;\ b)]+C_2$$

where C_1 and C_2 depend only on K(b) and a finite number of semi-norms of b and b_1 -see (4.8).

Let us remark that if, moreover, V(. ; b) verifies condition (T), we should have : $V(\lambda-\lambda^{1+\theta}\ ;\ b) - V(\lambda+\lambda^{1+\theta}\ ;\ b) \leqslant C\ \lambda^{\theta}\ V(\lambda\ ;\ b)$.

Proof

In order to link (4.9) and (3.21), we denote :

$P_0 = B_0^{-1}\ ;\ p = b^{-1}$.

Then, for $j \geqslant 1$, there is $p_j \in S(p(\phi\ \varphi)^{-j}\ ;\ \phi,\ \varphi)$ such that for each $N \geqslant 1$:

$$P_0 - Op^W(p + \sum_{j=1}^{N} p_j) = Op^W\ r_N \qquad \text{with} \qquad r_N \in S(p(\phi\ \varphi)^{-N-1}\ ;\ \phi,\ \varphi)$$

and each semi-norm of a p_j depend only on a finite number of semi-norms of b and b_1 - it is the mere construction of a parametrix for B_0.

Thus, we are in a slightly more general situation than in §3 : the symbol of P_0 is written as an asymptotic expansion (just like classical symbols). Nevertheless, the whole results may be adapted, by starting from :

$$q_{z\ ;\ j}^{(K)} = (p-z)^{-1}\ [p_j^{(K)} - \sum_{k=0}^{j-1}\ \sum_{\ell=0}^{j}\ \sum_{|\alpha+\beta|=j-k-\ell} \Gamma(\alpha,\beta)\ (\partial_\xi^\alpha\ D_x^\beta\ p_\ell)\ (\partial_\xi^\beta\ D_x^\alpha\ q_{z\ ;\ k}^{(K)})$$

instead of (3.3) (with $p_0 = p$)

Then, instead of (3.22), we arrive to :

$$|\tilde{N}(\lambda\ ;\ P_0) - \tilde{V}(\lambda,p)| \leqslant (2+C_1\ \lambda^{\theta-\delta})[\tilde{V}(\lambda+\lambda^{1-\theta}\ ;\ p)-\tilde{V}(\lambda-\lambda^{1-\theta}\ ;\ p)]+C_2$$

with C_1 and C_2 depending only on a finite number of semi-norms of $p,p_1,\ldots,p_N$, where N is a fixed integer. Now, let us remark that, for $\theta \leqslant \theta_0$, $\lambda^{\theta-\delta}$ may be replaced by $\lambda^{-\theta}$, and that, for any fixed constant $c > 0$, $\tilde{V}(\lambda+\lambda^{1-\theta}\ ;\ p) - \tilde{V}(\lambda-\lambda^{1-\theta}\ ;\ p)$ may be replaced by $\tilde{V}(\lambda+c\lambda^{1-\theta}\ ;\ p) - \tilde{V}(\lambda-c\lambda^{1-\theta}\ ;\ p)$.

On the other hand, we have, for $\mu = \lambda^{-1}$:

$N(\mu\ ;\ B_0) = \tilde{N}(\lambda\ ;\ P_0)\ ;\ V(\mu\ ;\ b) = \tilde{V}(\lambda\ ;\ p)$

and

$\tilde{V}(\lambda \pm c\lambda^{1-\theta}\ ;\ p) = V(\mu(1\pm c\mu^{\theta})^{-1}\ ;\ b)$

We choose $c=2^{\theta-1}$, and we have $(1+c\mu^{\theta})^{-1} > 1-\lambda^{\theta}$

$(1-c\mu^{\theta})^{-1} < 1+\lambda^{\theta}$.

From all that, we deduce (4.9). ∎

Lastly, here is a result looking like Garding's inequality. It consists of a relation between symbols positivity and operators positivity (that relation is not systematic as it is in anti-wick quantization - see [TU-SU]).

(4.10) Lemma

Let $b \in S(m\ ;\ \phi,\varphi)$ be a positive symbol. Let be $\varepsilon > 0$. Then, there is $d_\varepsilon \in S(m(\phi\,\varphi)^{-1}\ ;\ \phi,\varphi)$ such that :

$$0 < B + \varepsilon M + D_\varepsilon$$

and each semi-norm of d_ε depends only on a finite number of semi-norms of $(b+\varepsilon m)^{1/2}$.

Proof :

$c := (b+\varepsilon m)^{1/2} \in S(m^{1/2}\ ;\ \phi,\varphi)$. There is $d_\varepsilon \in S(m(\phi\,\varphi)^{-1}\ ;\ \phi,\varphi)$ such that :

$$Op^W(c^2) + Op^W d_\varepsilon = (Op^W c)^2 .$$

As $(Op^W c)^2$ is a positive operator, we get (4.10).

4.C Proof of proposition (4.1)

We shall denote by C various constants, independent of involved parameters $(\lambda,\varepsilon,\eta\ldots)$; and $C_j(\varepsilon)$ for $j=0,1,\ldots$, will be positive numbers depending only on ε and admitting a polynomial majorization with respect to ε^{-1}.

Here is the proof of (4.1) for (+) sign.

Let χ be a C^∞ function on $\mathbf{R}$, $\chi(\lambda) = 0$ if $\lambda < 1$ and $\chi(\lambda) = 1$ if $\lambda > 2$.

For $\varepsilon > 0$, we denote, $a_\varepsilon = a\,\chi(a/\varepsilon m) + \varepsilon m[1 - \chi(a/\varepsilon m)]$.

We then have the obvious following properties :

(4.11) $a_\varepsilon \in S(m\ ;\ \phi,\varphi)$

(4.12) $a_\varepsilon \geqslant \varepsilon m$

(4.13) a_ε coïncides with a in the region $\{a \geqslant 2\,\varepsilon m\}$

(4.14) $a \lesssim a_\varepsilon + \varepsilon m$.

Thanks to lemma (4.10), (4.14) gives that there is $d_{1,\varepsilon} \in S(m(\phi\,\varphi)^{-1}\ ;\ \phi,\varphi)$ such that :

$$A \lesssim A_\varepsilon + 2\,\varepsilon M + D_{1,\varepsilon} .$$

With lemmas (4.6) and (4.10), (4.12) gives that there is $d_{2,\varepsilon} \in S(m(\phi\,\varphi)^{-1}\ ;\ \phi,\varphi)$ such that :

$$\varepsilon M_0/2 \lesssim A_\varepsilon + D_{2,\varepsilon} .$$

Le b_ε be the summ : $a_\varepsilon + d_{2,\varepsilon}$. As M_0 is positive and invertible, so it is for B_ε. With $d_\varepsilon = d_{1,\varepsilon} - d_{2,\varepsilon}$ we then get :

(4.15) $A \lesssim B_\varepsilon + 2\,\varepsilon M + D_\varepsilon$.

We are going to estimate N^+ for each of the summands : at first the error terms $2\,\varepsilon M + D_\varepsilon$ and secondly the principal term B_ε ; then we will end the proof.

Step 1 : Error terms

(i) $2\,\varepsilon M$: Proposition (4.9) yields the rough estimate :

$$N_+(\lambda\ ;\ 2\,\varepsilon M) \leqslant C\ V(\lambda\ ;\ 2\,\varepsilon m)$$

with inequality (T') derived from condition (T), we get :

$$V(\lambda\ ;\ 2\,\varepsilon m) = V(\lambda/2\varepsilon\ ;\ m) \leqslant (2\,\varepsilon)^{\gamma_1}\ V(\lambda\ ;\ m)\ .$$

So, we have :

(4.16) $N_+(\lambda\ ;\ 2\,\varepsilon M) \leqslant C\ \varepsilon^{\gamma_1}\ V(\lambda\ ;\ m)$.

(ii) D_ε : as $d_\varepsilon \in S(m^{1+\gamma}\ ;\ \phi,\varphi)$, we draw from lemma (4.7) :

(4.17) $N(\lambda\ ;\ D_\varepsilon) \leqslant C_0(\varepsilon)\ \lambda^\tau\ V(\lambda\ ;\ m) \qquad \tau > 0$

for each semi-norm of d_ε is majorized by a polynomial function of ε^{-1}.

(iii) $2\,\varepsilon M + D_\varepsilon$: with lemma (4.3) for $\eta = 1/2$, (4.16) and (4.17) give us :

$$N(\lambda\ ;\ 2\,\varepsilon M + D_\varepsilon) \leqslant C(\varepsilon^{\gamma_1} + \lambda^\tau\ C_0(\varepsilon))\ V(\lambda/2\ ;\ m)$$

and, with the help of inequality (T')

(4.18) $N(\lambda\ ;\ 2\,\varepsilon M + D_\varepsilon) \leqslant C(\varepsilon^{\gamma_1} + \lambda^\tau\ C_0(\varepsilon))\ V(\lambda\ ;\ m)$.

Step 2 : Principal term

(i) B_ε is positive and invertible, with " main symbol " a_ε. As a_ε verifies (W) and (N) thanks to (4.11) and (4.12), proposition (4.9) yields :

(4.19) $N_+(\lambda\ ;\ B_\varepsilon) \leqslant V(\lambda\ ;\ a_\varepsilon) + (2 + C_1(\varepsilon)\ \lambda^\theta)\ [V(\lambda-\lambda^{1+\theta}\ ;\ a_\varepsilon) - V(\lambda+\lambda^{1+\theta}\ ;\ a_\varepsilon] + C_2(\varepsilon)$.

As the semi-norms of a_ε depends on ε^{-1} on a polynomial way, so it is for $C_1(\varepsilon)$ and $C_2(\varepsilon)$.

(ii) We are going to show how $V(\lambda\ ;\ a_\varepsilon)$ is close to $V_+(\lambda\ ;\ a)$.

(4.20) Lemma

$\gamma_0 = \min(\gamma_1/2, 1/2)$. $\exists C > 0, \forall \lambda < \lambda_0, \forall \varepsilon \in]0,1/2[$:

$$|V(\lambda\ ;\ a_\varepsilon) - V_+(\lambda\ ;\ a)| \leqslant C\ \varepsilon^{\gamma_0}\ V(\lambda\ ;\ m)$$

Proof

(1) Majorization of $V(\lambda\ ;\ a_\varepsilon)$. We have :

$$\int_{a_\varepsilon > \lambda} \leqslant \underbrace{\int_{a > \operatorname{Sup}(\lambda, 2\varepsilon m)}}_{I_1} + \underbrace{\int_{(a_\varepsilon > \lambda) \cap (a < 2\varepsilon m)}}_{I_2}$$

Now : $I_1 \leqslant V_+(\lambda\ ;\ a)$ and $I_2 \leqslant V(\lambda\ ;\ 2\varepsilon m) \leqslant C\ \varepsilon^{\gamma_1}\ V(\lambda\ ;\ m)$ by (4.16).

Therefore :

$$V(\lambda\ ;\ a_\varepsilon) \leqslant V_+(\lambda\ ;\ a) + C\ \varepsilon^{\gamma_1}\ V(\lambda\ ;\ m)\ .$$

(2) Minoration of $V(\lambda ; a_\varepsilon)$. Thanks to (4.14) we have :

$$V(a_\varepsilon ; \lambda) \geqslant V_+(\lambda ; a - \varepsilon m) .$$

Thanks to lemma (4.3), we have for $\eta \in [0,1]$:

$$V_+(\lambda ; a - \varepsilon m) \geqslant V_+(\frac{\lambda}{\eta} ; a) - V_+(\frac{1-\eta}{\eta} \lambda ; \varepsilon m) .$$

For η such that $\frac{1-\eta}{\eta} = \varepsilon^{1/2}$, and thanks to (T') we get :

$$V_+(\frac{\lambda}{\eta} ; a) \geqslant \eta^{\gamma_2} V_+(\lambda ; a) \geqslant (1 - \gamma_2 \varepsilon^{1/2}) V(\lambda ; a)$$

$$V_+(\frac{1-\eta}{\eta} \lambda ; \varepsilon m) \leqslant \varepsilon^{\gamma_1/2} V(\lambda ; m).$$

So, we obtain the wanted minoration. ∎

(iii) Conclusion about B_ε.

(4.19) and (4.20) give us :

$$N_+(\lambda ; B_\varepsilon) \leqslant V_+(\lambda ; a) + C_2(\varepsilon) + C[1+C_1(\varepsilon) \lambda^\theta] [\varepsilon^{\gamma_0} V(\lambda ; m)+V_+(\lambda-\lambda^{1+\theta} ; a)-V_+(\lambda+\lambda^{1+\theta} ; a)].$$

The condition (T) about $V_+(\lambda ; a)$ and the majorization $a \leqslant C.m$ allow us to deduce :

(4.21) $N_+(\lambda ; B_\varepsilon) \leqslant V_+(\lambda ; a) + C[1+C_1(\varepsilon) \lambda^\theta] (\varepsilon^{\gamma_0} + \lambda^\theta) V(\lambda ; m) + C_2(\varepsilon) .$

Step 3 : End of the proof

(4.15) gives us :

$$N_+(\lambda ; A) \leqslant N_+(\lambda ; B_\varepsilon + 2 \varepsilon M + D_\varepsilon)$$

and with lemma (4.4) :

$$N_+(\lambda ; A) \leqslant N_+((1-\eta)\lambda ; B_\varepsilon) + N_+(\eta \lambda ; 2 \varepsilon M + D_\varepsilon) .$$

We take $\eta = \varepsilon^\alpha$, $\alpha > 0$; then (4.18) and (T') give :

$$N_+(\eta \lambda ; 2 \varepsilon M + D_\varepsilon) \leqslant C(\varepsilon^{\gamma_1} + \lambda^\tau \varepsilon^{\alpha\tau} C_0(\varepsilon)) \varepsilon^{-\alpha \gamma_2} V(\lambda ; m).$$

We then choose α such that $\gamma_1 - \alpha \gamma_2 \equiv \gamma_3 > 0$. So, we get from the last inequalities and (4.21) :

$$N_+(\lambda ; A) \leqslant V_+(\lambda ; a) + C(\varepsilon^{\gamma_0} + \varepsilon^{\gamma_3} + C_1(\varepsilon) \lambda^\theta + C_3(\varepsilon) \lambda^\tau) V(\lambda ; m) + C_2(\varepsilon) .$$

Now, the inequality $K m^\gamma \geqslant (\phi\varphi)^{-1}$ implies that there is some $\tau' > 0$ such that

$V(\lambda ; m) \geqslant C \lambda^{-\tau'}$ (with $C > 0$)

Therefore, with $\omega = \inf(\theta, \tau, \tau')$, $C_4 = C_1 + C_2 + C_3$ and $\gamma_4 = \inf(\gamma_0, \gamma_3)$, we get:

$$N_+(\lambda ; A) \leqslant V_+(\lambda ; a) + C(\varepsilon^{\gamma_4} + C_4(\varepsilon) \lambda^\omega) V(\lambda ; m).$$

Now, there is some positive integer N such that $C_4(\varepsilon) \leqslant \varepsilon^{-N}$ for $\varepsilon < 0$ small enough. Then, we choose :

$$\varepsilon = \lambda^\beta \text{ with } \beta = \frac{\omega}{\gamma_4 + N}$$

Thus, we get the proposition (4.1) with $\zeta = \gamma_4 \beta$.

4.D Proof of proposition (4.2)

In view of (4.1), it remains to be proved:

(4.22) $V(\lambda\ ;\ a) \leqslant N(\lambda\ ;\ A) + O(\lambda^{\zeta}\ V(\lambda\ ;\ m))$.

Step 1 : Reduction to the case when a is positive

Let us admit that (4.22) is true for $a \geqslant 0$. Then, for any symbol $a \in S(m\ ;\ \phi,\varphi)$, a^2 is positive and belongs to $S(m^2\ ;\ \phi,\varphi)$. We get from (4.22) :

(4.23) $V(\lambda\ ;\ a^2) \leqslant N(\lambda\ ;\ Op^W\ a^2) + O(\lambda^{\zeta'}\ V(\lambda\ ;\ m^2))$.

Now there is $d \in S(m^2(\phi\,\varphi)^{-1}\ ;\ \phi,\varphi)$ such that :

$Op^W\ a^2 - A^2 = Op^W\ d$.

Thanks to lemmas (4.4) and (4.7), we deduce from (4.23) that, by possibly decreasing ζ' :

(4.24) $V(\lambda\ ;\ a^2) \leqslant N(\lambda\ ;\ A^2) + O(\lambda^{\zeta'}\ V(\lambda\ ;\ m^2))$.

As $N(\lambda\ ;\ A^2) = N(\lambda^{1/2}\ ;\ A)$ and $V(\lambda\ ;\ a^2) = V(\lambda^{1/2}\ ;\ a)$, we deduce (4.22) from (4.24) with $\zeta = 2\,\zeta'$.

Step 2 : Proof of (4.22) in the case when a is positive

Let a_ε be $a + \varepsilon\, m$. Thanks to lemmas (4.6) and (4.10) we obtain that there is $d_\varepsilon \in S(m(\phi\,\varphi)^{-1}; \phi,\varphi)$ such that :

$B_\varepsilon \equiv A_\varepsilon + D_\varepsilon \geqslant \varepsilon M_0/2$.

We apply to B_ε with main symbol a_ε , the proposition (4.9) and we get :

(4.25) $V(\lambda\ ;\ a_\varepsilon) \leqslant N(\lambda\ ;\ B_\varepsilon) + (1+C_1(\varepsilon)\,\lambda^{\theta})\,(V(\lambda-\lambda^{1+\theta}\ ;\ a_\varepsilon) - V(\lambda+\lambda^{1+\theta}\ ;\ a_\varepsilon)) + C_2(\varepsilon)$.

Thanks to lemma (4.3) and the condition (T) for V(.,a), we obtain a similar statement to (4.20). At last :

(4.26) $N(\lambda\ ;\ B_\varepsilon) \leqslant N((1-\eta)\lambda\ ;\ A) + N(\eta\lambda\ ;\ \varepsilon M + D_\varepsilon)$.

We use (4.26) for $\eta = \varepsilon^{\alpha}$ with a suitable $\alpha > 0$ and we transfer the obtained estimate to (4.25) ; we then get (4.2) like (4.1). ∎

(4.27) Remark

For the minoration of $N(\lambda)$ it is possible to reduce to the case when a is positive and to approximate a **everywhere** by an elliptic regularization (and to get (4.26)). That would be impossible for $N_+(\lambda)$ alone when a is not of constant sign (a_ε in 4.C is far from a when a is negative).

Annex A : Weyl-Ky Fan inequalities

In order to prove these inequalities, we start giving a Ky Fan's type caracterization for the eigenvalues of a compact self-adjoint operator T in a Hilbert space H. We will deduce Weyl's inequalities for eigenvalues, then for counting functions $N_{\pm}(\lambda)$.

With the same convention as in §1 about the eigenvalues $\lambda_j^{\pm}(T)$ of T, we state:

(A.1) Proposition

$\lambda_j^+(T) = \mathrm{Inf}_{K,S}\|T-K-S\|$,

where the inf is computed for all self-adjoint operators K of rank $<j$, and for all compact non positive operators S.

$\lambda_j^-(T) = -\mathrm{Inf}_{K,S}\|T-K-S\|$,

with S non negative.

[GO-KR] give the corresponding statement for singular values : $s_j(T) = \mathrm{Inf}_K\|T-K\|$, with K of rank $<j$.

Proof

We study the (+) sign. We denote by (φ_k^+), (φ_k^-) an orthonormal eigenfunctions basis of H, associated to the (λ_k^+), (λ_k^-).

(i) As :

$$T = \sum_{k\geq 1} \lambda_k^+ \varphi_k^+ \otimes \varphi_k^+ + \sum_{k\geq 1} \lambda_k^- \varphi_k^- \otimes \varphi_k^-$$

by taking :

$$K = \sum_{k=1}^{j-1} \lambda_k^+ \varphi_k^+ \otimes \varphi_k^+ \ , \quad S = \sum_{k\geq 1} \lambda_k^- \varphi_k^- \otimes \varphi_k^-$$

we have suitable K and S, and we get : $\|T-K-S\| = \lambda_j^+$. So, $\lambda_j^+ \geq \mathrm{Inf}_{K,S}\|T-K-S\|$.

(ii) If j is such that φ_j^+ does not exist (if the positive part of T is of rank $<j$), we have $\mathrm{Inf}_{K,S}\|T-K-S\| = 0$, and thus $\mathrm{Inf}_{K,S}\|T-K-S\| = \lambda_j^+ = 0$.

If φ_j^+ exists, let K and S be self-adjoint operators : K of rank $<j$ and S negative. And let $u \neq 0$ be orthogonal to Rg(K) in the space with the basis $\varphi_1^+, \ldots, \varphi_j^+$. We have :

$$\begin{aligned}\|(T-K-S)u\| &\geq \langle (T-S-K)u,u\rangle \\ &\geq \langle Tu,u\rangle \text{ for } \langle Ku,u\rangle = 0 \text{ and } \langle -Su,u\rangle \geq 0 \\ &= \sum_{k=1}^{j} \lambda_k^+ \langle u,\varphi_k^+\rangle^2 \geq \lambda_j^+\|u\|^2 .\end{aligned}$$

Therefore $\|T-K-S\| \geq \lambda_j^+$.

(A.2) Corollary

$\lambda_{j+k+1}^+(T_1+T_2) \leq \lambda_{j+1}^+(T_1) + \lambda_{k+1}^+(T_2)$.

Proof

Let K_1, K_2, S_1, S_2 be self-adjoint operators sucht that, $\mathrm{rank}(K_1) \leq j$, $\mathrm{rank}(K_2) \leq k$, $S_1, S_2 \leq 0$, and :

$$\|T_1 - K_1 - S_1\| = \lambda^+_{j+1}(T_1) \; ; \; \|T_2 - K_2 - S_2\| = \lambda^+_{k+1}(T_2).$$

As, rank$(K_1+K_2) \leqslant j + k$ and $S_1 + S_2 \leqslant 0$, we get (A.2). ■

In order to prove lemma (4.4), it remains to state :

(A.3) Proposition

Let (u_n), (v_n), (w_n) be three decreasing sequences of positive numbers, tending to zero. We denote :

$N(\lambda \; ; u) = \# \{n \geqslant 1 \; / \; u_n \geqslant \lambda\}$, and the same for v,w.

Then (i) implies (ii) :

(i) $\quad j,k \in \mathbf{N} : u_{j+k+1} \leqslant v_{j+1} + w_{k+1}$

(ii) $\quad \lambda < 0, \quad \eta \in [0,1], N(\lambda \; ; u) \leqslant N(\lambda\eta \; ; v) + N((1-\eta)\lambda \; ; w)$

Indeed, (i) is equivalent to (ii).

Proof

Let $\lambda > 0$ and $\eta \in [0,1]$ be real numbers.

a) if $v_1 < \eta\lambda$ and $w_1 < (1-\eta)\lambda$, then :

$u_1 \leqslant v_1 + w_1 < \lambda.$

Thus $N(\lambda \; ; u) = 0$ and the inequality (ii) is true for λ,η.

b) If not, let us suppose that $v_1 \geqslant \eta\lambda$. Let j be the integer such that : $u_{j+1} < \lambda \leqslant u_j$ (if j does not exist, $N(\lambda \; ; u) = 0$, and (ii) is true). We have : $N(\lambda \; ; u) = j$.

b1) if $v_j \geqslant \eta\lambda$, then $N(\eta\lambda \; ; v) \geqslant j$ and (ii) is true.

b2) if not : $v_j < \eta\lambda$. Let $k < j$ be the integer such that $v_{k+1} < \eta\lambda \leqslant v_k$ (k exists because $v_1 \geqslant \lambda\eta$). We have : $N(\eta\lambda \; ; v) = k$.

Now, thanks to (i) :

$u_j \leqslant v_{k+1} + w_{j-k}$

As $u_j \geqslant \lambda$ and $v_{k+1} < \lambda\eta$, we get :

$w_{j-k} \geqslant \lambda(1-\eta).$

So : $N((1-\eta)\lambda \; ; w) \geqslant j-k.$

Therefore :

$N(\lambda \; ; u) = j = k + (j-k) \leqslant N(\eta\lambda \; ; v) + N((1-\eta)\lambda \; ; w).$ ■

In order to deduce lemma (4.4), we apply (A.3) to the sequences $u = (\lambda^+_j(A+B))$, $v = (\lambda^+_j(A))$, $w = (\lambda^+_j(B))$, in basing our argument on corollary (A.2) to have (i).

[BI-SO 2] gives without proof the statement corresponding to lemma (4.4) for the singular values.

Annexe B : Composition formula with precised remainder

Let (ϕ,φ) be weight functions verifying (H_1). Let, for $i = 1,2$, m_i be (ϕ,φ)-temperate weights and $a_i \in S(m_i\ ;\ \phi,\varphi)$. It is well known that the operator $(Op^W a_1) \circ (Op^W a_2)$ admit a Weyl's symbol a given by :

(B.0) $$a(x,\xi) = \exp\{\tfrac{i}{2}\,\sigma(D_z,D_\zeta\ ;\ D_y,D_\eta)\}\, a_1(z,\zeta)\, a_2(y,\eta) \Big|_{\substack{z=y=x\\ \zeta=\eta=\xi}}$$

where σ is the symplectic form :

$$\sigma(z,\zeta\ ;\ y,\eta) = \langle\zeta,y\rangle - \langle z,\eta\rangle .$$

With the help of Taylor's formula, we determine the principal part a_N of (B.0) at the order N, for each integer N :

$$a_N(x,\xi) = \sum_{j=0}^{N} \frac{1}{j!}\left(\frac{i}{2}\,\sigma(D_z,D_\zeta\ ;\ D_y,D_\eta)\right)^j a_1(z,\zeta)\, a_2(y,\eta) \Big|_{\substack{z=y=x\\ \zeta=\eta=\xi}}$$

and the remainder r_N at the order N :

$$r_N(a_1,a_2\ ;\ x,\xi) = a(x,\xi) - a_N(x,\xi) .$$

The aim of this annex is to state :

(B1) Proposition : <u>universal estimate of the remainder.</u>

There exist $\rho \geq 3/2$, $\rho' \geq 3/2$ depending only on (ϕ,φ) such that for each $N \geq 1$ and $q \geq 0$, there is $\gamma = \gamma(N,q,n)$ and $j(n)$ such that for all $(x,\xi) \in \mathbf{R}^{2n}$:

$$\sum_{|\alpha|+|\beta|\leq q} |\partial_\xi^\alpha\, \partial_x^\beta\, r_N\,(a_1,a_2)\,(x,\xi)| \leq$$

$$\gamma(q,N,n)\,\Big(\sum_{\substack{N<k<\rho N + M_2 + j(n) + q\\ N<\ell<\rho' N + M_1 + j(n) + q + M}} s_k(a_1\ ;\ m_1)\, s_\ell(a_2\ ;\ m_2)\Big)\, m_1(x,\xi)\, m_2(x,\xi)\, (\phi\varphi)^{-N}\,(x,\xi)$$

where M_i is the exponent attributed to m_i in the definition (1.1/2) of a temperate weight, and M depends only on ϕ,φ,M_1 and M_2.

(B.2) Remarks

(1) Under our hypothesis (H_1), we will show in the proof that we may take $\rho = 2$.

(2) If, moreover, $\phi = \varphi$, then we may take $\rho = 3/2$.

(3) We conjecture that (B.1) is still true with the relaxed hypothesis (H'_1) :

$$(H'_1)\quad \begin{cases} \phi\varphi \geq 1 \text{ on } \mathbf{R}^{2n} \\ \phi \text{ and } \varphi \text{ are } (\phi,\varphi)\text{-temperate} \end{cases}$$

Proof

The inversion Fourier's formula, and the Taylor's formula with integral remainder, give :

(B.3) $r_N(a_1,a_2\ ;\ x,\xi) = \frac{1}{N!}\int_0^1 (1-t)^{N-1}\ r_{N,t}(a_1,a_2\ ;\ x,\xi)\ dt$

with $r_{N,t}$ defined by :

(B.4) $r_{N,t}(a_1,a_2,x,\xi) = \int e^{i\frac{t}{2}\sigma(\tilde z,\tilde\zeta\ ;\ \tilde y,\tilde\eta)}\ \sigma(\tilde z,\tilde\zeta\ ;\ \tilde y,\tilde\eta)^N\ e^{i(\tilde z x+\tilde\zeta\xi+\tilde y x+\tilde\eta\xi)}\ \hat a_1(\tilde z,\tilde\zeta)\hat a_2(\tilde y,\tilde\eta)\ d\!\!\!^-\tilde z\, d\!\!\!^-\tilde\zeta\, d\!\!\!^-\tilde y\, d\!\!\!^-\tilde\eta$

where $\hat a$ denotes the total Fourier's transform of a, and $d\!\!\!^-z = (2\Pi)^{-n}\ dz$.

Thanks to the well-known result about Fourier's transform of gaussian functions :

$$\mathcal{F}\left(e^{i\frac{t}{2}\sigma(\tilde z,\tilde\zeta\ ;\ \tilde y,\tilde\eta)}\right)(z,\zeta\ ;\ y,\eta) = \left(\frac{2\Pi}{t}\right)^{2n}\ e^{-\frac{i}{2t}\sigma(z,\zeta\ ;\ y,\eta)}$$

and thanks to Plancherel's theorem, (B.4) yields for $t \neq 0$:

(B.5) $r_{N,t}(a_1,a_2\ ;\ x,\xi) =$

$$(2\Pi t)^{-2n}\int e^{-\frac{i}{2t}\sigma(z,\zeta\ ;\ y,\eta)}\ \sigma^N(D_z,D_\zeta\ ;\ D_y,D_\eta)\ (a_1(z+x,\zeta+\xi)a_2(y+x,\eta+\xi))\ dz d\zeta dy d\eta$$

We obviously have :

$$r_{N,0}(a_1,a_2\ ;\ x,\xi) = \sigma^N(D_z,D_\zeta\ ;\ D_y,D_\eta)\ a_1(z,\zeta)\ .\ a_2(y,\eta)\ \Big|_{\substack{z=y=x\\ \zeta=\eta=\xi}}$$

We are going to study the expression (B.5) of $r_{N,t}$, by a classical method consisting in isolating the critical point 0 of the phase. Let $\chi \in C_0^\infty(]-2,2[)$ be a cut off function, $\chi \equiv 1$ on $[-1,1]$. Let be $\varepsilon > 0$ be a small parameter. We denote :

$$b(z,\zeta,y,\eta\ ;\ x,\xi) = \sigma^N(D_z,D_\zeta\ ;\ D_y,D_\eta)(a_1(z+x,\zeta+\xi).a_2(y+x,\eta+\xi)) \text{ also denoted } \sigma^N(a_1,a_2)$$

$$b_1(z,\zeta,y,\eta\ ;\ x,\xi) = \chi\left(\frac{|y|^2+|z|^2}{\varepsilon^2\ \varphi^2(x,\xi)} + \frac{|\eta|^2+|\zeta|^2}{\varepsilon^2\ \phi^2(x,\xi)}\right)\ b(z,\zeta,y,\eta\ ;\ x,\xi) \text{ and } b_2 = b-b_1.$$

We are going to estimate the contribution of b_1 to (B.5) - lemma (B.6) - and then the contribution of b_2 - lemma (B.7). By integrating in $t \in [0,1]$, we shall get (B.1).

(B.6) Lemma

For $\varepsilon > 0$ small enough, there exists $\gamma_1(N,n) > 0$ and $j_1(n) \in \mathbf{N}$ such that for all $a_i \in S(m_i\ ;\ \phi,\varphi)$, $t \in]0,1]$ and $(x,\xi) \in \mathbf{R}^{2n}$:

$$\left|t^{-2N}\int e^{-\frac{i}{2t}\sigma(z,\zeta\ ;\ y,\eta)}\ b_1(z,\zeta,y,\eta\ ;\ x,\xi)\ dz\ d\zeta\ dy\ d\eta\right| \leqslant$$

$$t^{-2n/(2n+1)}\ \gamma_1(N,n)\ \Big(\sum_{\substack{N\leqslant k\leqslant N+j_1(n)\\ N\leqslant \ell\leqslant N+j_1(n)}} s_k(a_1\,;m_1)s_\ell(a_2\,;m_2)\Big)\ m_1(x,\xi)\ m_2(x,\xi)\ .\ (\phi\,\varphi)^{-N}(x,\xi)\ .$$

Proof

In b_1 support, we have, thanks to the (ϕ,φ)-continuity :

$$\phi^{-1}(x+z,\xi+\zeta) \lesssim \phi^{-1}(x,\xi) \quad \text{and} \quad \phi^{-1}(x+y,\xi+\eta) \lesssim \phi^{-1}(x,\xi)$$

and so it is for φ^{-1}, m_1 and m_2.

We cut out again b_1 support to control the behavior with respect to t : $b_1 = b'_1 + b''_1$ with b'_1 near to the critical point and b''_1 far from the critical point. Precisely, with $\varepsilon_0 = 1/(2n+1)$, we define :

$$b'_1(z,\zeta,y,\eta\ ;\ x,\xi) = \chi\left(\frac{|z|^2+|\zeta|^2+|y|^2+|\eta|^2}{t^{1-\varepsilon_0}}\right) b_1(z,\zeta,y,\eta\ ;\ x,\xi) \quad \text{and} \quad b''_1 = b_1 - b'_1 .$$

The volume of b'_1 support is a $O(t^{2n(1-\varepsilon_0)})$. That fact allows us to estimate b'_1 contribution.

We estimate b''_1 contribution by integration by parts $2n(2n+1)$-times with the differential operator :

$$L = \frac{2\,it}{|z|^2+|\zeta|^2+|y|^2+|\eta|^2}\left(\zeta\,.\,\frac{\partial}{\partial y} - \eta\,.\,\frac{\partial}{\partial z} + y\,.\,\frac{\partial}{\partial \zeta} - z\,.\,\frac{\partial}{\partial \eta}\right).$$

Indeed, we observe that each integration with L allows a gain of t^{ε_0}, and that the volume of b''_1 support is a $O((\phi\,\varphi)^{2n}\ (x,\xi))$. By developping the composition formula up to $N+2n+1$, we get the wanted estimate about r_N, which is given by :

$$r_N = \sum_{j=N+1}^{N+2n+1} a_j + r_{N+2n+1} .$$

(B.7) Lemma

There exists $\gamma_2(N,n) > 0$ and $j_2(n) \in \mathbf{N}$, such that for all $a_i \in S(m_i\ ;\ \phi,\varphi)$, $t \in\ ]0,1]$ and $(x,\xi) \in \mathbf{R}^{2n}$:

$$|t^{-2n}\,.\int e^{-\frac{i}{2t}\sigma(z,\zeta\ ;\ y,\eta)}\, b_2(z,\zeta,y,\eta\ ;\ x,\xi)\, dz\, d\zeta\, dy\, d\eta| \leqslant$$

$$\gamma_2(N,n)\,\Big(\sum_{\substack{N\leqslant k\leqslant 2N+M_2+j_2(n)\\ N\leqslant \ell\leqslant 2N+M_1+j_2(n)}} s_k(a_1\ ;\ m_1)\, s_\ell(a_2\ ;\ m_2)\Big)\, m_1(x,\xi)\, m_2(x,\xi)\, (\phi\,\varphi)^{-N}\,(x,\xi)\ .$$

Proof

On b_2 support, we have

$$\frac{|y|^2+|z|^2}{\varphi^2(x,\xi)} + \frac{|\eta|^2+|\zeta|^2}{\phi^2(x,\xi)} \geqslant \varepsilon^2 .$$

We divide that region into three zones Z_1,Z_2,Z_3 defined by :

$$Z_1\ :\ \frac{|z|^2}{\varphi^2(x,\xi)} + \frac{|\zeta|^2}{\phi^2(x,\xi)} \leqslant \frac{\varepsilon^2}{2} \quad \text{and} \quad \frac{|y|^2}{\varphi^2(x,\xi)} + \frac{|\eta|^2}{\phi^2(x,\xi)} \geqslant \frac{\varepsilon^2}{2}$$

$$Z_2\ :\ \frac{|z|^2}{\varphi^2(x,\xi)} + \frac{|\zeta|^2}{\phi^2(x,\xi)} \geqslant \frac{\varepsilon^2}{2} \qquad \frac{|y|^2}{\varphi^2(x,\xi)} + \frac{|\eta|^2}{\phi^2(x,\xi)} \leqslant \frac{\varepsilon^2}{2}$$

$$Z_3 : \frac{|z|^2}{\varphi^2(x,\xi)} + \frac{|\zeta|^2}{\phi^2(x,\xi)} \geqslant \frac{\varepsilon^2}{2} \qquad \frac{|y|^2}{\varphi^2(x,\xi)} + \frac{|\eta|^2}{\phi^2(x,\xi)} \geqslant \frac{\varepsilon^2}{2}$$

We deduce from that partition, a breaking up of b_2 into three parts, $b_2^{(j)}$, the support of which being close to Z_j for $j=1,2,3$.

At first we are going to study $b_2^{(1)}$ in lemma (B.8). We note that $b_2^{(2)}$ is like $b_2^{(1)}$. At last, we will study $b_2^{(3)}$ in lemma (B.10).

(B.8) Lemma

For all $a_i \in S(m_i ; \phi,\varphi)$, for all $t \in [0,1]$, $(x,\xi) \in \mathbf{R}^{2n}$, we have :

$$|t^{-2n} \int e^{-\frac{i}{2t}\sigma(z,\zeta ; y,\eta)} b_2^{(1)}(z,\zeta,y,\eta ; x,\xi)\, dz\, d\zeta\, dy\, d\eta| \leqslant$$

$$\gamma(N,n) \sum_{N\leqslant k\leqslant 2N + M_2+j(n)} s_k(a_1 ; m_1)\, s_N(a_2 ; m_2)\, m_1(x,\xi)\, m_2(x,\xi)\, (\phi\,\varphi)^{-N}(x,\xi)\,.$$

Proof :

We integrate by parts with the help of operator L_1 :

$$\text{(B.9)} \quad L_1 = \frac{2it}{|y|^2\,\phi^2(x,\xi)+|\eta|^2\,\varphi^2(x,\xi)} \left(\phi^2(x,\xi)\, y \frac{\partial}{\partial\zeta} - \varphi^2(x,\xi)\,\eta\frac{\partial}{\partial z}\right).$$

Its acts on $a_1(x+z,\xi+\zeta)$. The condition about (z,ζ) which is imposed in the zone Z_1, implies that L_1 allows a gain consisting of

$$t(1+|y|^2\,\phi^2(x,\xi)+|\eta|^2\,\varphi^2(x,\xi))^{-1/2}$$

or :

$t(\phi\,\varphi)^{-1}(x,\xi)$ (thanks to condition about (y,η)).

We integrate $(2n+1)$-times to obtain the convergence of the integral with respect to (y,η) ; then we again integrate N-times to gain $(\phi\,\varphi)^{-N}(x,\xi)$. Here, we use the boundedness of ϕ^{-1} and φ^{-1} on $\mathbf{R}^{2n}$, in order to estimate $\sigma^N(a_1, a_2)$ - hypothesis (H_1) - . Thus we get (B.8).

(B.10) Lemma

There exists M_0 depending only on ϕ and φ, and $\gamma(n,N)$ such that :

$$|t^{-2n} \int e^{-\frac{i}{2t}\sigma(z,\zeta ; y,\eta)} b_2^{(3)}(z,\zeta,y,\eta ; x,\xi)\, dz\, d\zeta\, dy\, d\eta| \leqslant$$

$$\gamma(N,n) \sum_{\substack{N\leqslant k\leqslant 2N + M_2+2n + 1 \\ N\leqslant \ell\leqslant (M_0+1)N + M_0(M_2+2n+1) + M_1 + 2n + 1}} s_k(a_1 ; m_1)\, s_\ell(a_2 ; m_2)\, m_1(x,\xi)\, m_2(x,\xi)\, (\phi\,\varphi)^{-N}(x,\xi).$$

Proof

In that zone Z_3, we do use that ϕ^{-1} and φ^{-1} are (1,1) temperate. So, it exists $C, M_0>0$

such that :

$$\phi^{-1}(x+z,\xi+\zeta) \leqslant \phi^{-1}(x,\xi)\,(1+|z|+|\zeta|)^{M_0}$$

$$\varphi^{-1}(x+z,\xi+\xi) \leqslant \varphi^{-1}(x,\xi)\,(1+|z|+|\zeta|)^{M_0}.$$

Thus, each integration with the above operator L_1 - (B.9) - allows a gain consisting of :

$$t(1+|y|^2\,\phi^2(x,\xi) + |\eta|^2\,\varphi^2(x,\xi))^{-1/2}\,(1+|z|+|\zeta|)^{M_0}$$

or :

$$t(\phi\,\varphi)^{-1}\,(x,\xi)\,(1+|z|+|\zeta|)^{M_0/2}\,.$$

Thus we integrate by parts $(N+M_2+2n+1)$-times with the help of L_1. But we have a loss with respect to (z,ζ) variable. To overcome that, we integrate

$[M_0(N+M_2+2n+1) + M_1 + 2n + 1]$ - times

with the help of operator L_2 :

$$L_2 = \frac{2it}{|z|^2+|\zeta|^2}\,(z\frac{\partial}{\partial\eta} - \zeta\frac{\partial}{\partial y})\,.$$

Indeed, thanks to (H_1), L_2 allows a gain on Z_3 consisting of $(1+|z|^2+|\zeta|^2)^{-1/2}$. Thus we get (B.10).

Proof of remark (B.2/2)

Now, we suppose moreover that $\phi = \varphi$. We read again (B.8) and (B.10) proofs.

Lemma (B.8) : as $\sigma^N(a_1,a_2)$ is majorized by $\gamma\, m_2\, s_N(a_2\,;\,m_2)[m_1\, s_N(a_1\,;\,m_1)(\phi\,\varphi)^{-\frac{N}{2}}]$ $[N/2]+2n+2$ integrations by parts with L_1 are enough. So, in the estimate (B.8), we have $N \leqslant k \leqslant 3N/2 + M_2 + j(n)$.

Lemma (B.10) : here, we have the majorization :

$$|\sigma^N(a_1,a_2)(z,\zeta,y,\eta\,;\,x,\xi)| \leqslant \gamma m_2.s_N(a_2\,;\,m_2)\,.$$

$$[m_1.s_N(a_1\,;\,m_1)\,(\phi\,\varphi)^{-N/2}\,(x,\xi)\,(1+|z|+|\zeta|)^{M_0\,N}]\,.$$

Thus, we integrate $([N/2] + M_2 + 2n + 2)$-times with L_1 to have a suitable behavior with respect to (y,η) variables, and :

$M_0([3N/2] + M_2 + 2n + 2) + M_1 + 2n + 1$ - times

with L_2 for the behavior in (z,ζ) variables. Thus, in the estimate (B.10), we may sum on :

$N \leqslant k \leqslant 3N/2 + M_2 + j(n)$

$N \leqslant \ell \leqslant (3M_0/2 + 1)\,N + M_1 + j(n) + M$

where M depends only on m_1,m_2, ϕ and φ.

It is clear that those modifications yield the proof of (B.2/2).

Annex C : Computation for an example

We are going to compute the volume function of a symbol $a(x,\xi) = b(x)\,c(\xi)$ where b and c are C^∞ and homogeneous of order $\sigma < 0$, outside the ball $B(0,1/2)$ in $\mathbf{R}^n$. We have :

(C.1) $$V(a\,;\lambda) = \int_{a \geqslant \lambda} dx\,d\xi = V_1(a\,;\lambda) + V_2(a\,;\lambda)$$

where for i=1,2 :

$$V_i(a\,;\lambda) = \int_{J_i} dx\,d\xi$$

with :

$$J_1 = \{(x,\xi) \in \mathbf{R}^{2n} \,/\, (|x| \leqslant 1 \text{ or } |\xi| \leqslant 1) \text{ and } a(x,\xi) \geqslant \lambda\}$$
$$J_2 = \{(x,\xi) \in \mathbf{R}^{2n} \,/\, (|x| \geqslant 1 \text{ and } |\xi| \geqslant 1) \text{ and } a(x,\xi) \geqslant \lambda\}$$

V_1 may be computed just like the volume function of a symbol with one predominant homogeneity. So :

(C.2) $$V_1(a\,;\lambda) = \frac{\lambda^{n/\sigma}}{n} \int_{B^n \times S^{n-1}} b(x)^{-n/\sigma}\, c(\tilde{\xi})^{-n/\sigma}\, dx\, d\tilde{\xi} + \frac{\lambda^{n/\sigma}}{n} \int_{S^{n-1} \times B^n} c(\xi)^{-n/\sigma}\, b(\tilde{x})^{-n/\sigma}\, d\xi\, d\tilde{x} + 0(1)\ .$$

We have denoted :

$$\tilde{x} = \frac{x}{|x|}\ ,\ \tilde{\xi} = \frac{\xi}{|\xi|}\ ,\ B^n = \{x \in \mathbf{R}^n,\ |x| \leqslant 1\} \quad \text{or} \quad \{\xi \in \mathbf{R}^n,\ |\xi| \leqslant 1\}\ .$$

Let us compute V_2 :

$$V_2(a\,;\lambda) = \int_{|x|^\sigma b(\tilde{x})|\xi|^\sigma c(\tilde{\xi}) \geqslant \lambda,\ |x| \geqslant 1,\ |\xi| \geqslant 1} |x|^{n-1}\, |\xi|^{n-1}\, d|x|\, d|\xi|\, d\tilde{x}\, d\tilde{\xi}$$
$$= \int_{|x|^\sigma b(\tilde{x})\, c(\tilde{\xi}) \geqslant \lambda,\ |x| \geqslant 1} |x|^{n-1}\, v(x,\tilde{\xi})\, d|x|\, d\tilde{x}\, d\tilde{\xi}$$

where :

$$v(x,\tilde{\xi}) = \int_1^{(\lambda b^{-1}(\tilde{x})\, c^{-1}(\tilde{\xi}))^{1/\sigma} |x|^{-1}} r^{n-1}\, dr$$
$$= \frac{1}{n} \lambda^{n/\sigma}\, b(\tilde{x})^{-n/\sigma}\, c(\tilde{\xi})^{-n/\sigma}\, |x|^{-n} - \frac{1}{n}\ .$$

Thus :

$$V_2(a\,;\lambda) = \int_{b(\tilde{x})\, c(\tilde{\xi}) \geqslant \lambda} w(\tilde{x},\tilde{\xi})\, d\tilde{x}\, d\tilde{\xi}$$

where :

$$w(\tilde{x},\tilde{\xi}) = \int_1^{(\lambda b^{-1}(\tilde{x})\, c^{-1}(\tilde{\xi}))^{1/\sigma}} \frac{1}{n} \lambda^{n/\sigma}\, b^{-n/\sigma}(\tilde{x})\, c^{-n/\sigma}(\tilde{\xi})\, \frac{1}{r} - \frac{r^{n-1}}{n}\, dr\ .$$

We have :

$$w = \frac{1}{n\sigma}\, \lambda^{n/\sigma}\, (bc)^{-n/\sigma}\, \mathrm{Log}\, \frac{\lambda}{bc} - \frac{\lambda^{n/\sigma}}{n^2}\, (bc)^{-n/\sigma} + \frac{1}{n^2}$$

Thus :

$$V_2(a\ ;\lambda) = \frac{\lambda^{n/\sigma}}{n^2} \int_{bc > \lambda} (bc)^{-n/\sigma} [\frac{n}{\sigma} \text{ Log } \lambda + \text{Log}(bc)^{-n/\sigma} - 1]\, d\tilde{x}\, d\tilde{\xi} + O(1)$$

Now :

$$\int_{bc \leq \lambda} (bc)^{-n/\sigma} \text{ Log}(bc)^{-n/\sigma}\, d\tilde{x}\, d\tilde{\xi} = O(\lambda^{-n/\sigma} \text{ Log } \frac{1}{\lambda}) .$$

Therefore, we have :

(C.3) $$V_2(a;\lambda) = \frac{\lambda^{n/\sigma}}{n\sigma} \text{ Log } \lambda \int_{S^{n-1} \times S^{n-1}} (bc)^{-n/\sigma}\, d\tilde{x}\, d\tilde{\xi}$$

$$+ \frac{\lambda^{n/\sigma}}{n^2} \int_{S^{n-1} \times S^{n-1}} (bc)^{-n/\sigma} \text{ Log } \frac{(bc)^{-n/\sigma}}{e}\, d\tilde{x}\, d\tilde{\xi}$$

$$+ O(\text{Log } \frac{1}{\lambda})$$

So, (C.1), (C.2) and (C.3) give the asymptotics, when $\lambda \to 0$:

$$V(a\ ;\lambda) = \lambda^{n/\sigma} (\alpha \text{ Log } \frac{1}{\lambda} + \beta) + O(\text{Log } \frac{1}{\lambda})$$

with α given by (C.3) and β given by both (C.2) and (C.3).

REFERENCES

[BE] R. BEALS :
A general calculus of pseudodifferential operators. Duke Math. J. 42, 1-42 (1975)

[BO-KO] K. Kh. BOITMATOV, A.G. KOSTYUCHENKO :
The distribution of the eigenvalues of the equation $Au = \lambda Bu$ in the whole space. Soviet Math. Dokl. 30 (1) 245-248 (1984)

[BI-SO 1] M.S. BIRMAN, M.Z. SOLOMJAK :
Asymptotics of the spectrum of pseudodifferential operators with anisotropic-homogeneous symbols. Vestnik Leningrad Univ. Math. 10 (1982) 237-247 et 12 (1980) 155-161

[BI-SO 2] M.S. BIRMAN, M.Z. SOLOMJAK :
Compact operators whose singular numbers have powerlike asymptotics. J. Soviet Math. 27 (1) (1984) 2442-2447

[BI-SO 3] M.S. BIRMAN, M.Z. SOLOMJAK :
Asymptotic behavior of the spectrum of differential equations. J. Soviet Math. 12 (3) (1979) 247-283

[CV] Y. COLIN DE VERDIERE :
Une formule de trace pour l'opérateur de Schrödinger dans $\mathbf{R}^3$. Ann. E.N.S. 14 (1) 27-39 (1981).

[DA-RO] M. DAUGE, D. ROBERT :
Formule de Weyl pour une classe d'opérateurs pseudodifférentiels d'ordre négatif sur $L^2(\mathbf{R}^n)$. Note C.R. Acad. Sc. Paris 302 Série I, (5) 175-178 (1986)

[FE] V.I. FEIGIN :
Sharp estimates of the remainder in the spectral asymptotic for pseudodifferential operators in $\mathbf{R}^n$. Funk. Anal. Ego Prilozheniya. 16 (3), 88-89 (1982)

[FL-LA] J. FLECKINGER, M.L. LAPIDUS :
Eigenvalues of elliptic boundary value problems with an indefinite weight function. Preprint May 1985

[GO-KR] I.C. GOHBERG, M.G. KREIN :
Introduction à la théorie des opérateurs linéaires non autoadjoints. Dunod (1972)

[GR] G. GRUBB :
Singular Green operators and their spectral asymptotics. Duke Math. J. 51 (3) (1984) 477-528

[GU] L. GUILLOPE :
Une formule de trace pour l'opérateur de Schrödinger dans $\mathbf{R}^n$. Thèse de 3ème cycle. Univ. de Grenoble (1981)

[HE-RO 1] B. HELFFER, D. ROBERT :
Calcul fonctionnel par la transformation de Mellin et opérateurs admissibles. Journal of Functional Analysis 53 (3) (1983) 246-268

[HE-RO 2] B. HELFFER, D. ROBERT :
Propriétés asymptotiques du spectre d'opérateurs pseudodifférentiels sur $\mathbf{R}^n$. Comm. in Partial Differential Equations 7 (7) 795-882 (1982)

[HO 1] L. HÖRMANDER :
The Weyl Calculus of pseudodifferential operators. CPAM 32, 359-443 (1979)

[HO 2] L. HÖRMANDER :
On the asymptotic distribution of the eigenvalues of pseudodifferential operators in $\mathbf{R}^n$. Arkiv för Mathematik 17 (2) (1979) 297-313

[KO-SU] S.M. KOZLOV, M.A. SUBIN :
On the structure of functions of class S of self-adjoint elliptic operators on $\mathbf{R}^n$. Comm. of the Moscov Math. Soc. Russian Math. Surveys. 37 (2) 221-222 (1982)

[MA] A. MARTIN :
Bound states in the Strong Coupling Limit. Helv. Phys. Acta 45 (1972) 140-148

[MO] A. MOHAMED :
Etude spectrale des opérateurs pseudodifférentiels hypoelliptiques. Thèse de Doctorat. Univ. de Nantes 1983

[RO] D. ROBERT :
Propriétés spectrales d'opérateurs pseudodifférentiels. Comm. in Partial Differential Equations 3 (9) (1978) 755-826

[SI] B. SIMON :
On the number of bound states of two body Schrödinger operators. A review. Studies in Math. Physics. Princeton University Press 1976

[TU-SU] V.N. TULOVSKII, M.A. SUBIN :
On asymptotic distribution of eigenvalues of pseudodifferential operators in $\mathbf{R}^n$. Math. USSR Sbornik 21 (4) (1973) 565-583

NORMAL SOLVABILITY OF BOUNDARY VALUE PROBLEMS IN HALF SPACE

A.K. Erkip(*)
Department of Mathematics
METU
Ankara, Turkey

INTRODUCTION: We investigate the normal solvability of the half space boundary value problem

$$(*)\qquad \begin{aligned} Pu &= f && \text{in } \mathbb{R}^{n+1}_+ \\ B^k u &= g^k && \text{on } \partial\mathbb{R}^{n+1}_+ , \quad k=1,2,\dots,r \end{aligned}$$

with data in appropriate weighted Sobolev spaces. Roughly speaking P, B^k are differential operators in the normal direction with tangential pseudo-differential operator coefficients. Our main result is Theorem 4.3 establishing a sufficient condition, namely the m-d ellipticity of the system ((3.2), (3.3) and (4.1)). This coincides with m-d ellipticity in [5] and [6] for the non-weighted version. On the other hand our choice of P and B^k covers the case of differential operators with smooth coefficients of polynomial growth.

Our sufficient condition locally implies that (*) is an elliptic boundary value problem in the usual case. On the other hand m-d ellipticity brings more restrictions on the behaviour at infinity. Namely (3.2) says that the symbol of p satisfies $|p(x, \zeta)| \geq C(1+|x|)^{\ell}\,(1+|\zeta|)^{2r}$ for large $|x| + |\zeta|$, which is uniform ellipticity plus a similar restriction at infinity. Similary (3.3) and (4.1) are m-d versions of proper ellipticity and the Lopatinski-Shapiro (covering) condition.

Our approach follows the idea of Calderon [3], reducing (*) to a system of pseudo-differential equations on the boundary. The reduction is done as in Hörmander [8], [9] and Seeley [11]. While they get local results, by using the pseudo-differential operator calculus in Cordes [4] we are able to obtain global results. In fact the machinery in [8] can be easily adapted to the "global" calculus. A

(*) This work was partially supported by the Scientific and Technical Research Council of Turkey.

similar approach can be found in [2]. An alternative way would be to globalize the setup in Seeley [11] ([6]).

Following the preliminaries, in §2 we look at a constant coefficient boundary value problem on the half line, more or less on the lines of [10]. In §3 the Calderon projector is derived. In §4 we first obtain a left and a partial right inverse for the reduced system on the boundary using the pointwise results of §2. This in turn yields the a priori estimate (Theorem 4.1), which is the weighted version of Agmon, Douglis, Nirenberg estimate [1], and finally the normal solvability by establishing an explicit Fredholm inverse (Theorem 4.3).

We finally want to mention some possibilities of extending our results. Via partitions of unity and local coordinates a priori estimates and Fredholm inverses can be carried over to unbounded domains. In [7] we made some attempt in that direction. The main obstacle in such an approach is the fact that m-d ellipticity is not invariant under a change of coordinates. Locally there is no problem but the conditions at infinity stay out of control. On the other hand this suggests a new problem; namely, given an elliptic boundary value problem can we find the "right" change of coordinates such that it becomes m-d elliptic.

§1. *PRELIMINARIES*

We denote points in $\mathbb{R}^{n+1}$ by $x = (x', y)$, $\zeta = (\zeta', \eta)$ where $y, \eta \in \mathbb{R}$. The half-space is $\mathbb{R}^{n+1}_+ = \{(x', y): y \geq 0\}$ with boundary $\partial\mathbb{R}^{n+1}_+ = \{(x', 0)\} \cong \mathbb{R}^n$. With the same notation $D = (D', D_y) = (D_1, \dots, D_n, D_y)$, $D_j = -i\partial/\partial x_j$.

We will frequently use the function $\lambda(x) = (1 + |x|^2)^{-\frac{1}{2}}$ and the operator $\Lambda = \lambda(D) = (1-\Delta)^{-\frac{1}{2}}$. We will be working with weighted L^2-Sobolev spaces;

$$H_{(s,t)} = H_{(s,t)}(\mathbb{R}^k) = \{u: \lambda^{-t}(x)\Lambda^{-s}u \in L^2\} .$$

It follows from Cordes [4] that the order of $\lambda(x)$ and Λ can be reversed; in other words $\|\lambda^{-t}(x)\Lambda^{-s}u\|_{L^2}$ and $\|\Lambda^{-s}\lambda^{-t}(x)u\|_{L^2}$ are equivalent. Without discriminating either one we will denote the norm in $H_{(s,t)}$ by $\| \ \|_{(s,t)}$. The induced spaces $H^t_{(s,t)}$ of the half space are defined in the usual way, namely:

$$H^+_{(s,t)} = H_{(s,t)}(\mathbb{R}^{n+1}_+) = \{u: \exists\, v \in H_{(s,t)}(\mathbb{R}^{n+1}), \ v=u \ \text{ on } \ \mathbb{R}^{n+1}_+\}$$

$$\|u\|^+_{(s,t)} = \inf\{\|v\|_{(s,t)} : \ v=u \ \text{ on } \ \mathbb{R}^{n+1}_+\} .$$

We note that $\|u\|^+_{(s,t)} \cong \|\lambda^{-t}(x)u\|^+_{(s,0)}$ and the $(s, 0)$ space is the usual Sobolev space; hence the trace theorem applies, namely:

Lemma 1.1. For $s > j + \frac{1}{2}$ *there exists* c *such that:*

$$\|D_y^j u(x', 0)\|_{(s-j-\frac{1}{2}, t)} \leq c\|u\|^t_{(s,t)} \quad \textit{for all} \quad u\varepsilon S(\mathbb{R}^{n+1}).$$

As usual γ_j denotes the *trace operator;* i.e. the extension of the map $u \to D_y^j u(x', 0)$ to $H^+_{(s,t)}$.

We also want to note that the $H_{(\sigma,\tau)}$ spaces of Hörmander [8] can similarly be modified to the weighted case, $H_{(\sigma,\tau),t}$ spaces. The corresponding results in [8] apply in the weighted case.

We have the inclusion relations:

$$H_{(s',t')} \subset H_{(s,t)}, \quad H^+_{(s',t')} \subset H_{(s,t)} \quad \text{for} \quad s \leq s', \quad t \leq t'.$$

Moreover the inclusions are compact if $s < s'$ and $t < t'$ [4].

We will use the pseudo-differential operator (ψdo) calculus of Cordes [4]. We list some of the main points that we will need. We refer to [4] for the details.

For $m = (m_1, m_2)$, $\rho = (\rho_1, \rho_2)$ the symbol class $SS^{m,\rho,\delta}$ consists of $C^\infty(\mathbb{R}^{2k})$ functions satisfying the estimate:

$$D_x^\beta D_\zeta^\alpha q(x, \zeta) = O(1 + |\zeta|)^{m_1 - \rho_1|\alpha| + \delta|\beta|} (1 + |x|)^{m_2 - \rho_2|\beta|}.$$

For such q the ψdo of order m is defined as:

$$Qu(x) = q(m, D)u(x) = (2\pi)^{-k} \int e^{ix\zeta} q(x, \zeta)\hat{u}(\zeta)d\zeta \quad \text{for} \quad u\varepsilon S(\mathbb{R}^k).$$

For $\rho_j > 0$, $0 \leq \delta < \rho_1 < 1$ Q extends to a bounded operator of $H_{(s,t)}$ into $H_{(s-m_1, t-m_2)}$. Thus for $m_i < 0$ $i=1,2$ Q is compact on $H_{(s,t)}$.

A ψdo $p(M, D)$ of order m is said to be m-d elliptic on an open set $\Omega \subset \mathbb{R}^k$ if there exists c such that $|p(x, \zeta)| \geq c(1 + |\zeta|)^{m_1}(1 + |x|)^{m_2}$ for all sufficiently large $|x| + |\zeta|$, with $x\varepsilon\Omega$, $\zeta\,\varepsilon\,\mathbb{R}^k$. For an m-d elliptic P one can construct a (local) K-parametric Q of order $-m$ satisfying:

$$PQ = I + K_1, \quad QP = I + K_2 \quad \text{on} \quad \Omega$$

where K_1, K_2 are ψdo's of order $(-\infty, -\infty)$, hence integral operators with $S(\mathbb{R}^{2k})$ kernels.

We finally list some Lemmas that are generalizations of the ones of Hörmander [8], the generalization being quite straightforward after introducing $H_{(\sigma,\tau),t}$ spaces mentioned above. (See [6], [7] for details).

Lemma 1.2. Let K *be a* ψdo *of order* $(-\infty, -\infty)$. *For* $u \in H^+_{(s,t)}$ *we define* u^0 *as* $u^0 = u$ *for* $y \geq 0$, $u^0 = 0$ *for* $y < 0$. *Then for* $s' \geq 0$, t', *there exists* c *such that:*

$$\|Ku^0\|^+_{(s,t)} \leq c\|u\|^+_{(s',t')} \quad \text{for all} \quad u \in H^+_{(s,t)} .$$

Lemma 1.3. Let P *be* m-d *elliptic of order* m *of the particular form (3.1) on some neighborhood of* $\mathbb{R}^{n+1}_+$, Q *a K-parametrix for* P. *For* $s \geq 0$, t *there exists* c *such that:*

$$\|Qu^0\|^+_{(s+m_1,t+m_2)} \leq c\|u\|^+_{(s,t)} \quad \text{for all} \quad u \in H^+_{(s,t)} .$$

Lemma 1.4. Let Q *be a* ψ*do of order* $-m$ *whose symbol* $q(x, \zeta)$, *in some neighborhood of* $\mathbb{R}^{n+1}_+$, *has an asymptotic expansion* $\sum_{j\geq 0} q_j$ *satisfying:*

(i) $q_j(x, \zeta', \eta)$ *is a rational function of* η *for large* $|x| + |\zeta|$, $j=0,1,2,\ldots$

(ii) *For large* $|x| + |\zeta'|$ *the poles of* $q_j(z) = q_j(x, \zeta', z)$ *are not on the real line and all lie in some disc of radius* $O(|\zeta'|)$.

Then for $u \in S(\mathbb{R}^n)$ *with* $\delta^j = D^j_y\delta$ *the map* $u \to \lim_{y\to 0+} D^k_y(u \otimes \delta^j)$ *defines a* ψ*do* Q^{kj} *of order* $(j+k+1-m_1, -m_2)$. *Moreover the leading term in the asymptotic expansion of the symbol of* Q^{kj} *is (for large* $|x'| + |\zeta'|$):

$$\frac{1}{2\pi} \int_{\Gamma(\zeta')} q_0(0, x', \zeta', \eta)\eta^{k+j}\, d\eta$$

where $\Gamma(\zeta')$ *is any contour in the upper half plane enclosing the poles of* $q_0(z)$ *there.*

§2. THE CALDERON PROJECTOR AND BOUNDARY VALUE PROBLEMS ON THE HALF LINE

Let $\quad p(\zeta) = \sum_{j=0}^{2r} p_j\zeta^j \quad$ satisfy:

(2.1) $p(\zeta) \neq 0$ *for* $\zeta \in \mathbb{R}$

(2.2) $p(\zeta) = p_{2r}\, p^-(\zeta) p^+(\zeta)$ *where* $p^+(p^-)$ *has exactly* r *roots with positive (negative) imaginary parts.*

For $u \in S(\mathbb{R}^+)$ we set $u_m = \lim_{y \to 0+} D_y^m u(y)$. Via Laplace transform we see that (see [6])

(2.3) $p(D)u = 0$ for $y > 0$
if and only if

(2.4) $$u(y) = \sum_{m=0}^{2r-1} \sum_{j=0}^{2r-m-1} p_{j+m+1}\, u_m \frac{1}{2\pi i} \int_\Gamma e^{iy\eta} \frac{\eta^j}{p(\eta)}\, d\eta \quad \text{for } y > 0$$

where Γ is a contour in the upper half plane eclosing the zeroes of p^+. Differentiating and taking limit as $y \to 0+$ we get:

(2.5) $$u_k = \sum_{m=0}^{2r-1} \sum_{j=0}^{2r-m-1} p_{j+m+1}\, u_m \frac{1}{2\pi i} \int_\Gamma \frac{\eta^{j+k}}{p(\eta)}\, d\eta \qquad k=0,1,\ldots,2r-1 .$$

We set $\gamma u = (u_0, u_1, \ldots, u_{2r-1})$ and express (2.5) as the matrix equation:

(2.6) $\gamma u = Q \gamma u$

Thus γu is the Cauchy data of $u \in S(\mathbb{R}^+)$ satisfying (2.3) if and only if (2.6) holds. It is easy to see that Q projects $\mathbb{C}^{2r}$ onto the r-dimensional subspace corresponding to the Cauchy data of the solution space of $p^+(D)u = 0$. Q corresponds to the Calderon projector in one dimension.

On the other hand the $S(\mathbb{R}^+)$ solutions of (2.3) can be represented as:

(2.7) $$u(y) = \sum_{m=0}^{r-1} \sum_{j=0}^{r-m-1} p^+_{j+m+1}\, u_m \frac{1}{2\pi i} \int_\Gamma \frac{e^{iy\eta}\, \eta^j}{p^+(\eta)}\, d\eta$$

where $p^+(\zeta) = \sum_{m=0}^{r} p_j^+ \zeta^j$. We note that (2.7) only involves the Dirichlet data $(u_0, \ldots, u_{r-1})$ and the correspondence $u(y) \longleftrightarrow (u_0, \ldots, u_{r-1})$ is one to one.

We now consider the boundary value problem

(2.8) $$\begin{aligned} p(D)u &= 0 \quad \text{for } y > 0 \\ b^k(D)u &= c_k \quad \text{at } y=0,\ k=1,2,\ldots,r \end{aligned}$$

with $b^k(\zeta) = \sum_{j=0}^{r_k} b_j^k \zeta^j$, $r_k < 2r$ satisfying the *covering condition:*

(2.9) *The polynomials* $b^1, \ldots, b^r$ *are linearly independent modulo* p^+.
Setting $B = ((b_j^k))$ $k=1,2,\ldots,r$, $j=0,2,\ldots,2r-1$ with $b_j^k = 0$ for $j > r_k$,

γu is the Cauchy data of $u \varepsilon S(\mathbb{R}^+)$ satisfying (2.8) if and only if

$$(2.10) \qquad \begin{bmatrix} I - Q \\ \\ B \end{bmatrix} \gamma u = \begin{bmatrix} 0 \\ \vdots \\ 0 \\ c_1 \\ \vdots \\ c_r \end{bmatrix}$$

Now it follows from [10] that (2.8) has a unique $S(\mathbb{R}^+)$ solution if and only if the covering condition (2.9) holds, the solution given by (2.7) with

$$(2.11) \qquad \begin{bmatrix} u_0 \\ \vdots \\ u_{r-1} \end{bmatrix} = ((r_j^k))^{-1} \begin{bmatrix} c_1 \\ \vdots \\ c_r \end{bmatrix}$$

where $r^k(\zeta) = \sum_{j=0}^{r-1} r_j^k \zeta^j$ is the residue of $b^k(\zeta)$ modulo p^+. This first implies that the $3r \times 2r$ matrix in (2.10) is injective hence has rank $2r$, and that its range contains the subspace $0 \times \mathbb{R}^r$. Moreover computing the Cauchy data γu from (2.7) with (2.11) gives us a partial right inverse T, whose entries can be computed in terms of b_j^k 's and p_j^+, satisfying:

(2.12) $(I-Q)T = 0$ and $BT = I_r$

It is easy now to check that $S(u, w) = T(Bw-(1-Q)u)+(1-Q)u$ is a left inverse, that is:

$$(2.13) \qquad S \begin{bmatrix} I - Q \\ \\ B \end{bmatrix} = I_{2r}$$

§3. *THE CALDERON PROJECTOR FOR* $\mathbb{R}_+^{n+1}$

We now let $P = p(M, D)$ with $p \varepsilon \dot{S}S^{(2r,\ell),\rho,\delta}$ such that in some neighborhood Ω of $\mathbb{R}_+^{n+1}$

(3.1) $p(x, \zeta) = \sum_{j=0}^{2r} p_j(x, \zeta')\zeta^j$

We assume that:

(3.2) P *is m-d elliptic on* Ω

(3.3) $p_{(x',\zeta')}(z) = \lambda^{2r}(\zeta')\lambda^{\ell}(x')p(0, x', \zeta', \lambda^{-1}(\zeta')z)$ *has exactly* r *roots* $\tau_j(x',\zeta')$ *with positive imaginary part for large* $|x'| + |\zeta'|$.

(Note that $p_{(x',\zeta')}(z)$ cannot have real zeroes for large $|x'|+|\zeta'|$). *We set*

$$p^{+}_{(x',\zeta')}(z) = \prod_{j=1}^{r} (z-\tau_j(x',\zeta')).$$

We now take $u\varepsilon S(\mathbb{R}^{n+1}_{+})$. We let u^0 denote the zero extension of Lemma 1.2, and $u_m = \lim_{y\to 0+} D^m_y u$. Then as in Hörmander [8]

$$Pu^0 = (Pu)^0 + i^{-1} \sum_{m=0}^{2r-1} \sum_{j=0}^{2r-m-1} p_{j+m+1}(M, D')(u_m \otimes \delta^j)$$

Since P is m-d elliptic on Ω, there is a K-parametrix Q satisfying $PQ = I+K_1$, $QP = I+K_2$ on Ω; applying Q we get:

$$(3.4)\quad u^0 + K_2u^0 = Q(Pu)^0 + i^{-1} \sum_{m=0}^{2r-1} \sum_{j=0}^{2r-m-1} Qp_{j+m+1}(M, D')(u_m \otimes \delta^j)$$

By Lemmas 1.2, 1.3 K_2u^0 and $Q(Pu)^0$ are sufficiently smooth. Moreover by construction of the K-parametrix for large $|x| + |\zeta|$, $x\varepsilon\Omega$ the symbol of $Qp_{j+m+1}(M, D')$ has an asymptotic expansion consisting of rational functions of η with only powers of $p(x, \zeta)$ appearing in the denominator. On the other hand by m-d ellipticity $|p_{2r}(x, \zeta')| \geqq c|x|^{\ell}$ for large $|x|$, hence root estimates imply that the roots of $p(x, \zeta', z) = 0$ are of $O(|\zeta'|)$, thus Lemma 1.4 applies and we can take traces in (3.4).

$$(3.5)\quad u_k = \gamma_k(Q(Pu)^0 - K_2u^0) + i^{-1} \sum_{m=0}^{2r-1} \sum_{j=0}^{2r-m-1} (Qp_{j+m+1}(M, D'))^{kj}u_m$$

the sum on the right hand side being the Calderon projector on the trace $(u_0, u_1, \dots, u_{2r-1})$.

From the representation (3.4) and Lemmas 1.2, 1.3 for any $0 \leqq s' < s$, $t' < t$ we obtain the estimate:

$$(3.6)\quad \|u\|^{+}_{(s,t)} \leqq \|Q(Pu)^0\|^{+}_{(s,t)} + \|K_2u^0\|^{+}_{(s,t)} + \sum_{m=0}^{2r-1} \sum_{j=0}^{2r-m-1} \|Qp_{j+m+1}(u_m \otimes \delta^j)\|^{+}_{(s,t)}$$

$$\leqq C\{ \|Pu\|^{+}_{(s-2r,t-\ell)} + \|u\|^{+}_{(s',t')} + \sum_{m=0}^{2r-1} \|u_m\|_{(s-m-1/2,\, t)}\}$$

the last estimate following from the weighted version of Hörmander's $H_{(\sigma,\tau)}$ estimates mentioned in §1 (see [8] and [7] for the weighted version). Clearly (3.6) remains valid for $u\varepsilon H^{+}_{(s,t)}$ with $s \geqq 2r$.

§4. THE BOUNDARY VALUE PROBLEM

We let $B^k = \sum_{j=0}^{r_k} b_j^k(M', D')D_y^j$, $b_j^k \varepsilon SS^{(r_k-j,\ \ell_k),\rho,\delta}$, $r_k < 2r$ for $k=1,2,\ldots,r$ be the boundary operators. With P satisfying the assumptions of §3 we impose the *covering condition:*

(4.1) *The polynomials* $b^k_{(x',\zeta')}(z) = \sum_{j=0}^{r_k} b_j^k(x',\zeta')\lambda^{r_k-j}(\lambda')\lambda^{\ell_j}(x')z^j$ $k=1,2,\ldots,r$ *are linearly independent modulo* $P^+_{(x',\zeta')}$ *uniformly for large* $|x'| + |\zeta'|$.

By uniformity we mean that the determinant of the coefficients $r_j^k(x', \zeta')$ (as in §2) of the residues of the b^k's modulo p^+ stays bounded away from zero for large $|x'| + |\zeta'|$. The system $(P, B^1, \ldots, B^r)$ satisfying (3.1)-(3.3) and (4.1) will be called *m-d elliptic.*

Theorem 4.1. Let $(P, B^1, \ldots, B^r)$ *be an m-d elliptic system. There exists* $\varepsilon_i > 0$, *such that for all* (s, t), $s \geq 2r$ *there is* c *satisfying for all* $u\varepsilon H^+_{(s,t)}$:

$$\|u\|^+_{(s,t)} \leq c\{ \|Pu\|^+_{(s-2r,t-\ell)} + \sum_{k=1}^{r} \|\gamma_0 B^k u\|_{(s-r_k-1/2,\ t-\ell_k)} + \|u\|^+_{(s-\varepsilon_1,s-\varepsilon_2)}\}$$

Proof: By (3.6) it suffices to estimate $\|u_m\|_{(s-m-1/2,t)}$ in terms of the right hand side. For this we consider the $3r \times 2r$ system of pseudo-differential equations, $2r$ of them from (3.5) and r from the boundary equations. We first scale down this system by setting:

$$U_m = \Lambda^m u_m, \qquad G^k = \lambda^{\ell_k}(x')\Lambda^{r_k}\gamma_0 B^k u,$$

$$F^k = \Lambda^k \gamma_k(Q(Pu)^0 - K_2 u^0) \text{ with } \Lambda = \Lambda' = \lambda(D') \text{ to get:}$$

$$(4.2) \qquad \begin{bmatrix} I - \overline{Q} \\ \\ \overline{B} \end{bmatrix} \begin{bmatrix} U_0 \\ \vdots \\ U_{2r-1} \end{bmatrix} = \begin{bmatrix} \vdots \\ F^k \\ \vdots \\ G^k \\ \vdots \end{bmatrix} \quad \text{where}$$

$$\overline{Q} = ((i^{-1} \sum_{j=0}^{2r-m-1} \Lambda^k(Qp_{j+m+1}(M, D'))^{kj}\,\Lambda^{-m}))_{k,m}$$

$$\overline{B} = ((\lambda^{\ell_k}(x')\Lambda^{r_k}\, b_m^k(M', D')))_{k,m} \text{ with } b_m^k = 0 \text{ for } r_k < m < 2r$$

Due to normalization the entries of $\overline{Q}$, $\overline{B}$ are all ψdos of order (0, 0). The symbol of the matrix in (4.1) $\sigma(x', \zeta')$ can be computed asymptotically for large $|x'| + |\zeta'|$ via Lemma 1.4, and the leading ((0, 0) order) term turns out to be

$$\sigma_0(x', \zeta') = \begin{bmatrix} I - \overline{q}_0(x', \zeta') \\ \\ \overline{b}_0(x', \zeta') \end{bmatrix} \quad \text{with}$$

$$\overline{q}_0(x', \zeta') = ((\sum_{j=0}^{2r-m-1} \lambda^{k-m}(\zeta')p_{j+m+1}(x',\zeta') \frac{1}{2\pi i} \int_\Gamma q_0(0, x', \zeta', \eta)\eta^{k+I}d\eta))$$

$$\overline{b}_0(x', \zeta') = ((\lambda^{\ell_k}(x')\lambda^{r_k}(\zeta')b^k_m(x', \zeta')))$$

Noting that for large $|x'| + |\zeta'|$ $q_0(x, \zeta) = (p(x, \zeta))^{-1}$ and changing the variable η to $\lambda^{-1}(\zeta')\eta$ in the integral above we see that $\sigma_0(x', \zeta')$ is nothing but the matrix in (2.10) for the ordinary differential operators $p_{(x',\zeta')}(D)$, $b^k_{(x',\zeta')}(D)$. Due to the covering condition (4.1), (2.9) is satisfied hence the matrices $S = s(x', \zeta')$, $T = t(x', \zeta')$ of §2 exist satisfying:

(4.3) $\qquad s\sigma_0 = I_{2r}, \quad (1-\overline{q})t = 0, \quad \overline{b}t = I_r \quad$ for large $|x'| + |\zeta'|$.

On the other hand the m-d ellipticity of P and the $O(|\zeta'|)$ bound on the roots of $p(x, \zeta', z) = 0$ force $p^+(M, D)$ to be m-d elliptic of order (0,0). Then the construction of $s(x', \zeta')$ and $t(x', \zeta')$, the uniformity of the covering condition (4.1) and estimates similar to the ones in the proof of Lemma 1.4 (cf [6]) show that the entries of $s(x', \zeta')$ and $t(x', \zeta')$ are symbols of order (0, 0). Setting $S = s(M', D')$ $T = t(M', D')$ the product formulas for matrices and ψdos and (4.2) yields

$$(4.4) \qquad S\begin{bmatrix} I - \overline{Q} \\ \\ \overline{B} \end{bmatrix} = I + D_1, \qquad [I-\overline{Q}]T = D_2 \qquad \overline{B}T = I + D_3$$

with the ψdo matrices D_i all of negative order. Now applying S to (4.1) we obtain:

$$(4.5) \qquad \begin{bmatrix} U_0 \\ \vdots \\ U_{2r-1} \end{bmatrix} = S\begin{bmatrix} \vdots \\ F^k \\ \vdots \\ G^k \\ \vdots \end{bmatrix} - D_1 \begin{bmatrix} U_0 \\ \vdots \\ U_{2r-1} \end{bmatrix}$$

Estimating U_m's and hence u_m's from (4.4) one obtains:

$$\|u_m\|_{(s-m-1/2,t)} \leq c\{\|Pu\|^+_{(s-2r,t-\ell)} + \sum_{k=1}^{r} \|\gamma_0 B^k u\|_{(s-r_k-1/2,t-\ell_k)} + \|u\|^+_{(s-\varepsilon_1,t-\varepsilon_2)}\}$$

where $-(\varepsilon_1, \varepsilon_2)$ is the highest order of the entries of D_1. This concludes the proof.

The compactness of the inclusion $H^+_{(s,t)} \subset H^+_{(s-\varepsilon_1, t-\varepsilon_2)}$ and the a priori estimate of Theorem 4.1 shows that:

Corollary 4.2. Let $(P, B^1, \ldots, B^r)$ *be an m-d elliptic system. Then the boundary value problem:*

$$Pu = 0 \quad \textit{in} \quad \mathbb{R}^{n+1}_+$$

$$B^k u = 0 \quad \textit{on} \quad \partial\mathbb{R}^{n+1}_+ \qquad k=1,2,\ldots,r$$

has finite dimensional solution space in $H^+_{(s,t)}$ *for any* $s \geq 2r$, t.

We now construct a solution of the non-homogeneous boundary value problem with the operator T of (4.3). First for $\psi = (\psi_1, \ldots, \psi_r)$, $\psi_i \varepsilon S(\mathbb{R}^n)$ we set:

$$W\psi = i^{-1} \sum_{m=0}^{2r-1} \sum_{j=0}^{2r-m-1} Qp_{j+m+1}(M, D')(\Lambda^{-m}(T\psi)_m \otimes \delta^j)$$

$(T\psi)_m$ denoting the m-th component of $T\psi$. From (3.6),

$$(4.5) \qquad \|W\psi\|^+_{(s,t)} \leq c \sum_{m=0}^{2r-1} \|\Lambda^{-m}(T\psi)_m \otimes \delta^j\|_{(s-m-1/2,\ t)}$$

$$\leq c \sum_{k=1}^{r} \|\psi_k\|_{(s-1/2,\ t)}.$$

Since $p_{j+m+1}(M, D')(\Lambda^{-m}(T\psi)_m \times \delta^j)$ vanishes for $y > 0$;

$$(4.6) \quad PW\psi = K_1\{-i \sum_{m=0}^{2r-1} \sum_{j=0}^{2r-m-1} p_{j+m+1}(M, D')(\Lambda^{-m}(T\psi)_m \times \delta^j)\} \quad \text{in} \quad \mathbb{R}^{n+1}_+ .$$

To apply the boundary operators to $W\psi$, we compute the traces by (3.5) and use (4.3)

$$(4.7) \qquad \begin{bmatrix} B_1 W\psi \\ \vdots \\ B_r W\psi \end{bmatrix} = ((\lambda^{-p_k}(x')\Lambda^{-r_k}\delta_{km}))\cdot\overline{B}\,\overline{Q}T\psi .$$

But $\overline{B}\,\overline{Q}T = \overline{B}(T-D_2) = I + D_3 - \overline{B}\,D_2 = I + D$.

Now D is a ψdo matrix of negative order, hence a compact operator of $\prod_{k=1}^{r} H_{(s-1/2,t)}$ into itself, thus has finite codimentional range. By $(I+D)^{-1}$ we denote a right inverse on the range somehow extended to the whole space.

Theorem 4.3. Let $(P, B^1, \ldots, B^r)$ be m-d elliptic. Then for $s \geq 2r$, the operator

$$(P, B^1, \ldots, B^r)\colon H^+_{(s,t)} \to H^+_{(s-2r,\ t-\ell)} \times \prod_{k=1}^{r} H_{(s-r_k-1/2,\ t-\ell_k)} \quad \text{is Fredholm.}$$

Proof: By Corollary 4.2 the kernel is finite dimensional. For $(f, g^1, \ldots, g^r)$

belonging to the right hand side we define:

$$A(f, g^1, \dots, g^r) = Qf^0 + W(I + D)^{-1}\psi \quad \text{with}$$

$$\psi_k = \Lambda^{r_k}\lambda^{\ell_k}(x')(g^k - B^kQf^0).$$

First, (4.5) shows that A is a bounded operator into $H^+_{(s,t)}$. Then since K_1 is of order $(-\infty, -\infty)$ by (4.6)

$$PA(f, g^1, \dots, g^r) = f + K(f, g^1, \dots, g^r) \quad \text{in } \mathbb{R}^{n+1}_+$$

where K is compact. Finally for ψ in the range of $I+D$ by (4.7)

$$B^kA(f, g^1, \dots, g^r) = g^k \quad \text{on } \partial\mathbb{R}^{n+1}_+, \quad k=1,2,\dots,r.$$

Corollary 4.4. Let $(P, B^1, \dots, B^r)$ be m-d elliptic. Then for $f\varepsilon H^+_{(s-2r,t-\ell)}$, $g^k\varepsilon H_{(s-r_k-1/2,\ t-\ell_k)}$, $u\varepsilon H^+_{(s,t)}$, $s \geq 2r$ the boundary value problem

$$Pu = f \quad \text{in} \quad \mathbb{R}^{n+1}_+$$

$$B^ku = g^k \quad \text{on} \quad \partial\mathbb{R}^{n+1}_+ \quad k=1,2,\dots,r$$

is normally solvable.

REFERENCES

[1] S. AGMON, A. DOUGLIS, L. NIRENBERG: Estimates near the boundary for solutions of elliptic partial differential equations satisfying general boundary conditions.I. *Comm. Pure Appl. Math.* 12 (1959) 623-727.

[2] A. BOVE, B. FRANCHI, E. OBRECHT: A boundary value problem for elliptic equations with polynomial coefficients in a half space I, II, and, Elliptic equations with polynomially growing coefficients in a half space: *Boll. Un. Math. Ital.* B (5) 18 (1981) 25-45, 355-380.

[3] A.P. CALDERON: Boundary value problems for elliptic equations. *Outlines Joint Symp. Partial Differential Equations, Novosibirsk* (1963) 303-304.

[4] H.O. CORDES: A global parametrix for pseudo-differential operators over $\mathbb{R}^n$, with applications. Preprint No. 90, SFB72, University of Bonn.

[5] H.O. CORDES, A.K. ERKİP: The N-th order elliptic boundary problem for noncompact boundaries. *Rocky Mountain J. Math.* 10 (1980) 7-24.

[6] A.K. ERKİP: The elliptic boundary problem on the half space. *Comm. in P.D.E.* 4 (1979) 537-554.

[7] A.K. ERKİP: The N-th order elliptic boundary value problem on noncompact domains: Dissertation, University of California, Berkeley 1979.

[8] L. HÖRMANDER: Pseudo-differential operators and non-elliptic boundary problems. *Ann. Math.* 2:83 (1966) 129-209.

[9] L. HÖRMANDER: *The Analysis of Linear Partial Differential Operators III.* Springer-Verlag, Berlin, 1985.

[10] J.L. LIONS, E. MAGENES: *Non-Homogeneous Boundary Value Problems and Applications.* Vol. 1. Springer-Verlag, Berlin, 1972.

[11] R.T. SEELEY: Singular integrals and boundary value problems. *Am. J. Math.* 88 (1966) 781-809.

A Remark on Taniguchi-Kumanogo Theorem for Product of Fourier Integral Operators

By

Daisuke Fujiwara

Department of Mathematics, Tokyo Institute of Technology

Ohokayama, Meguroku, Tokyo 152, Japan

Summary: Taniguchi-Kumanogo estimate for a product of Fourier integral operators is stated in a form slightly different from the original one. Our estimate is rather sharp if Fourier integral operators are close to the identity.

§ 1 Introduction. We consider Fourier integral operator of the form

$$I(\nu;\phi,a)u(x) = \left(\frac{i\nu}{2\pi t}\right)^n \int a(x,y)\, e^{i\nu\phi(t,x,y)} u(y)\, dy,$$

where $\nu>1$ and t are real parameters, and $\int dy$ is the abbreviation of $\int_{\mathbb{R}^n} dy$. We assume that the phase function ϕ is of the form

$$\phi(x,y) = \frac{|x-y|^2}{2t} + t\,\omega(x,y)$$

and the second and higher derivatives of $\omega(x,y)$ are uniformly bounded on $\mathbb{R}^n\times\mathbb{R}^n$. We assume also that the amplitude function $a(x,y)$ belongs to the function space $\mathcal{B}(\mathbb{R}^n\times\mathbb{R}^n)$ of Schwartz. For any integer $m\geq 0$ we have

$$\| a \|_m = \sum_{|\alpha|+|\beta|\leq m} \sup_{x,y} \left|\left(\frac{\partial}{\partial x}\right)^\alpha \left(\frac{\partial}{\partial y}\right)^\beta a(x,y)\right| < \infty.$$

The phase function ϕ is a generating function of a canonical transformation

$$\chi : \mathbb{R}^n\times\mathbb{R}^n \ni \left(y, -\frac{\partial}{\partial y}\phi(t,x,y)\right) \longrightarrow \left(x, \frac{\partial}{\partial x}\phi(t,x,y)\right) \in \mathbb{R}^n\times\mathbb{R}^n.$$

If t is small, χ is close to the identity.

As the phase function $\phi(x,y)$ is not a homogeneous function, the Fourier integral operator $I(\nu;\phi,a)$ is slightly different from the standard ones of Hörmander [4]. That type of operator appears when one descusses the fundamental solution of the Schrödinger equation (cf. [3]).

Proving convergence of the Feynman path integral in [3], we needed to treat the product of many Fourier integal operators

$$I(\nu;\phi_k,a_k)I(\nu;\phi_{k-1},a_{k-1})\ldots\ldots I(\nu;\phi_2,a_2)I(\nu;\phi_1,a_1),$$

where $\phi_1, \phi_2,\ldots, \phi_k$ are phase functions of the form

$$\phi_j(x,y) = \frac{|x-y|^2}{2t_j} + t_j\ \omega(x,y),$$

and $a_1, a_2,\ldots, a_k$ are amplitude functions. We set $T_l = t_l+t_{l-1}+\ldots+t_1$. If each $|T_l|$, $l=1,2,\ldots,k$, is small then we can write this product as one Fourier integral operator. Each of $\phi_j(x,y)$ is the generating function of a canonical tranformation χ_j which is close to the identity. The composite $\chi_l\cdot\chi_{l-1}\cdot\ldots\cdot\chi_1$, $l=1,2,\ldots,k$, is a canonical tranformation with the generating function, which we denote by ψ_l. There exists an amplitude function $a(x,y)$ which may depends on t_j's and on ν such that

$$I(\nu;\psi_k,a) = I(\nu;\phi_k,a_k)I(\nu;\phi_{k-1},a_{k-1})\ldots\ldots I(\nu;\phi_k,a_k).$$

Kumanogo [6] [7] treated product of Fourier integral operators of standard type and Kitada-Kumanogo [5] discussed operators similar to ours. By Kumanogo's notation we may write $\psi_l=\phi_l\#\phi_{l-1}\#\ldots\#\phi_1$. Taniguchi-Kumanogo [8] and Taniguchi [9] gave the estimate of norm $\| a \|_m$ of the amplitude $a(x,y)$ in terms of those of $a_1, a_2,\ldots,$ a_k. Their estimate applied to our case gives the following

<u>Taniguchi-Kumanogo Theorem.</u> Assume that $|T_l|$ is sufficiently small for each $l=1,2,\ldots,k$. Then for any integer $m\geq 0$, there exist a positive integer $M(m)$ and a positive constant $C(m)$ independent of k such that

$$\| a \|_m \leqq C(m)^k \prod_{j=1}^{k} \| a_j \|_{M(m)}.$$

Taniguchi-Kumanogo theorem is very useful in constructing fundamental solutions of Schrödinger equation. cf. [3]. However this is not good enough in the following point: Keeping T fixed and let k go to ∞, the right hand side does not necessarily give finite bound even if each a_j is close to 1.

Assume that the amplidude function a_j is close to 1. Then we expect the constant C(m) is close to 1. To be more precise, we introduce semi-norms for a phase funtion

$$\phi(t,x,y) = \frac{|x-y|^2}{t} + t\,\omega(x,y).$$

For integer $m \geq 2$, we put

$$\varkappa_m(\phi) = \sup_{2 \leqq k \leqq m} \sum_{|\alpha|+|\beta|=k} \sup_{x,y} \left| \left(\frac{\partial}{\partial x}\right)^{\alpha} \left(\frac{\partial}{\partial y}\right)^{\beta} \omega(x,y) \right|.$$

Our aim in this note is the following

<u>Theorem.</u> Let $\phi_j(t_j,x,y)$, j=1,2,...,k, be phase functions of the form

$$\phi_j(t_j,x,y) = \frac{|x-y|^2}{t_j} + t_j\,\omega(x,y).$$

Assume that for each m=2,3,...

$$\varkappa_m = \sup_j \varkappa_m(\phi_j) < \infty.$$

Then there exist a positive constant τ with the following properties: If $|T_l|<\tau$ for each l=1,2,...,k, then there exists a function a(x,y) in $\mathcal{B}(\mathbb{R}^n \times \mathbb{R}^n)$ such that

$$I(\nu;\phi_k\#\ldots\#\phi_1,1+a) = I(\nu;\phi_k,1+a_k)I(\nu;\phi_{k-1},1+a_{k-1})\ldots.I(\nu;\phi_1,1+a_1).$$

For any integer m =0,1,2...., there exist a positive constant C(m) and a positive integer M(m) such that

(1) $$1+ \| a \|_m \leqq \prod_{j=1}^{k} (1+C(m)\| a_j \|_{M(m)}) \,(1+ C(m)\,|t_j|).$$

Assume that $\sum |t_j|<\infty$ and $\sum | a_j|_m <\infty$. Then the right hand side of (1) remains finite if we let k go to ∞.

Remark. We may put parameter t_j inside a_j and ω_j but it does not play essential role.

Skipping tedious technical lemmata, we will give a proof of the above theorem in §2. Technical lemmata will be proved in the rest of the paper.

§2. Outline of the proof.

Let $a(x,\xi,y) \in \mathcal{B}(\mathbb{R}^n\times\mathbb{R}^n\times\mathbb{R}^n)$. The pseudo-differential operator with symbol a is denoted by $J(\nu;a)$, i.e.,

$$J(\nu;a)\ u(x)= (\frac{\nu}{2\pi})^n \iint a(\ x,\xi,y)\ e^{i\nu(x-y)\xi}\ u(y)\ dyd\xi,$$

where $\int dy$ and $\int d\xi$ stand for integral over $\mathbb{R}^n$. Similar abbreviation will be made in the followings.

Proof of the theorem. We may assume without losing generality that all t_j's are positive. Let τ be so small that

$$4\kappa_2\ \tau^2 < \frac{1}{5}.$$

Let $\chi_1, \chi_2,\dots,\chi_k$ be canonical transformations with the generating functions $\phi_1, \phi_2,\dots,\phi_k$. Then the composite canonical transformations $\chi_2\cdot\chi_1$, $\chi_3\cdot\chi_2\cdot\chi_1$, ..., $\chi_k\cdot\chi_{k-1}\cdot\dots\cdot\chi_2\cdot\chi_1$ have generating functions $\psi_2 =\phi_2\#\phi_1$, $\psi_3=\phi_3\#\phi_2\#\phi_1$, ..., $\psi_k= \phi_k\#\phi_{k-1}\#\dots\#\phi_2\#\phi_1$, respectively. (cf. Theorem A2 of appendix II). And ψ_j is of the form

$$\psi_j(x^j,\ x^0) = \frac{|x^j-x^0|^2}{2T_j} + T_j\ \omega_{T_j}(x^j,\ x^0), \qquad j=1,2,\dots,k.$$

We know from Theorem A2 that

$$\kappa_2(\psi_j) < 14\ \kappa_2 \quad \text{and} \quad \kappa_l(\psi_j) \leq C_l(\kappa_2,\kappa_3,\dots,\kappa_l), \qquad l\geq 2,$$

where $C_1(\varkappa_2,\varkappa_3,\ldots,\varkappa_l)$ is a positive constant depending only on $\varkappa_2$, $\varkappa_3,\ldots,\varkappa_l$. We take τ still smaller in relation to $C_l(\varkappa_2,\varkappa_3,\ldots,\varkappa_l)$, $l=1,2,\ldots,3n+8$, if necessary. Since $|T_j|<\tau$, Lemma 3.9 asserts that the Fourier integral operator $I(\nu;\psi_j,1)$ has its inverse $I(\nu;\psi_j,1)^{-1}$. We have

$$I(\nu;\phi_j,1+a_j) = I(\nu;\phi_j,1+a_j)\ I(\nu;\psi_{j-1},1)\ I(\nu;\psi_{j-1},1)^{-1},$$

for $j=2,3,\ldots,k$. For each j, we can find an amplitude function $b_j(x,y)$ such that

$$(2.1)\qquad I(\nu;\phi_j,1+a_j)\ I(\nu;\psi_{j-1},1) = I(\nu;\psi_j,1+b_j),\quad j=2,3,\ldots,k.$$

There exists an amplitude function $p_j(x,\xi,y)$ such that

$$(2.2)\qquad I(\nu;\psi_j,1+b_j) = I(\nu;\psi_j,1)\ J(\nu;1+p_j),\quad j=2,3,\ldots,k.$$

For j=1, this reduces to

$$(2.3)\qquad I(\nu;\phi_1,1+a_1) = I(\nu;\phi_1,1)\ J(\nu,1+p_1).$$

Combining these, we have the expression

$$(2.4)\qquad I(\nu;\phi_j,1+a_j) = I(\nu;\psi_j,1)\ J(\nu;1+p_j)\ I(\nu;\psi_{j-1},1)^{-1}.$$

for $k \geqq j \geqq 2$. Therefore

$$I(\nu;\phi_k,1+a_k)\ I(\nu;\phi_{k-1},1+a_{k-1})\ldots\ldots I(\nu;\phi_1,1+a_1)$$
$$= I(\nu;\psi_k,1)\ J(\nu;1+p_k)J(\nu;1+p_{k-1})\ldots\ldots J(\nu;1+p_1).$$

We can find $p(x,\xi,y)$ such that

$$(2.5)\qquad J(\nu;1+p) = J(\nu;1+p_k)\ J(\nu;1+p_{k-1})\ldots\ldots J(\nu;1+p_1).$$

Finally we have

$$(2.6)\qquad I(\nu;\psi_k,1+a) = I(\nu;\psi_k,1)\ J(\nu;1+p).$$

The discussion above is nothing but the original discussion of Taniguchi [9] and Kumanogo-Taniguchi [8]. cf also [5], [6] and [7]. Our improvement lies in the following estimates: For each m=0,1,2,3,... there exist a positive constant C(m) and a positive integer M(m) such that

$$(2.7)\qquad 1+\|b_j\|_m \leq (1+C(m)\ \|a_j\|_{M(m)})\ (1+C(m)|t_j|\ |\frac{T_j}{T}|)$$

$$(2.8)\qquad \|p_j\|_m \le C(m)\,\|b_j\|_{M(m)},$$

$$(2.9)\qquad 1+\|p\|_m \le \prod_{j=1}^{k} (1+C(m)\,\|p_j\|_{M(m)}),$$

$$(2.10)\qquad 1+\|a\|_m \le (1+C(m)\,\|p\|_{M(m)})\,(1+C(m)|T|).$$

Combining all these estimates, we have

$$\begin{aligned} 1+\|a\|_m &\le (1+C(m)\,\|p\|_{M(m)})\,(1+C(m)|T|) \\ &\le (1+C(m)|T|)\prod_{j=1}^{k} (1+C(m)\,\|p_j\|_{M(m)}) \\ &\le (1+C(m)|T|)\prod_{j=1}^{k} (1+C(m)\,\|b_j\|_{M(m)}) \\ &\le (1+C(m)|T|)\prod_{j=1}^{k} (1+C(m)\,\|a_j\|_{M(m)})\,(1+C(m)|t_j|\,|\tfrac{T_j}{T}|). \end{aligned}$$

This prove the theorem.

The next section is devoted to prove existence of $I(\nu;\psi_j,1)^{-1}$, the proof of equalities (2.1), (2.2), (2.3), (2.5), (2.6), and estimates (2.7), (2.8), (2.9) and (2.10):

Existence of $I(\nu;\psi_j,1)^{-1}$ will be proved in Lemma 3.9.

(2.1) and estmate (2.7) will be proved in Lemma 3.13.

(2.2),(2.3) and estimate (2.8) will be proved in Lemma 3.10.

(2.5) and estimate (2.9) will be proved in Corollary 3.6.

(2.6) and estimate (2.10) will be proved in Lemma 3.8.

§3 Proof of Lemmata.

Lemma 3.1. Let $a,\ b\in\mathcal{B}(\mathbb{R}^n\times\mathbb{R}^n)$ and $t\varkappa_2(\phi) \le \frac{1}{4}$. Then there exist functions c_1 and $c_2\in\mathcal{B}(\mathbb{R}^n\times\mathbb{R}^n)$ such that

$$(3.1)\qquad I(\nu;\phi,a)\,I(\nu;\phi,b)^* = J(\nu,c_1),$$

$$(3.2)\qquad I(\nu;\phi,b)^*\,I(\nu;\phi,a) = J(\nu,c_2).$$

For any m=0, 1,... , there exist a positive constant C(m) and a

For any m=0, 1,... , there exist a positive constant C(m) and a positive integer M(m) such that

(3.3) $$\| c_1 \|_m \leq C(m) \| a \|_{M(m)} \| b \|_{M(m)},$$

(3.4) $$\| c_2 \|_m \leq C(m) \| a \|_{M(m)} \| b \|_{M(m)}.$$

Proof. We have by definition

$$I(\nu;\phi,a)\, I(\nu;\phi,b)^* u(x) =$$
$$= (\frac{\nu}{2\pi})^n \iint e^{i\nu(\phi(x,y)-\phi(z,y))}\, a(x,y)\, \overline{b(z,y)}\, u(z)\, dz\, dy.$$

We put $\phi(x,y) - \phi(z,y) = (x-z)\cdot\xi(x,y,z)$ with

$$\xi(x,y,z) = \int_0^1 \frac{\partial}{\partial x}\phi(sx+(1-s)z,y)\, ds.$$

Since $\frac{\partial\xi}{\partial y} = t^{-1}\{ -I + t^2 \int_0^1 \frac{\partial^2}{\partial x\partial y} \omega(sx+(1-s)z,y)\, ds\}$, we have the estimate $\left| \frac{\partial\xi}{\partial y} + \frac{1}{t} \right| \leq t\,\varkappa_2$, where $|\ \ |$ denotes the norm of an $n\times n$ matrix as a linear map of R^n and $\varkappa_2$ stands for $\varkappa_2(\phi)$. If $t^2\varkappa_2 < 1$, then $(\frac{\partial\xi}{\partial y})^{-1}$ exists and

(3.5) $$\left| \frac{1}{t} (\frac{\partial\xi}{\partial y})^{-1} + I \right| \leq t\,\varkappa_2\, (1 - t\varkappa_2)^{-1}.$$

This and Hadamard's global implicit function theorem prove that the correspondence $y \longrightarrow \xi$ is a global diffeomorphism of $\mathbb{R}^n$ for fixed (x,z). Hence we can represent y as a function of (x,ξ,z), which we denote by $y=y(x,\xi,z)$. By this change of variables we have

$$I(\nu;\phi,a)I(\nu;\phi,b)^*u(x) = (\frac{\nu}{2\pi})^n \iint e^{i\nu(x-z\cdot\xi)} c_1(x,\xi,z)u(z)\, dzd\xi,$$

where

(3.6) $$c_1(x,\xi,z) = t^{-n}\, a(x,y)\, \overline{b(z,y)}\, (\det \frac{\partial\xi}{\partial y})^{-1} \Big|_{y=y(x,\xi,z)}.$$

Estimates (3.3) follows easily from this. (3.2) and (3.4) will be proved similarly. We have proved Lemma 3.1.

To our purpose we need a slight modification of Lemma 3.1.

Lemma 3.2. Let $t^2\, \varkappa_2(\phi) \leq \frac{1}{4}$ and $a_1, a_2 \in \mathcal{B}(\mathbb{R}^n\times\mathbb{R}^n)$. Then there exist functions b_1 and b_2 in $\mathcal{B}(\mathbb{R}^n\times\mathbb{R}^n\times\mathbb{R}^n)$ such that

$$I(\nu;\phi,1+a_1)\, I(\nu;\phi,1+a_2)^* = J(\nu,1+b_1),$$

$$I(\nu;\emptyset,1+a_2)^* \; I(\nu;\emptyset,1+a_1) = J(\nu,1+b_2).$$

For any $m=0,1,2,\ldots$, there exist positive constant $C(m)$ and a positive integer $M(m)$ such that for $i=1,2$

$$(3.7) \qquad 1+\|b_i\|_m \le (1+C(m)\|a_1\|_{M(m)})(1+C(m)\|a_2\|_{M(m)})(1+C(m)|t|^2).$$

<u>proof</u> Just as in Lemma 3.1, we have

$$I(\nu;\emptyset,1+a_1) \; I(\nu;\emptyset,1+a_2)^* \; u(x) =$$

$$= (\tfrac{\nu}{2\pi})^n \iint e^{i\nu(x-z\cdot\xi)}(1+b_1(x,\xi,z))u(z) \, dz \, d\xi,$$

where

$$1+b_1(x,\xi,z)= t^{-n} \; (1+a_1(x,y))(1+\overline{a_2(z,y)})(\det \tfrac{\partial\xi}{\partial y})^{-1} \Big|_{y=y(x,\xi,z)}.$$

It follows from (3.5) that $t^{-n}|\det (\frac{\partial\xi}{\partial y})^{-1}| = 1+ t^2 \; p(x,y,z)$, and $p(x,y,z)$ satisfies the following estimate

$$|(\tfrac{\partial}{\partial x})^{\alpha} \; (\tfrac{\partial}{\partial y})^{\beta} \; (\tfrac{\partial}{\partial z})^{\gamma} \; p(x,y,z)| \le C_{\alpha\beta\gamma}$$

for any multi-indices α, β and γ. The constant $C_{\alpha\beta\gamma}$ depends only on κ_m with $m=|\alpha|+|\beta|+|\gamma|$. Since the diffeomorphism $\mathbb{R}^n \ni y \longrightarrow \xi\in\mathbb{R}^n$ induces a linear isomorphism of the Frechet spaces $\mathcal{B}(\mathbb{R}^n_x\times\mathbb{R}^n_y\times\mathbb{R}^n_z)$ and $\mathcal{B}(\mathbb{R}^n_x\times\mathbb{R}^n_\xi\times\mathbb{R}^n_z)$, we have the estimate (3.7).

The following Lemma is due to Kumanogo [7].

<u>Lemma 3.3 (Kumanogo)</u> Let $a_1(x,\xi,y)$, $a_2(x,\xi,y),\ldots,a_k(x,\xi,y)$ be a sequence of functions in $(\mathbb{R}^n_x\times\mathbb{R}^n_\xi\times\mathbb{R}^n_z)$. Then there exists a function $b(x,\xi,y)$ such that

$$J(\nu,a_k) \; J(\nu,a_{k-1}) \ldots J(\nu,a_1) = J(\nu,b).$$

And for any $m=0,1,2,\ldots,$, we have the estimate

$$(3.8) \qquad \|b\|_m \le C^k \sum_{m_1+m_2+\ldots+m_k=m} \prod_{j=1}^{k} \|a_j\|_{3n+6+m_j}.$$

The constant C is independent of m, k and of a_j. A fortiori

$$(3.9) \qquad \|b\|_m \le (2C)^{m+k-1} \prod_{j=1}^{k} \|a_j\|_{3n+6+m}$$

with the same constant C as in (3.8).

Remark 3.4. Since $m_j \leq m$, (3.8) implies that

$$\| b \|_m \leq C^k \frac{(m+k-1)!}{m!\,(k-1)!} \prod_{j=1}^{k} \| a_j \|_{3n+6+m}.$$

This and the inequality $\frac{(m+k-1)!}{m!\,(k-1)!} < 2^{m+k-1}$ prove (3.9).

Corollary 3.5. There exists a positive constant τ_0 such that for any $a \in \mathcal{B}(\mathbb{R}^n \times \mathbb{R}^n)$ satisfying

$$\| a \|_{3n+6} < \tau_0 \tag{3.10}$$

we can find a function $b \in \mathcal{B}(\mathbb{R}^n \times \mathbb{R}^n)$ with the property

$$J(\nu,1+a)\, J(\nu,1+b) = J(\nu,1+b)\, J(\nu,1+a) = I.$$

Moreover for any integer m = 0, 1, 2,.... we have

$$\| b \|_m \leq C(m)\, \| a \|_{3n+6+m}, \tag{3.11}$$

where the constant C(m) is independent of a if (3.10) holds.

Proof. Let τ_0 be so small that

$$C\, \tau_0 = \delta_1 < 1, \tag{3.12}$$

where C is the positive constant of Lemma 3.3. Then we construct $J(\nu,1+a)^{-1} = J(\nu,1+b)$ by the Neuman series. We put

$$J(\nu,b_k) = (-1)^k\, J(\nu,a)^k.$$

Then for each integer m=0,1,2....

$$\| b_k \|_m \leq C^k \sum_{m_1+m_2+\ldots+m_k=m} \prod_{j=1}^{k} \| a \|_{3n+6+m_j}.$$

Since $\| a \|_{3n+6+m_j} \leq \| a \|_{3n+6+m}$, we have, for $k \geq m$,

$$\| b_k \|_m \leq C^k \frac{(m+k-1)(m+k-2)\ldots k}{m\,!} \| a \|_{3n+6}^{\,k-m} \| a \|_{3n+6+m}^{\,m}$$

$$\leq \delta_1^{\,k-m} \frac{(m+k-1)(m+k-2)\ldots k}{m!} C^m \| a \|_{3n+6+m}^{\,m}.$$

It follows from this and the condition (3.11) that the series $b(x,\xi,y) = \sum_{k=1}^{\infty} b_k(x,\xi,y)$ converges in $\mathcal{B}(\mathbb{R}^n \times \mathbb{R}^n \times \mathbb{R}^n)$. We have

$$\| b \|_m \leq \sum_{k=1}^{m-1} C^k \frac{(k+m-1)!}{m!\,(k-1)!} \| a \|_{3n+6+m}^{\,k}$$

$$+ C^m \delta_1^{-m} \left(\frac{\partial}{\partial \delta_1}\right)^{m-1} \left(\delta_1^{\,2m-1} (1-\delta_1)^{-1}\right) \| a \|_{3n+6+m}^{\,m},$$

with the same constant C as in (3.8). We have proved the corollary.

The next Corollary gives proof of (2.5) and (2.9) of §2.

Corollary 3.6. Let $a_1(x,\xi,y)$, $a_2(x,\xi,y),\dots,a_k(x,\xi,y)$ be a sequence of functions of $\mathcal{B}(\mathbb{R}^n_x\times\mathbb{R}^n_\xi\times\mathbb{R}^n_z)$. Then there exists a function $b(x,\xi,y)$ such that

$$J(\nu,1+a_k)\,J(\nu,1+a_{k-1})\,\dots\,J(\nu,1+a_1) = J(\nu,1+b). \tag{3.13}$$

And for any $m=0,1,2,\dots,$, there holds the estimate

$$1+\|b\|_m \le \prod_{j=1}^{k} (1+C(m)\,\|a_j\|_{M(m)}). \tag{3.14}$$

The constant $C(m)$ is independent of k and a_j.

proof. (3.13) holds if we put

$$b = \sum_{l=1}^{k} \sum_{k\ge j_l>j_{l-1}>\dots>j_1\ge 1} b_{j_l j_{l-1}\dots j_1},$$

where $b_{j_l j_{l-1}\dots j_1}$ is the symbol function such that

$$J(\nu;b_{j_l j_{l-1}\dots j_1}) = J(\nu;a_{j_l})J(\nu;a_{j_{l-1}})\dots J(\nu;a_{j_1}).$$

It follows from (3.9) that for any $m=0,1,2,\dots$

$$\|b_{j_l j_{l-1}\dots j_1}\|_m \le (2C)^{m+l-1}\prod_{i=1}^{l}\|a_{j_i}\|_{3n+6+m}$$

$$\le (2C)^{(m+1)l}\prod_{i=1}^{l}\|a_{j_i}\|_{3n+6+m}.$$

Putting $C(m)=(2C)^{m+1}$, we can prove (3.14) as follows:

$$1+\|b\|_m \le 1+\sum_{l=1}^{k}\sum_{k\ge j_l>j_{l-1}>\dots>j_1\ge 1}\|b_{j_l j_{l-1}\dots j_1}\|_m$$

$$\le 1+\sum_{l=1}^{k}\sum_{k\ge j_l>j_{l-1}>\dots>j_1\ge 1} C(m)^l\prod_{i=1}^{l}\|a_{j_i}\|_{3n+6+m}$$

$$\le \prod_{j=1}^{k}(1+C(m)\,\|a_j\|_{3n+6+m}).$$

Lemma 3.7. For any $a\in\mathcal{B}(\mathbb{R}^n\times\mathbb{R}^n)$ and $p\in\mathcal{B}(\mathbb{R}^n\times\mathbb{R}^n\times\mathbb{R}^n)$ we have

$$J(\nu,p)\,I(\nu;\emptyset,a) = I(\nu;\emptyset,b_1),\quad I(\nu;\emptyset,a)\,J(\nu,p) = I(\nu;\emptyset,b_2).$$

For any $m=0,1,2,\dots,$ there exist a positive constant $C(m)$ and a positive integer $M(m)$ such that

(3.15) $$\| b_i \|_m \le C(m) \| a \|_{M(m)} \| p \|_{M(m)},$$

for i=1, 2. C(m) depends polynomially on $\varkappa_2, \varkappa_3, \ldots\ldots,$.

Proof. By definition

$$J(\nu,p)I(\nu;\phi,a)u(x)$$
$$= (\frac{\nu}{2\pi})^n (\frac{i\nu}{2\pi t})^{n/2} \iiint p(x,\xi,z)a(y,z)e^{i\nu((x-y)\xi+\phi(y,z))}u(z)dzdyd\xi.$$
$$= (\frac{i\nu}{2\pi t})^{n/2} \int b(x,z)\, e^{i\nu\phi(x,z)}\, u(z)\, dz,$$

where

$$b(x,z)= (\frac{\nu}{2\pi})^n \iint p(x,\xi,y)a(y,z)e^{i\nu((x-y)\xi+\phi(y,z)-\phi(x,z))}dy\, d\xi.$$

Change of variables y=x-w gives

$$b(x,z)=(\frac{\nu}{2\pi})^n \iint p(x,\xi,x-w)a(x-w,z)e^{i\nu(w\xi+\phi(x-w,z)-\phi(x,z))}dw\, d\xi.$$

The critical point of the phase is $w=0$, $\xi = \frac{\partial}{\partial x}\phi(x,z)$. The Hessian determinant is given by $\det H(x,w,\xi,z) = (-1)^n$. We can apply the stationary phase method of Appendix I to b(x,z) and we have $b(x,z)\in\mathcal{B}(\mathbb{R}^n \times \mathbb{R}^n)$. We also have the estimate (3.15) for i=1. The other half of Lemma will be proved similarly.

The next Lemma 3.8 proves (2.6) and (2.10) of §2.

Lemma 3.8. Assume that $a\in\mathcal{B}(\mathbb{R}^n\times\mathbb{R}^n)$. Then there exist b_1 and $b_2\in\mathcal{B}(\mathbb{R}^n\times\mathbb{R}^n)$ such that

$$J(\nu,1+a)\ I(\nu;\phi,1) = I(\nu;\phi,1+b_1),\quad I(\nu;\phi,1)\ J(\nu,1+a) = I(\nu;\phi,1+b_2).$$

For any integer $m\geq 0$ there exist a positive constant C(m) and a positive integer M(m) such that

$$\| b_i \|_m \le C(m) \| a \|_{M(m)}, \qquad i=1, 2.$$

Proof Since $J(\nu,1)$ = identity, we have

$$J(\nu,1+a)\ I(\nu;\phi,1) = I(\nu;\phi,1) + J(\nu,a)I(\nu;\phi,1).$$

This together with the previous lemma proves Lemma 3.8.

Existence of $I(\nu;\psi_j,1)^{-1}$, $j=1,2,\ldots,k$, of §2 follows from the next Lemma 3.9.

Lemma 3.9. There exists a positive constant τ_1 depending only on $\varkappa_2(\phi),\varkappa_3(\phi),\ldots,\varkappa_{3n+8}(\phi)$ such that if $t\leq\tau_1$ then $I(\nu;\phi,1)^{-1}$ exists and is of the form

$$I(\nu;\phi,1)^{-1} = I(\nu;\phi,1+a)^* \tag{3.16}$$

with an amplitude function $a(x,y)\in\mathcal{B}(\mathbb{R}^n\times\mathbb{R}^n)$. For any $m=0,1,2\ldots$ there exists a positive constant $C(m)$ such that

$$\| a \|_m \leq C(m)\, \varkappa_{m+3n+8}\; t^2. \tag{3.17}$$

Proof. We choose τ_1 so small that

$$\varkappa_2(\phi)\tau_1{}^2 \leq 4^{-1} \tag{3.18}$$

and that

$$C(3n+6)\tau_1{}^2 < \tau_0, \tag{3.19}$$

where $C(m)$ is the constant of Lemma 3.2 and τ_0 is the constant in Corollary 3.5. Since $C(3n+6)$ depends only on $\varkappa_2(\phi)$, $\varkappa_3(\phi),\ldots$, $\varkappa_{3n+8}(\phi)$, so does τ_1. From Lemma 3.2 and (3.18) we have

$$I(\nu;\phi,1)^*\, I(\nu;\phi,1) = J(\nu,1+p).$$

For any $m = 0,1,2\ldots$, we have $1 + \| p \|_m \leq 1 + C(m)\, t^2$. Since (3.19) holds, we can apply Corollary 3.5 and we can construct $J(\nu,1+p)^{-1} = J(\nu,1+q)$. Therefore

$$J(\nu,1+q)\, I(\nu;\phi,1)^*\, I(\nu;\phi,1) = I.$$

We define b_1 by the equality $I(\nu;\phi,1)J(\nu,1+\bar{q}\ ) = I(\nu;\phi,1+ b_1)$. Then (3.16) holds. The estimate (3.17) follows from Lemma 3.8 and this.

The next Lemma 3.10 proves (2.2), (2.3) and estimate (2.8) of §2.

Lemma 3.10. Assume the conditions of Lemma 3.9. Then for any $a\in\mathcal{B}(\mathbb{R}^n\times\mathbb{R}^n)$ there exist p_1 and $p_2\in\mathcal{B}(\mathbb{R}^n\times\mathbb{R}^n\times\mathbb{R}^n)$ such that

$$I(\nu;\phi,1+a) = I(\nu;\phi,1)\, J(\nu,1+p_1)= J(\nu,1+p_2)\, I(\nu;\phi,1).$$

And we have, for any $m = 0, 1,\ldots.,$

$$\| p_j \|_m \leq C(m) \| a \|_{M(m)} \quad , j=1,2.$$

Proof. Let $I(\nu;\phi,1)^{-1} = I(\nu;\phi,1+b)^*$. We have only to define p_1 by the equality $J(\nu,p_1) = I(\nu;\phi,1+b)^* I(\nu;\phi,a)$. The estimate for $\| p_j \|_m$ follows from Lemma 3.1 and Lemma3.9.

Lemma 3.11. Let ϕ_1 and ϕ_2 be two phase functions of the form

$$\phi_1(x,y) = \frac{|x-y|^2}{2t_1} + t_1\omega_1(x,y) \text{ and } \phi_2(x,y) = \frac{|x-y|^2}{2t_2} + t_2\omega_2(x,y).$$

Let a_1 and $a_2 \in \mathcal{B}(\mathbb{R}^n \times \mathbb{R}^n)$ be two amplitude functions and $\varkappa_m$, m=2,3,... be such constants as $\varkappa_m(\phi_i) \leqq \varkappa_m$ for i=1,2. Assume that $|t_1|,|t_2|<1$ and that

$$(3.20) \qquad \varkappa_2 \ (|t_1|+|t_2|)^2 < 1.$$

Then there exists a function $b \in \mathcal{B}(\mathbb{R}^n \times \mathbb{R}^n)$ such that

$$I(\nu;\phi_2,a_2)\ I(\nu;\phi_1,a_1) = I(\nu;\phi_2 \# \phi_1,b).$$

For any integer $m \geq 0$, there exists positive constant C(m) and a positive integer M(m) such that

$$\| b \|_m \leq C(m) \| a_1 \|_{M(m)} \| a_2 \|_{M(m)}.$$

Proof. For any $u(x) \in C_0^\infty(\mathbb{R}^n)$, we have

$$I(\nu;\phi_2,a_2)\ I(\nu;\phi_1,a_1)\ u(x)$$

$$=(\frac{\nu i}{2\pi t_1})^{n/2}(\frac{\nu i}{2\pi t_2})^{n/2}\iint a_2(x,y)a_1(y,z)e^{i\nu(\phi_2(x,y)+\phi_2(y,z))}u(z)dzdy.$$

We put

$$(3.21) \qquad J =\iint a_2(x,y)a_1(y,z)e^{i\nu(\phi_2(x,y)+\phi_2(y,z))}dy.$$

Let $\mu = \nu(\frac{1}{t_1}+\frac{1}{t_2})$ and $\tau = \frac{1}{t_1}+\frac{1}{t_2} = \frac{t_1+t_2}{t_1 t_2}$. Then

$$J =\int a_2(x,y)a_1(y,z)e^{i\mu\Phi(x,y,z)}\ dy,$$

where

$$\Phi(x,y,z)= \frac{1}{(t_1+t_2)}(t_2\frac{|x-y|^2}{2}+t_1t_2{}^2\omega_2(x,y)+t_1\frac{|y-z|^2}{2} +t_1{}^2t_2\omega_1(y,z)).$$

We want to apply the stationary phase method to (3.21). We have

$$\left|\frac{\partial^2}{\partial y \partial y}\Phi(x,y,z) - I\right| \leq \frac{1}{\tau}(|t_1|+|t_2|)\varkappa_2 \leq |t_1\ t_2|\ \varkappa_2 \leq \frac{1}{4}.$$

Therefore $\det \frac{\partial^2}{\partial y \partial y}\Phi(x,y,z) \geq (\frac{3}{4})^n$. The critical point of the phase is the solution $y=y(x,z)$ of the system of equations

$$\frac{\partial}{\partial y}\ \phi_2(x,y) + \frac{\partial}{\partial y}\ \phi_1(y,z) = 0.$$

The solution of this is unique, which we denote by $y=y(x,z)$.

Applying Theorem A1.2 of Appendix I, we have

$$J = \left(\frac{2\pi}{\mu}\right)^{n/2} e^{-\pi i n/4}\ H(x,z)^{-1/2}\ e^{i\mu\Phi(x,y(x,z),z)}$$

$$\times\{\ a_2(x,y(x,z))a_1(y(x,z),z) + \mu^{-1}\ p(\mu,x,z)\ \},$$

where

$$H(x,z) = \det \frac{\partial^2}{\partial y \partial y}\ \Phi(x,y(x,z),z).$$

Note that $\mu\ \Phi(x,y(x,z),z) = \nu\ \phi_2\#\phi_1(x,z)$ and

$$\left(\frac{i\nu}{2\pi t_1}\right)^{n/2} \left(\frac{i\nu}{2\pi t_2}\right)^{n/2} \left(\frac{2\pi}{\mu}\right)^{n/2} e^{-\pi n i/4} = \left(\frac{i\nu}{2\pi T}\right)^{n/2},$$

where $T = t_1 + t_2$. Consequently we have

$$I(\nu;\phi_2,a_2)I(\nu;\phi_1,a_1) = I(\nu;\phi_2\#\phi_1,b)$$

with

$$b(x,z)= H(x,z)^{-1/2}\{a_2(x,y(x,z))a_1(y(x,z),z) + \mu^{-1}p(\mu,x,z)\ \}.$$

For $m=0,1,2,\ldots$, we have

$$\|\ p(\mu,x,z)\|_m \leq C(m)\ \|\ a_2\ a_1\|_{M(m)}.$$

with some $C(m)$ and $M(m)$. Lemma is proved.

<u>Lemma 3.12</u>. Let ϕ_1 and ϕ_2 be the same phase function as in the previous lemma. Then we can find a function $a\,\mathcal{B}(\mathbb{R}^n\times\mathbb{R}^n)$ such that

$$I(\nu;\phi_2,1)\ I(\nu;\phi_1,1) = I(\nu;\phi_2\#\phi_1,1+a).$$

For any $m=0,1,2,\ldots$, we have

$$\|\ a\ \|_m \leq C(m)\ \frac{t_1 t_2}{t_1+t_2}.$$

The constant $C(m)$ depends polynomially on $\varkappa_2, \varkappa_3, \ldots$.

<u>Proof</u>. In Lemma 3.11 we put $a_1=a_2= 1$. Using notation of

Lemma 3.11, we have from (3.22)

$$J = \left(\frac{2\pi}{\mu}\right)^{n/2} \det\left[\frac{\partial^2}{\partial y \partial y}\Phi(x,y(x,z),z)\right]^{-1/2} e^{i\mu\Phi(x,y(x,z),z)}\{1+\mu^{-1}p(\mu,x,z)\}.$$

We can write

$$\det\left[\frac{\partial^2}{\partial y \partial y}\Phi(x,y(x,z),z)\right] = \det(I+\tau^{-1}A),$$

where

$$A = t_2 \frac{\partial^2}{\partial y \partial y}\omega_2(x,y(x,z)) + t_1 \frac{\partial^2}{\partial y \partial y}\omega_1(x,y(x,z)).$$

By the assumption (3.20), we get the estimate $\tau^{-1} | A | \leq \frac{1}{4}$. All entries of the matrix A belong to $\mathcal{B}(\mathbb{R}^n \times \mathbb{R}^n)$. Therefore

$$\det\left[\frac{\partial^2}{\partial y \partial y}\Phi(x,y(x,z),z)\right]^{-1/2} = 1 + \tau^{-1} q(x,z),$$

where q satisfies the estimate: $\| q \|_m \leqq C(m)$, $m=0,1,2,\ldots$.

This implies that

$$J = \left(\frac{2\pi}{\mu}\right)^{n/2} e^{\pi n i/4} e^{i\nu(\phi_2 \# \phi_1)(x,z)} \left\{ 1 + \frac{1}{\nu\tau} a(x,z)\right\},$$

where $a(x,z) = q(x,z)\{1+ \frac{1}{\nu\tau} p(\mu,x,z)\}$. Lemma is proved.

The next Lemma 3.13 proves (2.1) and estimate (2.7) of §2.

<u>Lemma 3.13.</u> Let ϕ_1 and ϕ_2 be the same phase functions as in Lemma 3.11. Assume that a_1 and a_2 belong to the space $\mathcal{B}(\mathbb{R}^n \times \mathbb{R}^n)$. Then there exists a function $b \in \mathcal{B}(\mathbb{R}^n \times \mathbb{R}^n)$ such that

$$I(\nu;\phi_2,1+a_2)\, I(\nu;\phi_1,1+a_1) = I(\nu;\phi_2 \# \phi_1,1+b).$$

For any $m=0,1,2,\ldots$, there exists a positive constant $C(m)$ and a positive integer $M(m)$ such that

$$1+ \| b \|_m \leq (1+C(m)\| a_1 \|_{M(m)})(1+C(m)\| a_2 \|_{M(m)})(1+C(m)\frac{t_1 t_2}{t_1+t_2}).$$

<u>Proof</u> .

$$\begin{aligned} I(\nu;\phi_2,1+a_2)\, I(\nu;\phi_1,1+a_1) &= \\ = I(\nu;\phi_2,1)\, I(\nu;\phi_1,1) &+ I(\nu;\phi_2,1)\, I(\nu;\phi_1,a_1) \\ + I(\nu;\phi_2,a_2)\, I(\nu;\phi_1,1) &+ I(\nu;\phi_2,a_2)\, I(\nu;\phi_1,a_1). \end{aligned}$$

Applying Lemma 3.11 and Lemma 3.12 to this, we can prove Lemma 3.13.

All the facts assumed in §2 have now been proved.

§ Appendix I

Usually the stationary phase method is formulated in localized form. (For instance, Fedryuk [2] or Hörmander [4]). We need a global version of it, which we state here.

Consider the following oscillatory integral

(A1.1) $$I(x,\nu) = \int_{\mathbb{R}^n} a(x,y)\, e^{i\nu\phi(x,y)}\, dy,$$

where $\nu \geqq 1$ is a parameter and $x \in \mathbb{R}^m$. The real phase function $\phi(x,y)$ and the amplitude function $a(x,y)$ are defined for $(x,y) \in \mathbb{R}^m \times \mathbb{R}^n$ and of class C^∞. We put further assumptions:

(A1.2) $0 < K_0 = \inf_{x,y} \left| \det\left(\frac{\partial^2}{\partial y \partial y} \phi(x,y)\right)\right|$.

(A1.3) For any integer $k \geqq 2$,

$$\sum_{2 \leq |\alpha|+|\beta| \leq k} \sup_{x,y} \left|\left(\frac{\partial}{\partial x}\right)^\alpha \left(\frac{\partial}{\partial y}\right)^\beta \phi(x,y)\right| = K_k < \infty.$$

(A1.4) For any integer $k \geqq 0$,

$$\| a \|_k = \sum_{|\alpha|+|\beta| \leq k} \sup_{x,y} \left|\left(\frac{\partial}{\partial x}\right)^\alpha \left(\frac{\partial}{\partial y}\right)^\beta a(x,y)\right| < \infty.$$

Let $\xi = \mathrm{grad}_y \phi(x,y)$. Then the map $\mathbb{R}^n \ni y \longrightarrow \xi \in \mathbb{R}^n$ is a global diffeomorphism for any fixed x because of (A1.2),(A1.3) and Hadamard's global implicit function theorem. Consequently the equation

$$\mathrm{grad}_y\, \phi(x,y) = 0$$

with respect to y has a unique solution $y = y(x)$. It is clear that the function $y(x)$ is of class C^∞ in x.

Let $H(x)$ be the Hessian matrix $H(x) = \frac{\partial^2}{\partial y \partial y} \phi(x,y)\big|_{y=y(x)}$. Then we denote $h(x,y) = \phi(x,y) - \phi(x,y(x)) - \frac{1}{2} \langle y-y(x), H(x)(y-y(x))\rangle$, which is the error term of the Taylor expansion up to order 2 of $\phi(x,y)$ around $y=y(x)$. Using the inverse matrix $H(x)^{-1} = (h'_{jk}(x))$ of $H(x)$, we define a differential operator of order 2

$$\langle H(x)D_y, D_y\rangle = \sum_{jk} h'_{jk}(x) \left(\frac{\partial}{\partial y_j}\right)\left(\frac{\partial}{\partial y_k}\right).$$

Theorem A1.1. Assume that (A1.2), (A1.3) and (A1.4) hold. Then for any integer $N\geq 0$, we have

$$I(x,\nu) = (\tfrac{2\pi}{\nu})^{n/2} |\det H(x)|^{-1/2} \exp i\tfrac{\pi}{4}(n-2\mathrm{Ind}(H(x))) \exp i\nu\phi(x,y(x))$$

$$\{\sum_{k=0}^{N-1} \frac{1}{k!}((-\frac{i}{2\nu}\langle H(x)^{-1}D_y, D_y\rangle)^k a(x,y) \exp i\nu h(x,y)|_{y=y(x)}\}$$

$$+ \exp i\nu\phi(x,y(x))\, p_N(x,\nu).$$

Here Ind $H(x)$ is the maximum dimension of negative subspaces of $H(x)$ and $p_N(x,\nu)$ satifies the following estimate: For any multi-index α there exists a positive constant C and a positive integer $l(\alpha,N)$ such that

$$|(\frac{\partial}{\partial x})^\alpha p_N(x,\nu)| \leq C\, \nu^{-N} \sum_{\beta\leq\alpha} \sum_{|\gamma|\leq l(\alpha,N)} \sup_{x,y} |(\frac{\partial}{\partial x})^\beta (\frac{\partial}{\partial y})^\gamma a(x,y)|.$$

Here $C=C(K_0^{-1}, K_{l(\alpha,N)})$ is bounded if K_0^{-1} and $K_{l(\alpha,N)}$ are bounded.

In particular, the case N=0 of Theorem A1.1 reads

Theorem A1.2. Under assumption of Theorem A1.1 we have

$$I(x,\nu) = e^{i\nu\phi(x,y(x))}\, q(x,\nu).$$

For any multi-index α there exists a positive constant C and a positive integer $l(\alpha)$ such that

$$|(\frac{\partial}{\partial x})^\alpha \gamma(x,\nu)| \leq C \sum_{\beta\leq\alpha} \sum_{|\gamma|\leq l(\alpha)} \sup_{x,y} |(\frac{\partial}{\partial x})^\alpha (\frac{\partial}{\partial y})^\gamma a(x,y)|.$$

The constant $C=C(K_0^{-1},K_{l(\alpha)})$ is bounded if K_0^{-1} and $K_{l(\alpha)}$ are bounded.

§ Appendix II. Phase functions.

Let $\phi_1(x,y), \phi_2(x,y), \ldots, \phi_k(x,y)$ be phase functions as in §1:

$$\phi_j(x,y) = \frac{|x-y|^2}{2\, t_j} + t_j\, \omega_j(x,y), \quad j = 1,2,\ldots,k. \tag{A2.1}$$

Then the function $\phi_j(x,y)$ is a generating function of the canonical transformation χ_j: $\mathbb{R}^n\times\mathbb{R}^n \ni (x^{j-1},\xi^{j-1}) \longrightarrow (x^j,\xi^j) \in \mathbb{R}^n\times\mathbb{R}^n$, where $\xi^{j-1} = -\frac{\partial}{\partial x^j}\phi_j(x^j,x^{j-1})$ and $\xi^j = -\frac{\partial}{\partial x^j}\phi_j(x^j,x^{j-1})$, $j=1,2,\ldots,k$. Let $\chi = \chi_k\cdot\chi_{k-1}\cdot\ldots\cdot\chi_1$ be the composite of these mappings. It is well known that χ has the generating function if $t_1+t_2+\ldots+t_k = T$ is very small. But how small T should be ? We state here the following

Theorem A2. Let $\phi_1(x,y)$, $\phi_2(x,y),\dots,\phi_k(x,y)$ be phase functions of the form

$$\phi_j(x,y) = \frac{|x-y|^2}{2t_j} + t_j\,\omega_j(x,y), \qquad \text{for } j=1,2,\dots,k.$$

Assume that

(A2.2) $$|\omega_j(x,y)| < \varkappa_0\,(1+|x|+|y|)^2,$$

(A2.3) $$|\tfrac{\partial}{\partial x}\omega_j(x,y)|+|\tfrac{\partial}{\partial y}\omega_j(x,y)| \le \varkappa_1\,(1+|x|+|y|),$$

(A2.4) $$\varkappa_m(\phi_j) \le \varkappa_m \quad \text{for } m \geq 2,$$

where $\varkappa_m$, m=0,1,2..., are positive constants. We further assume that

(A2.5) $$T=t_1+t_2+\dots+t_k < (20\varkappa_2)^{-1/2}.$$

Then the composite mapping $\chi=\chi_k\circ\chi_{k-1}\circ\cdots\circ\chi_1$ of canonical mappings $\chi_k, \chi_{k-1},\dots,\chi_1$ has the generating function $\psi(x,y)$ of the form

(A2.6) $$\psi(x,y) = \frac{|x-y|^2}{2T} + T\,\omega(x,y).$$

$\omega(x,y)$ may depend on T and satisfies the estimates (A2.2), (A2.3) and (A2.4) with $\varkappa_m$ replaced by another constant $\varkappa_m'$ which depends polynomially on $\varkappa_0, \varkappa_1,\dots,\varkappa_m$ but not on k. In particular we have $\varkappa_2' \le 14\,\varkappa_2$. And $\varkappa_m'$, $m \geq 2$, depends only on $\varkappa_2, \varkappa_3,\dots,\varkappa_m$.

References

[1] Asada, K. and Fujiwara, D., On some oscillaltory integral transformations in $L^2(\mathbb{R}^n)$, Japan J. Math. vol.4 (1978), 299-361.

[2] Fedryuk, M.V., The stationary phase methods and pseudo-differential operators, Russian Math. Survey, vol.26 (1971), 65-115.

[3] Fujiwara, D., A remark on convergence of Feynman path integrals, Duke Math. J. vol.47 (1980), 559-600.

[4] Hörmander, L., Fourier integral operators I, Acta Math. vol.127 (1971), 77-183.

[5] Kitada.H, and Kumanogo, H., A family of Fourier integral operators and the fundamental solution for a Schrödinger equation, Osaka J. Math. vol.8 (1981) 291-360.

[6] Kumanogo, H., Fundamental solution for a hyperbolic system with diagonal part, Comm. Partial diff. equation vol.4 (1979), 959-1015.

[7] Kumanogo, H., Pseudo-differential operators. MIT press, Cambridge and London, 1982.

[8] Kumanogo,H.-Taniguchi,K., Fourier integral operators of multi-phase and the fundamental solution for a hyperbolic system, Funkcial. Ekvac. vol.22 (1979),161-196.

[9] Taniguchi,K., Multi-products of Fourier integral operators and the fundamental solution for a hyperbolic system with involutive characteristics, Osaka J. Math. vol.21 (1984) 169-224.

ON THE ANALYTIC REGULARITY OF WEAK SOLUTIONS OF ANALYTIC SYSTEMS OF CONSERVATION LAWS WITH ANALYTIC DATA

Paul Godin
Département de Mathématique
Université Libre de Bruxelles
Campus de la Plaine, C.P.214
Boulevard du Triomphe
B-1050 Bruxelles
Belgium

1. Introduction.

We consider a hyperbolic system of conservation laws

$$\frac{\partial u}{\partial t} + \sum_{j=1}^{N} \frac{\partial}{\partial x_j} F_j(u) = 0 , \tag{1.1}$$

where (x,t) belongs to an open subset V of $\mathbb{R}^N \times \mathbb{R}$. For $j=1,\dots,N$, F_j is a sufficiently smooth mapping of an open subset G of $\mathbb{R}^m$ to $\mathbb{R}^m$ ($m \in \mathbb{Z}^+ \setminus \{0\}$), and $u={}^t(u_1,\dots,u_m)$ is a mapping of V to G. If $u \in L^\infty(V,G)$, one says that u is a weak solution of (1.1) provided the relation

$$\iint \left(\frac{\partial \psi}{\partial t} u + \sum_{j=1}^{N} \frac{\partial \psi}{\partial x_j} F_j(u)\right) dx\, dt = 0 \tag{1.2}$$

holds for all $\psi \in C_0^1(V)$. Assume that $(x_o,t_o) \in V$ and let S be a C^1 hypersurface through (x_o,t_o), transversal to the planes t=constant. Since we shall only be concerned with local problems, we may as well assume that S is given by an equation $x_N=\varphi(x',t)$, where $x=(x',x_N)$ and $\varphi \in C^1$. We take V so small that $V=V^- \cup (S \cap V) \cup V^+$, where $V^\pm=\{(x,t) \in V,\ x_N \gtrless \varphi(x',t)\}$. Below we shall mainly be concerned with weak solutions of (1.1) such that $u^\pm=u|_{V^\pm} \in C^1(\bar{V}^\pm)$. (For $0 < \rho \leq \infty$ and $E \subset \mathbb{R}^q$, we say that $f \in C^\rho(E)$ if there exist an open neighborhood E' of E in $\mathbb{R}^q$ and a function $f' \in C^\rho(E')$ such that $f'|_E=f$). If $u \in C^1(\bar{V}^\pm)$, it follows by integration by parts that

$$\frac{\partial u^\pm}{\partial t} + \sum_{j=1}^{N} A_j(u^\pm) \frac{\partial u^\pm}{\partial x_j} = 0 \tag{1.3}$$

in $V^\pm$, where $A_j = \frac{\partial F_j}{\partial u}$, whilst

$$\varphi_t(u^+-u^-) + \sum_{j=1}^{N-1} \varphi_{x_j}(F_j(u^+)-F_j(u^-)) - (F_N(u^+)-F_N(u^-)) = 0 \tag{1.4}$$

on S. ((1.4) is of course a necessary and sufficient condition for $u^{\pm} \in C^1(\bar{V}^{\pm})$ to be the restriction to $V^{\pm}$ of a weak solution in V). In this paper we shall assume that $V=\{(x,t) \in W, t > 0\}$, where W is an open neighborhood of (0,0) in $\mathbb{R}^N \times \mathbb{R}$, and that $S=\{(x',\varphi(x',t),t),(x',t) \in U \times]0,T[\}$, where $\varphi \in C^1(U \times [0,T])$, and U is an open neighborhood of 0 in $\mathbb{R}^{N-1}$. Put $V_o=\{(x,t) \in W, t=0\}$, $V_o^{\pm}=\{(x,0) \in V_o, x_N \gtrless \varphi_o(x')\}$, $S_o=\{(x,0) \in V_o, x_N=\varphi_o(x')\}$, where φ_o is a C^1 function defined in a neighborhood of 0 in $\mathbb{R}^{N-1}$ with $\varphi_o(0)=0$. We may assume that $V_o=V_o^- \cup S_o \cup V_o^+$. In this paper we study solutions u^+,u^-,φ of (1.3), (1.4) which also satisfy the initial conditions

$$u^{\pm}|_{V_o^{\pm}} = u_o^{\pm} \quad , \quad \varphi|_{t=0}=\varphi_o, \tag{1.5}$$

where $u_o^{\pm}$ and φ_o are given functions. When N=1, weak solutions of systems of conservation laws have been much studied (see e.g.[11] and the references given there). For $N > 1$, the first general existence results were obtained by Majda [7],[8],[9], in case u^-,u^+,φ define a shock front solution, namely when S is not characteristic for either $\frac{\partial}{\partial t} + \sum_{j=1}^{N} A_j(u^+)\frac{\partial}{\partial x_j}$ or $\frac{\partial}{\partial t} + \sum_{j=1}^{N} A_j(u^-)\frac{\partial}{\partial x_j}$. Besides hyperbolicity, Majda assumed uniform stability and entropy conditions, and compatibility of the initial data. (For further existence results, see [10]). In [5], propagation of analyticity of shock fronts satisfying Majda's conditions was proved, assuming the F_j and the initial data to be analytic. In the present paper we prove a unique continuation principle from which we can in particular obtain an analytic regularity result in a situation where S is characteristic.

Our paper is organized as follows. In Section 2 we prove the unique continuation principle announced above (Theorems 1 and 2). In Section 3 we show how the results of Section 2 can be applied to the proof of analytic regularity of some weak solutions with analytic data.

2. A unique continuation principle for some weak solutions with given Cauchy data.

Let $W,V,V^{\pm},V_o,V_o^{\pm},\varphi,S$ be as in Section 1 (recall that we assume that $\varphi(0,0)=0$). Denote by ρ a real number which does not belong to $\mathbb{Z}^+$ and is strictly larger than 1. Assume that $u^{\pm},\tilde{u}^{\pm} \in C^{\rho}(\bar{V}^{\pm})$, $\varphi \in C^{\rho+1}(U\times[0,T])$, $F_j \in C^{\rho+1}(G)$ for $1 \leq j \leq N$, and that

$$\frac{\partial u^{\pm}}{\partial t} + \sum_{j=1}^{N} A_j(u^{\pm}) \frac{\partial u^{\pm}}{\partial x_j} = 0 \tag{2.1}$$

in $V^{\pm}$,

$$\varphi_t(u^+-u^-)+\sum_{j=1}^{N-1}\varphi_{x_j}(F_j(u^+)-F_j(u^-))-(F_N(u^+)-F_N(u^-))=0 \tag{2.2}$$

on S,

$$\frac{\partial\tilde{u}^{\pm}}{\partial t}+\sum_{j=1}^{N} A_j(\tilde{u}^{\pm})\frac{\partial\tilde{u}^{\pm}}{\partial x_j}=0 \tag{2.3}$$

in $V^{\pm}$,

$$\varphi_t(\tilde{u}^+-\tilde{u}^-)+\sum_{j=1}^{N-1}\varphi_{x_j}(F_j(\tilde{u}^+)-F_j(\tilde{u}^-))-(F_N(\tilde{u}^+)-F_N(\tilde{u}^-))=0 \tag{2.4}$$

on S. We assume that $\frac{\partial}{\partial t}+\sum_{j=1}^{N} A_j(g)\frac{\partial}{\partial x_j}$ is strictly hyperbolic in the direction of dt for all $g \in G$. One purpose of this section is to prove the following result.

<u>Theorem 1</u>. Assume that $u^-=\tilde{u}^-$ on S. Then:

(i) if $u^+=\tilde{u}^+$ on V_o^+, it follows that $u^+=\tilde{u}^+$ in a neighborhood of 0 in $\bar{V}^+$;

(ii) if $u^-=\tilde{u}^-$ on V_o^-, it follows that $u^-=\tilde{u}^-$ in a neighborhood of 0 in $\bar{V}^-$.

Remark that no hypothesis is made on the fact that S is or is not characteristic. For reasons of symmetry it is also clear that the assumption that $u^-=\tilde{u}^-$ on S may be replaced by the condition $u^+=u^+$ on S.

We start with some preparations for the proof of Theorem 1(i). Part (ii) is actually simpler and will be dealt with afterwards. First we are going to straighten S. Put $v^+(x,t)=u^+(x',x_N+\varphi(x',t),t)$, $\tilde{v}^+(x,t)=\tilde{u}^+(x',x_N+\varphi(x',t),t)$. If $g',g'' \in G$ and $\theta g'+(1-\theta)g'' \in G$ for all $\theta \in [0,1]$, define $H_j(g',g'')=\int_0^1 A_j(g''+\theta(g'-g''))d\theta$, $1 \leq j \leq N$. Write $E_r=\{(x,t) \in \mathbb{R}^N \times \mathbb{R},\ x_N > 0,\ t > 0, |x|+t < r\}$. If r is small enough, $z^+=v^+-\tilde{v}^+$ belongs to $C^{\rho}(\bar{E}_r)$. Since $u^+=\tilde{u}^+$ on V_o^+, the composite function $H_j(v^+,\tilde{v}^+)$ is well defined for small r and also belongs to $C^{\rho}(\bar{E}_r)$. It is convenient to put $\tilde{B}_j=H_j(v^+,\tilde{v}^+)$ if $1 \leq j \leq N-1$, $\tilde{B}_N=H_N(v^+,\tilde{v}^+)-\varphi_t-\sum_{j=1}^{N-1}\varphi_{x_j}H_j(v^+,\tilde{v}^+)$.

Notice that

$$\sum_{j=1}^{N-1} (\tilde{B}_j - A_j(v^+))\frac{\partial v^+}{\partial x_j} + (\tilde{B}_N - (A_N(v^+) - \varphi_t - \sum_{j=1}^{N-1} \varphi_{x_j} A_j(v^+)))\frac{\partial v^+}{\partial x_N}$$

$$= F(v^+, \tilde{v}^+, \nabla v^+)z^+, \tag{2.5}$$

where F is a $m \times m$ matrix depending in a $C^{\rho-1}$ way on its arguments. From (2.1) and (2.5), it follows that in E_r (r small):

$$\frac{\partial v^+}{\partial t} + \sum_{j=1}^{N} \tilde{B}_j \frac{\partial v^+}{\partial x_j} = F(v^+, \tilde{v}^+, \nabla v^+)z^+. \tag{2.6}$$

In the same way we also get

$$\frac{\partial \tilde{v}^+}{\partial t} + \sum_{j=1}^{N} \tilde{B}_j \frac{\partial \tilde{v}^+}{\partial x_j} = G(v^+, \tilde{v}^+, \nabla \tilde{v}^+)z^+ \tag{2.7}$$

in E_r (r small), where G is an expression similar to F.
From (2.6) and (2.7) we obtain that, in E_r:

$$\frac{\partial z^+}{\partial t} + \sum_{j=1}^{N} \tilde{B}_j \frac{\partial z^+}{\partial x_j} = F , \tag{2.8}$$

where F has the bound

$$|F| \leq C|z^+| \tag{2.9}$$

with a constant C depending on $v^+, \tilde{v}^+$ only through $\sup|v^+| + \sup|\nabla v^+| + \sup|\tilde{v}^+| + \sup|\nabla\tilde{v}^+|$. Since on the other hand $u^- = \tilde{u}^-$ on S, (2.2) and (2.4) imply that $\varphi_t(v^+ - \tilde{v}^+) + \sum_{j=1}^{N-1} \varphi_{x_j}(F_j(v^+) - F_j(\tilde{v}^+)) - (F_N(\tilde{v}^+) - F_N(\tilde{v}^+)) = 0$ on $E'_r = \{(x,t) \in \mathbb{R}^N \times \mathbb{R},\ x_N = 0,\ t > 0,\ |x| + t < r\}$, r small, so that we obtain from Taylor formula that

$$\tilde{B}_N z^+ = 0 \tag{2.10}$$

on E'_r . Also we may assume that r is so small that

$$z^+ = 0 \tag{2.11}$$

if $t=0$, $x_N > 0$, $|x| < r$.
It is convenient to extend the coefficients $\tilde{B}_j$ in a suitable way. So let $\zeta(x,t) \in C_0^\infty(\mathbb{R}^N \times \mathbb{R}, [0,1])$ be equal to 1 for $|x|+|t| < r_1$ and to 0 for $|x|+|t| > r_2$, where $0 < r_1 < r_2 < r$, and put

$B_j(x,t)=\zeta(x,t)\tilde{B}_j(x,t)+(1-\zeta(x,t))\tilde{B}_j(0,0)$, $1 \leq j \leq N$, $P=\frac{\partial}{\partial t}+\sum_{j=1}^{N} B_j\frac{\partial}{\partial x_j}$. Define $\Omega=\{x \in \mathbb{R}^N, x_N > 0\}$. If r is small enough, P is strictly hyperbolic in the direction of dt at each point of $\bar{\Omega} \times \bar{\mathbb{R}}^+$ (of course $\mathbb{R}^+=\{t \in \mathbb{R}, t > 0\}$). Let $\gamma \in C^1(\bar{\Omega})$ (with values in $\mathbb{R}$) satisfy $\sup|\nabla\gamma| \leq 1$, and put $h(x,t)=t+\beta\gamma(x)$, where $\beta > 0$ is a small parameter. The basic estimate which will lead to Theorem 1 is contained in Proposition 1 below. We formulate this proposition in such a way that it will also have another interesting consequence (Theorem 2 below). Let $\mu=\bar{\Omega} \to \bar{\mathbb{R}}^+$ be a C^1 function. Define $D=\{(x,t) \in \Omega \times \mathbb{R}^+, t > \mu(x)\}$ and denote by $C^1_{(o)}(\bar{D})$ the set of restrictions to D of $C^1_o(\mathbb{R}^N_x \times \mathbb{R}_t)$ functions. Then we have the following proposition, where $\|\ \|_D$ stands for the $L^2(D)$ norm.

Proposition 1. One can find $\eta_o > 0$, $C > 0$, $\beta_o > 0$, such that for all $\eta \geq \eta_o$, $\beta \leq \beta_o$, and all $w \in C^1_{(o)}(\bar{D})$ satisfying $B_N w=0$ when $x_N=0 < t-\mu(x)$, $w-\sum_{j=1}^{N} \mu_{x_j} B_j w=0$ when $t-\mu(x)=0 < x_N$, the estimate

$$\eta\|e^{-\eta h}w\|_D \leq C\|e^{-\eta h}Pw\|_D \tag{2.12}$$

holds.

Before proving Proposition 1, let us show how it can be used to complete the proof of Theorem 1. Choose $\varepsilon > 0$, $\beta \leq \beta_o$ and put $h(x,t)=t+\beta(|x|^2+\varepsilon^2)^{1/2}$. Let $\nu > \beta\varepsilon$ be so close to $\beta\varepsilon$ that $\{(x,t) \in \Omega \times \mathbb{R}^+, h(x,t) \leq \nu\}$ is contained in E_{r_1}. Choose ν_1 such that $\varepsilon\beta < \nu_1 < \nu$ and $\omega(s) \in C^\infty(\mathbb{R},[0,1])$ such that $\omega(s)=1$ if $s < \nu_1$ and $\omega(s)=0$ if $s > \nu$. Put $\chi(x,t)=\omega(h(x,t))$, $w=\chi z^+$. It follows from (2.10) that $B_N w=0$ when $x_N=0 < t$, and (2.11) implies that $w=0$ if $t=0 < x_N$. Hence we may apply (2.12) with the choice $\mu \equiv 0$, and writing $P\chi.z^+=\chi Pz^+ + [P,\chi]\, z^+$, we obtain for $\eta \geq \eta_o$:

$$\eta\|e^{-\eta h}\chi z^+\|_D \leq C(\|e^{-\eta h}\chi F\|_D+\|e^{-\eta h}[P,\chi]\, z^+\|_D), \tag{2.13}$$

since $Pz^+=F$ on the support of χ. If η is large enough, we therefore deduce from (2.9) that

$$\eta\|e^{-\eta h}\chi z^+\|_D \leq C\|e^{-\eta h}[P,\chi]\, z^+\|_D. \tag{2.14}$$

Since $\chi=1$ when $g \leq \nu_1$, the right-hand side of (2.14) is $O(e^{-\eta\nu_1})$ as $\eta \to +\infty$. Hence it follows from (2.14) that $z^+=0$ if $g \leq \nu_1$. This proves Theorem 1 (i). The proof of Theorem 1 (ii) is similar; actually the situation is somewhat simpler since if we put $z^-=v^- -\tilde{v}^-$ and argue as above, we obtain that z^- satisfies an equation similar to (2.8) and an initial condition as (2.11) (this time for negative values of x_N), but now we have $z^-=0$ on E_r' instead of (2.10). The conclusion follows just like for z^+. It therefore remains to prove Proposition 1 in order to conclude the proof of Theorem 1.

Proposition 1 will be proved by symmetrizing P. Below we define this symmetrization and in Lemma 1 we list some useful estimates. After the proof of Lemma 1 we finally complete the proof of Proposition 1. If $\lambda \in \mathbb{R}$, denote by $S^{\lambda,\rho}$ the set of mappings a: $\mathbb{R}_x^N \times \mathbb{R}_\xi^N \to \mathbb{C}^{m\times m}$ such that

$$\|D_\xi^\alpha a(.,\xi)\|_{C_x^\rho} \leq C_\alpha(1+|\xi|)^{\lambda-|\alpha|} \tag{2.15}$$

for all $\alpha \in (\mathbb{Z}^+)^N$ and $\xi \in \mathbb{R}^N$. (Here of course $\| \|_{C_x^\rho}$ means the $C^\rho(\mathbb{R}_x^n)$ norm.) If $\nu \in \mathbb{Z}^+$, we put $N^{\lambda.\rho,\nu}(a)= \sum_{|\alpha|\leq\nu} C_\alpha$, where C_α are the smallest constants such that (2.15) holds. We equip $S^{\lambda,\rho}$ with the topology defined by the norms $N^{\lambda,\rho,\nu}$. Define $\Sigma_{hom}^{\lambda,\rho}=\{a \in S^{\lambda,\rho}$, a is positively homogeneous of degree λ with respect to ξ if $|\xi|$ is large and independent of x if $|x|$ is large$\}$ and endow it with the topology induced by $S^{\lambda,\rho}$.

Write $b(t,x,\xi)=\sum_{j=1}^N \xi_j B_j(x,t)$, $B(t)=i^{-1}\sum_{j=1}^N B_j(x,t)\partial_j$. (where i is the imaginary unit). If $\sigma < \rho$, it is easy to check that $b \in C(\bar{\mathbb{R}}_t^+,\Sigma_{hom}^{0,\sigma}) \cap C^1(\bar{\mathbb{R}}_t^+,\Sigma_{hom}^{0,\sigma-1})$. A standard construction using strict hyperbolicity gives a symmetrizer $r \in C(\bar{\mathbb{R}}_t^+,\Sigma_{hom}^{0,\sigma}) \cap C^1(\bar{\mathbb{R}}_t^+, \Sigma_{hom}^{0,\sigma-1})$ for b; namely $r(t,x,\xi)$ is a real symmetric $m \times m$ matrix for all t,x,ξ, with the following properties :

$$r(t,x,\xi) \geq \delta I \tag{2.16}$$

for some $\delta \in \mathbb{R}^+$ and all (t,x,ξ) with $|\xi| \geq 1$ (here I is the unit $m \times m$ matrix);

$$r(t,x,\xi)b(t,x,\xi)=b^*(t,x,\xi)r(t,x,\xi) \tag{2.17}$$

for all t,x,ξ, where $*$ denotes the adjoint matrix. Write (,) for the usual $L^2(\mathbb{R}_x^N)$ scalar product, $\| \|$ for the corresponding norm, and $\| \|_s$ for the usual $H^s(\mathbb{R}_x^N)$ norm. Fix $1 < \sigma < \rho$. Then choose $1 < \sigma' < \sigma$ and put $R(t)=r(t,x,D_x)$, $Q(t)=(R(t)+R^*(t))/2+C(1+|D_x|^2)^{-\sigma'/2}$, where $*$ means $L^2(\mathbb{R}_x^N)$ adjoint and C is a large constant (actually $C \geq C_4$ where C_4 is defined in Lemma 1(iii) below). Finally write $J(t)=[Q(t),\partial_t]$, $K(t)=Q(t)B(t)-B^*(t)Q(t)$.

Then the following estimates hold (where C_1,C_2,C_3,C_4,C_5 denote various strictly positive constants independent of t (and z)).

<u>Lemma 1.</u> For all $z \in L^2(\mathbb{R}^N_x)$, we have :

(i) $\|R(t)z\| \leq C_1 \|z\|$,

(ii) $\|J(t)z\| \leq C_2 \|z\|$,

(iii) $\mathrm{Re}(R(t)z,z) \geq C_3\|z\|^2 - C_4\|z\|^2_{-\sigma'/2}$,

(iv) $\|K(t)z\| \leq C_5\|z\|$.

<u>Proof of Lemma 1</u>. We shall only prove (iii) and (iv) since (i) and (ii) are well known (and at any rate can be proved with the method we use below for proving (iii) and (iv)). If b was a C^∞ function (or even a C^k function with k large), one could choose r with the same regularity and Lemma 1 would be an almost trivial consequence of the calculus of classical pseudodifferential operators. But here ρ may be close to 1. We shall therefore use Bony's paradifferential calculus [2] (and the slight extension described in [4]). Recall that if $a \in S^{\lambda,\sigma}$, one defines the paradifferential operator T_a by $T_aZ(x) = (2\pi)^{-N}\int e^{ix\xi}a'(x,\xi)\cdot \hat{Z}(\xi)d\xi$, $Z \in C_o^\infty(\mathbb{R}^N)$, where $a'(x,\xi) = \int \check{\chi}(x-y,\xi)a(y,\xi)dy$. Here $\hat{Z}$ is the Fourier transform of Z and $\check{\chi}$ denotes the inverse Fourier transform of χ with respect to the first variable; $\chi(\omega,\xi)$ is a C^∞ function, equal to 1 when $|\omega| \leq \varepsilon_1|\xi|$ and $|\xi| \geq 1$ and to 0 when $|\omega| \geq \varepsilon_2|\xi|$ or $|\xi| < 1/2$ ($\varepsilon_1,\varepsilon_2$ are strictly positive numbers with $\varepsilon_1 < \varepsilon_2$). $\chi(\tilde{\lambda}\omega,\tilde{\lambda}\xi) = \chi(\omega,\xi)$ if $\tilde{\lambda} \geq 1$ and $|\omega|+|\xi| \geq \varepsilon'$, where ε' is a small positive number. We shall need to compare a and T_a (actually only for $a \in \Sigma^{\lambda,\sigma}_{hom}$). Therefore we shall use the following property, which is a direct consequence of the results and methods of [2]:

If $a \in C([0,T],\Sigma^{\lambda,\sigma}_{hom})$ and $0 \leq -\tau \leq \sigma' < \sigma$, one can find $C > 0$ independent of $t \in [0,T]$ such that the estimate

$$\|(a(t,x,D_x)-T_{a(t)})z\|_{\tau+\sigma'} \leq C\,\|z\|_{\tau+\lambda} \tag{2.18}$$

holds for all $z \in H^{\tau+\lambda}(\mathbb{R}^N_x)$. (Using the results of [4], one can see that a similar inequality holds if $a \in C([0,T],S^{\lambda,\sigma})$, but we shall not need this fact).

<u>Proof of (iii)</u>. Let $\ell \in C(\bar{\mathbb{R}}^+_t,\Sigma^{o,\sigma}_{hom})$ be such that the relations $\ell(t) = \ell^*(t)$, $\ell^2(t) = r(t)$ and $\ell(t) \geq \delta^{1/2}I$ hold for $|\xi| \geq 1$. Choose $\chi \in C_o^\infty(\mathbb{R}^N,[0,1])$ such that $\chi(\xi) = 0$ for $|\xi| \leq 2$ and $\chi(\xi) = 1$ for $|\xi| \geq 3$. Put $\tilde{\ell}(t) = \chi\ell + (1-\chi)\delta^{1/2}I$, $\tilde{r}(t) = \tilde{\ell}^2(t)$, $\tilde{R}(t) = \tilde{r}(t,x,D_x)$. We may write

$$\tilde{R}(t)-T^*_{\tilde{\ell}(t)}T_{\tilde{\ell}(t)}=W_1(t)+W_2(t)+W_3(t), \tag{2.19}$$

where $W_1(t)=\tilde{R}(t)-T_{\tilde{r}(t)}$, $W_2(t)=T_{\tilde{\ell}^2(t)}-T^2_{\tilde{\ell}(t)}$, $W_3(t)=(T_{\tilde{\ell}(t)}-T^*_{\tilde{\ell}(t)})T_{\tilde{\ell}(t)}$. But we certainly have

$$|(W_j(t)z,z)| \leq C\|z\|^2_{-\sigma'/2}, \tag{2.20}$$

for j=1 because of (2.18), and for j=2,3 because of the paradifferential calculus ([2]). From the paradifferential calculus it also follows that, in particular,

$$\|z\|^2 \leq C(\|T_{\tilde{\ell}(t)}z\|^2+\|z\|^2_{-\sigma'/2}). \tag{2.21}$$

Since $\tilde{r}(t)=r(t)$ for $|\xi| \geq 3$, (iii) follows readily from (2.19),(2.20),(2.21).

Proof of (iv). It follows the same lines as that of (iii). Put $q(t)=r(t)+\tilde{C}(1+|\xi|^2)^{-\sigma'/2}$ (with the same $\tilde{C}$ as in the definition of Q(t)). We have

$$\begin{aligned}Q(t)B(t)-B^*(t)Q(t)&=(Q(t)-T_{q(t)})B(t)+T_{q(t)}(B(t)-T_{b(t)})+\\&(T_{q(t)}T_{b(t)}-T_{q(t)b(t)})+T_{q(t)b(t)-b^*(t)q(t)}+\\&(T_{b^*(t)q(t)}-T_{b^*(t)}T_{q(t)})+T_{b^*(t)}(T_{q(t)}-Q(t))+(T_{b^*(t)}-B^*(t))Q(t).\end{aligned} \tag{2.22}$$

Using (2.18) and the results of [2],[4] about the calculus of paradifferential operators, it is easy to show that each term of the right-hand side of (2.22) is uniformly bounded in $L^2(R^N_x)$. (iv) follows at once. The proof of Lemma 1 is complete.

From Lemma 1(iii), it follows immediately that

$$(Q(t)z,z) \geq C_3\|z\|^2, \tag{2.23}$$

for all $z \in L^2(R^N_x)$ and all $t \in \bar{R}^+$. Write $M(t)=I+\beta\sum_{j=1}^N \frac{\partial\gamma}{\partial x_j}(.)B_j(.,t)$. If β_o is small enough and $\beta \leq \beta_o$, (2.23) implies that for all $z \in L^2(R^N_x)$:

$$\mathrm{Re}(Q(t)M(t)z,z) \geq C_6\|z\|^2 \tag{2.24}$$

with $C_6 > 0$ independent of t and z.

We are now ready to prove Proposition 1 using the energy integral method. If $g \in C(\bar{D},\mathbb{C}^m)$, we define $\tilde{g}:\mathbb{R}^N_x \times \mathbb{R}^+_t \to \mathbb{C}^m$ by $\tilde{g}=g$ in $\bar{D}$, $\tilde{g}(x,t)=0$ if $(x,t) \notin \bar{D}$. We define Q by $(Qf)(x,t)=Q(t)f(x,t)$ and use the same convention for M and B.

Proof of Proposition 1. Put $f=e^{-\eta h}w$, $P_\eta=e^{-\eta h}Pe^{\eta h}$. Of course $P_\eta=P+\eta M$. Denote by $((\ ,\ ))$ (resp. $((\ ,\ ))_D$) the usual $L^2(\mathbb{R}^N_x \times \mathbb{R}^+_t)$ (resp. $L^2(D)$) scalar product. We may write

$$2\mathrm{Re}((\tilde{f},QP_\eta f))=T_1+T_2, \tag{2.25}$$

where $T_1=2\mathrm{Re}((\tilde{f},Q\widetilde{Pf}))$, $T_2=2\eta\mathrm{Re}((\tilde{f},Q\widetilde{Mf}))$. We have $\widetilde{Mf}=M\tilde{f}$, so (2.24) gives that

$$T_2 \geq C\eta\|f\|^2_D. \tag{2.26}$$

To estimate T_1 we adapt the ideas of [3,Lemma 3.1] to our situation. Choose $\underline{f} \in C^1_o(\mathbb{R}^N_x \times \mathbb{R}_t)$, $\underline{\mu} \in C^1(\mathbb{R}^N_x)$ whose restrictions to D are equal to f, μ respectively and put $f_\varepsilon(x,t)=\zeta(x_N/\varepsilon)\zeta((t-\underline{\mu}(x))/\varepsilon)\underline{f}(x,t)$, where $\varepsilon > 0$ and $\zeta(s) \in C^\infty(\mathbb{R},\mathbb{R})$ is equal to 0 if $s < -1$ and to 1 if $s > -1/2$. To simplify the writing put $S_1=\{(x,t) \in \mathbb{R}^{N+1},\ x_N=0 < t-\mu(x)\}$, $S_2=\{(x,t) \in \mathbb{R}^{N+1},\ t-\mu(x)=0 < x_N\}$. Denote by $d\sigma_j$ the canonical surface measure on S_j and by $n^j=(n^j_1,\dots,n^j_N,n^j_t)$ the unit outer normal to D on S_j. Since $Q=Q^*$, we have

$$((f_\varepsilon,Q\widetilde{Pf}))=((P^*Qf_\varepsilon,\tilde{f}))+\sum_{j=1}^{2}\int_{S_j} Qf_\varepsilon\ .\ \overline{B^jf}\, d\sigma_j\ , \tag{2.27}$$

where in the integral over S_j, . means scalar product of vectors of $\mathbb{C}^m$ and $B^j=n^j_t+\sum_{k=1}^{N}n^j_kB_k$. (In (2.27), $Qf_\varepsilon \in H^1$ and formula (2.27) is easily justified by approximating Qf_ε with C^1 functions). Actually $B^jf=0$ on S_j so the integrals in (2.27) disappear. From (2.27) it then follows that

$$((f_\varepsilon,Q\widetilde{Pf}))+((QPf_\varepsilon,\tilde{f}))=(((J+iK)f_\varepsilon,\tilde{f})), \tag{2.28}$$

where J and K are as in Lemma 1 above. Write $D'=(\mathbb{R}^N_x \times \mathbb{R}^+_t) \setminus D$. Since $B^jf=0$ on S_j, an explicit computation of Pf_ε using the definition of f_ε shows that $\iint_{D'} Pf_\varepsilon.\overline{Q\tilde{f}}\, dx\, dt \to 0$ as $\varepsilon \to 0$. Since $Pf_\varepsilon=Pf$ in D, we have $((Pf_\varepsilon,Q\tilde{f}))_D=((\widetilde{Pf},Q\tilde{f}))$. Hence

$$((QPf_\varepsilon,\tilde{f})) \to ((Q\widetilde{Pf},\tilde{f})) \tag{2.29}$$

as $\varepsilon \to 0$. Now when $\varepsilon \to 0$, we have $f_\varepsilon \to \tilde{f}$ in $L^2(\mathbb{R}^N_x \times \mathbb{R}^+_t)$; it follows that $((Jf_\varepsilon,\tilde{f})) \to ((J\tilde{f},\tilde{f}))$ because of Lemma 1(ii) and that $((Kf_\varepsilon,\tilde{f})) \to ((K\tilde{f},\tilde{f}))$ because of Lemma 1(iv).
Hence if we let $\varepsilon \to 0$ in (2.28) and use (2.29), we obtain

$$2\mathrm{Re}((\tilde{f},Q\widetilde{Pf}))=(((J+iK),\tilde{f},\tilde{f})). \tag{2.30}$$

From (2.30) and Lemma 1(ii) and (iv), it follows that

$$T_1 \geq -C\ \|f\|^2_D. \tag{2.31}$$

Writing $|((\tilde{f},Q\widetilde{P_\eta f}))| \leq (\text{large constant})\dfrac{\|Q\widetilde{P_\eta f}\|^2}{\eta} + (\text{small constant})\cdot \eta\|f\|^2_D$, where $\|\ \|$ is the $L^2(\mathbb{R}^N_x \times \mathbb{R}^+)$ norm, we obtain from (2.25), (2.26),(2.31) that

$$\eta\|f\|_D \leq C\ \|P_\eta f\|_D$$

for η large. If we replace f by $e^{-\eta h}w$, Proposition 1 follows at once. Therefore the proof of Theorem 1 is also complete.

Using Proposition 1 we can also obtain a unique continuation result for weak solutions of (1.1) which are not smooth across at most two prescribed hypersurfaces (Theorem 2 below). Let V,V_o,U,φ_o be as in Section 1 and in the beginning of this section (in particular $\varphi_o(0)=0$). Assume that $\varphi^-,\varphi^+ \in C^1(U \times [0,T])$ satisfy $\varphi^\pm(x',0)=\varphi_o(x')$, $\varphi^-_t(x',t) < \varphi^+_t(x',t)$ if $(x',t) \in U \times [0,T]$. Write $V^\pm=\{(x,t) \in V, x_N \gtrless \varphi^\pm(x',t)$, $V^\#=\{(x,t) \in V,\varphi^-(x',t) < x_N < \varphi^+(x',t)\}$, $S^\pm=\{(x',\varphi^\pm(x',t),t),\ (x',t) \in U \times [0,T]\}$. Let $u,\tilde{u}$ be weak solutions of (1.1) in V and for $*=+,-,\#$, write $u^*=u|_{V^*}$, $\tilde{u}^*=\tilde{u}|_{V^*}$. As before, let $\rho \in \mathbb{R}^+ \setminus \mathbb{Z}^+$ be larger than 1 and assume that $u^*,\tilde{u}^* \in C^\rho(\bar{V}^*)$ for $*=+,-,\#$. Write as before $V^\pm_o=\{(x,t) \in \bar{V}^\pm,t=0\}$. Then we have:

Theorem 2. If $u^\pm=\tilde{u}^\pm$ on $V^\pm_o$ and on $S^\pm$, it follows that $u^*=\tilde{u}^*$ in a neighborhood of 0 in $\bar{V}^*$ if $*=+,-,\#$.

Proof of Theorem 2. It follows immediately from the proof of Theorem 1 (ii) that $u^\pm=\tilde{u}^\pm$ in a neighborhood of 0 in $\bar{V}^\pm$. For $*=+,-,\#$, put $v^*(x,t)=u^*(x',x_N+\varphi^-(x',t),t)$, $\tilde{v}^*(x,t)=\tilde{u}^*(x',x_N+\varphi^-(x',t),t)$, $z^*=v^*-\tilde{v}^*$; define also $\psi(x',t)=\varphi^+(x',t)-\varphi^-(x',t)$. We now argue as we did to obtain (2.10) above. We have $\varphi^-_t(v^\#-\tilde{v}^\#)+\sum_{j=1}^{N-1}\varphi^-_{x_j}(F_j(v^\#)-F_j(\tilde{v}^\#))-(F_N(v^\#)-F_N(\tilde{v}^\#))=0$ when $x_N=0$ and $(x',t) \in U \times [0,T]$, and

and $\varphi_t^+(v^\#-\tilde{v}^\#)+\sum_{j=1}^{N-1}\varphi_{x_j}^+(F_j(v^\#)-F_j(\tilde{v}^\#))-(F_N(v^\#)-F_N(\tilde{v}^\#))=0$ when $x_N=\psi(x',t)$ and $(x',t)\in U\times[0,T]$. Hence if we put $z^\#=v^\#-\tilde{v}^\#$, $\tilde{B}_j^\#=H_j(v^\#,\tilde{v}^\#)$, $1\le j\le N-1$, $\tilde{B}_N^\#=H_N(v^\#,v^\#)-\varphi_t-\sum_{j=1}^{N-1}\varphi_{x_j}H_j(v^\#,v^\#)$, where $H_j(g',g")=\int_0^1 A_j(g"+\theta(g'-g"))d\theta$ as above, we readily obtain that

$$\tilde{B}_N^\# z^\# = 0 \tag{2.32}$$

when $x_N=0$, $(x',t)\in U\times[0,T]$, while

$$(\tilde{B}_N^\#-\psi_t-\sum_{j=1}^{N-1}\psi_{x_j}\tilde{B}_j^\#)z^\#=0 \tag{2.33}$$

when $x_N=\psi(x',t)$, $(x',t)\in U\times[0,T]$. Since $\psi_t>0$, it follows from the implicit function theorem that $\psi(x',t)-x_N=\Phi(x,t)(t-\Psi(x))$ close to 0, where Φ is continuous, $\Psi\in C^1$ and $\Phi>0$. Then (2.33) implies at once that $z^\#-\sum_{j=1}^{N}\Psi_{x_j}\tilde{B}_j^\# z^\#=0$ on $t=\Psi(x)$ in a neighborhood of 0. Hence, cutting $z^\#$ off outside a neighborhood of 0, we may apply Proposition 1 as we did for the proof of Theorem 1, but now with μ replaced by a suitable extension of Ψ. Theorem 2 follows easily.

3. Analytic regularity via unique continuation.

If $E\subset\mathbb{R}^q$, let us say that $f\in A(E)$ if there exist an open neighborhood E' of E in $\mathbb{R}^q$ and a function f' analytic in E' such that $f'|_E=f$. Let $V,V_0,V^\pm,V_0^\pm,U,u,\varphi,S$ be as in the beginning of Section 2. We assume as before that the system of conservation laws in (1.1) is strictly hyperbolic in the direction of dt. The following result is an easy consequence of Theorem 1 above.

<u>Corollary 1</u>. Assume that $F_j\in A(G)$ for $1\le j\le N$, and that $u\in C^\infty(\bar{V}^\pm)$ and $\varphi\in A(U\times[0,T])$ satisfy (2.1), (2.2). Assume also that $u^-|_S\in A(\bar{S})$ Then the following holds :

(i) If $u^+|_{V_0^+}\in A(\bar{V}_0^+)$, then u^+ is analytic in a neighborhood of 0 in $\bar{V}^+$.

(ii) If $u^-|_{V_0^-}\in A(\bar{V}_0^-)$, then u^- is analytic in a neighborhood of 0 in $\bar{V}^-$.

<u>Proof of Corollary 1</u>. For $\varepsilon>0$ small, put $W_\varepsilon^\pm=\{(x,0)\in V_0,\ |x'|<\varepsilon,\ x_N \gtrless \varphi_0(x')\mp\varepsilon\}$. If ε is small enough, one can find $\tilde{u}_0^+\in A(\overline{W_\varepsilon^+})$ such that $\tilde{u}_0^+|_{V_0^+}=u|_{V_0^+}$. Decreasing ε if necessary and using the Cauchy-Kowalewski theorem, one can obtain $\tilde{u}^+$ analytic in an open neighborhood X^+ of 0 in $\mathbb{R}_x^N\times\mathbb{R}_t$ such that

$$\frac{\partial \tilde{u}^+}{\partial t} + \sum_{j=1}^{N} A_j(\tilde{u}^+) \frac{\partial \tilde{u}^+}{\partial x_j} = 0 \tag{3.1}$$

in X^+, and

$$\tilde{u}^+ = \tilde{u}_o^+ \tag{3.2}$$

on W_ε^+. Put $v^+(x,t)=u^+(x',x_N+\varphi(x',t),t)$, $\tilde{v}^+(x,t)=\tilde{u}^+(x',x_N+\varphi(x',t),t)$. Since (2.2) holds, one has

$$\varphi_t(v^+-v^-)+\sum_{j=1}^{N-1} \varphi_{x_j}(F_j(v^+)-F_j(v^-))-(F_N(v^+)-F_N(v))=0 \tag{3.3}$$

in a neighborhood of 0 when $x_N=0 \leq t$. Define $\Lambda(x',t)=(\varphi_t(\tilde{v}^+-v^-)+\sum_{j=1}^{N-1} \varphi_{x_j}(F_j(\tilde{v}^+)-F_j(v^-))-(F_N(\tilde{v}^+)-F_N(v^-)))|_{x_N=0}$; Λ is analytic close to 0 when $t \geq 0$. Now it follows from (3.1) and (3.2) that $\tilde{u}^+$ and u^+ have the same Taylor expansion at 0; hence (3.3) implies that Λ has zero Taylor expansion at 0. By analyticity Λ therefore must vanish in a neighborhood of 0 when $t \geq 0$. Choose $\tilde{u} \in L^\infty(V,G)$ such that $\tilde{u}=u^-$ in V^-, $\tilde{u}=\tilde{u}^+$ in V^+. Decreasing V in necessary, we see that $\tilde{u}$ is a weak solution of (1.1) in V because of the vanishing property of Λ. Since Theorem 1(i) shows that $u^+=\tilde{u}^+$ in a neighborhood of 0 in $\bar{V}^+$, Corollary 1(i) follows immediately.
To prove Corollary 1(ii), we apply once more the Cauchy-Kowalewski theorem to find $\tilde{u}^-$ analytic in a neighborhood X^- of 0 in $\mathbb{R}^{N+1}$, such that

$$\frac{\partial \tilde{u}^-}{\partial t} + \sum_{j=1}^{N} A_j(\tilde{u}^-) \frac{\partial \tilde{u}^-}{\partial x_j} = 0 \tag{3.4}$$

in X^-, and

$$\tilde{u}^- = \tilde{u}_o^- \tag{3.5}$$

in W_ε^-. From (3.4),(3.5), it follows that $\tilde{u}^-$ and u^- have the same Taylor expansion at 0. Hence the analytic functions $\tilde{u}^-|_{\bar{S}}$ and $u^-|_{\bar{S}}$ also have the same Taylor expansion at 0. Therefore $\tilde{u}^-=u^-$ on S close to 0. Theorem 1(ii) then shows that $\tilde{u}^-=u^-$ in a neighborhood of 0 in V , which implies Corollary 1 at once.

As an application of Corollary 1, we consider so called contact discontinuities. Let $\lambda_1(u,\xi) < \lambda_2(u,\xi)<\ldots< \lambda_m(u,\xi)$ be the eigenvalues of $\sum_{j=1}^{N} \xi_j A_j(u)$, $\xi \in \mathbb{R}^N \setminus 0$. (Recall that we assume that

$\frac{\partial}{\partial t} + \sum_{j=1}^{N} A_j(u) \frac{\partial}{\partial x_j}$ is strictly hyperbolic in the direction of dt).

A solution u^+, u^-, φ of (1.3),(1.4) is called a contact discontinuity corresponding to λ_k if the surface $x_N=\varphi(x',t)$ is a characteristic surface of the k-th family with respect to both u^+, u^-, namely

$$\varphi_t(x',t)=\lambda_k(u^{\pm}(x',\varphi(x',t),t),(-\nabla_{x'}\varphi(x',t),1)). \tag{3.6}$$

We have the following result.

<u>Theorem 3</u>. Assume that $u^{\pm} \in C^{\infty}(\bar{V}^{\pm})$ and $\varphi \in C^{\infty}(U \times [0,T])$ are a C^{∞} contact discontinuity in a neighborhood of 0, corresponding to λ_k with k=1 or m. If $F_j \in A(G)$ and $u_o^{\pm} \in A(\bar{V}_o^{\pm})$, $\varphi_o \in A(U)$, then $u^{\pm}$ is analytic in a neighborhood of 0 in $\bar{V}^{\pm}$, and φ is analytic in a neighborhood of 0 in $U \times [0,T]$.

<u>Proof of Theorem 3</u>. We may assume that k=1, since the proof for k=m is similar. Using a change of variables which transforms the characteristic surfaces of $\frac{\partial}{\partial t} + \sum_{j=1}^{N} A_j(u^-)\frac{\partial}{\partial x_j}$ corresponding to λ_1, with initial data t=0, $x_N=\varphi_o(x')$+constant (where |constant| is small), into pieces of hyperplanes x_N=constant, we can easily check that if $\delta > 0$ is small, the following holds: one can find an open neighborhood Δ_δ of 0 in $R_x^N \times R_t$, with diameter $\leq \delta$, such that $D_\delta=\{(x,t) \in \Delta_\delta, x_N < \varphi(x',t), t > 0\}$ is connected and such that for each $(\bar{x},\bar{t}) \in D_\delta$, $(\bar{x},\bar{t})$ is in the domain of influence of V_o^-. This last property means that for each $(\bar{x},\bar{t}) \in D_\delta$, one can find a bounded open subset Σ_o of V_o^- with C^{∞} boundary $\partial\Sigma_o \subset V_o^-$ and a continuous family of C^{∞} functions ψ_μ, $\mu \in [0,1]$, defined on $\bar{\Sigma}_o$, such that $\psi_o \equiv 0$, $\psi_\mu(x) < \psi_{\mu'}(x)$ if $\mu < \mu'$ and $x \in \Sigma_o$, $\psi_\mu|_{\partial\Sigma_o}=0$, and such that the following holds if $\Sigma_\mu=\{(x,\psi_\mu(x)), x \in \Sigma_o\}$:

(i) $\Sigma_\mu \subset D_\delta$ for all $\mu \in]0,1]$;

(ii) Σ_μ is spacelike for $\partial_t+\sum_{j=1}^{N} A_j(u^-)\partial_{x_j}$ for all $\mu \in [0,1]$ (that is all normals to Σ_μ are directions of strict hyperbolicity for $\partial_t+\sum_{j=1}^{N} A_j(u^-)\partial_{x_j}$);

(iii) $(\bar{x},\bar{t}) \in \Sigma_\mu$ for some $\mu \in]0,1[$.

It follows from Theorem 1.2.2 of [1] that u^- is analytic in D_δ. Now define W_ε^- as in the proof of Corollary 1 above and let $\tilde{u}_o^- \in A(W_\varepsilon^-)$ be such that $\tilde{u}_o^-|_{V_o^-}=u_o^-$. By the Cauchy-Kowalewski theorem one can find $\tilde{u}^-$ analytic in a neighborhood X^- of 0 in R^{N+1} such that

$\frac{\partial \tilde{u}^-}{\partial t} + \sum_{j=1}^{N} A_j(\tilde{u}^-)\frac{\partial \tilde{u}^-}{\partial x_j} = 0$ in X^-, $\tilde{u}^-|_{W_\varepsilon^-} = \tilde{u}_o^-$. The uniqueness part of the Cauchy-Kowalewski theorem implies that $\tilde{u}^- = u^-$ in a non empty open subset of D_δ; hence the analytic functions $\tilde{u}^-, u^-$ must be equal in the whole of D_δ (if δ is small enough); so $u^- \in A(\bar{D}_\delta)$. Since $\varphi_o \in A(U)$, it therefore follows from (3.6) and the analyticity of λ_1 that one can find an open neighborhood $\tilde{U}$ of 0 in $\mathbb{R}_x^N$ and a number $\tilde{T} > 0$ such that $\varphi \in A(\bar{\tilde{U}} \times [0,\tilde{T}])$. Define $S = \{(x',\varphi(x',t),t),(x',t) \in \tilde{U} \times]0,\tilde{T}[\}$. If $\tilde{U}$ is small and $\tilde{T}$ is close to 0, we have $u^-|_S = \tilde{u}^-|_S$ and so $u^-|_S \in A(\bar{S})$. From Corollary 1(i) we therefore obtain that u^+ is analytic in a neighborhood of 0 in $\bar{V}^+$. The proof of Theorem 3 is complete.

In the same way one can obtain analyticity results for solutions of (1.1) which present two contact discontinuities associated with the extreme eigenvalues. Let $V, S^\pm, V_o^\pm, V^*$, $* = +,-,\#, \varphi_o$, be as in Theorem 2. Assume that $u^* \in C^\rho(\bar{V}^*)$ for $* = +,-,\#$ and that u is a weak solution in V of a system of type (1.1), strictly hyperbolic in the direction of dt , with $u|_{V^*} = u^*$, $* = +,-,\#$ and $u|_{V_o^\pm} = u_o^\pm \in A(\bar{V}_o^\pm)$. Assume that $\varphi_o \in A(U)$ and that S^-, S^+ are surfaces of contact discontinuity with respect to λ_1, λ_m respectively, that $F_j \in A(G)$ for $1 \le j \le N$, and that the following holds:

The Cauchy problem $\frac{\partial \tilde{u}}{\partial t} + \sum_{j=1}^{N} \frac{\partial}{\partial x_j} F_j(\tilde{u}) = 0$ in V, $\tilde{u}|_{V_o^\pm} = u_o^\pm$ has a weak solution $\tilde{u} \in L^\infty(V,G)$ which presents contact discontinuities across surfaces $\tilde{S}^\mp$ given by equations $x_N = \tilde{\varphi}^\mp(x,t)$, $x' \in U, t \in [0,T]$, with $\tilde{S}^-, \tilde{S}^+$ associated to λ_1, λ_m respectively. Furthermore $\tilde{u}|_{\tilde{V}^*} = \tilde{u}^* \in A(\bar{\tilde{V}}^*)$ for $* = +,-,\#$, where $\tilde{V}^\pm = \{(x,t) \in V, x_N \gtrless \tilde{\varphi}^\pm(x',t)\}$, $\tilde{V}^\# = \{(x,t) \in V,$ $\varphi^-(x',t) < x_N < \tilde{\varphi}^+(x',t)\}$ and $\tilde{\varphi}^\pm = \varphi_o$ for $t=0$. (3.7)

Let $W_\varepsilon^\pm$, $\varepsilon > 0$ small, be as in the proof of Corollary 1. Extend $u_o^\pm$ to $\tilde{u}_o^\pm \in A(\bar{W}_\varepsilon^\pm)$ and denote by $w^\pm$ the solution of the Cauchy problem $\frac{\partial w^\pm}{\partial t} + \sum_{j=1}^{N} A_j(w^\pm)\frac{\partial w^\pm}{\partial x_j} = 0$ in a neighborhood $X^\pm$ of 0 in $\mathbb{R}_x^N \times \mathbb{R}_t$, $w^\pm = u_o^\pm$ in $W_\varepsilon^\pm$. Comparing $u^\pm$ and $\tilde{u}^\pm$ with $w^\pm$ (exactly as we did when we compared the functions called $\tilde{u}^-$ and u^- in the proof of Theorem 3), we easily obtain that $\tilde{V}^* = V^*$ close to 0 for $* = +,-,\#$ and that $\tilde{u}^\pm = u^\pm$ close to 0. From Theorem 2 it follows that $\tilde{u}^\# = u^\#$ close to 0. Hence u^* is analytic in a neighborhood of 0 in $\bar{V}^*$ for $* = +,-,\#$.

When $|u_o^+ - u_o^-|$ is small, Harabetian [6] has obtained general results which as a very special case give sufficient conditions for (3.7) to hold. (For small $|u_o^+ - u_o^-|$ his results can also be applied to avoid the construction of $\tilde{u}$ we made in Corollary 1 above as a step towards the proof of Theorem 3).

REFERENCES.

[1] S.ALINHAC-G.METIVIER, Propagation de l'analyticité des solutions d'équations hyperboliques non linéaires, Invent.Math. 75 (1984) 189-204.

[2] J.M.BONY, Calcul symbolique et propagation des singularités pour les équations aux dérivées partielles non linéaires, Ann.Scient. Ec.Norm.Sup.4e série, 14 (1981), 209-246.

[3] K.O.FRIEDRICHS-P.D.LAX, Boundary value problems for first order operators, Comm.Pure Appl.Math. 18 (1965), 355-388.

[4] P.GODIN, Subelliptic non linear oblique derivative problems, Amer.J.Math. 107 (1985), 591-615.

[5] P.GODIN, Analytic regularity of unformly stable shock fronts with analytic data, preprint.

[6] E.HARABETIAN, Convergent series expansions for hyperbolic systems of conservation laws, Trans.Amer.Math.Soc. 294,2 (1986), 383-424.

[7] A.MAJDA, The stability of multi-dimensional shock fronts, Mem. Amer.Math.Soc. 275 (1983).

[8] A.MAJDA, The existence of multi-dimensional shock fronts, Mem. Amer.Math.Soc. 281 (1983).

[9] A.MAJDA,"Compressible fluid flows and systems of conservation laws in several space variables", Springer, New York, Berlin Heidelberg, Tokyo, 1984.

[10] G.METIVIER, Interaction de deux chocs pour un système de deux lois de conservation, en dimension deux d'espace , to appear in Trans.Amer.Math.Soc.

[11] J.SMOLLER, "Shock waves and reaction-diffusion equations", Springer, New York, Berlin, Heidelberg, Tokyo, 1983.

COMPLEX POWERS OF PSEUDO-DIFFERENTIAL BOUNDARY VALUE PROBLEMS WITH THE TRANSMISSION PROPERTY

Gerd Grubb
Mathematics Department of Copenhagen University
Universitetsparken 5, DK-2100 Copenhagen, Denmark

1. Introduction.

Complex powers of differential operators and of pseudo-differential operators have been studied in numerous works. On one hand, there has been a long development of the theory of fractional powers of the one-dimensional differentiation operator (two different stages of this development are described in Hille-Phillips [17] and Oldham-Spanier [23]), and on the other hand, the advent of pseudo-differential operators has allowed a fine analysis of powers of multi-dimensional elliptic and hypoelliptic operators. Here the complex powers A^z can of course be *defined* via the spectral resolution when A is a selfadjoint positive operator in a Hilbert space, and they can be defined more generally by use of the Cauchy integral formula

$$A^z = \frac{i}{2\pi} \int_C \lambda^z (A-\lambda)^{-1} d\lambda \ , \tag{1.1}$$

when the resolvent $(A-\lambda)^{-1}$ exists for λ in a suitable region of the complex plane and satisfies appropriate estimates on C, where C is a curve in $\mathbb{C}$ going around the spectrum in the positive direction. However, the point is not merely to make these definitions possible, but rather to *analyse* the resulting operators, to find their detailed structure and special properties in terms of differential and integral operator calculi. - It is also of interest to study other functions of the operator, such as the exponential function $\exp(-tA)$ and other analytic functions, or C^∞ functions in general.

For operators on manifolds without boundary, there are many studies, beginning with Seeley [28], who showed that when A is a classical pseudo-differential operator (with a certain ellipticity property) then so is A^z for any $z \in \mathbb{C}$. This was followed by extensions by Nagase-Shinkai [22], Hayakawa-Kumanogo [15], Strichartz [31], Dunau [5], Robert [26], Helffer-Robert [16], Widom [32], Iwasaki-Iwasaki [20],... (where the tendency of the results is that $f(A)$ is more general, the more general A or f is).

Much less has been said about operators on manifolds with boundary, where the boundary conditions give special complications. Here the complex powers of elliptic differential operator realizations were studied in Seeley [29] and later Laptev [21], whereas pseudo-differential generalizations were considered in Grubb [10] (sketching a treatment of Boutet de Monvel [3] operators, based on [9]) and Rempel-Schulze [25] (treating a larger class than that of [3], with less assumptions and conclusions on the x_n-behavior). The exponential function ("heat operator") was considered in Greiner [7] for the differential operator case and in Grubb [12] for pseudo-differential boundary problems, and Widom [33] treated certain truncated ps.d.o.s of order ≤ 0. Cordes [4] worked out an abstract approach.

The basic step in most of these calculi is to analyse the *resolvent* $(A-\lambda)^{-1}$; then the other functions are developed from that. Here a marked difference between the differential operator and the pseudo-differential operator cases is felt, namely that in the differential operator case the symbols are polynomial and the spectral parameter can be included as just another polynomial variable, whereas in the ps.d.o. case the homogeneous symbols generally have a certain non-smoothness at zero that gives trouble when the parameter λ is adjoined. For the latter case, we introduce the so-called "regularity number" ν, which measures the irregularity (roughly speaking, it indicates the amount of Hölder continuity the homogeneous principal symbols have at the zero section of the cotangent bundle). It plays a decisive rôle in the description of the symbol properties one obtains in the functional calculus.

The book [14], which is about to appear, gives an extensive account of the symbolic calculus needed for the resolvent construction (for pseudo-differential boundary problems having the transmission property), as a special case of a more general theory; and it describes with full details the development of exponential functions and power functions from this theory, as well as many other applications.

In the present paper, we explain the steps in the construction in plain words, aiming directly for the complex powers. This should be useful for the reader who is particularly interested in that subject, and it may in general serve as an illumination and introduction to the technicalities of the general theory. Section 2 explains the basic ingredients in the calculus of pseudo-differential boundary operators of [3], Section 3 contains the hypotheses and conclusions for the resolvent construction, and Section 4 goes through a variety of results for the complex powers.

In this paper, we moreover place a special emphasis on the Fourier integral operator structure (conormality) of the boundary terms (in relation to [19, Ch.18]); in particular we analyse this for the complex powers, going beyond [14].

The notation we use is as in standard texts on differential and pseudo-differential operators. Let us just mention that when $d \geq 0$ and $N \geq 1$ are integers, $H^d(\overline{\Omega})^N$ stands for the Sobolev space of N-vector valued functions u on $\overline{\Omega}$, whose derivatives up to order d are in $L^2(\overline{\Omega})^N$, with norm $\|u\|_d = (\Sigma_{|\alpha|\leq d} \|D^\alpha u\|^2_{L^2})^{\frac{1}{2}}$; and there are generalizations of this to the cases where $d \in \mathbb{R}$, and where the functions are replaced by sections in a vector bundle E over $\overline{\Omega}$; the space is then denoted $H^d(E)$.

2. Pseudo-differential boundary operators.

For some $n \geq 1$, we consider a compact n-dimensional C^∞ manifold $\overline{\Omega}$ with boundary $\partial\Omega$ and interior Ω. The operators we study act in spaces of C^∞ functions on $\overline{\Omega}$ or $\partial\Omega$ (possibly vector valued or sections in bundles), and they may be extended to suitable Sobolev spaces or distribution spaces over $\overline{\Omega}$ and $\partial\Omega$.

The operators belong to the Boutet de Monvel calculus [3] of pseudo-differential boundary operators, whose ingredients are the following:

1) Classical pseudo-differential operators (ps.d.o.s) P of integer order d defined on an open n-dimensional manifold Σ extending $\overline{\Omega}$; it is assumed that P has the transmission property [2] at $\partial\Omega$, which assures that the "restriction" of P to Ω

$$(2.1) \qquad P_\Omega u = r^+Pe^+u\,, \qquad u \in C^\infty(\overline{\Omega})\,,$$

(where e^+ extends by setting the function equal to zero on $\Sigma\setminus\overline{\Omega}$, and r^+ restricts back to Ω) maps $C^\infty(\overline{\Omega})$ into $C^\infty(\overline{\Omega})$.

2) Classical ps.d.o.s S of order $d \in \mathbb{R}$ acting in $C^\infty(\partial\Omega)$.

3) Trace operators T of order $d \in \mathbb{R}$ going from $C^\infty(\overline{\Omega})$ to $C^\infty(\partial\Omega)$, generally of the form (for some integer $r \geq 0$)

$$(2.2) \qquad T = \sum_{0\leq j<r} S_j\gamma_j + T'\,,$$

where the S_j are ps.d.o.s in $\partial\Omega$ of order $d-j$, γ_j is the j-th normal derivative

$$(2.3) \qquad \gamma_j : u(x',x_n) \sim (-i\partial/\partial x_n)^j u(x',0)$$

(here x' is a coordinate in $\partial\Omega$ and x_n is a normal coordinate), and T' is a certain kind of Fourier integral operator continuous from $L^2(\Omega)$ into $H^{-d-\frac{1}{2}}(\partial\Omega)$; here r is called the *class* of T, and T' has class 0.

4) Poisson operators K of order $d \in \mathbb{R}$ going from $C^\infty(\partial\Omega)$ to $C^\infty(\overline{\Omega})$; they can be characterized as the adjoints of the trace operators T' of class 0 and order $d-1$ (this strange order convention assures that the composition TK of a trace operator T of order d and a Poisson operator K of order d' is a ps.d.o. on $\partial\Omega$ of order $d+d'$).

5) Singular Green operators (s.g.o.s) G of order $d \in \mathbb{R}$ acting in $C^\infty(\overline{\Omega})$, generally of the form (for some integer $r \geq 0$)

$$(2.4) \qquad G = \sum_{0\leq j<r} K_j\gamma_j + G'\,,$$

where the K_j are Poisson operators of order $d-j$ and G' is a certain kind of Fourier integral operator continuous from $L^2(\Omega)$ to $H^{-d}(\overline{\Omega})$.

Altogether, these operators form systems

$$(2.5)\qquad A = \begin{pmatrix} P_\Omega + G & K \\ & \\ T & S \end{pmatrix} : \begin{matrix} C^\infty(\overline{\Omega})^N \\ \times \\ C^\infty(\partial\Omega)^M \end{matrix} \to \begin{matrix} C^\infty(\overline{\Omega})^{N'} \\ \times \\ C^\infty(\partial\Omega)^{M'} \end{matrix} ,$$

also called Green operators. (Here G is a singular Green operator, and it is sometimes advantageous to view also T, K and S as "singular" Green operators). The important point is now that these systems form an *algebra*, in the sense that the composition of two such systems (with matching vector dimensions) gives a third one, see Boutet de Monvel [2,3], Grubb [13,14], Rempel-Schulze [24]. Moreover, the composition rules are reflected in an associated symbolic calculus.

We now give some simple examples to illustrate the terminology.

Example 2.1. A differential operator P_0 with C^∞ coefficients on Σ is in particular a ps.d.o. having the transmission property at $\partial\Omega$. A differential trace operator $T = \gamma_0 P_0$ (which can also be written as $\sum_{0\leq j\leq d} S_j\gamma_j$ when P_0 is of order d, the S_j being differential operators on $\partial\Omega$) belongs to the trace operators we consider; here $T' = 0$. A system defining an elliptic boundary value problem

$$(2.6)\qquad A = \begin{pmatrix} P_\Omega \\ \\ T \end{pmatrix} : C^\infty(\overline{\Omega})^N \to \begin{matrix} C^\infty(\overline{\Omega})^N \\ \times \\ C^\infty(\partial\Omega)^M \end{matrix} ,$$

where P is an elliptic differential operator on Σ of order d, and T is a differential trace operator (generally vector valued, when $N > 1$ or $d > 2$), has a parametrix, belonging to the calculus, of the form

$$(2.7)\qquad \mathcal{B} = ((P^{-1})_\Omega + G \quad K) ,$$

where

$$(2.8)\qquad (P^{-1})_\Omega + G \text{ solves the problem } \begin{matrix} P_\Omega u = f , \\ Tu = 0 , \end{matrix}$$

and

$$(2.9)\qquad K \text{ solves the problem } \begin{matrix} P_\Omega u = 0 , \\ Tu = \varphi ; \end{matrix}$$

here P^{-1} is a pseudo-differential parametrix of P on Σ, G is a correction term adapted to the particular boundary condition (and $(P^{-1})_\Omega + G$ is often called the Green operator for the problem (2.8), G being the "singular" part); and K is a generalization of the usual Poisson operator solving the problem where $P = -\Delta$,

$T = \gamma_0$.

Example 2.2. Consider a matrix-formed problem

$$\begin{pmatrix} P_{11} & P_{12} \\ P_{21} & P_{22} \end{pmatrix}_\Omega \begin{pmatrix} u_1 \\ u_2 \end{pmatrix} = \begin{pmatrix} g_1 \\ g_2 \end{pmatrix} \quad \text{in } \Omega ,$$

(2.10)

$$\begin{pmatrix} T_{11} & T_{12} \\ T_{21} & T_{22} \end{pmatrix} \begin{pmatrix} u_1 \\ u_2 \end{pmatrix} = \begin{pmatrix} \psi_1 \\ \psi_2 \end{pmatrix} \quad \text{at } \partial\Omega ,$$

where the P_{ij} are differential operators and the T_{ij} are differential trace operators. If the system $\{P_{11,\Omega}, T_{11}\}$ is elliptic with inverse $(R_1 \quad K_1)$ (where $R_1 = (P_{11}^{-1})_\Omega + G_1)$, then we can solve the first rows for u_1 and insert in the second rows, reducing (2.10) to a problem for u_2 of the form

$$(2.11) \qquad \begin{aligned} (P_\Omega + G)u &= f \quad \text{in } \Omega , \\ Tu &= \varphi \quad \text{at } \partial\Omega , \end{aligned}$$

where $P = P_{22} - P_{21}\, P_{11}^{-1}\, P_{12}$, and G is a singular Green operator absorbing the contributions from G_1, from truncation (the replacement of $Q_\Omega Q'_\Omega$ by $(QQ')_\Omega$) and from K_1 and the trace operators; the composition rules assure that all these effects are of s.g.o. type. Also T is a pseudo-differential trace operator (generally with nontrivial T' as in (2.2)). Here, even if s.g.o. terms did not enter in the original problem, they certainly do so after the above manipulations.

Let us now explain in more detail what the operators look like. Since one can show that the operator classes are invariant under coordinate changes preserving the boundary, we need only describe the case where $\overline{\Omega}$ is replaced by $\overline{\mathbb{R}}^n_+ = \{x \in \mathbb{R}^n \mid x_n \geq 0\}$, with interior $\mathbb{R}^n_+$ and boundary $\mathbb{R}^{n-1}$; here $\mathbb{R}^n$ plays the rôle of Σ. Generally, we use this as local coordinates. The coordinate in $\mathbb{R}^{n-1}$ is denoted x'.

The *pseudo-differential operators* have the well-known description

$$(2.12) \qquad (Pu)(x) = (2\pi)^{-n} \int_{\mathbb{R}^{2n}} e^{i(x-y)\cdot\xi}\, p(x,\xi)u(y)dyd\xi \equiv OP(p)u(x) ,$$

where p is the symbol of P, assumed to lie in $S^d_{1,0}(\mathbb{R}^n, \mathbb{R}^n)$, i.e. for any $\alpha,\beta \in \mathbb{N}^n$,

$$(2.13) \qquad |D^\beta_x D^\alpha_\xi\, p(x,\xi)| \leq c(x)\, \langle\xi\rangle^{d-|\alpha|} .$$

(The function $c(x)$ always denotes a continuous function of x depending on the indices, $\langle\xi\rangle$ stands for $(1+|\xi|^2)^{\frac{1}{2}}$, and $\mathbb{N}$ denotes the nonnegative integers.)

P and p are moreover said to be *polyhomogeneous* when there is an asymptotic expansion

$$(2.14)\qquad p(x,\xi) \sim \sum_{\ell\in\mathbb{N}} p_{d-\ell}(x,\xi)$$

in C^∞ terms $p_{d-\ell}$ that are homogeneous in ξ of degree $d-\ell$ for $|\xi| \geq 1$, (2.14) being valid in the sense that $p - \Sigma_{\ell<M}\, p_{d-\ell} \in S^{d-M}_{1,0}$ for any $M \in \mathbb{N}$. Then p_d is called the principal symbol, also denoted p^0. The *transmission property* means that at $x_n = 0$, the inverse Fourier transforms in ξ_n of the symbol and its derivatives,

$$(2.15)\qquad \tilde{p}_{\alpha,\beta}(x',z_n,\xi') = F^{-1}_{\xi_n\to z_n}\, D^\beta_x D^\alpha_\xi\, p(x',0,\xi',\xi_n)\ ,$$

are C^∞ for $z_n \to 0+$ and for $z_n \to 0-$ (this does not exclude that $\tilde{p}$ contains distributions supported by $\{z_n = 0\}$, differential operator terms). For polyhomogeneous symbols, the transmission property may equivalently be formulated as the symmetry property (for all indices)

$$(2.16)\qquad D^\beta_x D^\alpha_\xi\, p_{d-\ell}(x,\xi) = e^{i\pi(d-\ell-|\alpha|)} D^\beta_x D^\alpha_\xi\, p_{d-\ell}(x,-\xi)\ , \quad \text{for } x_n = 0 \text{ and } \xi' = 0\,(|\xi|\geq 1).$$

Next, consider *singular Green operators of class* 0 (the term G' in (2.4)). A s.g.o. G of class 0 can be described by a formula

$$(2.17)\qquad (Gu)(x) = (2\pi)^{-n+1} \int_{\mathbb{R}^{2n-2}} e^{i(x'-y')\cdot\xi'}\, dy'd\xi' \int_0^\infty \tilde{g}(x',x_n,y_n,\xi')u(y)dy_n$$

$$\equiv \mathrm{OPG}(\tilde{g})u(x)\ , \qquad \text{for } u \in \mathscr{S}(\overline{\mathbb{R}}^n_+)[\equiv r^+\mathscr{S}(\mathbb{R}^n)]\ ,$$

where $\tilde{g}$ (the so-called *symbol-kernel* of G) is a C^∞ function of $(x',\xi') \in \mathbb{R}^{2n-2}$ and $(x_n,y_n) \in \overline{\mathbb{R}}_+ \times \overline{\mathbb{R}}_+$ that is rapidly decreasing in x_n, y_n. In fact it satisfies estimates

$$(2.18)\qquad |||\,\tilde{g}\,|||_{k,k',m,m',\alpha,\beta} \equiv \| x_n^k D_{x_n}^{k'} y_n^m D_{y_n}^{m'} D^\beta_{x'} D^\alpha_{\xi'}\, \tilde{g}(x',x_n,y_n,\xi')\|_{L^2_{x_n,y_n}}$$

$$\leq c(x')\langle\xi'\rangle^{d-k+k'-m+m'-|\alpha|} \qquad \text{for } |\xi'| \geq 1\ ,$$

for all indices $k,k',m,m' \in \mathbb{N}$ and $\alpha,\beta \in \mathbb{N}^{n-1}$. Symbol-kernels satisfying (2.18) are said to be of $S_{1,0}$ type, and they are moreover said to be *polyhomogeneous* if there is an expansion

$$(2.19)\qquad \tilde{g}(x',x_n,y_n,\xi') \sim \sum_{\ell\in\mathbb{N}} \tilde{g}_{d-1-\ell}(x',x_n,y_n,\xi')$$

where $\tilde{g} - \Sigma_{\ell<M}\, \tilde{g}_{d-1-\ell}$ satisfies the estimates (2.18) with d replaced by d-M, for any $M \in \mathbb{N}$, and the $\tilde{g}_{d-1-\ell}$ are C^∞ functions with the quasi-homogeneity

property

$$(2.20)\qquad \tilde g_{d-1-\ell}(x',x_n/s,y_n/s,s\xi') = s^{d+1-\ell}\,\tilde g_{d-1-\ell}(x',x_n,y_n,\xi')$$

$$\text{for } |\xi'| \geq 1,\ s \geq 1 .$$

The quasi-homogeneity is easier to accept when one considers the associated *symbol*

$$(2.21)\qquad g(x',\xi',\xi_n,\eta_n) = F_{x_n\to\xi_n}\,\overline{F}_{y_n\to\eta_n}\,[e^+_{x_n} e^+_{y_n} \tilde g(x',x_n,y_n,\xi')] \quad \text{(cf. (2.1)ff.)}$$

and it terms $g_{d-1-\ell}$; here (2.20) means precisely that $g_{d-1-\ell}$ is homogeneous in (ξ',ξ_n,η_n) of degree $d-1-\ell$ for $|\xi'| \geq 1$. Here g_{d-1} resp. $\tilde g_{d-1}$ (also denoted g^0 resp. $\tilde g^0$) are called the principal symbol resp. symbol-kernel.

There is another description of G that shows the invariant character more clearly, namely by reference to conormal distributions, as defined in Hörmander [19, Section 18.2]. First, it is possible to extend the symbol-kernel $\tilde g(x',x_n,y_n,\xi')$ to a function $\tilde g_1(x',x_n,y_n,\xi')$ of $(x',\xi') \in \mathbb{R}^{2n-2}$ and all $(x_n,y_n) \in \mathbb{R}^2$, so that estimates like (2.18) hold for the $L^2_{x_n,y_n}(\mathbb{R}^2)$-norm, by the method of Seeley [27]. (We note however, that this method does not preserve polyhomogeneity, because of the use of cut-off functions.) Then if we define the operator G_1 on $\mathbb{R}^n$ by a formula like (2.17) but with the integration in y_n extended to $\mathbb{R}$, we have that G is simply its restriction to $\mathbb{R}^n_+$:

$$(2.22)\qquad Gu = r^+G_1e^+u \qquad \text{for } u \in \mathscr{S}(\overline{\mathbb{R}}^n_+) .$$

Defining the associated *symbol* g_1 by $g_1(x',\xi',\xi_n,\eta_n) = F_{x_n\to\xi_n}\,\overline{F}_{y_n\to\eta_n}\,\tilde g_1(x',x_n,y_n,\xi')$, we can also express G_1 by the formula

$$(2.23)\qquad (G_1u)(x) = (2\pi)^{-n-1}\int_{\mathbb{R}^{2n+1}} e^{i(x'-y')\cdot\xi'+ix_n\xi_n-iy_n\eta_n} g_1(x',\xi',\xi_n,\eta_n)u(y)\,dy\,d\xi'\,d\xi_n\,d\eta_n .$$

Now the family of estimates like (2.18) for $\tilde g_1$ implies by use of the Parseval-Plancherel theorem and the standard estimate

$$(2.24)\qquad \sup_t|\varphi(t)|^2 \leq 2\,\|\varphi\|_{L^2}\,\|D_t\varphi\|_{L^2}$$

for functions on $\mathbb{R}$, that g_1 satisfies

$$(2.25)\qquad \begin{aligned}&|D^\beta_{x'}D^\alpha_{\xi'}D^k_{\xi_n}D^m_{\eta_n}\,g_1(x',\xi',\xi_n,\eta_n)| \leq \\ &\qquad c(x')\langle\xi'\rangle^{d-1-|\alpha|-k-m}\left(\frac{1+|\xi'|}{1+|\xi'|+|\xi_n|}\right)^{k'}\left(\frac{1+|\xi'|}{1+|\xi'|+|\eta_n|}\right)^{m'},\end{aligned}$$

for all indices, and then moreover

$$|D^{\beta}_{x'} D^{\theta}_{\xi', \xi_n, \eta_n} g_1| \leq c(x') \langle \xi' \rangle^{d-1-|\theta|} \left(\frac{1+|\xi'|}{1+|\xi'|+|\xi_n|+|\eta_n|} \right)^N ,$$

(2.26)

$$\text{for all } \beta \in \mathbb{N}^{n-1}, \ \theta \in \mathbb{N}^{n+1} \ \text{ and } \ N \in \mathbb{N} .$$

The system of estimates (2.26) is in fact equivalent with the system (2.25), and with the system (2.18) for $\tilde{g}_1$. For the passage from (2.25) to (2.26) one sets $k' = m' = N$ and uses that

$$\frac{1+|\xi'|}{1+|\xi'|+|\xi_n|} \, \frac{1+|\xi'|}{1+|\xi'|+|\eta_n|} \leq \frac{1+|\xi'|}{1+|\xi'|+|\xi_n|+|\eta_n|} . \tag{2.27}$$

(We take the opportunity to remark that one should use this argument instead of the one indicated in [14, Remark 2.4.12].)

This shows two properties of g_1:

(a) g_1 is a symbol in $S^{d-1}_{1,0}(\mathbb{R}^{n-1}, \mathbb{R}^{n+1})$ (i.e., satisfies estimates (2.13) with x resp. ξ replaced by x' resp. (ξ', ξ_n, η_n));

(b) g_1 is rapidly decreasing for $|\xi_n|$ and $|\eta_n| \to \infty$.

Conversely, (a) and (b) imply (2.26).

Now the operators of the form (2.23) whose symbols satisfy (a) (for some order d) are precisely the operators G_1 on $\mathbb{R}^n$ whose Schwartz kernel K_{G_1}, which is a distribution on

$$X = \mathbb{R}^n \times \mathbb{R}^n , \tag{2.28}$$

is *conormal* with respect to the submanifold

$$Y = \{(x,y) \in X \mid x' = y', \ x_n = 0, \ y_n = 0\} , \tag{2.29}$$

cf. Hörmander [19, Section 18.2]. And, as shown there, the property of being a conormal distribution with respect to $Y \subset X$ is invariant under coordinate changes in X preserving Y, and there are explicit rules for how the symbol is modified under such coordinate changes. And, the coordinate changes in X preserving Y, applied to the kernel K_{G_1}, correspond precisely to the transformation of G_1 under coordinate changes in $\mathbb{R}^n$ preserving $\{x_n = 0\}$ (applied to u and $G_1 u$ in (2.23)).

One furthermore has that the property (b) can be expressed invariantly as the rapid decrease of g_1 along $N(Y) \cap N(Z)$, where

$$Z = \{(x,y) \in X \mid x_n = 0, \ y_n = 0\} , \tag{2.30}$$

and $N(Y)$ resp. $N(Z)$ denote the normal bundles of Y resp. Z in X. Translating all this back to the manifold situation we find:

Proposition 2.3. *The singular Green operators* G *of class* 0 *on* $\bar{\Omega}$ *extend to operators* G_1 *on* Σ, *that are characterized as the operators whose Schwartz kernels* K_{G_1} *are distributions on* $X = \Sigma \times \Sigma$ *which are conormal with respect to* $Y = \mathrm{diag}(\partial\Omega \times \partial\Omega)$, *and whose symbols are rapidly decreasing along the intersection of the normal bundles* $N(Y) \cap N(Z)$, *where* $Z = \partial\Omega \times \partial\Omega$. *The class of such operators* G_1 *is defined independently of the coordinate changes in* Σ *preserving* $\partial\Omega$.

Further details are given in Grubb [14, Section 2.4], where it is also shown by direct calculation how the space of s.g.o.s G themselves is preserved, as well as the polyhomogeneous subspace. The *Poisson operators* K and the *trace operators* T' of class 0 have very similar characterizations, where X is replaced by $X_1 = \Sigma \times \partial\Omega$ resp. $X_2 = \partial\Omega \times \Sigma$, and Y and Z are restricted accordingly. (Thanks are due to L. Hörmander for helping us clear up these questions.)

3. Functional calculus and the resolvent.

The calculus, we are interested in, goes a step further than the mere "algebra" point of view, in that we want to study *functions* of the operators. More specifically, if B is the *realization in* $L^2(\Omega)$ associated with an elliptic system $\{P_\Omega + G, T\}$ with $P_\Omega + G$ of order $d \in \mathbb{N}$:

$$\text{(3.1)} \qquad Bu = (P_\Omega + G)u \,, \quad D(B) = \{u \in H^d(\bar{\Omega}) \mid Tu = 0\} \,,$$

then it is of interest to study the functions $f(B)$ of B (as bounded or unbounded operators in $L^2(\Omega)$) defined by a suitable functional calculus. One can also study functions $f(A)$ of the full systems of the form (2.5), provided that $N = N'$ and $M = M'$ (we find this less central than the questions for operators close to the differential operator case). An important example is the exponential function $\exp(-tB)$ (the "heat operator"), that solves the *evolution* problem for functions $u(x,t)$:

$$\text{(3.2)} \qquad \begin{aligned} \frac{\partial u}{\partial t} + P_\Omega u + Gu &= 0 \quad &&\text{for } x \in \Omega,\ t > 0 \,, \\ Tu &= 0 \quad &&\text{for } x \in \partial\Omega,\ t > 0 \,, \\ u|_{t=0} &= u_0 \quad &&\text{for } x \in \Omega \,; \end{aligned}$$

and $\exp(-tA)$ likewise solves the initial value problems for $\frac{\partial u}{\partial t} + Au = 0$. Another example is the complex power function B^z (resp. A^z) that is of interest for interpolation theory, for spectral theory and for many other purposes.

As indicated in the Introduction, we define such functions via the Cauchy integral formula

(3.3) $$f(B) = \frac{i}{2\pi}\int_C f(\lambda)(B-\lambda)^{-1}d\lambda ,$$

for which the study of the resolvent

(3.4) $$R_\lambda = (B-\lambda)^{-1}$$

(resp. $(A-\lambda)^{-1}$) is essential.

Since the methods for the operators B and A are quite similar, we concentrate on B. Generally, R_λ is found on the form

(3.5) $$R_\lambda = (P-\lambda)_\Omega^{-1} + G_\lambda ,$$

where $(P-\lambda)^{-1}$ is a parametrix of $P-\lambda$ on Σ (one can often obtain a true inverse, taking Σ compact) and G_λ is a λ-dependent family of s.g.o.s. Insertion in the Cauchy formula (3.3) then gives

(3.6) $$f(B) = \frac{i}{2\pi}\int_C f(\lambda)(P-\lambda)_\Omega^{-1}d\lambda + \frac{i}{2\pi}\int_C f(\lambda)G_\lambda d\lambda$$
$$= [f(P)]_\Omega + G_f ,$$

where $f(P)$ is the corresponding function of P (again of a pseudo-differential nature), and G_f, the contribution from G_λ, has an s.g.o.-like structure.

In the analysis of these operators, $f(P)$ is quite well known. Let us first mention the study of Seeley [28], where it is shown that when P is polyhomogeneous and satisfies a parameter-ellipticity condition for $\lambda \in \overline{\mathbb{R}}_-$, then the complex powers P^z exist for all $z \in \mathbb{C}$ and are again polyhomogeneous. Nagase-Shinkai [22] and Hayakawa-Kumanogo [15] extended this study to more general symbols, Strichartz [31] and Dunau [5] gave methods to include more general functions $f(t)$ (with f of $S_{1,0}$ type as a function of t, cf. (2.13)), and Widom [32] and Iwasaki-Iwasaki [20] analysed the exponential function $\exp(-tP)$ (see also Widom [33] for a study of $\exp(-tP_\Omega)$ under certain hypotheses at the boundary). There are moreover the studies of Robert [26] and Helffer-Robert [16],$\cdots$, concerning $f(P)$ when P is a ps.d.o. on $\mathbb{R}^n$ with certain global properties. Results on $[f(P)]_\Omega$ can be deduced from these results. (See also Schrohe [34].)

We shall then concentrate on G_f, on which much less has been said. For the differential operator case, Seeley [29] and later Laptev [21] obtained results on the power function, and Greiner [7] and others analysed the exponential function. For pseudo-differential problems having the transmission property, summaries of our studies (on resolvents, power functions and exponential functions) were published in [10,11,12] and full details are given in [14]; and cases without the transmission property were treated in Rempel-Schulze [25]. A Banach algebra point of view was introduced by Cordes [4].

The studies including boundary terms have so far mainly been concerned with specific analytic functions. As for C^∞ functions f, the method of Strichartz [31] for $S_{1,0}$ functions requires a certain commutativity, that the boundary contributions do not fulfill. The method of Helffer-Robert [16] takes a different path: using some very precise estimates of the complex powers P^{-s+it} (for $t \in \mathbb{R}$, s fixed > 0) one can define $f(P)$ for $S_{1,0}$ functions $f(t)$ supported in $]0,\infty[$ via the Mellin transform, when P is a selfadjoint positive ps.d.o. It would be interesting to apply this method to realizations B as well; this would require a further development of the resolvent estimates that we have obtained up to now (one needs a better knowledge of the behavior of G_λ in (3.5) for $\operatorname{Im}\lambda \to 0$, $\operatorname{Re}\lambda > 0$).

Let us now give an account of our present results (explained in all details in [14]), for the resolvent of B.

We consider a system

$$(3.7) \qquad A = \begin{pmatrix} P_\Omega + G \\ \\ T \end{pmatrix} : C^\infty(E) \quad \to \quad \begin{matrix} C^\infty(E) \\ \times \\ C^\infty(F) \end{matrix}$$

with the following ingredients:

- E is an N-dimensional C^∞ vector bundle over $\bar\Omega$, $N \geq 1$, and $C^\infty(E)$ denotes the space of C^∞ sections in E (locally, they are like functions in $C^\infty(\bar\Omega)^N$).
- P is a pseudo-differential operator of integer order $d \geq 0$ defined in a bundle $\tilde E$ extending E to a neighboring manifold Σ, and P has the transmission property at $\partial\Omega$.
- G is a singular Green operator in E of order d, its class r being $\leq d$ (cf. (2.4)).
- F is a direct sum $F = F_0 \oplus \cdots \oplus F_{d-1}$ of C^∞ vector bundles F_k over $\partial\Omega$ of dimensions $M_k \in [0,N]$ (the zero-dimensional bundles are included for notational convenience), and $M = M_0 + \cdots + M_{d-1}$; here $C^\infty(F) = C^\infty(F_0) \times \cdots \times C^\infty(F_{d-1})$ where $C^\infty(F_k)$ is locally like $C^\infty(\partial\Omega)^{M_k}$.
- T is a column vector of trace operators $T = \{T_k\}_{0\leq k\leq d-1}$, where T_k goes from $C^\infty(E)$ to $C^\infty(F_k)$, is of order k, and is of the form

$$(3.8) \qquad T_k = \sum_{0\leq j\leq k} S_{kj}\gamma_j + T'_k$$

(in particular, it is of class $k+1$, cf. (2.2)), the top coefficient S_{kk} being a morphism from $E|_{\partial\Omega}$ to F_k (locally it is an $M_k{\times}N$-matrix valued *function* $S_{kk}(x')$ on $\partial\Omega$).

With A is accociated the system of symbols: $p(x,\xi)$ (given in local coordinates over Σ), $g(x',\xi',\xi_n,\eta_n)$ and $t(x',\xi) = \{t_k(x',\xi)\}_{0\leq k\leq d-1}$ (given in local coordinates over $\partial\Omega$). We assume that the operators are polyhomogeneous, and we denote by p^h, g^h resp. t^h and t_k^h the functions coinciding with the principal symbols for $|\xi'| \geq 1$ and extended to be strictly homogeneous in the full region where $\xi' \neq 0$. At each x', ξ' (with $\xi' \neq 0$) they define the *boundary symbol operator*, i.e. the "model" operator on $\mathbb{R}_+$ obtained by applying the operator definitions with respect to the x_n-variable only:

$$(3.9) \quad a^h(x',\xi',D_n) = \begin{pmatrix} p^h(x',0,\xi',D_n)_{\mathbb{R}_+} + g^h(x',\xi',D_n) \\ t^h(x',\xi',D_n) \end{pmatrix} : \mathcal{S}(\overline{\mathbb{R}}_+)^N \to \begin{matrix} \mathcal{S}(\overline{\mathbb{R}}_+)^N \\ \times \\ \mathbb{C}^M \end{matrix} ,$$

it extends to a continuous operator from $H^d(\overline{\mathbb{R}}_+)^N$ to $L^2(\mathbb{R}_+)^N \times \mathbb{C}^M$.

Definition 3.1. *Let* $\theta \in [0,2\pi[$. *The system* $A_\lambda = \{P_\Omega + G - \lambda I, T\}$ *is said to be* *parameter-elliptic for* λ *on the ray* $\lambda = re^{i\theta}$ $(r \in \overline{\mathbb{R}}_+)$, *when* (I)-(III) *hold:*

(I) *For all* x *and all* $|\xi| + |\lambda| \neq 0$, *the interior symbol* $p^h(x,\xi) - \lambda I : \mathbb{C}^N \to \mathbb{C}^N$ *is invertible.*

(II) *For all* x', *all* $\xi' \neq 0$, *all* λ, *the boundary symbol operator*

$$(3.10) \quad a^h(x',\xi',\lambda,D_n) = \begin{pmatrix} p^h(x',0,\xi',D_n)_{\mathbb{R}_+} + g^h(x',\xi',D_n) - \lambda \\ t^h(x',\xi',D_n) \end{pmatrix} : H^d(\overline{\mathbb{R}}_+)^N \to \begin{matrix} L^2(\mathbb{R}_+)^N \\ \times \\ \mathbb{C}^M \end{matrix}$$

is bijective.

(III) *For all* x', *all* $\lambda \neq 0$, *the boundary symbol operator* $a^h(x',\xi',\lambda,D_n)$ *converges for* $\xi' \to 0$ *to a bijective operator* $a^h(x',0,\lambda,D_n) : H^d(\overline{\mathbb{R}}_+)^N \to L^2(\mathbb{R}_+)^N \times \mathbb{C}^M$.

The first two conditions are fairly natural, and the third condition is written out explicitly (rather than being included in (II)) since there do exist interesting problems where (I) and (II) but not (III) are satisfied (see e.g. Remark 3.2.16 in [14]). The convergence for $\xi' \to 0$ is a convergence in the symbol norms we use for these operators; it implies convergence in the norm of operators from $H^d(\overline{\mathbb{R}}_+)^N$ to $L^2(\mathbb{R}_+)^N \times \mathbb{C}^M$.

When (I) and (II) hold, condition (III) implies that the boundary condition is *normal*, i.e., the coefficient $S_{kk}(x')$ of γ_k in T_k (cf. (3.8)) is *surjective* for each k (since the term $S_{kk}(x')\gamma_k$ is all that survives in $t_k^h(x',\xi',D_n)$ when $\xi' \to 0$). *When* $N = 1$ (and more generally, when e.g. $\dim M_k = N$ or 0 for each k, and $p^h(x,\xi)$ is diagonalizable for $x_n = 0$, $\xi' = 0$, cf. [14, Proposition 1.5.9]), *condition* (III) *is simply equivalent with normality*, when (I) and (II) hold.

A closer study shows that the continuity (in symbol norm) of a^h at $\xi' = 0$ is actually a Hölder continuity with exponent ν, where ν is an integer or half-integer in the interval

$$(3.11) \qquad \nu \in [\tfrac{1}{2}, d]$$

(which can be improved to $\nu = +\infty$ in purely differential operator problems). This number ν, the so-called *regularity number*, plays an important rôle in the symbol calculus.

Example 3.2. Obvious examples of ps.d.o.s P satisfying condition (I) are the *strongly elliptic* ps.d.o.s (those for which $[p^h(x,\xi) + p^h(x,\xi)^*]$ is positive definite for $\xi \neq 0$), where the parameter-ellipticity of $p - \lambda$ holds on all rays $\lambda = re^{i\theta}$ with $\theta \in [\pi/2, 3\pi/2]$. The transmission property here implies that $d = 2m$, m integer. For such operators, one can show that the system with the Dirichlet condition $\{P_\Omega - \lambda, \gamma\}$ (where $\gamma = \{\gamma_0, \gamma_1, \cdots, \gamma_{m-1}\}$) satisfies all three conditions (I-III) ([14, Section 1.7]). The parameter-ellipticity is stable under small perturbations, so it will also hold when P_Ω and γ are replaced by $P_\Omega + G$ resp. $\gamma + T'$, when G and T' are small. (The perturbed problems need not satisfy the Gårding inequality, although $\{P_\Omega, \gamma\}$ does.)

As a particularly simple example, we mention the case where $P = -\Delta$ on Ω, with one of the boundary conditions

$$\text{(3.12)}\qquad \begin{aligned} &\text{(i)} && \gamma_0 u + T_0' u = 0\,, \\ &\text{(ii)} && \gamma_1 u + T_1' u = 0\,, \text{ or} \\ &\text{(iii)} && \gamma_1 u + S_1 \gamma_0 u + T_1' u = 0\,, \end{aligned}$$

where T_0' (of order 0) and T_1' (of order 1) are of class 0 and S_1 is a ps.d.o. in $\partial\Omega$ of order 1 (all suitably small). It is found here that $\nu = \frac{1}{2}$ for (i), $\nu = 3/2$ for (ii) and $\nu = 1$ for (iii). Further details and examples are given in [14, Sections 1.5-7].

The main results are the following ([14, Section 3.3]):

Theorem 3.2. *When* $A_\lambda = \{P_\Omega + G - \lambda, T\}$ *is parameter-elliptic on the ray* $\lambda = re^{i\theta}$, *then there is an* $r_0 \geq 0$ *so that the inverse*

$$\text{(3.13)}\qquad B_\lambda = A_\lambda^{-1} = (R_\lambda \quad K_\lambda)$$

exists for all $\lambda = re^{i\theta}$, $r \geq r_0$. *In particular, the resolvent* $R_\lambda = (B-\lambda)^{-1}$ *of the realization* B *(cf. (3.1)) exists and satisfies the following estimates, uniformly in* $\lambda = re^{i\theta}$ $(r \geq r_0)$,

$$\text{(3.14)}\qquad \left(\langle\lambda\rangle^{1+s/d}\, \|R_\lambda f\|_0 + \|R_\lambda f\|_{s+d}\right) \leq C_s\left(\langle\lambda\rangle^{s/d}\, \|f\|_0 + \|f\|_s\right)$$

for any $s \geq 0$; *and related estimates hold for* K_λ.

Here R_λ is the resolvent of the realization B since it solves the semi-homogeneous problem, as in (2.8).

Theorem 3.3. *The resolvent* R_λ *is the sum of a ps.d.o. and a s.g.o. (depending on the parameter* λ*)*

$$R_\lambda = Q_{\lambda,\Omega} + G_\lambda \,, \tag{3.15}$$

where Q_λ *is a parametrix of* $P-\lambda$. *Denoting* $|\lambda|^{1/d}$ *by* μ, *we have (for* λ *on the ray* $\lambda = re^{i\theta}$, $r \geq 0$*):*

1^0 *The symbol* $q(x,\xi,\mu,\theta)$ *of* Q_λ *has an asymptotic expansion* $q \sim \Sigma_{\ell\in\mathbb{N}}\, q_{-d-\ell}$, *where each* $q_{-d-\ell}(x,\xi,\mu,\theta)$ *is homogeneous of degree* $-d-\ell$ *in* (ξ,μ) *for* $|\xi| \geq 1$, *and the expansion holds in the sense that, for all indices* $\alpha,\beta \in \mathbb{N}^n$ *and* $j,M \in \mathbb{N}$:

$$|D_x^\beta D_\xi^\alpha D_\mu^j[q - \sum_{\ell<M} q_{-d-\ell}]| \leq c(x)\Big(\langle\xi\rangle^{d-|\alpha|-M}\langle(\xi,\mu)\rangle^{-2d-j} + \langle(\xi,\mu)\rangle^{-d-|\alpha|-M-j}\Big). \tag{3.16}$$

2^0 *The symbol* $g(x',\xi',\xi_n,\eta_n,\mu,\theta)$ *of* G_λ *has an expansion* $g \sim \Sigma_{\ell\in\mathbb{N}}\, g_{-d-1-\ell}$, *where* $g_{-d-1-\ell}(x',\xi',\xi_n,\eta_n,\mu,\theta)$ *is homogeneous of degree* $-d-1-\ell$ *in* (ξ',ξ_n,η_n,μ) *for* $|\xi'| \geq 1$, *and the expansion holds in the following sense, most easily expressed in terms of the associated symbol-kernels* $\tilde g$ *(where* $\tilde g = F^{-1}_{\xi_n\to x_n} \overline{F}^{-1}_{\eta_n\to y_n} g$*):*

$$\begin{aligned} &\| x_n^k D_{x_n}^{k'} y_n^m D_{y_n}^{m'} D_{x'}^\beta D_{\xi'}^\alpha D_\mu^j [\tilde g - \sum_{\ell<M} \tilde g_{-d-1-\ell}] \|_{L^2_{x_n,y_n}(\mathbb{R}_+\times\mathbb{R}_+)} \\ &\leq c(x')\Big(\langle\xi'\rangle^{\nu-[k-k']_+-[m-m']_+-|\alpha|-M}\langle(\xi',\mu)\rangle^{-d-\nu+[k-k']_-+[m-m']_--j} \\ &\qquad + \langle(\xi',\mu)\rangle^{-d-k+k'-m+m'-|\alpha|-M-j}\Big) \end{aligned} \tag{3.17}$$

holds for all indices $\alpha,\beta \in \mathbb{N}^{n-1}$, $k,k',m,m',j,M \in \mathbb{N}$; *here we have denoted*

$$a_\pm = \max\{0\,, \pm a\}\,. \tag{3.18}$$

There are related statements for the Poisson operator part K_λ of $\mathcal{B}_\lambda$, cf. (3.13).

As we see, the regularity number ν plays the rôle in the estimates (3.17) of limiting the improvement of the estimates with respect to μ, when x_n^k, y_n^m or $D_{\xi'}^\alpha$ are applied. In fact, the term containing powers of $\langle\xi'\rangle$ is dominating as soon as the exponent on $\langle\xi'\rangle$ is negative, and therefore the expressions are never better than $O(\mu^{-d-\nu-j})$ for $\mu\to\infty$. Similarly, the symbols in (3.16) are never better than $O(\mu^{-2d-j})$ for $\mu\to\infty$ (this is a special effect of having pseudo-differential and not just differential problems).

One can show by computation that the class of symbols satisfying the estimates (3.17) is invariant under coordinate changes preserving the boundary (cf. [Theorem 2.4.11]), but we have not found a simple characterization by conormality and related properties, as in the parameter-independent case (Proposition 2.3 above).

4. Complex powers.

To give a concrete example of the functional calculus that the resolvent study allows, we now consider complex powers, defined by the formula

$$B^z = \frac{i}{2\pi}\int_C \lambda^z R_\lambda d\lambda = (P^z)_\Omega + G^{(z)} , \tag{4.1}$$

where

$$P^z = \frac{i}{2\pi}\int_C \lambda^z Q_\lambda d\lambda \quad \text{and} \quad G^{(z)} = \frac{i}{2\pi}\int_C \lambda^z G_\lambda d\lambda . \tag{4.2}$$

Here $\{P_\Omega + G-\lambda, T\}$ is assumed to be parameter-elliptic on the ray $\overline{\mathbb{R}}_-$ (where $\lambda = -\mu^d$, $\mu \geq 0$), and then C is taken as a Laurent loop (the boundary of the set $\mathbb{C} \setminus (\mathbb{R}_- \cup \{|\lambda| \leq \delta\})$ for some $\delta > 0$). We assume for simplicity that $\{P_\Omega + G-\lambda, T\}$ is invertible for all $\lambda \in \overline{\mathbb{R}}_-$, (the finite part of the spectrum occurring on $\mathbb{R}_-$ can otherwise be avoided by a slight shift of C). Since the parameter-ellipticity necessarily holds also for neighboring rays $\lambda = e^{i\theta}\mu^d$ with $\theta \in]\pi-\varepsilon, \pi+\varepsilon[$, the Laurent loop can be replaced by a keyhole-shaped contour, with a larger opening the larger ε is. In fact, if B is positive selfadjoint (this is the case where one might hope to extend the methods of Helffer-Robert [16]), the two rays in C can be taken arbitrarily close to the *positive* real axis.

When $\operatorname{Re} z < 0$, the integrals in (4.1) and (4.2) converge in the operator norm in $L^2(E)$, so the complex powers are well-defined by this formulation, and moreover, the resulting operators can be represented by Fourier integral formulas where the symbols are obtained by applying the Cauchy integral formula to the resolvent symbols. We now go through some results for which most of the details are found in Grubb [14, Sections 4.4 and 4.5].

Before getting into the symbol properties, let us note that one easily finds by use of (3.14) that B^z maps $L^2(E)$ continously into $H^{d\theta}(E)$ for $\operatorname{Re} z < -\theta \leq 0$. One would here expect B^z to map into $H^{d|\operatorname{Re} z|}(E)$, and this is the case at least when the following result applies:

Theorem 4.1. *If* B *is selfadjoint* > 0, *and the coefficients* S_{kj} *in the formulas (3.8) for* T_k *are differential operators, then for* $0 < \theta < 1$ *with* $d\theta - \frac{1}{2} \notin \mathbb{Z}$,

$$R(B^{-\theta}) = D(B^\theta) = \{u \in H^{d\theta}(E) \mid T_k u = 0 \quad \text{for} \quad k < d\theta - \tfrac{1}{2}\} . \tag{4.3}$$

If $d\theta - \frac{1}{2} \in \mathbb{Z}$, *then* $D(B^\theta) \subset H^{d\theta}(E)$.

This is a generalization of results of Grisvard [8] and Seeley [30] for the differential operator case, and it is actually proved by a reduction to their result (we also get a detailed description of $D(B^\theta)$ when $d\theta - \frac{1}{2}$ is integer).

Now let us consider the terms in B^z, cf. (4.1),(4.2).

For the pseudo-differential part P^z it was shown already in Seeley [28] that it is a polyhomogeneous ps.d.o. of order $d' = d\,\mathrm{Re}\,z$, and we can just add that its symbol $p^{(z)}(x,\xi) \sim \Sigma_{\ell\in\mathbb{N}} p^{(z)}_{d'-\ell}$ has the symmetry property (for all indices)

$$(4.4)\quad D^\beta_x D^\alpha_\xi p^{(z)}_{d'-\ell}(x,\xi) = e^{i\pi(-d-\ell-|\alpha|)} D_x D_\xi p^{(z)}_{d'-\ell}(x,-\xi) \text{ for } x_n = 0 \text{ and } \xi' = 0 \ (|\xi| \geq 1)$$

(implied by a corresponding symmetry property for the resolvent $(P-\lambda)^{-1}$), and this coincides with the transmission property for P^z if and only if $d'+d$ is an even integer (cf. (2.16)). Note for example that noninteger powers of $-\Delta$ do not have the transmission property, but square roots of fourth order operators do.

For the part $G^{(z)}$ the resolvent estimates imply:

<u>Theorem 4.2.</u> *Let* $\mathrm{Re}\,z < 0$, *and let* $d' = d\,\mathrm{Re}\,z$. $G^{(z)}$ *is a generalized s.g.o. of order* d', *in the sense that it is defined from a polyhomogeneous symbol-kernel* $\tilde g^{(z)}(x',x_n,y_n,\xi')$ *by the formula (2.17), where* $\tilde g^{(z)}$ *(that is derived by insertion of* $\tilde g(x',x_n,y_n,\mu,\theta)$ *from Theorem 3.3 2° in the Cauchy formula) satisfies* <u>*some*</u> *of the usual estimates (cf. (2.18)):*

$$(4.5)\quad \| D^\beta_{x'} x_n^k D^{k'}_{x_n} y_n^m D^{m'}_{y_n} D^\alpha_{\xi'} [\tilde g^{(z)} - \sum_{\ell<M} \tilde g^{(z)}_{d'-1-\ell}] \|_{L^2_{x_n,y_n}} \leq c(x') \langle\xi'\rangle^{d'-k+k'-m+m'-|\alpha|-M}$$

holds for the indices satisfying

$$(4.6)\quad \begin{aligned} k'+m' &< |d'| + |\alpha| + k + m + M, \\ [k'-k]_+ + [m'-m]_+ &< |d'| + \nu, \end{aligned}$$

here ν *is the regularity number (cf. (3.11)ff.), recall also (3.18). In particular, the estimates are preserved under application of* $(x_n D_{x_n})^N$ *and* $(y_n D_{y_n})^{N'}$, *all* N, N'.

The second line in (4.6) disappears when complex powers of *differential* operators are considered (they have regularity $\nu = +\infty$), but the first line alone causes a marked difference between $G^{(z)}$ and the original singular Green operators.

Let us find out, to what extent the considerations of Section 2 in connection with conormality can be extended: It is not hard to extend $\tilde g^{(z)}$ (by use of Seeley [27]) to a function $\tilde g^{(z)}_1$ defined for all $x_n, y_n \in \mathbb{R}$ and satisfying a similar system of estimates (4.5)-(4.6), such that $G^{(z)} = r^+ G^{(z)}_1 e^+$, where $G^{(z)}_1$ is defined from $\tilde g_1$ by (2.17) with the integration over y_n extended to $\mathbb{R}$. Let $g^{(z)}_1(x',\xi',\xi_n,\eta_n) = F_{x_n\to\xi_n} \overline{F}_{y_n\to\eta_n} \tilde g^{(z)}_1$, then the estimates like (4.5) give rise to the system of estimates

$$(4.7)\quad \| D^k_{\xi_n} \xi_n^{k'} D^m_{\eta_n} \eta_n^{m'} D^\alpha_{\xi'} D^\beta_{x'} g^{(z)}_1 \|_{L^2_{\xi_n,\eta_n}} \leq c(x') \langle\xi'\rangle^{d'-k+k'-m+m'-|\alpha|},$$

for all indices satisfying (4.6) with $M = 0$.

Again, the use of (2.24) (with respect to ξ_n as well as η_n) leads to a system of pointwise estimates

$$|D^{\beta}_{x'} D^{\alpha}_{\xi'} D^{k}_{\xi_n} D^{m}_{\eta_n} g_1^{(z)}(x',\xi',\xi_n,\eta_n)|$$

$$(4.8) \qquad \leq c(x')\langle\xi'\rangle^{d'-1-|\alpha|-k-m}\left(\frac{1+|\xi'|}{1+|\xi'|+|\xi_n|}\right)^{k'}\left(\frac{1+|\xi'|}{1+|\xi'|+|\eta_n|}\right)^{m'}$$

for all indices satisfying (4.6) with $M=0$,

but this family of estimates does not give back the full system (4.7) when one passes from sup-norms to L^2-norms, because of the restrictions on k' and m' (note that the identity

$$(4.9) \qquad \|(1+|\xi'|+|\xi_n|)^{-a}\|_{L^2_{\xi_n}} = c(1+|\xi'|)^{\frac{1}{2}-a}$$

"eats" some of the control over products with powers of ξ_n). Moreover, the limitations on k' and m' in (4.6) only allow us to obtain (2.26) for a restricted set of N. Here, rather than using (4.8) and (2.27) (that give too low values of N), we use that (4.7) implies

$$|(\xi_n+i\eta_n)^N D^k_{\xi_n} D^m_{\eta_n} D^{\beta}_{\xi'} D^{\alpha}_{x'} g_1^{(z)}| \leq c(x')\langle\xi'\rangle^{d'-1-k-m-|\alpha|+N},$$

when

$$(4.10) \qquad \begin{aligned} &N < |d'| + |\alpha| + k+m, \\ &\max\{[N-k]_+, [N-m]_+\} < |d'| + \nu; \end{aligned}$$

and then also

$$(4.11) \qquad |D^{\beta}_{x'} D^{\alpha}_{\xi'} D^{k}_{\xi_n} D^{m}_{\eta_n} g_1^{(z)}| \leq c(x')\langle\xi'\rangle^{d'-1-k-m-|\alpha|}\left(\frac{1+|\xi'|}{1+|\xi'|+|\xi_n|+|\eta_n|}\right)^{N}.$$

When ν is finite, we see how N is limited in relation to k, m and ν in the second line of (4.10), regardless of $|\alpha|$, whereas when $\nu=+\infty$, N can almost outweigh the exponent on $\langle\xi'\rangle$. In all cases there remains a negative exponent on $\langle\xi'\rangle$, and the estimates are not $S_{1,0}(\mathbb{R}^{n-1},\mathbb{R}^{n+1})$ estimates. So we do not get conormality as explained in Section 2, and certainly not a rapid decrease for $|(\xi_n,\eta_n)| \to \infty$, only a finite decrease linked to the differentiation orders. Let us formulate the results that we do get:

Theorem 4.3. *Let* $\mathrm{Re}\, z < 0$. *Then* $G^{(z)} = r^+ G_1^{(z)} e^+$, *where* $G_1^{(z)}$ *is an operator defined from a symbol* $g_1^{(z)}(x',\xi',\xi_n,\eta_n)$ *by the formula (2.23); here* $g_1^{(z)}$ *satisfies the estimates (4.7) and (4.8). Moreover it satisfies (4.11) for the indices fulfilling (4.10).*

In the case of differential operators, other methods than the above are available. For this case, Laptev [21] gets by use of more special properties of the resolvent kernel, that $G^{(z)}$ does actually have a kernel that is conormal in X with respect to Y (as defined in Section 2). But even in this case, one cannot expect $G^{(z)}$ to be a singular Green operator in the original sense, as the following example shows:

Example 4.4. Let B be the realization of the "biharmonic" operator $(1-\Delta)^2$ with Dirichlet boundary condition on a smooth bounded subset $\overline{\Omega}$ of $\mathbb{R}^n$:

$$D(B) = \{u \in H^4(\overline{\Omega}) \mid \gamma_0 u = \gamma_1 u = 0\} .$$

By the theorem of Grisvard (see Theorem 4.1 ff. above),

$$R(B^{-\frac{1}{2}}) = D(B^{\frac{1}{2}}) = \{u \in H^2(\overline{\Omega}) \mid \gamma_0 u = \gamma_1 u = 0\} .$$

Here $B^{-\frac{1}{2}} = (1-\Delta)_\Omega^{-1} + G^{(-\frac{1}{2})}$, since $[(1-\Delta)^2]^{-\frac{1}{2}} = (1-\Delta)^{-1}$. If $G^{(-\frac{1}{2})}$ were an ordinary s.g.o. (necessarily of order -2 and class 0), composition with $(1-\Delta)_\Omega$ would give

$$(1-\Delta)_\Omega\, B^{-\frac{1}{2}} = I + G' ,$$

for some s.g.o. G' of order 0 and class 0. On the boundary symbol level, one would then have, for each $\xi' \neq 0$,

$$\left(|\xi'|^2 + D_{x_n}^2\right) b^{(-\frac{1}{2})} = I + g'$$

with a Hilbert-Schmidt operator g' in $L^2(\mathbb{R}_+)$. Then $I + g'$ would be a Fredholm operator in $L^2(\mathbb{R}_+)$ with index 0, but this contradicts the fact that the operator on the left is a bijection of $L^2(\mathbb{R}_+)$ onto a space of codimension 1 (for $b^{(-\frac{1}{2})}$ maps $L^2(\mathbb{R}_+)$ bijectively onto $\{u \in H^2(\overline{\mathbb{R}}_+) \mid u(0) = D_{x_n}u(0) = 0\}$, whose image by $|\xi'|^2 + D_{x_n}^2$ has codimension 1 in $L^2(\mathbb{R}_+)$). Stated in another way, $D(B^{\frac{1}{2}})$ contains too many boundary conditions in order for $B^{\frac{1}{2}}$ to be a realization associated with $1-\Delta$. Altogether, the term $G^{(-\frac{1}{2})}$ cannot be a true s.g.o.

We note that in the calculus of Cordes [4], the square roots do indeed belong to the algebra, which we see as a sign that it is structurally more coarse than the algebra we study here.

At any rate, the estimates on the symbols are strong enough to imply some interesting spectral estimates:

Theorem 4.5. *Let* $\operatorname{Re} z < -(2d)^{-1}$. *Then the characteristic values* $s_k(G^{(z)})$ *(the eigenvalues of* $|G^{(z)}| = (G^{(z)*}G^{(z)})^{\frac{1}{2}}$) *satisfy*

(4.12) $$s_k(G^{(z)}) \le C_\delta k^{-d|\mathrm{Re}\, z|/(n-1)+\delta} \quad \text{for } k \to \infty ,$$

where $\delta = 0$ *if* $n \geq 3$ *and* δ *is any positive number if* $n = 2$.

The method of proof is the same as in Grubb [13], when one uses that it suffices to have estimates of a certain finite set of the seminorms $|||\tilde{g}^{(z)}|||_{\dots}$ (cf. (2.18)); also C_δ can be estimated by these seminorms. Actually, we trust that the restrictions $\delta > 0$ for $n = 2$, and $\mathrm{Re}\, z < -(2d)^{-1}$, are due to the method only. Note the

Corollary 4.6. *The operator* $G^{(z)}$ *is of trace class when* $\mathrm{Re}\, z < -(n-1)/d$.

Further corollaries and applications to spectral theory are developed in [14, Chapter 4].

Let us now turn to another classical problem, namely the question of how the function $\mathrm{tr}\, G^{(z)}$ (the trace of $G^{(z)}$), that is well-defined and holomorphic for $\mathrm{Re}\, z < -(n-1)/d$ in view of the above calculus, extends as a meromorphic function for larger values of $\mathrm{Re}\, z$. Here we find the following result where the regularity ν (recall (3.11)ff.) plays an important rôle:

Theorem 4.7. *The trace of* $G^{(z)}$ *extends to a meromorphic function on the region*

(4.13) $$\{z \in \mathbb{C} \mid \mathrm{Re}\, z < (\nu - {}^1\!/\!{}_4)/d\}$$

(note that $(\nu - {}^1\!/\!{}_4)/d > 0$*), with simple poles at the points*

(4.14) $$z = \frac{1-n}{d}, \frac{2-n}{d}, \dots, \frac{-1}{d}, \frac{1}{d}, \frac{2}{d}, \dots, \frac{\nu'}{d},$$

where ν' *is the largest integer less than* ν.

The theorem follows essentially from a decomposition of $\mathrm{tr}\, G_\lambda$ into a number of exact terms $s_j(-\lambda)^{-1+(n-j)/d}$ (giving the poles) plus a remainder (defining a holomorphic term in (4.13)).

It is well known from Seeley [28], that $\mathrm{tr}\, P^z$ on compact manifolds Σ, and hence also $\mathrm{tr}(P^z)_\Omega$, extends to a meromorphic function on $\mathbb{C}$ with simple poles at the points $(j-n)/d$ for $j \in \mathbb{N}$, except for $j = n$, where there is no pole. (Note that here $-n/d$ is included as a pole, in contrast with (4.14).) Then we find for $B^z = (P^z)_\Omega + G^{(z)}$, using some more explicit formulas explained in [14, Section 4.4]:

Theorem 4.8. $\mathrm{tr}\, B^z$ *is defined as a meromorphic function on the region (4.13), with simple poles at the points*

(4.15) $$z = \frac{-n}{d}, \frac{1-n}{d}, \ldots, \frac{-1}{d}, \frac{1}{d}, \frac{2}{d}, \ldots, \frac{\nu'}{d}$$

(ν'= *largest integer less than* ν). *In particular,* B^z *is analytic at* $z = 0$, *and the value* $c_0(B)$ *is determined from the symbols*

$$D_x^\alpha D_\xi^\beta p_{d-\ell}(x',0,\xi) \quad \textit{for} \quad \ell = 0,1,\cdots,n; \quad |\alpha+\beta| \le n-\ell ;$$

(4.16) $$D_{x'}^\alpha D_{\xi'}^\beta g_{d-\ell}(x',\xi,\eta_n) \quad \textit{for} \quad \ell = 0,1,\cdots,n-1; \quad |\alpha+\beta| \le n-1-\ell ;$$

$$D_{x'}^\alpha D_{\xi'}^\beta t_{d-\ell}(x',\xi) \quad \textit{for} \quad \ell = 0,1,\cdots,n-1; \quad |\alpha+\beta| \le n-1-\ell ,$$

given in local coordinates.

The study of Rempel-Schulze [25] of complex powers of realizations $\underline{A}_T$ of pseudo-differential operators not necessarily having the transmission property, and of order $d > 1$, obtains the meromorphic extendability of $(\underline{A}_T)^z$ into the region $\{\operatorname{Re} z < -(n-\frac{1}{2})/d\}$, with a simple pole at $-n/d$.

There is an application of the statements on the value of $\operatorname{tr} B^z$ at $z = 0$ that we shall likewise mention. When B_1 is a realization defined, within the Boutet de Monvel calculus, from an elliptic boundary problem $\{P_{1,\Omega} + G_1, T_1\}$, where $P_{1,\Omega} + G_1$ has order ≥ 0 and T_1 is *normal*, then the adjoint B_1^* (as unbounded operators in $L^2(E)$) is a realization of the same type, and $B = B_1^* B_1$ is also a realization (associated with $P_1^* P_1$) defined by a normal boundary condition, this is proved in [14, Theorems 1.4.6 and 1.6.9]. Here B is selfadjoint ≥ 0 and satisfies in particular the parameter-ellipticity condition for $\theta \in]0,2\pi[$; the same holds for $B' = B_1 B_1^*$. Then one can use some elementary observations on the index in Atiyah-Bott-Patodi [1] to show the following result:

Corollary 4.9. *When* B_1 *is a normal elliptic realization, defined from a system* $\{P_{1,\Omega} + G_1, T_1\}$, *then the index of* B_1 *is given by the formula*

(4.17) $$\operatorname{index} B_1 = c_0(B_1^* B_1) - c_0(B_1 B_1^*) ,$$

where c_0 *is as defined in Theorem 4.8. The index of* B_1 *equals the index of the full operator* $\{P_{1,\Omega} + G_1, T_1\}$.

The formula (4.17) is of course rather complicated when it comes to actual computation. Let us just note that general properties of the index show that it will depend only on the principal symbols (since the lower order parts give rise to relatively compact perturbations); moreover, the number of derivatives of the symbols of G_1 and T_1 that are used (cf. (4.16)) is one less than what enters in the "coarse" formulas of Rempel-Schulze [24] for general (not necessarily normal) elliptic problems. It would be interesting to study refinements of the formula, as done for other formulas by Fedosov [6], Hörmander [18], Rempel-Schulze [24].

We shall end by making some further comments concerning the region (4.13) of analytic continuation, limited by the regularity number ν. For the function $\operatorname{tr} P^z$, where P is in general of finite regularity d, one does not have this restriction. For, one can for example use the fact that in the calculation of the resolvent, $(P-\lambda)^{-1}$ may be replaced by the right hand side of the formula

$$(4.16) \qquad (P-\lambda)^{-1} = -\lambda^{-1} - \lambda^{-2}P - \dots - \lambda^{-m}P^{m-1} - \lambda^{-m}P^m(P-\lambda)^{-1},$$

where the first m terms drop out in the Cauchy formula for P^z ($\operatorname{Re} z < 0$), and the product $P^m(P-\lambda)^{-1}$ has a much better regularity than $(P-\lambda)^{-1}$ for large m (because of factors of the type $|\xi|^{md}$ that kill singularities at $\xi = 0$). Then P^z can be extended meromorphically to all of $\mathbb{C}$. Unfortunately, this argument does not work for B^z. For, in the analogous formula for $R_\lambda = (B-\lambda)^{-1}$,

$$(4.17) \qquad (B-\lambda)^{-1} = \lambda^{-1} - \dots - \lambda^{-m}B^{m-1} - \lambda^{-m}B^m(B-\lambda)^{-1},$$

the composition formulas give terms in $B^m(B-\lambda)^{-1}$ that are not more regular than $(B-\lambda)^{-1}$ itself, e.g. terms of the type $D_{x_n}^{md} G_\lambda$ (where (3.17) shows that $D_{\xi'}^\alpha D_{x_n}^{md} \tilde{g}$ has a dominating factor $\langle\xi'\rangle^{-1}$ if $|\alpha| \geq \nu + 1$, just like $D_{\xi'}^\alpha \tilde{g}$ itself).

We do not know whether (4.13) can actually be improved. It may well be that the loss of $^1/_4$ is merely technical. One may also speculate that at least in problems generated from elliptic differential operator systems (hence with rational symbols) there ought to be meromorphic extendability to $\mathbb{C}$, in view of the analytic properties of the whole problem. Or, a meromorphic extension should exist if only P is freely pseudo-differential, but G and T are generated from differential operators.

There is a closely related question of giving an asymptotic series expansion for the trace of the "heat" operator $\exp(-tB)$ for $t \to 0+$ (when there is parameter-ellipticity on *all* rays $\lambda = re^{i\theta}$ for $\theta \in [\pi/2, 3\pi/2]$); here ν puts a similar limitation on the number of terms we can describe, see [14, Section 4.2]. In this type of question, Widom [33] expects the existence of a full expansion (with logarithmic terms) when the boundary conditions have a simple nature.

BIBLIOGRAPHY

[1] M.F. Atiyah, R. Bott and V.K. Patodi: On the heat equation and the index theorem, Inventiones Math. 19 (1973), 279-330.

[2] L. Boutet de Monvel: Comportement d'un opérateur pseudo-différentiel sur une variête à bord, I-II, J. d'Analyse Fonct. 17 (1966), 241-304.

[3] L. Boutet de Monvel: Boundary problems for pseudo-differential operators, Acta Math. 126 (1971), 11-51.

[4] H.O. Cordes: *Elliptic Pseudo-Differential Operators - An Abstract Theory*. Lecture Note 756, Springer Verlag, Berlin 1979.

[5] J. Dunau: Fonctions d'un opérateur elliptique sur une variéte compacte. J. Math. pures et appl. 56 (1977), 367-391.

[6] B.V. Fedosov: Analytical formulas for the index of elliptic operators, Trudy Mosk. Mat. Obsv. 30 (1974), 159-241 = Trans. Moscow Math. Soc. 30 (1974), 159-240.

[7] P. Greiner: An asymptotic extension for the heat equation, Arch. Rat. Mech. Anal. 41 (1971), 163-218.

[8] P. Grisvard: Caractérisation de quelques espaces d'interpolation, Arch. Rat. Mech. Anal. 25 (1967), 40-63.

[9] G. Grubb: On Pseudo-Differential Boundary Problems. Reports no.2 (1979), 1, 2, 7 and 8 (1980), Copenhagen University Math. Dept. Publication Series.

[10] G. Grubb: A resolvent construction for pseudo-differential boundary value problems, with applications, 18th Scandinavian Congr. Math. Proceedings, 1980, Birkhäuser, Boston, 307-320.

[11] G. Grubb: Problèmes aux limites dépendant d'un paramètre, C.R. Acad. Sci. Paris (Sér.I) 292 (1981), 581-583. La résolvante d'un problème aux limites pseudo-differentiel elliptique, C.R. Acad. Sci. Paris (Sér.I) 292 (1981), 625-627.

[12] G. Grubb: The heat equation associated with a pseudo-differential boundary problem, Seminar Analysis 1981-82, Akad. Wiss. Berlin, 27-41. [Also available as Preprint no.2, 1982, Copenhagen Univ. Math. Dept.]

[13] G. Grubb: Singular Green operators and their spectral asymptotics, Duke Math. J. 51 (1984), 477-528.

[14] G. Grubb: *Functional Calculus of Pseudo-Differential Boundary Problems*, Progress in Mathematics vol.65, Birkhäuser, Boston 1986.

[15] K. Hayakawa and H. Kumano-go: Complex powers of a system of pseudo-differential operators, Proc. Japan Acad. 47 (1971), 359-364.

[16] B. Helffer and D. Robert: Calcul fonctionnel par la transformation de Mellin et opérateurs admissibles, J. Functional Analysis 53 (1983), 246-268.

[17] E. Hille and R. Phillips: *Functional Analysis and Semi-Groups*. Amer. Math. Soc. Colloq. Publ. 31, Providence 1957.

[18] L. Hörmander: The Weyl calculus of pseudo-differential operators, Comm. Pure Appl. Math. 32 (1979), 359-443.

[19] L. Hörmander: *The Analysis of Linear Partial Differential Operators*, III. Springer Verlag, Berlin 1985.

[20] C. Iwasaki and N. Iwasaki: Parametrix for a degenerate parabolic equation and its application to the asymptotic behavior of spectral functions for stationary problems, Publ. R.I.M.S. Kyoto 17 (1981), 577-655.

[21] A.A. Laptev: Spectral asymptotics of a class of Fourier integral operators, Trudy Mosk. Mat. Obsv. 43 (1981), 92-115 = Trans. Moscow Math. Soc. 1983, 101-127.

[22] M. Nagase and K. Shinkai: Complex powers of non-elliptic operators, Proc. Japan Acad. 46 (1970), 779-783.

[23] K.B. Oldham and J. Spanier: *The Fractional Calculus*, Academic Press, New York 1974.

[24] S. Rempel and B.-W. Schulze: *Index Theory of Elliptic Boundary Problems.* Akademie-Verlag, Berlin 1982.

[25] S. Rempel and B.-W. Schulze: Complex powers for pseudo-differential boundary problems I, Math. Nachr. 111 (1983), 41-109. Complex powers for pseudo-differential boundary problems II, Math. Nachr. 116 (1984), 269-314.

[26] D. Robert: Propriétés spectrales d'opérateurs pseudo-différentiels, Comm. Part. Diff. Equ. 3 (1978), 755-826.

[27] R. Seeley: Extension of C^∞ functions, Proc. Amer. Math. Soc. 15 (1964), 625-626.

[28] R. Seeley: Complex powers of an elliptic operator, Amer. Math. Soc. Proc. Symp. Pure Math. 10 (1967), 288-307.

[29] R. Seeley: Analytic extension of the trace associated with elliptic boundary problems, Amer. J. Math. 91 (1969), 963-983.

[30] R. Seeley: Interpolation in L^p with boundary conditions, Studia Math. 44 (1972), 47-60.

[31] R.S. Strichartz: A functional calculus for elliptic pseudo-differential operators, Amer. J. Math. 94 (1972), 711-722.

[32] H. Widom, A complete symbolic calculus for pseudo-differential operators, Bull. Sc. Math. 104 (1980), 19-63.

[33] H. Widom: *Asymptotic Expansions for Pseudodifferential Operators on Bounded Domains.* Lecture Note 1152, Springer Verlag, Berlin 1985.

[34] E. Schrohe: Complex powers of elliptic pseudodifferential operators, Int. Eq. Op. Th. 9 (1986), 337-354.

Some Spectral Properties of Periodic Potentials

V. Guillemin and A. Uribe

§1. Introduction

In this article we will describe some asymptotic features of the spectra of the Laplace operator on line bundles over a 2n-torus. These results are similar to the results described in [GU] for the Laplace operator on line bundles over $\mathbb{C}P^n$.

Let Ω be the standard symplectic form on $\mathbb{R}^{2n}$ and let Γ be a lattice subgroup of $\mathbb{R}^{2n}$. We will say that Γ is integral if $\Omega(v,w) \in \mathbb{Z}$ for $v,w \in \Gamma$. This property can be formulated in terms of the torus

$$T_\Gamma = \mathbb{R}^{2n}/\Gamma$$

and the symplectic form, Ω_Γ, on T_Γ associated with Ω. Namely let ι be the natural mapping of $H^2(T_\Gamma,\mathbb{Z})$ into $H^2(T_\Gamma,\mathbb{R})$. Then Γ is integral if and only if the DeRham class $[\Omega_\Gamma]$ is integral, i.e. if and only if there exists a $c \in H^2(T_\Gamma,\mathbb{Z})$ such that

$$\iota(c) = [\Omega_\Gamma].$$

We will henceforth assume this is the case. Then, having made the choice of a c in the equation above, there exists a circle bundle

$$\pi: M_\Gamma \to T_\Gamma$$

with Chern class equal to c. Moreover there exists, as we will see in §2, a more or less canonical connection on M_Γ whose curvature form is $2\pi i\Omega_\Gamma$. Now, for every integer e, let γ_e be the character of the circle group: $\gamma_e(\omega) = \omega^e$, and consider the line bundle

$$L_e = M \times_{\gamma_e} \mathbb{C}.$$

Corresponding to the connection on M is a connection on this line bundle which we will denote by ∇_e.

Let us make T_Γ into a Riemannian manifold by equipping it with the Riemannian metric associated with the standard quadratic form, $x_1^2 + \ldots + x_{2n}^2$, on $\mathbb{R}^{2n}$. Let Δ_e be the Laplace operator on the space of global sections of L_e. We will denote by $\lambda_{e,k}$, $k = 1,2,3\ldots$ the eigenvalues of Δ_e, arranged in increasing order, and by $N_{e,k}$, $k = 1,2,3\ldots$ their multiplicities. In §3 we will show that $N_{e,k} \to \infty$ as e and $k \to \infty$.

Now let V be a smooth function on T_Γ. By Rellich's theorem the eigenvalues of the Schroedinger operator, $\Delta_e + V$, are made up of bands

$$\lambda_{e,k,i}, \qquad i = 1,\ldots\ldots,N_{e,k} \tag{1.1}$$

with

$$|\lambda_{e,k} - \lambda_{e,k,i}| \leq \sup|V|.$$

Let $\delta(t)$ be the Dirac δ-function, and, for e and $E = \lambda_{e,k}$, let

$$\mu_{e,E} = \frac{1}{N_{e,k}} \sum_i \delta(t - (\lambda_{e,k,i} - \lambda_{e,k})) . \tag{1.2}$$

Our main result has to do with the asymptotic behavior of this measure as E and e tend to infinity. To state this result we need some notation. Let us make the standard identification:

$$\mathbb{R}^{2n} \cong \mathbb{C}^n$$

so that the standard Riemannian form and the standard symplectic form on $\mathbb{R}^{2n}$ are the real and imaginary parts respectively of the standard Hermetian form on $\mathbb{C}^n$. Let S^{2n-1} be the sphere of radius 1, $\{v \in \mathbb{C}^n, |v| = 1\}$, in $\mathbb{C}^n$. Given a smooth function, f, on $\mathbb{R}^{2n}$ (i.e. on $\mathbb{C}^n$) we define a smooth function, f_r, on $\mathbb{C}^n \times S^{2n-1}$ by the formula

$$f_r(u,v) = \frac{1}{r} \int_0^{2\pi} f(u + \frac{re^{is}v}{2\pi})ds .$$

If f is periodic with respect to Γ, so is f_r; so the formula above defines a transformation, $f \to f_r$, from smooth functions on T_Γ to smooth functions on $T_\Gamma \times S^{2n-1}$. Denote by σ_1 and σ_2 the standard probability measures on T_Γ and S^{2n-1} and by σ the product measure $\sigma_1 \times \sigma_2$.

Theorem 1. Let r^2 be a rational multiple of 2π. Let E and e tend to infinity so that

(1.3) $$E = e^2r^2 + 2\pi ne .$$

Then

(1.4) $$\mu_{e,E} = (V_r)_*\sigma$$

We will prove this theorem in §§5 and 6 after reviewing (in §2,..) some standard facts about discrete subgroups of the 2n+1-dimensional Heisenberg group and sub-Laplacians on the corresponding locally homogeneous spaces. In §7 we will discuss some of its implications.

§2. Discrete subgroups of the Heisenberg group.

Let $\mathcal{H}_n$ be the 2n+1-dimensional Heisenberg group. We recall that

$$\mathcal{H}_n = \mathbb{R}^{2n} \times \mathbb{R}$$

equipped with the multiplication law

$$(v,s) \circ (w,t) = (v+w, s+t + \tfrac{1}{2}\Omega(v,w)).$$

The projection

$$\pi: \mathcal{H}_n \to \mathbb{R}^{2n}, \quad \pi(v,t) = v$$

is a group homomorphism and its kernel, $\mathbb{R}$, is the center of $\mathcal{H}_n$.

As in §1 let Γ be a lattice subgroup of $\mathbb{R}^{2n}$ which is integral with respect to the symplectic form, Ω; and let us try to determine all closed subgroups of $\mathcal{H}_n$ which have Γ as homomorphic image. Obviously the group $\Gamma \times \mathbb{R}$ has this property, and it is not hard to show that all other subgroups with this property are discrete normal subgroups of $\Gamma \times \mathbb{R}$. Without loss of generality we can restrict our attention to subgroups, Γ', having the two properties below:

$$(*) \qquad \pi(\Gamma') = \Gamma$$
and
$$\{(o,n),\ n \in \mathbb{Z}\} = \text{kernel } \pi\colon \Gamma' \to \Gamma.$$

The fundamental structure theorem for integral sublattices of $\mathbb{R}^{2n}$ says that one can find a basis, $e_1,\ldots,e_n,\ f_1,\ldots,f_n$ of $\mathbb{R}^{2n}$ and a sequence of positive integers $m_1,\ldots,m_n$ such that:

i. $e_1,\ldots,f_n$ is a symplectic basis of $\mathbb{R}^{2n}$, i.e.

$$(2.1) \qquad \Omega(e_i,e_j) = \Omega(f_i,f_j) = 0 \quad \text{and} \quad \Omega(e_i,f_i) = \delta_{ij}\ .$$

ii. Each term in the sequence $m_1,\ldots,m_n$ is divisible by the term preceding it.

iii. The vectors $e_1,\ldots,e_n,\ m_1f_1,\ldots,m_nf_n$ are generators of Γ over $\mathbb{Z}$.

(See, for instance, [W], lemma A3.2.) Now let

$$F\colon \Gamma \to \mathbb{Z}$$

be the map:

$$F(\Sigma\ a_ie_i + b_im_if_i) = \Sigma\ m_ia_ib_i\ .$$

<u>Proposition 2.1</u>.

i. Let α be a linear functional on $\mathbb{R}^{2n}$. The map χ_α: $\Gamma \times \mathbb{R} \to S^1$ defined by

$$(2.2) \qquad \chi_\alpha(v,t) = (\exp 2\pi it)(\exp \pi i)(F(v) + \alpha(v))$$

is a character of $\Gamma \times \mathbb{R}$.

ii. Let Γ_α be the kernel of χ_α. Then Γ_α has the properties (*). Moreover every subgroup with these properties is of this form.

iii. $\Gamma_\alpha = \Gamma_\beta$ if and only if the difference $\gamma = \alpha-\beta$ maps Γ into $2\mathbb{Z}$.

Consider now the coset space $\mathcal{H}_n/\Gamma_\alpha$. From the action of the center, $\mathbb{R}$, of $\mathcal{H}_n$ on $\mathcal{H}_n$ one obtains a free action of the circle group $S^1 = \mathbb{R}/\mathbb{Z}$ on $\mathcal{H}_n/\Gamma_\alpha$. Moreover, since the homomorphism π: $\mathcal{H}_n \to \mathbb{R}^{2n}$ maps Γ_α onto Γ, it maps the coset space of Γ_α onto the coset space of Γ, i.e. it defines a map

(2.2) $$\pi\colon \mathcal{H}_n/\Gamma_\alpha \to \mathbb{R}^{2n}/\Gamma = T_\Gamma .$$

Proposition 2.2. The map (2.2) makes $\mathcal{H}_n/\Gamma_\alpha$ into a principal S^1 bundle over T_Γ.

The action of $\mathcal{H}_n$ on $\mathcal{H}_n/\Gamma_\alpha$ is locally free; so, at every point p of $\mathcal{H}_n/\Gamma_\alpha$, there is a canonical identification of the tangent space of p with the Lie algebra of $\mathcal{H}_n$:

(2.3) $$T_p \cong h_n = \mathbb{R}^{2n} \oplus \mathbb{R} .$$

We will call the first summand the horizontal part of T_p and the second summand the vertical part. It is clear that this decomposition is S^1 invariant and is, therefore, the defining data for a connection.

Proposition 2.3. The curvature form of the connection associated with (2.3) is $2\pi i\Omega_\Gamma$.

Remark. Let $c_\alpha \in H^2(T_\Gamma,\mathbb{Z})$ be the Chern class of the circle bundle (3.2). Let ι be the inclusion map of $H^2(T_\Gamma,\mathbb{Z})$ into $H^2(T_\Gamma,\mathbb{R})$. By the proposition, $\iota_* c_\alpha = [\Omega_\Gamma]$. However $c_\alpha = c_\beta$ if and only if $\Gamma_\alpha = \Gamma_\beta$. One can show that every circle bundle over T_Γ with the property $\iota_* c = [\Omega_\Gamma]$ is isomorphic to (2.2) for some Γ_α.

We will conclude this section with a few words about the L^2-theory of the coset space $\mathcal{H}_n/\Gamma_\alpha$. For every integer, k, let $L^2(\mathcal{H}_n/\Gamma_\alpha)_k$ be the subspace of $L^2(\mathcal{H}_n/\Gamma_\alpha)$ consisting of L^2-functions, ϕ, which transform under the action of S^1 according to the formula

$$\phi(e^{2\pi i\theta}x) = e^{2\pi ik\theta}\phi(x).$$

It is clear that

(2.4) $$L^2(\mathcal{H}_n/\Gamma_\alpha) = \bigoplus_k L^2(\mathcal{H}_n/\Gamma_\alpha)_k$$

(Hilbert space direct sum). Moreover, (2.4) is a decomposition of $L^2(\mathcal{H}_n/\Gamma_\alpha)$ into $\mathcal{H}_n$ invariant subspaces. We will describe how each of these invariant subspaces decomposes into irreducibles. First of all we recall that the irreducible unitary representations of $\mathcal{H}_n$ consist of the Stone-Von Neumann representations, ρ_λ, $\lambda \in \mathbb{R}-0$, and the trivial one-dimensional representations of the quotient group, $\mathcal{H}_n/\text{Center} \cong \mathbb{R}^{2n}$. We recall also how the Stone-Von Neumann representations are defined. Fix a symplectic basis $e_1,\dots,e_n,f_1,\dots,f_n$ of $\mathbb{R}^{2n}$ (e.g. the basis (2.1)) and let V and W be the subspaces of $\mathbb{R}^{2n}$ spanned by the e's and f's. Then the underlying space on which the representation, ρ_λ, acts can be taken to be $L^2(V)$, and the representation itself is defined by the formulas

$$(2.5)\qquad \begin{aligned} (\rho_\lambda(e)\phi)(v) &= \phi(v+e) \\ (\rho_\lambda(f)\phi(v) &= ((\exp 2\pi i\lambda)\Omega(v,f))\phi(v)) \\ (\rho_\lambda(s)\phi)(v) &= (\exp 2\pi i\lambda s)\Phi(v) \end{aligned}$$

for $\phi \in L^2(V)$, $e \in V$, $f \in W$ and $s \in \mathbb{R} = \text{Center of } \mathcal{H}_n$. It is clear from (2.5) that the representation of $\mathcal{H}_n$ on $L^2(\mathcal{H}_n/\Gamma_\alpha)_k$ is a direct sum of a finite number of copies of the representation, ρ_k. Therefore, to describe the representation of $\mathcal{H}_n$ on $L^2(\mathcal{H}_n/\Gamma_\alpha)$ we only have to determine the multiplicity with which ρ_k occurs for each k.

<u>Proposition 2.4</u>. The representation of $\mathcal{H}_n$ on $L^2(\mathcal{H}_n/\Gamma_\alpha)_k$ is a direct sum of N_k copies of the representation ρ_k, N_k being k^m times the symplectic volume of the torus T_Γ.

"Proof": For simplicity assume $\alpha = 0$. By Frobenius reciprocity the multiplicity with which ρ_k occurs in $L^2(\mathcal{H}_n/\Gamma_0)$ is equal to the multiplicity with which the trivial representation of Γ_0 occurs in the restriction of ρ_k to Γ_0. Since Γ_0 is generated by the elements, $(e_i,0)$, $(m_if_i,0)$ and $(0,n)$, $n \in \mathbb{Z}$, it follows from (2.5) that this multiplicity is equal to the dimension of the space of functions, ϕ, on V with the properties

$$(2.6)\qquad \begin{gathered} \phi(v+e_j) = \phi(v) \quad \text{for all } j \\ \text{and} \\ \phi(v) = (\exp 2\pi i\, m_j k\, \Omega(v,f_j))\phi(v) \quad \text{for all } j. \end{gathered}$$

The first condition implies that ϕ is period with respect to the lattice

$$\Gamma \cap V = \{r_1 e_1 + \cdots + r_n e_n,\ r_i \in \mathbb{Z}\},$$

i.e. is of the form

$$\phi(v) = \Sigma\ c(f) \exp 2\pi i\ \Omega(v,f)$$

the sum taken over the dual lattice $\{r_1 f_1 + \cdots + r_n f_n\}$. The second condition says that in this expansion

$$c(f) = c(f + m_i k f_i) \tag{2.7}$$

for all f in this lattice and all i. It is clear that the set of equations (2.7) has exactly $m_1 \cdots m_k k^n$ independent solutions.

Remark. Cartier has shown (see [C]) that this "proof" can be made rigorous. For a more elaborate treatment of the material in this section see [L-V] or [W].

§3. The sub-Laplacian on $\mathcal{H}_n/\Gamma_\alpha$.

The Lie algebra, h_n, of the Heisenberg group is $\mathbb{R}^{2n} \oplus \mathbb{R}$ equipped with the bracket operation

$$[(v,s),(w,t)] = (0,\Omega(v,w)).$$

In particular if $e_1,\ldots,e_n,f_1,\ldots,f_n$ is the standard symplectic basis of $\mathbb{R}^{2n}$ then

$$[(e_i,0),(f_i,0)] = (0,1)$$

and all other brackets are zero. Let $X_1,\ldots,X_n$, $Y_1,\ldots,Y_n$ and T be the right invariant vector fields on $\mathcal{H}_n$ corresponding to the basis vectors $(e_1,0),\cdots,(e_n,0)$, $(f_1,0),\cdots,(f_n,0)$, $(0,1)$ of h_n. By abuse of notation we will also denote by X_1, etc. the vector fields on $\mathcal{H}_n/\Gamma_\alpha$ associated with these basis vectors. The second order differential operator

$$\Box = -(X_1^2 + \cdots + X_n^2 + Y_1^2 + \cdots + Y_n^2) \tag{3.1}$$

on $\mathcal{H}_n/\Gamma_\alpha$ (or on $\mathcal{H}_n$) is called the sub-Laplacian. It can be extended to a self-adjoint operator on $L^2(\mathcal{H}_n/\Gamma_\alpha)$; and this operator clearly commutes with the representation of the circle group on $L^2(\mathcal{H}_n/\Gamma_\alpha)$ and, therefore, preserves the decomposition (2.4). The sub-Laplacian is of interest for us because of the following fact (which is an easy corollary of Proposition 2.3).

Proposition 3.1. The restriction of $\square$ to $L^2(\mathcal{H}_n/\Gamma_\alpha)_e$ is unitarily equivalent to the Laplace operator, Δ_e, defined in §1.

Let $x_1,\ldots,x_n$ and $y_1,\ldots,y_n$ be the coordinate functions associated with the symplectic basis vectors $e_1,\ldots,e_n$ and $f_1,\ldots,f_n$ of $\mathbb{R}^{2n}$. Under the Stone-Von Neumann representation, ρ_e, the right-invariant vector fields $X_1,\ldots,X_n$ and $Y_1,\ldots,Y_n$ correspond to the operators

$$(3.2)\qquad \frac{\partial}{\partial x_i},\ i = 1,\ldots,n \quad\text{and}\quad 2\pi i e x_i,\ i = 1,\ldots,n,$$

on $L^2(\mathbb{R}^n)$. (See 2.5.) Therefore, the operator $-(X_1^2 +\cdots+ X_n^2 + Y_1^2 +\cdots+ Y_n^2)$ corresponds to the harmonic oscillator on $L^2(\mathbb{R}^n)$:

$$(3.3)\qquad \sum \frac{\partial}{\partial x_i^2} +(2\pi e)^2 x_i^2 .$$

The eigenvalues of this operator are the numbers

$$(3.4)\qquad 2\pi e(2k+n),\qquad k = 0,1,2,\ldots$$

and the multiplicity M_k of the k-th eigenvalue is

$$(3.4)\qquad M_k = \binom{k+n-1}{n-1} .$$

By Proposition 2.4, $L^2(\mathcal{H}_n/\Gamma_\alpha)$ consists of N_e copies of $L^2(\mathbb{R}^n)$ and the sub-Laplacian is equal to (3.3) on each of these copies; therefore, Proposition 3.1 implies:

Proposition 3.2. The eigenvalues of Δ_e are

$$\lambda_{e,k} = 2\pi e(2k+n),\qquad k = 0,1,2,\ldots$$

and the multiplicity with which $\lambda_{e,k}$ occurs is

$$N_{e,k} = \text{Volume}(T_\Gamma)\, e^n \begin{bmatrix} k+n-1 \\ n-1 \end{bmatrix} .$$

§4. The bicharacteristics of the sub-Laplacian

Since $\mathcal{H}_n$ is the universal covering space of $\mathcal{H}_n/\Gamma_\alpha$, to compute the bicharacteristics of the sub-Laplacian on $\mathcal{H}_n/\Gamma_\alpha$ it suffices to compute them on $\mathcal{H}_n$ itself. Let $e_1,\dots,e_n,f_1,\dots,f_n$ be the standard symplectic basis of $\mathbb{R}^{2n}$ and let $x_1,\dots,x_n,y_1,\dots,y_n,t$ be the coordinates on $\mathcal{H}_n = \mathbb{R}^{2n} \times \mathbb{R}$ associated with this basis. Let $\xi_1,\dots,\xi_n$, $\eta_1,\dots,\eta_n$ and τ be the corresponding dual cotangent coordinates. In these coordinates

$$X_i = \frac{\partial}{\partial x_i} + y_i\frac{\partial}{\partial t},\quad Y_i = \frac{\partial}{\partial y_i} \quad\text{and}\quad T = \frac{\partial}{\partial t}.$$

Therefore the symbol of the sub-Laplacian is

$$\sum (\xi_i + \tau y_i)^2 + \eta_i^2 \tag{4.1}$$

and the Hamilton-Jacobi equations are

$$\dot{x}_i = 2(\xi_i + \tau y_i),\quad \dot{y}_i = 2\eta_i,\quad \dot{t} = 2\sum(\xi_i + \tau y_i)y_i \tag{4.2}$$

and

$$\dot{\xi}_i = 0,\qquad \dot{\eta}_i = -2(\xi_i + \tau y_i)\tau,\qquad \dot{\tau} = 0$$

In particular, ξ_i, τ and $\gamma_i = \eta_i + \tau x_i$ are conserved quantities. Notice now that for $\tau \neq 0$ the equations for $\dot{x}_i$ and $\dot{y}_i$ can be written

$$\dot{x}_i = 2\tau\left[y_i + \frac{\xi_i}{\tau}\right],\quad \dot{y}_i = -2\tau\left[x_i - \frac{\gamma_i}{\tau}\right].$$

In other words if we set: $x_i^{\#} = x_i - \frac{\gamma_i}{\tau}$ and $y_i^{\#} = y_i + \frac{\xi_i}{\tau}$ these equations say that

$$\left(\frac{d}{ds}\right)(x_i^{\#} + \sqrt{-1}\, y_i^{\#}) = -2\tau\sqrt{-1}(x_i^{\#} + y_i^{\#}).$$

Thus

$$(4.3) \qquad (x_i + \sqrt{-1}\, y_i)^{\#} = \mu_i\, e^{-\sqrt{-1}\, 2\tau s}$$

where s is the time parameter along the integral curve of the system and μ_i is some fixed complex number. Moreover,

$$H = \tau^2 \sum_{i=1}^{n} (x_i^{\#})^2 + (y_i^{\#})^2 = \tau^2 \sum |\mu_i|^2$$

i.e. the μ_i's satisfy:

$$(4.4) \qquad \sum_{i=1}^{n} |\mu_i|^2 = H/\tau^2 .$$

Notice finally that the equation for $\dot{t}$ in line one of (4.2) can be rewritten:

$$\begin{aligned} \dot{t} &= 2\tau \sum \left[y_i + \frac{\xi_i}{\tau}\right] y_i \\ &= 2\tau \sum y_i^{\#}\left[y_i^{\#} - \frac{\xi_i}{\tau}\right] \\ &= 2\tau \sum (y_i^{\#})^2 - 2\tau \sum \frac{\xi_i}{\tau}\, y_i^{\#} . \end{aligned}$$

Therefore, if we integrate s from 0 to π/τ the second term drops out and we obtain

$$(4.5) \qquad \begin{aligned} \Delta t &= 2\tau \sum \int_0^{\pi/\tau} (y_i^{\#})^2 ds \\ &= \tau \sum \int_0^{\pi/\tau} \left[(x_i^{\#})^2 + (y_i^{\#})^2\right] ds \\ &= \tau(\pi/\tau) \sum |\mu_i|^2 = \pi(H/\tau^2) \end{aligned}$$

Summarizing we have proved

<u>Proposition 4.1</u>.

a) The quantities H, τ, ξ_i and $\gamma_i = \eta_i + \tau x_i$ are constant along bicharacteristics of H.

b) Identify $\mathbb{R}^{2n}$ with $\mathbb{C}^n$ via the map: $(x_1,\dots,x_n,y_1,\dots,y_n) \to (x_1 + \sqrt{-1}y_1,\dots,x_n + \sqrt{-1}y_n)$. Then for fixed $\tau \neq 0$ the bicharacteristics of H are circles of radius $\sqrt{H/\tau}$ and period π/τ lying on affine planes of the form $p + V$ in $\mathbb{R}^{2n}$, p being an arbitrary point of $\mathbb{R}^{2n}$ and V a real two-dimensional subspace of $\mathbb{R}^{2n}$ corresponding to a complex one-dimensional subspace of $\mathbb{C}^n$.

c) As each bicharacteristic of H makes one complete circuit in $\mathbb{R}^{2n}$, the t coordinate of the bicharacteristic increases by the quantity $\pi(H/\tau^2)$.

§5. Time-averaged perturbations.

Let $\Pi_{e,E}$ be orthogonal projection of $L^2(\mathcal{H}_n/\Gamma_\alpha)$ onto the space

$$\{f \in L^2(\mathcal{H}_n/\Gamma_\alpha),\ \Box f = Ef\} . \tag{5.1}$$

The first step in the proof of Theorem 1 will be to show that the Schroedinger operator, $\Box + V$, has the same asymptotic behavior as the "time-averaged" Schoedinger operator, $\Box + V_{av}$, where

$$V_{av} = \sum_E \sum_{e\neq 0} \Pi_{e,E} V \Pi_{e,E} \tag{5.2}$$

Indeed let $\Delta_e = \pi_e \Box \pi_e$ (see Proposition 3.1). By Proposition 3.2 the eigenvalues of Δ_e are the integer multiples of $2\pi e$:

$$E = \lambda_{e,k} = 2\pi e(2k+n), \quad k = 0,1,\dots \tag{5.3}$$

and occur with multiplicities, $N_{e,E}$. If we add the perturbative term, V or V_{av}, to Δ_e this eigenvalue breaks up into an eigen-band

$$\lambda_{e,k,i}, \quad i = 1,\dots,N_{e,E}$$

or

$$\lambda^{av}_{e,k,i}, \quad i = 1,\dots,N_{e,E}$$

Now let $\mu_{e,E}$ be the measure (1.2) (which describes how the $\lambda_{e,k,i}$'s are distributed on this eigenband) and let $\nu_{e,E}$ be the corresponding measure for the $\lambda^{av}_{e,k,i}$'s:

$$(5.4)\qquad \nu_{e,E} = \frac{1}{N_{e,E}} \sum \delta(\lambda - (\lambda^{av}_{e,k,i} - \lambda_{e,k})) .$$

Proposition 5.1. As e and E tend to infinity along the parabola (1.3), the difference between $\mu_{e,E}$ and $\nu_{e,E}$ tends weakly to zero.

Proof. The gaps between the successive $\lambda_{e,E}$'s goes to infinity as e and E go to infinity by (5.3); so the "averaging" lemma of [U], §1 is applicable here.

Now let D be the generator of the circle group action on $\mathcal{H}_n/\Gamma_\alpha$; i.e. in terms of the coordinates $x_1,\dots,x_n,y_1,\dots,y_n,t$ of §4

$$D = \frac{1}{2\pi i}\frac{\partial}{\partial t} .$$

We will define the operator $\square\circ D^{-1}$ abstractly by defining it to be

$$(5.5)\qquad \square\cdot D^{-1} = \sum_{E}\sum_{e\neq 0} (E/e)\pi_{e,E} .$$

It is easy to see that this is a self-adjoint operator on $L^2(\mathcal{H}_n/\Gamma_\alpha)$. As such it generates a one-parameter group of unitary operators; and, for the moment, we will define this group abstractly:

$$(5.6)\qquad U(t) = \sum_{E}\sum_{e\neq 0} \exp\left[\sqrt{-1}\ t\ \frac{E}{e}\right]\pi_{e,E} .$$

Notice that, by (5.3), U(t) is periodic of period one in t. Comparing (5.6) with (5.2) one obtains the formula

$$(5.7)\qquad V_{av} = \int_0^1 U(s)VU(-s)ds .$$

In the next section we will derive Theorem 1 from a trace formula for V_{av} which we will now describe. Let

$$(5.8)\qquad \Pi_r = \sum \Pi_{e,E}$$

the sum taken over all E and e for which

$$E = r^2e^2 + 2\pi ne \,, \tag{5.9}$$

i.e. over the parabola (1.3). Notice that, by definition,

$$\int \lambda^m d\nu_{e,E}(\lambda) = \operatorname{trace} \Pi_{e,E}(V_{av})^m \Pi_{e,E} \,.$$

Therefore, adding up the traces of

$$\Pi_{e,E} V_{av}^m \Pi_{e,E} \exp \sqrt{-1} \, \frac{E}{e} \, t$$

for all e and E satisfying (5.9), we obtain

$$\operatorname{trace} \Pi_r V_{av}^m U(t) \Pi_r = \sum_{e,E} \left[\int \lambda^m d\nu_{e,E}\right] \exp \sqrt{-1}(r^2 e + 2\pi n)t \,. \tag{5.10}$$

As before the sum on the right is over all e,E satisfying (5.9). In the next section we will show how to obtain Theorem 1 from (5.10). We will devote the remainder of this section to showing that the left hand side of this formula is a classical Fourier integral distribution in t. (This is, in fact, the central ingredient in the proof of Theorem 1.) To start with observe that on the set, $\tau \neq 0$, of the cotangent bundle of $\mathcal{H}_n/\Gamma_\alpha$, $\square \circ D^{-1}$ is micro-differential operator of order one with

$$\sigma(\square \circ D^{-1}) = 2\pi H/\tau \,. \tag{5.11}$$

Let Ψ_t, $-\infty < t < \infty$ be the one-parameter group of canonical transformations generated by the Hamiltonian (5.11). From Proposition 4.1 we read off:

<u>Proposition 5.2</u>. The trajectories of Ψ are circles of radius $\sqrt{H}/\tau$ and period ½ on T_Γ.

By Egorov's theorem $U(t)$ is a one-parameter group of quantized canonical transformations on $\tau \neq 0$ associated with Ψ. Therefore by the formula, (5.7), V_{av} is a micro-differential operator on $\tau \neq 0$. Indeed, by Egorov's theorem,

Proposition 5.3. The symbol of V_{av} at any point, p, in the set, $\tau \neq 0$, is the integral of V over the circle on T_Γ obtained by projecting onto T_Γ the trajectory of Ψ through p.

We will next turn our attention to the projection operator, π_r. The image of π_r is the set of all L^2 solutions of the hyperbolic differential equation

$$\square = r^2D^2 + 2\pi nD \; . \tag{5.12}$$

At the symbolic level, this equation reads

$$H = r^2(\tau/2\pi)^2 \; . \tag{5.13}$$

Let Σ_r be the locus of points in the punctured cotangent bundle of $\mathcal{H}_n/\Gamma_\alpha$ satisfying (5.13). Notice that Σ_r is contained entirely in the set $\tau \neq 0$ and is a codimension one co-isotropic manifold. The null-foliation of Σ_r is by the integral curves of the Hamiltonian vector field $\Xi_r = \Xi_{H_r}$ where

$$H_r = H - (r\tau/2\pi)^2 \; . \tag{5.14}$$

We will show below that these integral curves are periodic and, in fact, all of the same period. This will imply that the null-foliation of Σ_r is fibrating (and is even the foliation of Σ_r by the fibers of a principal S^1 bundle). Since r^2 is a rational multiple of 2π, there exist positive integers, p and q, such that

$$r^2 = 2\pi q/p \; . \tag{5.15}$$

We will choose p and q to be as small as possible, i.e. containing no common factor.

Proposition 5.4. The integral curves of Ξ_r on Σ_r are simply periodic of period $(\pi/\tau)p$.

Proof. By Proposition 4.2 these curves are periodic in the x-y-ξ-η coordinates of period π/τ. Moreover, by part c) of Proposition 4.2, as each integral curve makes a complete circuit in the space of x-y-ξ-η variables, the t variable increases by the amount

$$(\pi H/\tau^2) - \theta^2(d/d\tau)(\tau/2\pi)^2(\pi/\tau) = -(r^2/2\pi) = -q/p .$$

Hence as the integral curve of Ξ_r makes p complete circuits in x-y-ξ-η space it makes one complete circuit in Σ_r. Q.E.D.

Notice also that these null-bicharacteristics are circles of radius $\sqrt{H}/\tau = r/2\pi$ in x-y space. It is also clear from Proposition 4.2 that if we reduce Σ_r by identifying points on the same null-leaves, the space we obtain is

$$T_\Gamma \times S^{2n-1} \times \mathbb{R}^+ . \tag{5.16}$$

In other words, to summarize what we have shown above,

<u>Proposition 5.5</u>. Σ_r is a principal S^1-bundle over the space, (5.16), and its fibers are the leaves of the null-foliation.

Finally we remark that by the theory of symplectic reduction there exists a symplectic form, Ω_{reduced}, on the space (5.16) whose pull-back to Σ_r is the restriction to Σ_r of the standard symplectic form on $T^*(\mathcal{H}_n/\Gamma_\alpha)$. It is easy to see that Ω_{reduced} is the product of the symplectic form, Ω_Γ, on T_Γ and the standard $SO(2n)$-invariant symplectic-contact form on $S^{2n-1} \times \mathbb{R}^+$.

Lets return now to the projection operator, Π_r. This is an example of the type of projection operators which we studied extensively in the papers [GS,1], [GS,2] and [GS,3]. Its main feature is:

<u>Proposition 5.6</u>. Π_r is a quantized canonical transformation (of classical type) associated with the "idempotent" canonical relation

$$\Sigma_r \underset{\pi}{\times} \Sigma_r ,$$

π being the fibration of Σ_r over the space, (5.16), described above.

§6. <u>Proof of Theorem 1</u>.

Consider first the right hand side of (5.10). By (5.15) $r^2 = 2\pi q/p$ where p and q are positive mutually prime integers. Therefore, the e's occuring in the right hand sum are

$$e = pj, \qquad j = 1,2,\ldots$$

Letting $\nu_j = \nu_{e,E}$ with $e = qj$ and $E = r^2e^2 + 2\pi ne$ we can rewrite the right hand side as

$$(6.1) \qquad \sum_j \left(\int \eta^m d\nu_j\right)(\exp 2\pi\sqrt{-1}(qj+n)t) \ .$$

Next consider the left hand side of (5.10). It follows from the results of §5 that this is a classical Fourier integral distribution on $\mathbb{R}$, periodic of period one and non-singular except at the points $0, 1/q, \ldots, 1$. Moreover, as is clear by comparing the two sides of 5.10, it is Floquet-periodic of period $1/q$ with a Floquet multiplier: $\exp 2\pi\sqrt{-1}\ n/q$. As such it can be expanded in an asymptotic sum of the form

$$(6.2) \qquad \sum_{k=0} a_k \chi_k(t)$$

where

$$(6.3) \qquad \chi_k(t) = \sum_{j\neq 0} |j|^{-k} \exp 2\pi\sqrt{-1}(qj+n)t \ .$$

The coefficient, a_0, of the leading term of (6.2) can be computed by standard symbolic techniques. (See, for instance, §4 of [DG].) This computation, which we will spare the reader, shows that

$$a_0 = \int_{T_\Gamma \times S^{2n-1}} (V_r)^m d\sigma \ .$$

Lets now compare (6.1) and (6.2). In view of (6.3), the asymptotic formula,

$$\sum_j \left[\int \lambda^m d\nu_j\right]\left[\exp 2\pi\sqrt{-1}(qj+n)t\right] \sim \sum_{k=0}^{-\infty} a_k \chi_k(t) \ ,$$

translates into the somewhat simpler formula

$$\int \lambda^m d\nu_j \sim \sum_{k=0}^{-\infty} a_k |j|^{-k} \ .$$

In particular

$$\int \lambda^m d\nu_j \to a_o = \int (V_r)^m d\sigma$$

as $j \to \infty$.

Finally, by Proposition 5.1, we can replace ν_j by μ_j in the expression on the left hand side.

§7. Some implications of Theorem 1.

Let us denote by μ_r the weak limit, (1.4), of $\mu_{e,E}$ as e and E tend to infinity along the parabola, $E = r^2 e + 2\pi ne$. Without loss of generality we can assume that the integral of the potential function, V, over T_Γ is zero. This implies that for all r

$$\int \lambda d\mu_r(\lambda) = 0 \ . \tag{7.1}$$

The main object which we will be looking at in this section is the quantity

$$E(r) = \int \lambda^2 d\mu_r(\lambda) \tag{7.2}$$

which we propose as a plausible measure of how "dispersed" the eigenvalues of $\Delta_e + V$ are on the high energy bands. We will derive an explicit formula for $E(r)$ in terms of the Fourier coefficients of V and then discuss to what extent one can determine the Fourier coefficients themselves from $E(r)$.

Parenthetical Remark. $E(r)$ is expressible as a spectral invariant of V only for those r's for which r^2 is a rational multiple of 2π. However, its clear that if we know $E(r)$ for these r we know it for all r.

Our starting point is a formula for V_r in terms of the Fourier coefficients of V: Let Γ^* be the dual of the lattice, Γ, and let

$$V(x) = \sum_{\omega\in\Gamma^*} c_\omega e^{i(\omega,x)} \ . \tag{7.3}$$

Proposition 7.1. For $x \in T_\Gamma$ and $v \in S^{2n-1}$

$$V_r(x,v) = \sum_{\omega\in\Gamma^*} c_\omega J_0\left[(r/2\pi)\,|\langle\omega,v\rangle|\right] e^{i(\omega,x)} \ . \tag{7.4}$$

<u>Notation</u>. We make the usual identification, $\mathbb{R}^{2n} = \mathbb{C}^n$. The square bracket, $\langle\ ,\ \rangle$, is the Hermitian inner product in $\mathbb{C}^n$ and the round bracket, $(\ ,\ ) = \mathrm{Re}\langle\ ,\ \rangle$ is the Euclidean inner product in $\mathbb{R}^{2n}$. J_0 is the standard Bessel function of order zero.

<u>Proof</u>. First of all we will prove the two-dimensional version of this formula. More explicitly, given $\omega \in \mathbb{R}^2$, we will compute the mean of $e^{i(\omega,x)}$ over the circle of radius $r/2\pi$ centered at the origin in $\mathbb{R}^2$. It is clear that this mean depends only on $|\omega|$, so we can assume $\omega = (|\omega|,0)$. In polar coordinates this is

$$(1/2\pi)\int_0^{2\pi} e^{i(r/2\pi)|\omega|\cos\theta}\,d\theta$$

or

$$J_0(r|\omega|/2\pi)$$

by a well-known formula for the zeroth order Bessel function ([WW], page 362). We will now prove (7.4). Without loss of generality we can assume $x = 0$ and $v = e_1 = (1,0,\ldots,0)$ in $\mathbb{C}^n$. By the computation we have just made

$$V_r(0,e_1) = \sum c_\omega J_0\left[(r/2\pi)|\omega^\#|\right]$$

where $|\omega^\#|$ is the Euclidean length of ω, restricted to the two-dimensional real subspace (one-dimensional <u>complex</u> subspace) of $\mathbb{C}^n$ spanned by $v = e_1$ and $\sqrt{-1}v = f_1$; i.e. if $\omega = (a_1,\ldots,a_n,b_1,\ldots,b_n)$, $|\omega^\#| = \sqrt{a_1^2 + a_1^2}$. On the other hand the Hermitian inner product of ω and $e_1 = v$ is $a_1 + ib_1$, so $|\langle\omega,v\rangle| = \sqrt{a_1^2 + a_1^2}$. Q.E.D.

From (7.4) we deduce that

$$E(r) = \sum_{\omega\in\Gamma^*} |c_\omega|^2 G\left[(r/2\pi)|\omega|\right] \tag{7.5}$$

where

$$G(s) = \int_{\mathbb{C}P^{n-1}} J_0^2(s|z_1|)\,d\kappa\ , \tag{7.6}$$

κ being the standard $SU(n)$-invariant probability measure on $\mathbb{C}P^{n-1}$. Notice that, when $n = 1$,

(7.7) $$G(s) = J_0(s)^2 .$$

For $n \geq 2$ the expression, (7.6), can be somewhat simplified. Since J_0 is even, there exists an analytic function, H_0, such that

$$J_0(s)^2 = H_0(s^2) ..$$

Substituting this into (7.6) the expression for $G(s)$ can be rewritten

(7.8) $$\int_{\mathbb{C}P^{n-1}} H_0(s^2|z_1|^2)d\kappa$$

where $\langle(z_1,\ldots,z_n)\rangle \in \mathbb{C}P^{n-1} \to |z_1|^2$ is the moment map for the symplectic action:

$$e^{i\theta} \circ z \to (e^{i\theta}z_1, z_2, \ldots, z_n)$$

of S^1 on $\mathbb{C}P^{n-1}$. By the Duistermaat-Heckman theorem the "push-forward" of κ by this moment mapping is the polynomial measure

$$(1 - \lambda)^{n-2}d\lambda$$

on the interval, [0,1], of the real axis. Therefore, we can rewrite (7.8):

$$\int_0^1 H_0(s^2\lambda)(1 - \lambda)^{n-2}d\lambda ,$$

or, finally setting $\lambda = t^2$, and using again (7.7),

(7.9) $$G(s) = 2\int_0^1 J_0(st)^2(1-t^2)^{n-2}tdt \quad \text{for} \quad n \geq 2 .$$

The main facts which we will need about $G(s)$ can be deduced equally well from (7.9) or (7.6):

<u>Proposition 7.2</u>. $G(s)$ is a real-analytic function of s, is everywhere positive and is of order $O(1/s)$ for large s. ([WW], page 368.)

Proof. The last assertion follows from the fact that

$$J_0(s) \sim (2/\pi s)^{1/2} \cos (s - \pi/4)$$

for s large.

Now let α be a number between zero and one. Multiplying (7.5) by $r^{-\alpha}$ and integrating from zero to infinity we obtain

$$\int_0^\infty r^{-\alpha}E(r)dr = \sum |c_\omega|^2 \int_0^\infty r^{-\alpha}G(|\omega|r/2\pi)dr ,$$

and, hence,

$$\frac{1}{\gamma_\alpha}\int_0^\infty r^{-\alpha}E(r)dr = \sum |c_\omega|^2|\omega|^{-1+\alpha} \tag{7.10}$$

where $0 < \alpha < 1$ and

$$\gamma_\alpha = \int_0^\infty s^{-\alpha}G(s/2\pi)ds . \tag{7.11}$$

Notice, by the way, that the integral (7.11) makes sense since the integrand is of order $O(s^{-\alpha})$ at zero and of order $O(s^{-1-\alpha})$ at infinity.

Since the right hand side of (7.10) is analytic in α for all α we deduce:

Theorem 2. The Sobolev norms

$$\|V\|_s^2 = \sum_{\omega\in\Gamma^*} |c_\omega|^2|\omega|^{-2s}$$

are spectral invariants of V for all real values of s.

Suppose now that the lattice, Γ, has the following property:

(SLS) If $\omega_1, \omega_2 \in \Gamma^*$ and $|\omega_1| = |\omega_2|$, then $\omega_1 = \pm\omega_2$.

Remark. It is easy to see that this property is a generic property, not only for sublattices of $\mathbb{R}^{2n}$ in general ([ERT]), but also for the restricted class of lattices we are considering here, i.e. integral lattices.

Example. For $n = 1$. Let Γ be the lattice generated by $\omega_1 = (1,a)$ and $\omega_2 = (b,1+ab)$ where a and b are algebraically independent over the rationals.

Now consider $\|V\|^2_{-s}$ as a function of s for s large. Clearly

$$\|V\|^2_{-s} = |c_{\omega_0}|^2 |\omega_0|^{-2s}(1 + o(1))$$

for s large, where $\pm\omega_0$ is the non-zero element of shortest length in Γ. Similarly

$$\|V\|^2_{-s} - |c_{\omega_0}|^2|\omega_0|^{-2s} = |c^2_{\omega_1}||\omega_1|^{-2s}(1 + o(1))$$

$\pm\omega_1$ being the element of next shortest length, and so on. Continuing, we obtain

Theorem 3. If Γ has the property (SLS) then the norms, $|c_\omega|^2$, of the Fourier coefficients of V are spectral invariants of V.

If V is even, i.e. if $V(x) = V(-x)$, the Fourier cofficients of V are real, so in particular:

Corollary. If V is even, the c_ω's themselves are, up to factors of ± 1, spectral invariants of V.

It is interesting to contrast these results with the much more subtle inverse spectral results for "band-invariants" of periodic potentials described in [ERT].

Concluding Remarks. For $n = 1$ and for certain Γ's the $\lambda_{e,k,i}$'s in (1.1) are the energy levels of an electron on a flat metallic plate in the presence of a uniform magnetic field pointing in the direction perpendicular to the plate. Our results seem to contradict what has been observed physically about these energy levels: i.e. the experimentally observed distribution of the $\lambda_{e,k,i}$'s about their Landau levels (the $\lambda_{e,k}$'s) seems to depend on the parameter, α, in an essential way. (This dependence on α is called "band" dependence.) On the other hand the distribution of the $\lambda_{e,k,i}$'s predicted by (1.4) is band independent! The only explanation we can see for this discrepancy is that the magnetic forces required to produce the distribution (1.4) may simply be too great to be realized experimentally.

Acknowledgements. We are grateful to Richard Melrose for several stimulating discussions, and to Jochen Bruning for pointing out to us the simple relationship between E(r) and the Sobolev norms of V described above.

Bibliography

[C] P. Cartier, "Quantum mechanical commutation relations and theta functions," Proc. of Symp. in Pure Math. IX, Amer. Math. Soc. Providence, R.I. 361-386 (1966).

[DG] J.J. Duistermaat and V. Guillemin, "The spectrum of positive elliptic operators and periodic bicharacteristics," Invent. Math. 29, 39-79 (1975).

[GS,1] V. Guillemin and S. Sternberg, "Some problems in integral geometry and some related problems in micro-local analysis," Am. J. Math. 101, 915-955 (1979).

[GS,2] V. Guillemin and S. Sternberg, "The metaplectic representation, Weyl operators and spectral theory," J. of Funct. Analysis, Vol. 42, 128-225 (1981).

[GS,3] V. Guillemin and S. Sternberg, "A generalization of the notion of polarization" (to appear).

[GU] V. Guillemin and A. Uribe, "Clustering theorems with twisted spectra," Math. Ann. 273, 479-506 (1986).

[LV] G. Lion and M. Vergne, The Weil representation, Maslov index and theta series. Progress in Mathematics, Birkhauser, Boston, Basel, Stuttgart (1980).

[U] A. Uribe, "Band asymptotics with non-smooth potentials" (to appear).

[W] N. Wallach, Symplectic geometry and Fourier analysis. Math. Sci. Press, 53 Jordan Road, Brookline, MA 02146 (1977).

[WW] E.T. Whittaker and G.N. Watson, A course in modern analysis. Cambridge U. Press, (fourth edition) Cambridge (1927).

[ERT] G. Eskin, J. Ralston and E. Trubowitz, "On isospectral periodic potentials in $\mathbb{R}^n$", Comm. P. Appl. Math. 37, 647-676 (1984).

The lifespan of classical solutions of non-linear hyperbolic equations

Lars Hörmander

P r e f a c e

During the last decade a number of results have been obtained both on global existence and on "blowup" of solutions of hyperbolic differential equations. In the two dimensional case John [3] showed that blowup always occurs in the "genuinely non-linear case" for small data of compact support. In fact, his methods allow one to determine the time of blowup asymptotically. This will be shown in Chapter I where we give a self contained and somewhat simplified exposition of these methods. (For systems of conservation laws DiPerna and Majda [1] have determined the limit of global entropy solutions after appropriate rescaling. Their results suggest but do not prove where blowup occurs before passage to the limit.)

Much less is known in the case of several variables. For radial solutions of the non-linear wave equation

$$u''_{tt} = c(u'_t)^2 \Delta u$$

in 3 space dimensions John [4] has proved an upper bound for the lifespan by applying the methods of [3] to the equation expressed in polar coordinates. A slightly modified version of his arguments will be given in Section 2.2. The rest of

Chapter II is devoted to proving existence theorems which in the radially symmetric case prove that John's upper bound for the lifespan actually gives the lifespan asymptotically. It seems highly plausible that the lower bound which we establish for arbitrary data of compact support is asymptotically correct. This is supported by the fact that the limit of the solutions after appropriate rescaling blows up at that time. However, just as the results of DiPerna and Majda [1] do not imply those of Chapter I, we can make no such rigorous conclusions. To prove the existence theorems we first construct approximate solutions and rewrite the Cauchy problem as a Cauchy problem for the difference between the exact and the approximate solution. The approximate solution is sufficiently accurate to allow existence to be proved by the methods of John-Klainerman [5] and Klainerman [6] in the form given them in Hörmander [2]. The limitation in the existence theorem comes from the blowing up of the approximate solution which is obtained by solving essentially a scalar first order differential equation. F. John [4a] has independently proved the same existence theorem using a combination of the methods of [4] and [5].

Chapter I

The two dimensional case.

1.1. Scalar first order operators. For later reference we shall here recall some basic and elementary facts concerning the Cauchy problem

(1.1.1) $\partial u/\partial t + a(u)\partial u/\partial x = 0; \; u(0,x) = u_0(x);$

where $a \in C^1$ and $u_0 \in C_0^1$. (All functions are assumed real valued.) If u is a C^1 solution for $0 \leqq t < T$, then $u(t,x) = u_0(y)$ on the straight characteristic line

$$x = y + ta(u_0(y)). \tag{1.1.2}$$

We have $\partial x/\partial y = 1 + ta'(u_0(y))u_0'(y)$, so the map $y \mapsto x$ is invertible if this is always positive. Then

$$\partial u/\partial x = u_0'(y)/(\partial x/\partial y) \tag{1.1.3}$$

which is unbounded if $a'(u_0)u_0'$ takes a negative minimum $-M$ at y and $t \to 1/M$. Thus we conclude that the largest T such that a C^1 solution exists for $0 \leqq t < T$ is given by

$$1/T = \max -a'(u_0(y))u_0'(y); \tag{1.1.4}$$

$T = +\infty$ if the maximum is 0. In case $u \in C^2$ we could also argue by differentiating the equation (1.1.1) with respect to x. With $w = \partial u/\partial x$ we obtain

$$Lw = -a'(u)w^2 \tag{1.1.5}$$

where $L = \partial/\partial t + a(u)\partial/\partial x$ denotes differentiation with respect to t along the characteristics (1.1.2). Thus

$$1/w - 1/w(0,y) = a'(u_0(y))t$$

which is just another way of writing (1.1.3). However, the preceding argument can be adapted to systems where the explicit integration is not available. In that case we shall only consider small initial data. Note that if we replace u_0 by $u_0(\varepsilon,.)$ where ε is small and $u_0(0,.)=0$, and replace T by T_ε, then

$$1/(\varepsilon T_\varepsilon) = \max\,(-a'(u_0(\varepsilon,y))\varepsilon^{-1}\partial u_0(\varepsilon,y)/\partial y) \to \max\,(-a'(0)\partial^2 u_0(0,y)/\partial\varepsilon\partial y), \text{ as } \varepsilon\to 0. \tag{1.1.4$'$}$$

In case a' vanishes of order N at 0 we obtain a better result,

$$1/(\varepsilon^{N+1}T_\varepsilon) \to \max\,(-a^{(N+1)}(0)\partial^2 u_0(0,y)/\partial\varepsilon\partial y(\partial u_0(0,y)/\partial\varepsilon)^N/N \tag{1.1.4$''$}$$

However, in the genuinely non-linear case where $a'(0) \neq 0$ the right-hand side of (1.1.4)' is not equal to zero, so (1.1.4)' gives precise information on the lifespan of solutions with small initial data.

The purpose of the subsequent sections is to extend (1.1.4)' to more general situations.

So far we have only discussed the lifespan of the solution. The explicit solution above shows that for the solution u_ε with initial data εu_0 we have

$$u(t,x) = \varepsilon u_0(y) \text{ when } x=y+ta(\varepsilon u_0(y)).$$

It follows easily that when $\varepsilon \to 0$

$$\varepsilon^{-1} u(t/\varepsilon, x+ta(0)/\varepsilon) \to u_0(y), \text{ if } x=y+tu_0(y)a'(0).$$

This means that the limit is the solution of Burgers' equation

$$\partial U/\partial t + a'(0)U\partial U/\partial x = 0, \quad U(0,x) = u_0(x),$$

as long as a solution without singularities exists.

Let us now return to the solution of (1.1.1). It is easy to estimate the L^1 norm of $w(t,.)$ for $t < T$. Indeed, let τ be a C^1 arc where $0 \leqq t < T$, such that τ intersects the orbits of L transversally. Then

$$\int_\tau w(dx - a(u)dt) = \int_\tau du = u_0(\beta) - u_0(\alpha),$$

$$\int_\tau |w||dx-a(u)dt| = \int_\alpha^\beta |du_0|$$

if the end points of τ are on the orbits of L starting at $(0,\alpha)$ and $(0,\beta)$ respectively. This is clear since $dx-a(u)dt = du$ and u is constant along the orbits of L. If t is constant on τ we obtain in particular a bound for the L^1 norm of $w(t,.)$, and if $dx = \lambda dt$ on τ, we obtain a bound for $\int_\tau |\lambda - a(u)||w|dt$, hence a bound for $\int_\tau |w|dt$, if $|\lambda - a(u)|$ is bounded from below on τ.

1.2. Generalities on first order systems. We shall now discuss the Cauchy problem for the hyperbolic system

(1.2.1) $\partial u/\partial t + a(u)\partial u/\partial x = 0$

where $u = (u_1, \ldots, u_N)$ and $a(u) = (a_{jk}(u))_{j,k=1}^{N}$ is an $N \times N$ matrix, with C^∞ entries for the sake of simplicity. We assume that $a(0)$ has real distinct eigenvalues. This implies that for all u in a neighborhood of 0 the matrix $a(u)$ has real eigenvalues

$$\lambda_1(u) < \ldots < \lambda_N(u)$$

and eigenvectors $r_1(u), \ldots, r_N(u)$ depending smoothly on u.

The equation (1.2.1) is well defined if u takes its values in a manifold M of dimension N and a is a section of Hom(TM, TM), that is, $a(u)$ is a linear map in T_uM which depends smoothly on u. It is useful to keep this in mind later on when we change the dependent variables u. (Note that this is legitimate since we only consider C^1 solutions.) However, for the time being we shall assume that u takes its values in $\mathbf{R}^N$.

Definition 1.2.1. A C^1 solution of (1.2.1) is called simple at (t_0, x_0) if the range belongs to a C^1 curve.

The definition is taken from John [3, p. 383]. As there we observe that the definition means that $u(t,x) = U(s(t,x))$ where $s \mapsto U(s)$ is a parametrization of the curve and $s \in C^1$. The equation (1.2.1) can now be written

$$\partial s/\partial t\, U' + \partial s/\partial x\, a(U)U' = 0$$

Thus U' is an eigenvector of $a(U)$, so for some j we have $a(U)U' = \lambda_j(U)U'$ and

(1.2.2) $\partial s/\partial t + \lambda_j(U(s))\partial s/\partial x = 0$

which is a scalar equation of the form discussed in Section 1.1. Conversely, let $s \mapsto U(s)$ be an integral curve of the eigenvector field $r_j(u)$, thus $U'(s) = r_j(U(s))$, and let s

satisfy (1.2.2). Then $u(t, x) = U(s(t,x))$ satisfies (1.2.1), and we shall say that u is j-simple. We can find a j-simple C^∞ solution locally with u prescribed at any point.

Let u be a C^2 solution of (1.2.1) taking values in $\mathbf{R}^N$, and let $w = \partial u/\partial x$, thus $\partial u/\partial t = -a(u)w$. If f is any C^1 function in $\mathbf{R}^N$ then the derivatives

$$\partial f(u)/\partial x = f'(u)\partial u/\partial x = f'(u)w, \quad \partial f(u)/\partial t = -f'(u)a(u)w$$

are linear functions of w. Thus $\partial a(u)/\partial x$ is a linear function of w, so differentiation of (1.2.1) with respect to x gives

$$(1.2.3) \qquad \partial w/\partial t + a(u)\partial w/\partial x = \gamma(u, w)$$

where γ is a quadratic form in w.

(1.2.3) is best interpreted if one splits w into its components along the eigenvectors, so we write

$$(1.2.4) \qquad \partial u(t, x)/\partial x = w(t, x) = \sum w_i(t, x) r_i(u(t, x)).$$

Since $ar_i = \lambda_i r_i$ and the derivatives of r_i are linear in w, we obtain from (1.2.3)

$$(1.2.3)' \qquad L_i w_i = \sum \gamma_{ijk}(u) w_j w_k, \quad i = 1, \ldots, N,$$

where

$$(1.2.5) \qquad L_i = \partial/\partial t + \lambda_i(u(t,x))\partial/\partial x$$

denotes differentiation with respect to t along the ith characteristic for the solution u, and γ_{ijk} are smooth functions of u.

Now recall that $w_i(dx - \lambda_i(u)dt) = du_i$ when N=1. To obtain an analogue for general N we calculate the differential

$$(1.2.6) \quad d(w_i(dx - \lambda_i(u)dt)) = (\partial w_i/\partial t + \lambda_i(u)\partial w_i/\partial x + w_i \partial \lambda_i/\partial x) dt \wedge dx$$

$$= (\sum \gamma_{ijk}(u) w_j w_k + w_i \langle \partial \lambda_i/\partial u, \sum w_k r_k(u)\rangle dt \wedge dx = \sum \Gamma_{ijk}(u) w_j w_k dt \wedge dx$$

where the last equality defines Γ_{ijk}. In particular (1.2.6) holds for any j simple solution u. Then $w_i = 0$ for $i \neq j$, and $w_j(dx-\lambda_j(u)dt)$ is exact (equal to ds with the notation above), so it follows that

(1.2.7) $$\Gamma_{ijj} = 0 \text{ for } i,\ j = 1, \dots, N.$$

By (1.2.6) this means that

(1.2.8) $$\gamma_{ijj} = -\delta_{ij}\langle\partial\lambda_i(u)/\partial u,\ r_i(u)\rangle,$$

where the Kronecker symbol δ_{ij} is 0 when $i\neq j$ and 1 when $i=j$. (See also John [3, pp. 383 and 393-396] where essentially the same facts were proved by explicit calculations.) For later reference we sum up the important conclusions in a lemma:

Lemma 1.2.2. If u is a C^2 solution of (1.2.1) then the derivative of the eigenvector component w_i of $\partial u/\partial x$ defined by (1.2.4) satisfies the characteristic differential equation (1.2.3)' where γ_{ijj} is given by (1.2.8). If τ is a C^1 arc intersecting the orbits of L_i transversally and τ_i is the open set bounded by τ, the orbits of L_i through the end points and an interval τ_{i0} where $t = 0$, then

(1.2.9) $$\int_\tau |w_i(dx-\lambda_i(u)dt)| \leqq \int_{\tau_{i0}} |w_i|dx + \int_{\tau_i} |\sum \Gamma_{ijk}w_jw_k|dxdt$$

if u is a C^2 solution in τ_i. Here $\Gamma_{ijj} = 0$ for all i and j.

Proof. Only (1.2.9) remains to be proved. It is obvious by (1.2.6) and Stokes' formula if absolute values are replaced by appropriate signs. Hence (1.2.9) follows if w_i has a constant sign on τ. Thus (1.2.9) holds if we integrate over the open set where $w_i\neq 0$, which proves the lemma.

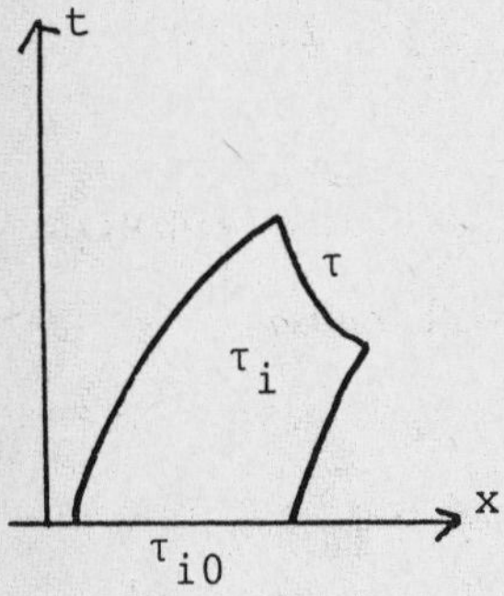

Note that γ_{jjj} depends on the normalization of r_j but is otherwise independent of the choice of u variables since it is just the derivative of λ_j along an orbit of r_j. The important point in (1.2.9) is that by (1.2.7) there are no squares in the right-hand side. The factors w_j will be concentrated in disjoint sets so at least one factor will be small in every term in the sum. Estimating the double integral will therefore cause no problems. We shall return to this in Section 1.4. Here we shall content ourselves with two simple consequences of (1.2.9).

Theorem 1.2.3. If $u \in C^2$ satisfies (1.2.1) and $\alpha \leqq x \leqq \beta$ in the support of the initial data, then

$$\alpha+\lambda_1(0)t \leqq x \leqq \beta+\lambda_N(0)t \text{ in supp } u, \ 0 \leqq t < T.$$

Proof. Let D be a "triangle" bounded by an interval where t=0 and the initial data of u vanish, and by integral curves of L_1 and L_N. Let

$$D_s = \{(t,x) \in D;\ t \leqq s\}\ ,\ \tau_s = \{x;\ (s,\ x) \in D\},$$

and set

$$(1.2.10) \qquad Y(s) = \sum_i \iint_{D_s} |w_i(t,\ x)|dxdt.$$

Then Y(0) = 0 and by (1.2.9) we obtain

$$0 \leqq Y'(s) = \sum_i \int_{\tau_s} |w_i(s,\ x)|dx \leqq CY(s).$$

Hence Y = 0 identically in D, and u is a constant in D which must be 0. The boundaries of D are therefore straight characteristic lines, which proves the theorem.

In this application we did not use (1.2.7) at all. However, it is important in the proof of the following well known result (see John [3, Section 3] or Lax [11]):

Theorem 1.2.4. If $u \in C^2$ satisfies (1.2.1) and $\alpha \leqq x \leqq \beta$ in the support of the initial data, then u is N simple to the right of the orbit of L_{N-1} through (0, β) and to the left of that of L_2 through $(0, \alpha)$.

Proof. Now let D be the set in the half plane $t > 0$ to the right of the orbit of L_{N-1} through $(0, \beta)$. We modify the definition (1.2.10) of $Y(s)$ so that the sum is only taken for $i < N$. Then

$$0 \leqq Y'(s) = \sum_{i<N} \int_{\tau_s} |w_i(s, x)|\, dx.$$

When we apply (1.2.9) to the terms in the sum, the interval on $t = 0$ which occurs lies to the right of β since $i \neq N$, and every term in the sum contains a factor w_j with $j \neq N$, by (1.2.7). Hence $Y'(s) \leqq C\,Y(s)$, and as before it follows that $Y(s) = 0$ identically. Thus u is N-simple in D. The other statement is proved in the same way.

When $N = 2$ it follows that u just consists of two simple solutions when t is large enough. In Section 1.3 we shall give a direct discussion of this simple case.

Using well-known arguments (cf. Lax [10, p.42]) one can deduce from (1.2.3)' a local existence theorem for the Cauchy problem for (1.2.1) when the Cauchy data u_0 are small. To do so we assume first that a C^2 solution is already known for $0 \leqq t < T$, say. If

$$M_1(t) = \sup_{j,x} |w_j(t, x)|$$

it follows from (1.2.3)', integrated along the orbits of L_j, that

$$M_1(t) \leqq M_1(0) + C_1 \int_0^t M_1(s)^2 ds.$$

Thus $M_1(t) < Y(t)$ if $Y(0) = M_1(0)$ and $Y'(t) > C_1 Y(t)^2$, for

$$M_1(t) - Y(t) < C_1 \int_0^t (M_1(s)^2 - Y(s)^2) ds$$

so we may conclude that $M_1(t) < Y(t)$ if t is the supremum of all s such that $M_1 \leqq Y$ in $[0, s]$. Hence

$$M_1(t) \leqq M_1(0)/(1 - C_1tM_1(0)), \; 0 \leqq t < 1/C_1M_1(0).$$

When $2C_1M_1(0)t < 1$ we obtain $M_1(t) \leqq 2M_1(0)$.

To estimate u itself we write

$$u = \sum u_jr_j(u)$$

and note that

$$a(u)\partial u/\partial x = \sum \partial u_j/\partial x \; \lambda_j(u)r_j(u) + \sum u_ja(u)\partial r_j(u)/\partial u \; w,$$

$$\partial u/\partial t = \sum \partial u_j/\partial t \; r_j(u) - \sum u_j\partial r_j(u)/\partial u \, aw.$$

Hence the equation (1.2.1) gives, for the component along r_j,

$$|L_ju_j| \leqq C|u||w|.$$

Set

$$M_0(t) = \sup_{j,x} |u_j(t,x)|.$$

Then

$$M_0(t) \leqq C_0 \int_0^t M_0(s)M_1(s)ds + M_0(0)$$

which implies

$$M_0(t) \leqq M_0(0) \exp(\int_0^t C_0M_1(s)ds) \leqq M_0(0)(1-C_1tM_1(0))^{-C_0/C_1}$$

$$\leqq M_0(0)2^{C_0/C_1} \text{ if } 2C_1M_1(0)t < 1.$$

If we apply a partial differentiation D^α to (1.2.3)' we obtain

$$L_iD^\alpha w_i + 2\sum \gamma_{ijk}w_jD^\alpha w_k = f_\alpha$$

where f_α is determined by derivatives of u of order $\leqq|\alpha|$. Hence we can estimate derivatives of u of order $\leqq 1+k$ when $2C_1M_1(0)t<1$ by a constant times a bound for them when t=0.

The preceding method for estimating a hypothetical solution easily gives an existence proof for given Cauchy data $u_0 \in C^\infty$ with all derivatives bounded. Indeed, put $u^0(t, x) = u_0(x)$ and define successively u^ν for $\nu \geqq 1$ by solving the linear Cauchy problem

$$\partial u^\nu/\partial t + a(u^{\nu-1})\partial u^\nu/\partial x = 0; \; u^\nu(0,x) = u_0(x).$$

Since the derivatives of a(u) only contributed to the right-hand side of our estimates above we conclude inductively that the estimates of M_1 and M_0 given for a hypothetical solution remain valid for all u^ν. We can then also get uniform bounds for $D^\alpha u^\nu$ for any α. What remains is just to prove that u^ν converges in the maximum norm. Set $v = u^\nu - u^{\nu-1}$ and $V = u^{\nu+1} - u^\nu$. Then $v = V = 0$ when $t = 0$, and from the equations

$$\partial u^{\nu+1}/\partial t + a(u^\nu)\partial u^{\nu+1}/\partial x = 0, \quad \partial u^\nu/\partial t + a(u^{\nu-1})\partial u^\nu/\partial x = 0$$

it follows that

$$\partial V/\partial t + a(u^\nu)\partial V/\partial x = (a(u^{\nu-1}) - a(u^\nu))\partial u^\nu/\partial x.$$

The right-hand side can be estimated by a constant times $|v|$. As in the estimate of M_0 above we obtain that if $|v(t,x)| \leqq m(t)$ when $2C_1M_1(0)t<1$, then

$$M(t) = \sup_x |V(t,x)|$$

satisfies in the same interval

$$M(t) \leqq \int_0^t (C_0 M(s) + Cm(s))M_1(s)ds.$$

Thus M is bounded by the solution of the Cauchy problem

$$Y'(t) = (C_0 Y(t) + Cm(t))M_1(t), \quad Y(0)=0.$$

Since $Y(t)$ is bounded by a constant times $\int_0^t m(s)ds$, it follows that

$$M(t) \leqq C' \int_0^t m(s)ds.$$

Inductively we conclude that

$$|u^{\nu+1} - u^\nu| \leqq C'^\nu t^{\nu-1}/(\nu-1)!$$

which proves the uniform convergence of the sequence u^ν, hence of all its derivatives. Thus we have proved that the Cauchy problem for (1.2.1) has a solution with given Cauchy data u_0. The uniqueness follows from the preceding argument by an obvious modification but was also proved in Theorem 1.2.3. Extending the a priori estimates at the beginning of the discussion to an estimate of the modulus of continuity, we find that it suffices to have $u_0 \in C^k$. Thus we have proved:

Theorem 1.2.5. For any $u_0 \in C^k$, $k \geqq 1$, with bounded derivatives and $\sup |D^\alpha u_0|$ sufficiently small when $|\alpha| \leqq 1$, the equation (1.2.1) has a C^k solution with $u(0,x) = u_0(x)$ for $0 \leqq t < T$ provided that

$$T \sup |u_0'| \leqq c$$

where c is a constant depending only on a; we have $\sup|u| \leqq C \sup|u_0|$.

In the preceding result we may allow u_0 and u to depend on a parameter ε (or several parameters). The solution is in C^k as a function of (x, ε) if the data are. In fact, to obtain a priori estimates we just differentiate (1.2.1) with respect to ε which gives

$$\partial u_\varepsilon'/\partial t + a(u)\partial u_\varepsilon'/\partial x + a'(u)u_\varepsilon' w = 0.$$

Writing

$$u_\varepsilon' = \sum \psi_j r_j(u)$$

we obtain equations of the form

$$L_j \psi_j + c_j \psi_j = 0$$

where c_j denotes a smooth function of u and w. Thus we obtain estimates of u_ε' in terms of C^1 bounds for the data, and by repeated differentiation we obtain C^k estimates. Now these a priori estimates can also be used in the existence proofs above to show that the solution constructed becomes a C^k function of (x, ε).

1.3. First order systems with two unknowns. When discussing simple wave solutions in Section 1.2 we observed the importance of the orbits of the vector fields r_j. (More precisely, these are line fields since only the directions are well defined.) In the two dimensional case one can change u coordinates in a neighborhood of the origin so that they become parallel to the coordinate axes. It suffices to choose the coordinates so that u_2 (resp. u_1) is constant along the orbits of r_1 (resp. r_2). Then a takes diagonal form, so (1.2.1) reduces to

$$\partial u_j/\partial t + \lambda_j(u)\partial u_j/\partial x = 0;\ j = 1, 2; \tag{1.3.1}$$

where $u = (u_1, u_2)$ and $\lambda_1(0) < \lambda_2(0)$. Let $u^0 \in C_0^\infty(\mathbf{R}^2, \mathbf{R}^2)$, assume that $u^0(0, x) = 0$ identically, and pose the initial condition

$$u(0,x) = u^0(\varepsilon,x) \tag{1.3.2}$$

where ε is small. From Theorem 1.2.5 we know that there is a solution for $t < c/\varepsilon$, such that $|u(t,x)| \leqq C'\varepsilon$. If ε is small enough it follows that with $c_1=(2\lambda_1(0)+\lambda_2(0))/3$ and $c_2=(\lambda_1(0)+2\lambda_2(0))/3$ we have

$$\lambda_1(u) \leqq c_1 < c_2 \leqq \lambda_2(u).$$

If $\operatorname{supp} u^0 \subset \mathbf{R} \times [a,b]$ we conclude that $u_1(t,x) = 0$ for $x>b+c_1t$ and that $u_2(t,x) = 0$ for $x<a+c_2t$. Since $a+c_2t \leqq x \leqq b+c_1t$ implies

$$t \leqq (b-a)/(c_2-c_1) = t_0,$$

it follows that u_1 and u_2 have disjoint supports for $t>t_0$, so u will then consist of two simple wave solutions with disjoint supports. They will therefore remain small as long as they exist, and the lifespan T_ε can be computed from (1.1.4)' if we replace T_ε by $T_\varepsilon-t_0$ and u_0 by $u_j(t_0,x)$, which vanishes when $\varepsilon=0$ and has the derivative $\partial u_j^0(0,x-\lambda_j(0)t_0)/\partial\varepsilon$ with respect to ε when $\varepsilon=0$. Hence we have

$$1/(\varepsilon T_\varepsilon) \to \max_{j,y} (-\partial\lambda_j(0)/\partial u_j \partial^2 u_j^0(0,y)/\partial\varepsilon\partial y). \tag{1.3.3}$$

From the discussion in Section 1.1 it follows also that $u_j(t/\varepsilon,x+t\lambda_j(0)/\varepsilon)/\varepsilon$ converges to the solution of Burger's equation $\partial U_j/\partial t+\partial\lambda_j(0)/\partial u_j\, U_j\partial U_j/\partial x = 0$ with initial value

$\partial u_j^0(0,.)/\partial\varepsilon$.

We can interpret (1.3.3) invariantly in the following way. $\partial u^0(0,.)/\partial\varepsilon$ can be regarded as a map from **R** to the tangent space at 0 of the manifold M where u takes its values. This vector space is the direct sum of the one dimensional eigenspaces, and $\partial\lambda_j(0)/\partial u_j\ \partial u_j^0(0,y)/\partial\varepsilon$ is the differential of λ_j in the direction of the component of $\partial u^0(0,y)/\partial\varepsilon$ along the jth eigenspace. Thus we have

<u>Theorem 1.3.1.</u> The largest T_ε, such that the Cauchy problem (1.3.1), (1.3.2) with $u^0 \in C_0^\infty(\mathbf{R}^2, M)$ and $u_0(0,.) = 0$ has a C^2 solution for $0 \leqq t < T_\varepsilon$, is given asymptotically by

(1.3.4) $$\lim_{\varepsilon\to 0} 1/\varepsilon T_\varepsilon = \max_{j,y} -\frac{d}{dy}\langle \lambda_j'(0),\ \pi_j \partial u^0(0,y)/\partial\varepsilon\rangle,$$

where π_j is the projection in the tangent space of M at 0 on the jth eigenspace along the other one.

In this statement there is no longer any condition on the choice of u coordinates; we have just functions with values in a manifold M with a point marked as 0.

1.4. <u>Systems with more than two unknowns.</u> We shall now return to the discussion of (1.2.1) when u takes its values in $\mathbf{R}^N$, N>2. With $u^0 \in C_0^\infty(\mathbf{R}^2, \mathbf{R}^N)$, $u^0(0,x)\equiv 0$, we pose the initial condition (1.3.2) where ε is small. From Theorem 1.2.5 we know that a smooth solution exists for $t \leqq c/\varepsilon$, and it is a C^∞ function of (t,x,ε). At first we shall emphasize the dependence on ε by using the notation $u(t,x,\varepsilon)$ for it, and we shall write

(1.4.1) $$u^{(\nu)}(t,x) = (\partial/\partial\varepsilon)^\nu u(t,x,\varepsilon)\big|_{\varepsilon=0},\ \nu = 1,2,\ldots$$

We may assume that the u variables are chosen so that

(1.4.2) the u_j axis is an orbit for the jth eigenvector field r_j when $j=1,\ldots,N$.

Explicitly this means that $a(te_j)e_j = \lambda_j(te_j)e_j$ if e_j is the jth basis vector. The advantage of such coordinates is obviously that they allow one to recognize j simple waves of compact support as solutions u for which the components other than u_j vanish. That such coordinates exist is clear. In fact, suppose that we already have an embedding φ of a neighborhood of 0 in $\mathbf{R}^k$ in a neighborhood of 0 in $\mathbf{R}^N$, mapping the coordinate axes in $\mathbf{R}^k$ to the first k eigenvector orbits through 0 in $\mathbf{R}^N$. If $k<N$ we extend it to a local embedding Φ of $\mathbf{R}^{k+1}$ so that $\Phi(v,0) = \varphi(v)$ and $\partial\Phi(v,s)/\partial s = r_{k+1}(\Phi(v,s))$ for (v,s) in a neighborhood of 0 in $\mathbf{R}^k\times\mathbf{R}$. After repeating this construction N times we get an embedding of $\mathbf{R}^N$ with the required properties. Moreover, the coordinates obtained are such that for every k the parallels of the u_k axis with $u_{k+1}=\ldots=u_N=0$ are orbits of r_k.

From (1.4.2) it follows if $a = (a_{jk})$ that

(1.4.3) $$a_{jk}(te_k) = \delta_{jk}\lambda_k(te_k).$$

In particular, $a_{jk}(0) = \delta_{jk}\lambda_k(0)$. Hence we obtain when $\varepsilon=0$ by differentiating (1.2.1) and using (1.4.1)

$$\partial_t u_j^{(1)} + \lambda_j(0)\partial_x u_j^{(1)} = 0,$$

hence

$$u_j^{(1)}(t,x) = \partial u_j^0(0,x-\lambda_j(0)t)/\partial\varepsilon.$$

Recursion formulas for determining $u^{(\nu)}$ when $\nu>1$ are obtained if u is replaced by the formal power series $\sum u^{(\nu)}\varepsilon^\nu/\nu!$ in (1.2.1). The coefficient of $\varepsilon^\nu/\nu!$ gives an equation of the form

$$\partial u^{(\nu)}/\partial t + a(0)\partial u^{(\nu)}/\partial x = f^\nu$$

where f^ν is a linear combination of products of components of $u^{(\mu)}$ and $\partial u^{(\mu)}/\partial x$ with $\mu<\nu$. If a term in f_j^ν only involves $u_k^{(1)}, \ldots, u_k^{(\nu-1)}$ and the x derivatives, but no other u components, then it must contain a factor $\partial^s a_{jk}(0)/\partial u_k{}^s$ with $s\neq 0$. If $j\neq k$ we know from (1.4.3) that $a_{jk}(te_k) = 0$ identically, so this derivative is 0. Hence the support of f_j is contained in the union of the support of $u_j^{(\mu)}$, $\mu<\nu$, and of the intersection of those of $u_k^{(\mu)}$ and $u_{k'}^{(\mu')}$ for some $\mu<\nu,\mu'<\nu$ and $k\neq k'$. Also note that the initial values of $u^{(\nu)}$ all vanish outside $[a,b]$ if $\operatorname{supp} u^0 \subset \mathbf{R}\times[a,b]$.

Suppose that $\lambda_1(0) < \lambda_2(0) < \ldots < \lambda_N(0)$, and let

$$c = \min_{0<j<N} (\lambda_{j+1}(0)-\lambda_j(0)).$$

Set

$$\Lambda_j = \{(t,\lambda_j(0)t),\ t\geqq 0\},\ K_1 = \{0\}\times[a,b].$$

Then $\operatorname{supp}\ u_j^{(1)} \subset K_1+\Lambda_j$, and we claim that $\operatorname{supp} u_j^{(\nu)} \subset K_\nu+\Lambda_j$ where

$$K_\nu = \{(t,x);\ a+\lambda_2(0)t \leqq x \leqq b+\lambda_{N-1}(0)t,\ 0\leqq t\leqq T_\nu\}, \tag{1.4.4}$$

$$T_\nu = (b-a)/c \sum_{0\leqq\mu<\nu-1} (1+(\lambda_{N-1}(0)-\lambda_2(0))/c)^\mu. \tag{1.4.5}$$

Let us assume that $\nu>1$ and that this has already been proved for smaller values of ν. We just have to show that the support of f_j^ν is contained in $K_\nu+\Lambda_j$. For terms involving a factor $u_j^{(\mu)}$ or $\partial u_j^{(\mu)}/\partial x$ the support is contained in $K_{\nu-1}+\Lambda_j \subset K_\nu+\Lambda_j$ by assumption. To cope with the other terms we must show that K_ν contains $(K_{\nu-1}+\Lambda_k)\cap(K_{\nu-1}+\Lambda_l)$ if $k\neq l$. Now

$$(K_\infty+\Lambda_k)\cap(K_\infty+\Lambda_l) \subset K_\infty \text{ if } k\neq l,$$

because $\lambda_k(0)$ and $\lambda_l(0)$ cannot both be equal to the same extreme eigenvalue $\lambda_1(0)$ or $\lambda_N(0)$. Hence

$$(K_{\nu-1}+\Lambda_k)\cap(K_{\nu-1}+\Lambda_l)\subset K_\infty.$$

Let $t \leqq T_{\nu-1}$ in $K_{\nu-1}$, thus

$$a+\lambda_2(0)T_{\nu-1} \leqq x \leqq b+\lambda_{N-1}(0)T_{\nu-1} \text{ when } t=T_{\nu-1}.$$

From a look at a figure it follows if $\lambda_k(0)<\lambda_l(0)$ that

$$t \leqq T_{\nu-1}+((b-a)+(\lambda_{N-1}-\lambda_2(0))T_{\nu-1})/(\lambda_l(0)-\lambda_k(0)) \leqq$$
$$\leqq T_{\nu-1}(1+(\lambda_{N-1}-\lambda_2(0))/c) + (b-a)/c$$

in $(K_{\nu-1}+\Lambda_k)\cap(K_{\nu-1}+\Lambda_1)$. Thus it follows inductively that T_ν can be estimated by the right-hand side of (1.4.5), which completes the proof.

Remark. From (1.4.5) it follows in particular that $u_k^{(\nu)} = 0$ when $k>1$ and $x < a+\lambda_2(0)t$. This is of course a consequence of Theorem 1.2.4. An analogous observation holds true when $x>b+\lambda_{N-1}(0)t$. Also note that T_ν grows linearly with ν if $N=3$ but grows exponentially otherwise.

The complete decoupling exploited in Section 1.3 does not occur when $N>2$. We shall therefore need two lemmas on solutions of equations of the form (1.1.5), with a' replaced by a function of t and error terms added which cover the interaction with other components.

Lemma 1.4.1. Let w be a solution in [0, T] of the ordinary differential equation

$$dw/dt = a_0(t)w^2 + a_1(t)w + a_2(t) \tag{1.4.6}$$

with a_j continuous and $a_0 \geqq 0$. Let

$$K = \int_0^T |a_2(t)|dt \exp(\int_0^T |a_1(t)|dt). \tag{1.4.7}$$

If $w(0) > K$ it follows that

$$\int_0^T a_0(t)dt \exp(-\int_0^T |a_1(t)|dt) < (w(0)-K)^{-1}. \tag{1.4.8}$$

Proof. Let us first assume that $a_1=0$, and introduce

$$w_2(t) = \int_0^t |a_2(s)|ds.$$

Then $w_2(0)=0$ and $w_2(T)=K$. Let w_1 be the solution of the differential equation

$$dw_1/dt = a_0(t)(w_1-K)^2 \; ; \; w_1(0) = w(0).$$

Thus

$$(w_1(t)-K)^{-1} - (w_1(0)-K)^{-1} = -\int_0^t a_0(s)\, ds,$$

so w_1 is increasing, and if w_1 exists in [0,T] then

(1.4.8)' $$\int_0^T a_0(s)ds < (w_1(0)-K)^{-1}.$$

Since

$$d(w_1-w_2)/dt = a_0(t)(w_1-K)^2-|a_2(t)| \leqq a_0(t)(w_1-w_2)^2+a_2(t)$$

and $w_1-w_2 = w$ when $t = 0$, we obtain $w_1-w_2 \leqq w$ in $[0,T]$ as long as w_1 exists. Thus w_1 cannot become infinite which proves that (1.4.8)' holds. For a general a_1 we just set

$$w(t) = W(t) \exp(\int_0^t a_1(s)\, ds).$$

This reduces (1.4.6) to

$$dW/dt = a_0(t) \exp(\int_0^t a_1(s)\, ds)W^2+ a_2(t) \exp(-\int_0^t a_1(s)\, ds),$$

and we just have to apply the special case of the lemma already proved.

Lemma 1.4.2. Let a_j be continuous functions in $[0,T]$, set $a_0^+ = \max(a_0,0)$, and define K by (1.4.7). If $0 \leqq w_0$ and

(1.4.9) $$\int_0^T a_0^+(t)dt \exp(\int_0^T |a_1(t)|dt) < (w_0+K)^{-1},$$

(1.4.10) $$\int_0^T |a_0(t)|dt \exp(\int_0^T |a_1(t)|dt) < K^{-1},$$

then (1.4.6) has a solution in $[0,T]$ with $w(0) = w_0$, and

(1.4.11) $$w(T)^{-1} \geqq (w_0+K)^{-1} - \int_0^T a_0^+(t)dt \exp(\int_0^T |a_1(t)|dt) \text{ if } w(T) \geqq 0,$$

(1.4.12) $$|w(T)|^{-1} \geqq K^{-1} - \int_0^T |a_0(t)|dt \exp(\int_0^T |a_1(t)|dt) \text{ if } w(T)<0.$$

Proof. Assume first that $a_1=0$ and let w_2 again be the integral of $|a_2|$ vanishing at 0. Thus $w_2(0) = 0$ and $w_2(T) = K$. Now let w_1 be the solution of

$$dw_1/dt = a_0^+(t)(w_1+K)^2, \quad w_1(0) = w_0,$$

that is,

$$(w_1(t)+K)^{-1} = (w_0+K)^{-1} - \int_0^t a_0^+(s)ds.$$

By (1.4.9) the increasing function w_1 exists in $[0, T]$. Since

$$d(w_1+w_2)/dt = a_0^+(t)(w_1+K)^2+|a_2(t)| \geqq a_0(t)(w_1+w_2)^2+a_2(t)$$

and $w_1+w_2=w$ at 0, we obtain $w \leqq w_1+w_2 \leqq w_1+K$ in $[0,T]$ if w exists, hence

$$1/w(T) \geqq (w_1(T)+K)^{-1} = (w_0+K)^{-1} - \int_0^T a_0^+(t)dt$$

if $w(T)>0$, which proves (1.4.11).

If on the other hand w has a zero in $[0, T]$ then we can apply (1.4.11) to $-w$, with w_0 replaced by 0 and to an interval starting at the zero of w. This gives (1.4.12). Now if we do not assume a priori that w exists in $[0, T]$, it follows that (1.4.11), (1.4.12) hold with T replaced by any smaller t such that a solution exists in $[0, t]$. Hence we have a fixed upper bound in any such interval, and it follows at once that a solution does exist in $[0, T]$, for the considered set of t values is both open and closed. Finally when a_1 is not identically 0 we can reduce to the case already studied just as in the proof of Lemma 1.4.1. The proof is complete.

We are now ready to prove an extension of Theorem 1.3.1.

Theorem 1.4.3. Assume that the eigenvalues $\lambda_1, \ldots, \lambda_N$ of $a(u)$ are distinct when $u = 0$, and let $u^0 \in C_0^\infty(\mathbf{R}^2, \mathbf{R}^N), u^0(0,.)=0$. Then the largest T_ε such that (1.2.1) has a C^2 (or C^∞) solution for $0 \leqq t < T_\varepsilon$ satisfying (1.3.2) is given asymptotically by (1.3.4).

Proof. We may assume that $\lambda_1(u) < \ldots < \lambda_N(u)$ for $|u| < \delta$ and that the u coordinate axes are orbits of the eigenvector fields. We shall denote by $[a, b]$ the smallest interval such that $\operatorname{supp} u^0 \subset \mathbf{R}\times[a, b]$.

If the right-hand side of (1.3.3) is positive, we must show that for any smaller positive number $1/M$ we have $T_\varepsilon < M/\varepsilon$ if ε is small enough. In addition we must prove that if $1/M$ is larger than the right-hand side of (1.3.3) then $T_\varepsilon \geqq M/\varepsilon$ for small ε. In the proof it is thus sufficient to discuss solutions defined in the strip $0 \leqq t \leqq M/\varepsilon$ for some fixed M.

Thus assume that we have a C^2 solution for $0 \leqq t \leqq T$ and that $\varepsilon T \leqq M$ for some fixed M. With a slight modification of the approach of John [3] we introduce the L^1 norm

$$J(t) = \sup_{0 \leqq s \leqq t} \sum_1^N \int |w_j(s,x)|dx$$

which is $O(\varepsilon)$ for fixed t, and the maximum norm

$$V(t) = \sup_{0 \leqq s \leqq t} \sup_i \sup_{(x,s) \notin R_i} |w_i(s,x)|.$$

Here R_i is the strip bounded by the integral curves of the vector fields L_i in (1.2.5), starting at (0, a) and (0, b), and by the lines $t = 0$ and $t = T$. If $|u| < \delta$, which will be justified later on, then $|\lambda_i - \lambda_j| \geqq c$ for some fixed $c > 0$, so the R_i are disjoint when $t > t_0 = (b-a)/c$. For fixed t we have seen that $w_i = O(\varepsilon^2)$ outside R_i. (In fact, we proved that the support of the first order term in w_i lies in the strip $K_1 + \Lambda_i$ which is at distance $O(\varepsilon)$ from R_i.) We wish to prove that these estimates remain valid uniformly when $t \leqq T$.

Let E be a constant such that $J(0) \leqq E\varepsilon/2$. We claim that for a suitable constant F

(1.4.13) $\qquad J(t) < E\varepsilon, \quad V(t) < F\varepsilon^2$ when $0 \leqq t \leqq T$.

It suffices to prove that if this is true when $t < T$ then the strict inequality is also true when $t = T$, for then the set of all $T' \leqq T$ such that (1.4.13) is valid with T replaced by T' is both open and closed, and it contains t_0.

To prove the first part of (1.4.13) we use (1.2.9), recalling the bound for supp u given in Theorem 1.2.3. This gives for $s \leqq t$

$$\sum \int |w_i(s,x)|dx \leqq \sum \int |w_i(0,x)|dx + CtV(t)J(t) + C\varepsilon^2,$$

for $w = O(\varepsilon)$ up to time t_0 while for $s > t_0$ and $j \neq k$ we can estimate

$$\int |w_j(s, x)w_k(s, x)|dx$$

by $J(t)V(t)$ since the strips R_j and R_k are disjoint then. Recalling that $j\neq k$ in the sum in (1.2.9) (by (1.2.7)), we obtain

$$J(t) \leqq E\varepsilon/2 + CM\varepsilon^{-1}F\varepsilon^2 J(t) + C\varepsilon^2$$

which implies

$$J(t) \leqq (E\varepsilon/2+C\varepsilon^2)/(1-CMF\varepsilon)$$

if $CMF\varepsilon < 1$. For small ε we conclude that $J(t) < 2E\varepsilon/3$, say.

Before passing to the estimate of $V(t)$ we observe that $J(t) \leqq E\varepsilon$ implies $|u(t, x)| \leqq CE\varepsilon$ since u has compact support for fixed t. If ε is so small that $CE\varepsilon < \delta$ this justifies our a priori hypothesis that $|u| \leqq \delta$.

To estimate $V(t)$ we shall apply (1.2.3)' on an orbit of L_i starting outside $[a, b]$. The initial data of w_i are 0 then. When $t \leqq t_0$ the sum in (1.2.3)' is $O(\varepsilon^2)$. For $t>t_0$ the sum when neither j nor k is equal to i is $O(F^2\varepsilon^4)$ except in an interval τ contained in R_j for some $j\neq i$, hence in no other R_k. By (1.2.9) again we have

$$c\int_\tau |w_j|dt \leqq E\varepsilon/2 + CMF\varepsilon J(t) + C\varepsilon^2 < E\varepsilon,$$

if ε is small enough. Hence it follows from (1.2.3)' that

$$|w_i(t,x)| \leqq C\varepsilon^2 + E\varepsilon F\varepsilon^2 + CF^2\varepsilon^4 M/\varepsilon,$$

where the three terms estimate the integral of $|L_i w_i|$ on the orbit when $t \leqq t_0$, in the intersections with R_j for $j \neq i$ and the rest of the orbit respectively. When ε is sufficiently small we obtain the second part of (1.4.13) if $F > C$.

Having verified (1.4.13) we shall now estimate T by applying Lemma 1.4.1 to an orbit T of L_i inside R_i. We can write

$$L_i w_i = \gamma_{iii}(u)w_i^2 + a_1 w_i + a_2;\ a_1 = 2\sum_{j\neq i}\gamma_{iij}w_j,\ a_2 = \sum_{j,k\neq i}\gamma_{ijk}.$$

By the estimates just given we have

$$\int_0^t |a_2(t)|\ dt < C\varepsilon^2,\ \int_0^t |a_1(t)|dt < C\varepsilon\ .$$

Hence

$$K = \int_0^t |a_2(s)|ds\ \exp\ (\int_0^t |a_1(s)|ds) < 2C\varepsilon^2$$

if ε is small. The integrals are of course taken along T. If $\gamma_{iii}(0) > 0$ and $w_i(0,x) > 2C\varepsilon^2$ at the initial point $(0,x) \in T$ it follows that

$$\int_0^T \gamma_{iii}(u)\ dt < e^{C\varepsilon}(w_i(0,x) - 2C\varepsilon^2)^{-1}.$$

Since $u = O(\varepsilon)$ it follows if we take $T = T_\varepsilon - 1$ and multiply by ε that

$$\overline{\lim_{\varepsilon\to 0}}\ \varepsilon T_\varepsilon \gamma_{iii}(0)\ \partial^2 u_i^0(0,x)/\partial x \partial\varepsilon \leqq 1.$$

By (1.2.8) we have $\gamma_{iii}(0) = -\partial\lambda_i(0)/\partial u_i$, so we obtain

$$\overline{\lim_{\varepsilon\to 0}}\ \varepsilon T_\varepsilon(-\partial\lambda_i(0)/\partial u_i)\partial^2 u_i^0(0,x)/\partial x\partial\varepsilon \leqq 1. \tag{1.4.14}$$

Changing the sign of u_i we have the same conclusion if $\partial^2 u_i^0(0,x)/\partial x\partial\varepsilon$ and $-\partial\lambda_i(0)/\partial u_i$ are both negative, and (1.4.14) is trivial when the signs are opposite. This completes the proof of the upper bound for T_ε implied by (1.3.4).

It remains to prove that if

$$\max_{i,x}\ (-\partial\lambda_i(0)/\partial u_i)\ \partial^2 u_i^0(0,x)/\partial x\partial\varepsilon < A \tag{1.4.15}$$

then $\varepsilon T_\varepsilon > 1/A$ for small ε. To do so we shall continue our estimates above to derive a uniform bound for w_i when $t \leqq T \leqq (A\varepsilon)^{-1}$ and ε is small. Once we have such a bound it is easy to deduce as in Section 1.2 that $u \in C^\infty$ for $t \leqq T$, hence that $T_\varepsilon > T$ and $T_\varepsilon > (A\varepsilon)^{-1}$. To estimate w_i in R_i we use the same differential equation for w_i on an orbit of L_i as before, but now we appeal to Lemma 1.4.2. Assume that the initial value w_{i0} of w_i is $\geqq 0$ for example. By (1.4.15) we have $\gamma_{iii}(u)w_{i0} < \varepsilon B$ on the orbit if B is the mean value between the two sides and ε is small. Thus

$$w_{i0} \int_0^T a_0^+ \, dt \leqq \varepsilon TB \leqq B/A < 1.$$

As before $K < 2C\varepsilon^2$, and

$$\int_0^T |a_1(t)|dt = O(\varepsilon), \quad \int_0^T |a_0(t)|\,dt = O(1/\varepsilon),$$

so (1.4.9) and (1.4.10) hold when ε is small. The right-hand side of (1.4.11) is at least $\frac{1}{2}(1-B/A)(w_{i0}+K)^{-1}$, and that of (1.4.12) is $\geqq (2K)^{-1}$. Hence we conclude that w_i is bounded by ε times a constant independent of T. By the remarks above the proof is now completed by establishing bounds for the higher order derivatives of u in a completely straightforward manner. We leave this for the reader.

Having proved Theorem 1.4.3 we shall make some remarks on additional information which can be obtained from the proof. The argument at the beginning of this section shows that for any μ we have $w_i = O(\varepsilon^\mu)$ if $t = t_\mu$ is large, except in some disjoint intervals moving to the right with i. Let R_i^μ now be the strips generated by these intervals. We can then show in the same manner as before that ε^2 can be replaced by ε^μ in (1.4.13). Thus the solution u will for small ε be very close to a j simple solution when we stay well away from the other strips R_k^μ. To study u closely in R_i^μ it is convenient to assume the u coordinates chosen so that all parallels of the u_i axis are integral curves of r_i. Since

$$\partial u/\partial x = \sum w_k r_k(u)$$

and r_i has no component except in the ith coordinate direction, it follows from (1.4.13), extended as just indicated, that when $t_\mu < t < T_\varepsilon$ we have

$$\partial u_j/\partial x = O(\varepsilon^\mu) \text{ in } R_i^\mu \text{ if } j\neq i \text{ while } \partial u_i/\partial x - w_i = O(\varepsilon^\mu).$$

Hence $u_j = O(\varepsilon^{\mu-1})$ in R_i^μ for any μ if $j\neq i$. From the equation

$$\partial u_i/\partial t + (a(u)\partial u/\partial x)_i = 0$$

we obtain for $t > t_\mu$ that in R_i^μ

$$\partial u_i/\partial t + \lambda_i(u)\partial u_i/\partial x = O(\varepsilon^\mu),$$

for

$$a(u)\partial u/\partial x = \sum w_k\lambda_k(u)r_k(u) = w_i\lambda_i(u)r_i(u) + O(\varepsilon^\mu) =$$
$$= \partial u_i/\partial x\lambda_i(u)r_i(u) + O(\varepsilon^\mu).$$

Hence $u_j = O(\varepsilon^{\mu-1})$ in R_i^μ for $j\neq i$, and the oscillation of u_i on the orbits of L_i is $O(\varepsilon^{\mu-1})$ when $t_\mu < t < T_\varepsilon$. As in Sections 1.1 and 1.3 we conclude that the nearly simple i wave approaches a solution of Burger's equation after appropriate rescaling. (See also DiPerna and Majda [1].)

1.5. Scalar operators of higher order. In this section we shall discuss quasilinear hyperbolic equations of the form

(1.5.1) $$\sum_0^m c_j(\partial^{m-1}u)\partial_t^{m-j}\partial_x^{j}u = 0$$

with the Cauchy boundary condition

(1.5.2) $$u - \varepsilon\varphi = O(t^m) \text{ as } t\to 0,$$

where $\varphi\in C_0^\infty(\mathbf{R}^2)$ is given and $\partial^{m-1} = \{\partial_t^{m-j}\partial_x^{j-1}\}_{1\leqq j\leqq m}$. We assume that $c_0 = 1$ identically and write

$$P(\partial^{m-1}u, \xi) = \sum c_j(\partial^{m-1}u)\, \xi_t^{m-j}\xi_x^{j},$$
$$p(\vartheta, \xi) = P(\vartheta^{m-1}, \xi)$$

when $\vartheta, \xi\in\mathbf{R}^2$; $\xi = (\xi_t, \xi_x)$. The polynomial $p(0, \xi)$ is assumed strictly hyperbolic which makes $p(\vartheta, \xi)$ strictly hyperbolic with respect to ξ for small ϑ.

We reduce (1.5.1) to a first order system by introducing

(1.5.3) $$U_j = \partial_t^{m-j}\partial_x^{j-1}u, 1 \leqq j \leqq m.$$

The equation (1.5.1) can then be written

$$\partial_t U_1 + \sum_1^m c_j(U)\partial_x U_j = 0,$$

and together with the compatibility conditions $\partial_t U_{j+1} = \partial_x U_j$, $1 \leqq j < m$, we obtain the system

$$\partial_t U + a(U)\partial_x U = 0, \tag{1.5.4}$$

where

$$a = \begin{pmatrix} c_1 & \cdots & c_m \\ -1 & 0 & .. \; 0 \\ 0 & -1 & . \; 0 \\ \cdots & \cdots & \cdots \\ 0 \ldots & & -1 \; 0 \end{pmatrix}. \tag{1.5.5}$$

The determinant of $a-\lambda I$ is $\sum(-\lambda)^{m-j}c_j$. If we denote the roots by $\lambda_1, \ldots, \lambda_m$ as in Section 1.4, it follows that

$$P(U, \xi) = \prod_1^m (\xi_t+\lambda_k\xi_x). \tag{1.5.6}$$

For the eigenvector of a corresponding to the eigenvalue λ_k we have $U_j + U_{j+1}\lambda_k = 0$, so u is proportional to

$$r_k = \{(-\lambda_k)^{m-j}\}_{j=1}^m.$$

We can calculate the derivative of λ_k at $U = 0$ in the direction r_k by differentiating (1.5.6). This gives, with $\xi = (-\lambda_k(0), 1)$,

$$\partial_\varepsilon P(\varepsilon r_k(0), \xi) = \prod_{\nu\neq k}(\lambda_\nu(0)-\lambda_k(0))\langle\lambda_k'(0), r_k(0)\rangle, \; \varepsilon = 0.$$

Equivalently, we have

$$\partial_\varepsilon p(\varepsilon\xi, \xi)_{|\varepsilon=0} = \prod_{\nu\neq k} (\lambda_\nu(0)-\lambda_k(0))\langle\lambda_k'(0), r_k(0)\rangle.$$

To apply (1.3.4) we must also when $t = 0$ write $U = \sum \gamma_k r_k(0)$. Let

$$Q_k = \prod_{\nu\neq k}(\partial_t+\lambda_\nu(0)\partial_x),$$

and note that if $U = r_j(0)$ at $(0, x)$ then

$$Q_k u = \prod_{\nu\neq k}(\lambda_\nu(0)-\lambda_j(0))$$

there. This is 0 if $j\neq k$. Hence we have when $t=0$

$$\gamma_k \prod_{\nu\neq k}(\lambda_\nu(0)-\lambda_k(0)) = Q_k u = \varepsilon\prod_{\nu\neq k}(\partial_t+\lambda_\nu(0)\partial_x)\varphi.$$

The right-hand side of (1.3.4) becomes

$$\max -\partial_\varepsilon p(\varepsilon\xi,\ \xi)_{|\varepsilon=0,\xi=(-\lambda_k,1)}\partial_x \prod_{\nu\neq k}(\lambda_\nu(0)-\lambda_k(0))^{-2}(\partial_t+\lambda_\nu(0)\partial_x)\varphi$$

when $t = 0$, so we have proved:

<u>Theorem 1.5.1.</u> Let the equation (1.5.1) be strictly hyperbolic with respect to t when $\partial^{m-1}u = 0$, and let $\varphi \in C_0^\infty(\mathbf{R}^2)$. Then the largest T_ε such that (1.5.1) has a solution with the Cauchy data (1.5.2) is given asymptotically by

$$(1.5.7)\quad 1/\varepsilon T_\varepsilon \to \max\ -\partial_\varepsilon p(\varepsilon\xi,\ \xi)_{|\varepsilon=0} q(\xi,\ \partial)\varphi_{|t=0},$$

where the maximum is taken for all real x and ξ with $\xi \neq 0$ and $p(0,\xi)=0$, and

$$q(\xi,\ \partial)=(\partial p(0,\xi)/\partial\xi_t)^{-2}\ \partial_x p(0,\partial)/(\xi_x\partial_t-\xi_t\partial_x).$$

Note that the quantity in the right-hand side of (1.5.7) is homogeneous in ξ of degree 0, so it is enough to take $\xi = (-\lambda_k(0),\ 1)$ for $k = 1, \ldots, m$. If we multiply all coefficients by the same function $c(\partial^{m-1}u)$, we see that the right-hand side does not change so it was not necessary to assume in the theorem that c_0 was normalized to be equal to 1 although we did so in the proof.

Theorem 1.5.1 shows that for general φ the lifespan of the solution is only $O(1/\varepsilon)$ unless $p(\xi,\ \xi)$ vanishes of order m+1 at 0 when $p(0,\xi) = 0$. One might call the equation genuinely non-linear if this does not happen.

Since Chapter II will be devoted entirely to second order operators in more than two variables, we shall make some additional comments here on the case $m = 2$ of the preceding results. They can of course be obtained directly from the simple observations in Section 1.3 and are independent of Section 1.4. Between the two strips where U is 1 or 2 simple we have $U = 0$, hence u is a constant. To determine this constant to the first approximation we just have to solve the linear problem

$$(\partial_t+\mu_1\partial_x)(\partial_t+\mu_2\partial_x)v = 0;\ v = \varphi\ ,\ \partial_t v = \partial_t\varphi \text{ when } t=0;\ \mu_j=\lambda_j(0).$$

This gives

$$u = \varepsilon|\lambda_2(0)-\lambda_1(0)|^{-1}\int \varphi_t'(0,x)dx + O(\varepsilon^2)$$

for the constant value of u between the strips where U is simple. Thus it is not necessarily equal to 0. Since $U = (\partial_t u,\ \partial_x u) = O(\varepsilon)$ and these strips have straight and parallel boundaries, it follows that $u = O(\varepsilon)$.

In the strip where U is k-simple (k = 1 or 2) we know that U lies on the curve through 0 defined by

$$dU_1/dU_2 = -\lambda_k(U),$$

and U is constant on lines with slope $\lambda_k(U)$ in the tx plane. Since on the curve

$$U_1 = -\lambda_k(0)U_2-(-\lambda_k(0)\partial\lambda_k(0)/\partial U_1+\partial\lambda_k(0)/\partial U_2)U_2^2/2+O(U_2^3),$$

the component of U along such a line is

$$U_1+\lambda_k(U)U_2 = (-\lambda_k(0)\partial\lambda_k(0)/\partial U_1+\partial\lambda_k(0)/\partial U_2)U_2^2/2+O(U_2^3).$$

In the genuinely non-linear case the leading term on the right-hand side is not 0, so the oscillation of u along the lines where U is constant (and u linear) is just $O(\varepsilon)$; we knew already that it could not be larger than that.

To determine the asymptotic behavior of the solution of (1.5.1) and (1.5.2) when m = 2 it is also convenient to parametrize the integral curve of r_k through 0 by U_2, that is, solve the equation

$$dU_1/dU_2 = -\lambda_k(U);\ U_1 = 0 \text{ when } U_2 = 0.$$

For U_2 we then obtain the scalar equation

$$\partial U_2/\partial t + \lambda_k(U_1,\ U_2)\ \partial U_2/\partial x = 0$$

in the k simple region; here

$$\lambda_k(U_1, U_2) = \lambda_k(0) + aU_2 + O(U_2^2);$$
$$a = (-\partial\lambda_k(0)/\partial U_1\lambda_k(0) + \partial\lambda_k(0)/\partial U_2).$$

From Section 1.1 we know therefore that

$$\varepsilon^{-1}U_2(t/\varepsilon, x+t\lambda_k(0)/\varepsilon) \to V$$

where with $j = 3-k$ (the other index among 1 and 2)

$$\partial V/\partial t + a V\partial V/\partial x = 0; \quad V = (\partial_t + \lambda_j(0)\partial_x)\varphi/(\lambda_j(0) - \lambda_k(0)), \ t=0.$$

Thus

$$\varepsilon^{-1}\partial u(t/\varepsilon, x+t\lambda_k(0)/\varepsilon)/\partial x \to V,$$

and we obtain the asymptotic properties of u itself by integrating with boundary value 0 at one side or the other of the strip.

It is easy and instructive to make (1.2.3), (1.2.6) explicit for the system derived from a second order scalar equation. Thus assume that $m = 2$ and let $c_0=1$, $c_1=\lambda_1(U)+\lambda_2(U)$, $c_2=\lambda_1(U)\lambda_2(U)$. The eigenvalues of a are now λ_1 and λ_2, and the eigenvectors are $(-\lambda_j(U), 1)$. The equations

$$\partial U/\partial x = \sum w_j(-\lambda_j(U), 1)$$

mean that

$$u''_{tx} = -w_1\lambda_1 - w_2\lambda_2, \quad u''_{xx} = w_1 + w_2,$$

hence

$$w_1(\lambda_2 - \lambda_1) = \lambda_2 u''_{xx} + u''_{xt} = L_2 u'_x$$

and similarly for w_2. Recall that $L_j = \partial_t + \lambda_j(U)\partial_x$. Thus

$$(\lambda_2-\lambda_1)L_1w_1 + w_1L_1(\lambda_2-\lambda_1) = (L_1\lambda_2)u''_{xx} + (\sum c_j\partial_t^{2-j}\partial_x^{j})u'_x =$$
$$= (L_1\lambda_2)u''_{xx} - (\lambda_1+\lambda_2)'_x u''_{xt} - (\lambda_1\lambda_2)'_x u''_{xx}, \text{ which}$$

gives

$$(\lambda_2-\lambda_1)L_1w_1 = -w_1L_1(\lambda_2-\lambda_1) + (w_1+w_2)(L_1\lambda_2 - (\lambda_1\lambda_2)'_x) +$$
$$+ (\lambda_1w_1 + \lambda_2w_2)(\lambda_1+\lambda_2)'_x = w_1(L_1\lambda_1 + (\lambda_1-\lambda_2)\lambda'_{1x}) + w_2L_2\lambda_2.$$

Now we have for any C^1 function $f(U)$

$$L_1f(U) = f'_1(U)L_1u'_t + f'_2(U)L_1u'_x,$$

$L_1 u'_x = u''_{xt}+\lambda_1 u''_{xx} = (\lambda_1-\lambda_2)w_2$; $L_1 u'_t = -\lambda_2 L_1 u'_x = -\lambda_2(\lambda_1-\lambda_2)w_2$.

Thus we have

$$L_1 f(U)=w_2(\lambda_2-\lambda_1)(\lambda_2\partial/\partial U_1-\partial/\partial U_2)f,$$

and a similar formula for $L_2 f(U)$, which gives

(1.5.8) $L_1 w_1=-w_1\partial\lambda_1/\partial x+w_1 w_2((\lambda_2\partial/\partial U_1-\partial/\partial U_2)\lambda_1-(\lambda_1\partial/\partial U_1-\partial/\partial U_2)\lambda_2)$,

$$\partial\lambda_1/\partial x = -(w_1\lambda_1+w_2\lambda_2)\partial\lambda_1/\partial U_1+(w_1+w_2)\partial\lambda_1/\partial U_2.$$

We get a particularly simple analogue of (1.2.6),

(1.5.9) $d(w_1(dx-\lambda_1 dt))=w_1 w_2((\lambda_2\partial/\partial U_1-\partial/\partial U_2)\lambda_1-(\lambda_1\partial/\partial U_1-\partial/\partial U_2)\lambda_2)dt\wedge dx$.

For the non-linear wave equation $u''_{tt}=c^2 u''_{xx}$ we have $\lambda_1=c=-\lambda_2$, so the right-hand side of (1.5.9) becomes $-2\, w_1 w_2 \partial c/\partial U_2 dt\wedge dx$.

It vanishes if c is just a function of $U_1 = \partial u/\partial t$; then we have

(1.5.8)' $L_1 w_1 = w_1(w_1-w_2)cc'$, $L_2 w_2 = w_2(w_2-w_1)cc'$,

(1.5.9)' $d(w_1(dx-cdt)) = 0$, $d(w_2(dx+cdt)) = 0$.

The work of John [4] on the rotationally symmetric wave equation with three space variables starts essentially from these simple formulas. We shall review his work in Section 2.2.

Chapter II

Second order operators with three or two space variables

2.1. The linear wave equation. In this section we shall

discuss some basic facts on the solution of the wave equation in $\mathbf{R}^{1+n}$, $n \geqq 1$,

$$\partial^2u/\partial t^2 = \Delta u, \tag{2.1.1}$$

with Cauchy data

$$u = f,\ \partial_t u = g \text{ when } t = 0, \tag{2.1.2}$$

where $f, g \in C_0^\infty(\mathbf{R}^n)$. If u_g is the solution when $f = 0$ then $\partial_t u_f$ is the solution when $g = 0$, for $\partial_t^2 u_f = \Delta u_f = 0$ when $t = 0$. Thus the general solution is $u_g + \partial_t u_f$, and it is enough to study u_g in what follows. It is given by convolution with the fundamental solution

$$E = \tfrac{1}{2}\pi^{\frac{1}{2}(1-n)}\chi_+^{\frac{1}{2}(1-n)}(t^2-|x|^2).$$

Here $x = (x_1,\dots,x_n)$ and

$$\chi_+^a(s) = s^a/\Gamma(a+1),\ s > 0;\ \chi_+^a(s) = 0,\ s \leqq 0, \text{ if } \operatorname{Re} a > -1,$$

$$d\chi_+^a/ds = \chi_+^{a-1} \text{ for all } a \in \mathbf{C}.$$

Hence $\chi_+^{-k} = \delta_0^{(k-1)}$, $k = 1, 2, \dots$, which is supported by the origin. In the sense of distribution theory we have

$$u_g(t, x) = \int E(t, x-y)g(y)dy,\ t > 0.$$

Set $x = r\omega$ where $r = |x|$ and $\omega \in S^{n-1}$. Then $r \leqq t+M$ in supp u_g if $|y| \leqq M$ in supp g, for $|x| \leqq |x-y|+|y| \leqq t+M$ in the support of the "integrand". When n is odd we also have $r \geqq t-M$, for $|x| \geqq |x-y|-|y| \geqq t-M$ (Huygens' principle). When n is even we note instead that $u_g(t, x) = O(t^{1-n})$ if $|x| < \tfrac{1}{2}t$ and $t \to \infty$, for E is homogeneous of degree $1-n$. The main contributions to u_g must therefore always occur when $r-t$ is not too large, so we set $r = t+\rho$ where $-r \leqq \rho \leqq M$, if $2r \geqq t \geqq r-M$. Then

$$t^2-|x-y|^2 = (r-\rho)^2-|r\omega-y|^2 = 2r(\langle\omega,y\rangle-\rho)+\rho^2-|y|^2,$$

and we obtain by the homogeneity of E

$$(2\pi r)^{\frac{1}{2}(n-1)}u_g=\tfrac{1}{2}\int \chi_+^{\frac{1}{2}(1-n)}(\langle\omega,y\rangle-\rho+(\rho^2-|y|^2)/2r)g(y)\,dy =$$
$$= \tfrac{1}{2}\int \chi_+^{\frac{1}{2}(1-n)}(s-\rho+\rho^2/2r)G(\omega,r^{-1},s)ds =$$
$$= \tfrac{1}{2}\int \chi_+^{\frac{1}{2}(1-n)}(s+(t^2-r^2)/2r)G(\omega,r^{-1},s)ds.$$

Here

$$G(\omega, z ,s) = \int \delta(s-\langle\omega,y\rangle+|y|^2 z/2)g(y)\,dy$$

is a C^∞ function in $S^{n-1}\times[0,\frac{1}{2M}]\times\mathbf{R}$ with $|s|\leqq 5M/4$ in the support and $G(\omega,0,s) = R(\omega,s;g)$,

$$(2.1.3) \qquad R(\omega,s;g) = \int \delta(s-\langle\omega,y\rangle)g(y)dy = \int_{\langle\omega,y\rangle=s} g(y)dS(y)$$

denoting the Radon transform of g. It follows at once that

$$(2.1.4) \qquad r^{\frac{1}{2}(n-1)}u_g(t,x) = F(\omega,r^{-1},\rho)$$

where

$$(2.1.5) \qquad F(\omega, z, \rho)=\tfrac{1}{2}(2\pi)^{\frac{1}{2}(1-n)}\int \chi_+^{\frac{1}{2}(1-n)}(s-\rho+\rho^2 z/2)G(\omega,z,s)ds$$

is a C^∞ function in $S^{n-1}\times[0,\frac{1}{2M}]\times\mathbf{R}$ with $\rho\leqq M$ in the support. This result is due to G. Friedlander who only assumed that u satisfies the wave equation for large $|x|$. However, when n is even we shall also need precise estimates for F and its derivatives for large negative ρ.

<u>Lemma 2.1.1</u> For all α, β, γ we have when $|z| \leqq 1/2M$

$$(2.1.6) \qquad |D_\omega^\alpha D_z^\beta D_\rho^\gamma F(\omega,z,\rho)| \leqq C_{\alpha\beta\gamma}(1+|\rho|)^{\frac{1}{2}(1-n)+|\beta|-|\gamma|},$$

where D_ω^α is defined by means of some local coordinates in S^{n-1}.

<u>Proof.</u> When ρ is bounded there is no information in (2.1.6) beyond the infinite differentiability of F, so we may assume that $\rho<-2M$, say, in the proof. The derivative we shall estimate is a finite linear combination of terms of the form

$$(2.1.7) \qquad \int \chi_+^a(s-\rho+\rho^2 z/2)z^b\rho^c G'(\omega,z,s)ds$$

where G' denotes some derivative of G with respect to ω and z, b and c are non-negative integers, and

$$(2.1.8) \qquad b+a \leqq \tfrac{1}{2}(1-n), \quad c+a-b \leqq |\beta|-|\gamma|+\tfrac{1}{2}(1-n).$$

This is clear when $|\beta|=|\gamma|=0$. If we differentiate with respect to z then $|\beta|$ increases by 1 and a,b,c remain unchanged, or b decreases by 1, or a decreases by 1 while c increases by 2. If we differentiate with respect to ρ then $|\gamma|$ increases by 1 and c decreases by 1 or a decreases by 1 while b-c is unchanged. Thus the claim follows by induction. The terms (2.1.7) can be estimated for $\rho < -2M$ by

$$C(|\rho|+\rho^2 z)^a z^b |\rho|^c < C(|\rho|+\rho^2 z)^{a+b} |\rho|^{c-2b} \leqq C|\rho|^{a+c-b}$$

which proves the lemma in view of (2.1.8).

In particular we find that

$$|u_g(t, x)| \leqq C(r+|t^2-r^2|)^{\frac{1}{2}(1-n)}, \tag{2.1.9}$$

for we have already seen that this is true when $r<\frac{1}{2}t$, and when $r>\frac{1}{2}t$ we have $|t^2-r^2|=|\rho|(t+r)<3|\rho|r$, so the estimate follows from (2.1.6). When $r\to\infty$ for fixed ω and ρ then $r^{\frac{1}{2}(n-1)}u_g(t,x)$ converges to the Friedlander radiation field $F_0(\omega,\rho) = F(\omega, 0, \rho)$,

$$F_0(\omega, \rho)=\tfrac{1}{2}(2\pi)^{\frac{1}{2}(1-n)} \int \chi_+^{\frac{1}{2}(1-n)}(s-\rho)R(\omega, s; g)ds. \tag{2.1.10}$$

For $r > \frac{1}{2}t$ we have

$$|u_g(t,x) - r^{\frac{1}{2}(1-n)}F_0(\omega,\rho)| \leqq Cr^{-2}(r+|t^2-r^2|)^{\frac{1}{2}(3-n)}, \tag{2.1.11}$$

for

$$|F(\omega, 0, \rho) - F(\omega, r^{-1}, \rho)| \leqq Cr^{-1}(1+|\rho|)^{\frac{1}{2}(1-n)+1}$$

by (2.1.6).

So far we have only examined the solution of (2.1.1), (2.1.2) when f = 0. However, in the general case we can choose $\psi\in C^\infty(\mathbf{R})$ equal to 1 in $[1,\infty)$ and 0 in $(-\infty,0)$ and obtain when $t > 1$

$$u = E*K, \quad K = \square(\psi(t)u) = 2\psi'\partial u/\partial t+\psi''u.$$

Thus $K \in C_0^\infty$, and

$$u(t, x) = \int ds \int E(t-s, x-y)K(s, y)\,dy$$

is a superposition of solutions of the form u_g just discussed.

For the solution $u_g(t-s,x)$ of (2.1.1) we have the radiation field $F_0(\omega,\rho+s)$ if F_0 is that of u_g, and (2.1.9), (2.1.11) hold as before. Hence we conclude that the radiation field of u exists and that (2.1.9) and (2.1.11) hold for u.

We shall also need estimates for the derivatives of u. It is obvious that $D^\alpha u = O(t^{1-n-|\alpha|})$ if $|x|<\frac{1}{2}t$ and $t\to\infty$. However, to state precise results also near the light cone we must consider the Lie algebra of vector fields spanned by the vector fields $\partial/\partial x_i$, $i = 0, \ldots, n$, and

(2.1.12) $$\sum_0^n x_i\partial/\partial x_i, \quad x_0\partial/\partial x_j+x_j\partial/\partial x_0, \quad x_j\partial/\partial x_k - x_k\partial/\partial x_j,$$

where $j,k = 1,\ldots,n$ and $x_0=t$. (This is the Lie algebra of the inhomogeneous conformal Lorentz group.) These vector fields play a crucial role in the work of Klainerman [6] (see also Hörmander [2]), and we shall return to his methods in Sections 2.3 and 2.4. We note already now that the vector fields all commute with $\square$ apart from the radial vector field,

$$[\square, \sum x_i\partial/\partial x_i] = 2\square.$$

When u satisfies (2.1.1) it follows that $Z^\alpha u$ is also a solution of (2.1.1) for any product Z^α of the vector fields (2.1.12) and $\partial/\partial x_i$. Hence (2.1.9) is also valid for $Z^\alpha u$, so we have proved most of the following theorem.

<u>Theorem 2.1.2.</u> For the solution of the Cauchy problem (2.1.1), (2.1.2) with f, $g\in C_0^\infty(\mathbf{R}^n)$, we have for any product Z^α of the vector fields (2.1.12) and $\partial/\partial x_i$, $i = 0, \ldots, n$,

(2.1.13) $$|Z^\alpha u(t, x)| \leqq C_\alpha(1+t+t\,||x|-t|)^{\frac{1}{2}(1-n)}, \quad t > 0.$$

The limit

(2.1.14) $$F_0(\omega, \rho) = \lim_{r\to\infty} r^{\frac{1}{2}(n-1)}u(r-\rho, r\omega)$$

exists and is a C^∞ function in $S^{n-1}\times\mathbf{R}$ vanishing for $\rho>M$, if $|x|\leqq M$ in $\operatorname{supp} f\cup\operatorname{supp} g$; we have

(2.1.15) $$F_0(\omega,\rho)=\tfrac{1}{2}(2\pi)^{\frac{1}{2}(1-n)}\chi_-^{\frac{1}{2}(1-n)}*(R(\omega,.;g)-R(\omega,.;f)')$$

with convolution and differentiation taken in the ρ variable. Thus F_0 has compact support if n is odd, and F_0 is a polyhomogeneous symbol of degree $\frac{1}{2}(1-n)$ in the ρ variable when n is even. When $|x|>\frac{1}{2}t>1$ we have

(2.1.16) $$|Z^\alpha(u(t,x)-|x|^{\frac{1}{2}(1-n)}F_0(x/|x|,|x|-t))|\leqq$$
$$\leqq C(1+t+t||x|-t|)^{\frac{1}{2}(3-n)}t^{-2}.$$

Proof. What remains is just to prove the last estimate when $\alpha\neq 0$, for this implies that the radiation field of $\partial u/\partial t$ is $-\partial F_0/\partial\rho$, so (2.1.15) follows from (2.1.10). To prove (2.1.16) we may assume that

$$u(t,x) = u_g(t,x) = r^{\frac{1}{2}(1-n)}F(\omega,r^{-1},\rho)$$

where F satisfies (2.1.6) and $F(\omega,0,\rho) = F_0(\omega,\rho)$. Thus the problem is to estimate

(2.1.17) $$Z^\alpha(r^{\frac{1}{2}(1-n)}(F(\omega,z,\rho)-F(\omega,0,\rho))),\ r=|x|=z^{-1},\omega=x/r,\rho=r-t.$$

When doing so we observe that application of the vector fields (2.1.12) to homogeneous functions preserves the degree of homogeneity while application of $\partial/\partial x_i$ lowers it. For the vector fields (2.1.12) we have

$$(\sum_0^n x_j\partial/\partial x_j)\rho = \rho,\quad (x_0\partial/\partial x_j+x_j\partial/\partial x_0)\rho = -\omega_j\rho,$$

while ρ is annihilated by the other vector fields (2.1.12).

i) Let us first consider terms in the expansion of (2.1.17) where no Z acts on the z variable. These are of the form

$$a(t,x)\,D_\omega^\alpha\rho^\nu D_\rho^\gamma(F(\omega,z,\rho)-F(\omega,0,\rho))$$

where a is a homogeneous function of degree $\leqq\frac{1}{2}(1-n)$ and $0\leqq\nu\leqq\gamma$. This follows at once by induction. Using (2.1.6) we can estimate it by

$$Cr^{\frac{1}{2}(1-n)}z(1+|\rho|)^{\frac{1}{2}(1-n)+1}$$

which is equivalent to the right-hand side of (2.1.16).

ii) Next we consider terms where some derivative falls on the z variable. These are of the form

$$a(t,x)D_\omega^{\alpha}D_z^{\beta}\rho^{\nu}D_\rho^{\gamma}F(\omega,z,\rho)$$

where $0\leqq\nu\leqq\gamma$ and a is homogeneous of degree $\leqq\frac{1}{2}(1-n)-|\beta|$, for each z differentiation of F is accompanied by a factor $Z_j z$ of degree $\leqq -1$. By (2.1.6) we can estimate such a term by

$$Cr^{\frac{1}{2}(1-n)-|\beta|}(1+|\rho|)^{\frac{1}{2}(1-n)+|\beta|}\leqq C'r^{\frac{1}{2}(1-n)-1}(1+|\rho|)^{\frac{1}{2}(1-n)+1}$$

since $1+|\rho|\leqq 2r$ and $|\beta|\geqq 1$. This completes the proof of (2.1.16) and of the theorem.

The case n=1 of the results above is somewhat exceptional and very elementary. Then $\omega=\pm 1$ and we have for t>M and $x\neq 0$

$$u(t,x) = F(\operatorname{sgn}x, |x|-t)$$

where F according to (2.1.15) is determined by

$$g(x)+f'(x)=-2F'(-1,-x),\ g(x)-f'(x)=-2F'(1,x).$$

We can rewrite (1.5.7) in terms of F instead of the Cauchy data if the equation (1.5.1) reduces to $\Box u = 0$ when $u' = 0$. When $\xi = (-1, \pm 1)$ we have with the notation there

$$q(\xi, \partial) = (\partial_x(\partial_t^2-\partial_x^2)/(\pm\partial_t+\partial_x))/4 = \partial_x(\pm\partial_t-\partial_x)/4,$$

$$q(\xi, \partial)(f+tg)_{|t=0} = (\pm g'-f'')/4 = -F''(\pm 1,\pm x)/2.$$

Hence (1.5.7) can be written

$$1/\varepsilon T_\varepsilon \to \max_{\xi,r} \sum \partial c_j(0)/\partial U_k\xi_k\xi_1^{2-j}\xi_2^{j}F''(\xi_2, r)/2,$$

where $\xi= (-1, \xi_2)$ and $\xi_2 = \pm 1$. We shall prove related results when n = 3 or n = 2.

Assume now that n>1. Then we shall prove

<u>Proposition 2.1.3.</u> Unless f and g are identically 0 the radiation field F_0 of the solution u of (2.1.1), (2.1.2) is not identically 0.

<u>Proof.</u> If F_0 is identically 0 it follows from the theorem of supports and (2.1.15) that $R(\omega,\rho;g) - R(\omega,\rho;f)' \equiv 0$.

This means that

$$R(-\omega,-\rho;\, g) - dR(-\omega,-\rho;f)/d\rho \equiv 0.$$

If we carry out the differentiation and replace $-\omega$ and $-\rho$ by ω and ρ afterwards, it follows that

$$R(\omega,\rho;g) + dR(\omega,\rho;f)/d\rho \equiv 0$$

so $R(\omega,\rho;g) \equiv 0$ and $dR(\omega,\rho;\, f) \equiv 0$. Hence f and g vanish identically.

Remark. The proof shows that the projection of the support of $R(\omega,\rho;\, g) - R(\omega,\rho;f)'$ on S^{n-1} cannot omit two antipodal points. In fact, if $R(\omega,\rho;\, g)$ vanishes for all ω in an open set then it vanishes identically since the Fourier transform of g will vanish in the open cone which it generates.

When $n = 3$ and f, g are functions of r only (hence C^∞ functions of r^2), then equation (2.1.1) can be written

$$(ru)''_{tt} - (ru)''_{rr} = 0.$$

Hence

$$u(t,\, x) = r^{-1}F_0(r-t),\ t > M,$$

so no limit needs to be taken in (2.1.14). It is easy to compute F_0, for

$$R(\omega,\rho;g) = \int_{|\rho|}^{\infty} g(t)d\pi(t^2-\rho^2) = \tfrac{1}{2}\int_{\rho}^{\infty} tg(t)dt$$

and similarly for $R(\omega,\rho;f)$. It follows that the radiation field which only depends on ρ is given by

(2.1.18) $\quad dF(\rho)/d\rho = (d(\rho f(\rho))/d\rho - \rho g(\rho))/2.$

The quantity on the right plays an important role in the results of F. John [4] discussed in Section 2.2.

2.2. The blowup of radial solutions. We shall now give an exposition of the work of John [4] in the spirit of Section 1.4 here. Thus we consider radial solutions of the non-linear wave equation with three space variables

(2.2.1) $$\partial^2 u/\partial t^2 = c^2 \Delta u$$

where c is a positive C^∞ function of $\partial u/\partial t$ only. (This is quite essential if one is to have radial solutions.) We prescribe Cauchy data

(2.2.2) $$u = \varepsilon f,\ \partial u/\partial t = \varepsilon g \text{ when } t = 0$$

where f and g are radial functions in C_0^∞. It is no restriction to assume that $c(0) = 1$ and that $c'(0) \geqq 0$, for these conditions can be achieved by a linear change of t variable or a change of sign for u. In the results below we shall assume genuine non-linearity in the sense that $c'(0)\neq 0$, thus $c'(0)>0$. By the existence theorems of John and Klainerman [5] and Klainerman [6] (see also Hörmander [2]) a unique C^∞ solution exists for $0 \leqq t < c_0/\varepsilon$ for some $c_0 > 0$ (depending on f and g). The solution is an even function of $r = \pm|x|$, so $v = ru$ is an odd function of r satisfying

(2.2.1)' $$v''_{tt} - c(v'_t/r)^2 v''_{rr} = 0,$$

(2.2.2)' $$v(0,r) = \varepsilon r f(r),\ \partial v(0,r)/\partial t = \varepsilon r g(r) \text{ when } t=0.$$

v is also a C^∞ function of ε, and $\partial_\varepsilon v_{|\varepsilon=0}$ is r times the solution of (2.1.1), (2.1.2), thus with $F\in C_0^\infty$ given by (2.1.18)

$$\partial_\varepsilon v_{|\varepsilon=0} = F(r-t) \text{ when } r \geqq 0 \text{ and } t > M,$$

where M is chosen so that the data vanish for $|x|>M$. Ignoring at first the fact that the argument of c is now v'_t/r and not v'_t (although this will be the reason for the drastically increased lifespan of the solution) we set as at the end of Section 1.5

(2.2.3) $$v''_{tr} = c(w_2-w_1),\quad v''_{rr} = w_1+w_2.$$

This means that with $L_1=\partial/\partial t+c\partial/\partial r$ and $L_2=\partial/\partial t-c\partial/\partial r$ we have

(2.2.3)' $\quad L_1v_t' = v_{tt}''+cv_{tr}'' = cv_{tr}''+c^2v_{rr}'' = cL_1v_r' = 2c^2w_2;$

a similar equation holds for w_1. Equivalently,

(2.2.4) $\quad L_1u_t' = u_{tt}''+cu_{tr}'' = c(2cw_2-u_t')/r,$

(2.2.4)' $\quad L_2u_t' = u_{tt}''-cu_{tr}'' = c(2cw_1+u_t')/r.$

In deriving equations corresponding to (1.5.8), (1.5.9) we start from the equation $2cw_1=cv_{rr}''-v_{tr}''$ and obtain

$$2cL_1w_1+2w_1L_1c = c^2v_{rrr}'''-v_{ttr}''' + (L_1c)v_{rr}'',$$

hence

(2.2.5) $\quad 2cL_1w_1=\partial_r(c^2v_{rr}''-v_{tt}'')-2c\partial c/\partial r(w_1+w_2)+(w_2-w_1)L_1c =$
$$=\partial_r(c^2v_{rr}''-v_{tt}'')+(w_2-w_1)\partial c/\partial t-(3w_1+w_2)c\partial c/\partial r.$$

So far we have neither used the equation for v nor the assumption that c only depends on $\partial u/\partial t$. Doing so we obtain

$$\partial c/\partial t = c'v_{tt}''/r = c'c^2(w_1+w_2)/r;$$

$$\partial c/\partial r = c'(v_{tr}''/r-v_t'/r^2) = c'(c(w_2-w_1)-u_t')/r.$$

Hence

(2.2.6) $\quad L_1w_1 = c'c\, w_1(w_1-w_2)/r + u_t'c'(3w_1+w_2)/2r,$

(2.2.6)' $\quad L_2w_2 = c'c\, w_2(w_2-w_1)/r - c'u_t'(3w_2+w_1)/2r,$

where the second equation follows from the first when we replace L_1,w_1,w_2,c by $L_2,w_2,w_1,-c$. (See also John [4].) This replaces (1.5.8)', and instead of (1.5.9)' we have

(2.2.7) $\quad d(w_1(dr-cdt)) = u_t'c'(w_1+w_2)/2r\; dt\wedge dr,$

(2.2.7)' $\quad d(w_2(dr-cdt)) = -u_t'c'(w_1+w_2)/2r\; dt\wedge dr.$

In fact, the left-hand side of (2.2.7) is equal to $(L_1w_1+w_1\partial c/\partial r)dt\wedge dr$.

We can now prove the following theorem of John [4] by arguments close to those of Section 1.4; in the course of the proof we shall get some improved information on the solutions which will be useful later on.

Theorem 2.2.1. If the Cauchy problem (2.2.1), (2.2.2), where $c = c(\partial u/\partial t)$ and $c(0) = 1$, has a C^2 solution for $0 \leqq t < T_\varepsilon$ then

$$\overline{\lim_{\varepsilon \to 0}} (\varepsilon \log T_\varepsilon) \max c'(0)F''(\rho) \leqq 1 \tag{2.2.8}$$

where F is defined by (2.1.18).

Note that F is a function of compact support which does not vanish identically unless f and g do, so F" takes both positive and negative values. The assertion (2.2.8) is void if $c'(0) = 0$. As remarked before we may therefore assume that $c'(0)>0$ (otherwise the signs of u, f, g can be changed). Let $0 < 1/A < \max (c'(0)F''(\rho))$; we have to prove that

$$T_\varepsilon \leqq \exp (A/\varepsilon) \text{ for small } \varepsilon.$$

Proof of Theorem 2.2.1. Let $M = \sup\{|x|;\ x \in \operatorname{supp} f \cup \operatorname{supp} g\}$, and let R be the part of $\{(t, r);\ t \geqq 0,\ 0 \leqq r \leqq t+M\}$ to the right of the bicharacteristic Γ_{-M} defined by

$$dr/dt = c(u_t'),\quad r = -M \text{ when } t = 0.$$

As in Section 1.4 we introduce

$$J(t) = \sup_{0<s<t} \int_{(s,r)\in R} |w_1(s, r)|dr, \tag{2.2.9}$$

$$U(t) = \sup_{2M<s<t}\ \sup_{(s,r)\in R} s|u_s'(s,r)|, \tag{2.2.10}$$

$$V(t) = \sup_{2M<s<t}\ \sup_{(s,r)\in R} s^2|w_2(s, r)|. \tag{2.2.11}$$

Choose E so that $J(2M) \leqq E\varepsilon/2$. Assuming that u is a solution at least in R when $t \leqq T \leqq \exp (A/\varepsilon)$ we shall prove that for small ε we have if $C > c'(0)$

$$J(t) < E\varepsilon,\quad U(t) < 2E\varepsilon,\quad V(t) < 3CE^2\varepsilon^2,\quad t \leqq T. \tag{2.2.12}$$

For reasons of continuity it suffices to show that these inequalities for $t < T$ imply strict inequality also when $t = T$ provided that ε is small enough.

Before starting on the verification of (2.2.12) we must make some geometrical observations concerning the bicharacteristics Γ_λ and C_μ defined by $dr/dt = \pm c$ with initial conditions $r = \lambda$ resp. $r = \mu$ when $t = 0$. In any compact set we have $|c-1| < C'\varepsilon$, and by (2.2.12) we have $|c-1|<2CE\varepsilon/s$ in R when $2M \leqq s \leqq t$. If $(t, r)\in\Gamma_\lambda$ and $|\lambda|\leqq M$ it follows that

$$|r - t - \lambda| \leqq 2C\varepsilon E\log t + 2MC'\varepsilon < 2CEA + 1 = C'$$

if ε is small enough. Similarly $|r+t-\mu|< C'$ if $(t,r)\in C_\mu\cap R$. For any other point $(t', r')\in C_\mu\cap R$ we obtain

$$2|t-t'|\leqq|t+r-t'-r'|+|t-r|+|t'-r'|\leqq 2M+4C',$$

hence $|t-t'|\leqq M+2C'$.

1) To estimate J(t) we use (2.2.7) as in the proof of Lemma 1.2.2. This gives for $t>2M$

$$\int_{(t,r)\in R} |w_1(t,r)|dr \leq \int_{(2M,r)\in R} |w_1(2M,r)|dr + \iint_{2M<s<t;(s,r)\in R} |\partial u/\partial s\, c'(w_1+w_2)/2r|dsdr.$$

The first term is by assumption at most $E\varepsilon/2$. We estimate u'_s by $2E\varepsilon/s$. The second term is then bounded by

$$E\varepsilon(CJ(t)\int_{2M}^{t} ds/s^2 + CV(t) \iint_{2M<s<t;(s,r)\in R} dsdr/2rs^2).$$

By the geometric observations above the integrals are bounded independently of t, so we obtain the bound $E\varepsilon/6$, say, when ε is small enough. Hence $J(t) < 2E\varepsilon/3$.

2) The estimate of U(t) follows from (2.2.4)' by integration along C_μ, but first we must estimate the integral of w_1 over $C_\mu\cap R$. To do so we observe that $dr-cdt =-2cdt$ when $dr=-cdt$. Hence the argument in 1) above shows that the integral of $|w_1|dt$ on $C_\mu\cap R$ can be estimated by $E\varepsilon/3$. For small ε we have $r>t/2$ in R when $t>2M$ so it follows that the integral of $2c^2|w_1|/r$ over $C_\mu\cap R$ to the right of (t,r) is at most $5E\varepsilon/3t$, say. Now the term cu'_t/r can be removed from (2.2.4)' by multiplication with an integrating factor with logarithm $O(\varepsilon/t)$, which only affects the constants in the estimates by a factor $1+O(\varepsilon)$, so we obtain $U(t) < 11E\varepsilon/6$ for small ε.

3) An estimate of w_2 is also obtained by integrating (2.2.6)' along C_μ. The right-hand side can be written aw_2+b where

$$a = (cc'(w_2-w_1)-\frac{3}{2}c'u_t')/r, \quad b = -c'u_t'w_1/2r.$$

It is clear that the integral of $|aw_2|$ over $C_\mu \cap R$ is $O(\varepsilon^3/t^2)$. Since $C > c'(0)$ we have

$$\int |b| ds < 2C\varepsilon^2 E^2/t^2$$

if we integrate on C_μ to the right of (t,r), estimating the integral of $|w_1|$ as in part 2) of the proof. This completes the verification of (2.2.12).

With (2.2.12) proved it is now easy to get an upper bound for T by applying Proposition 1.4.1 to the differential equation (2.2.6) on Γ_λ. With the notation in that lemma we have $a_2 = u_t'c'w_2/2r$ which we can estimate by $2E\varepsilon t^{-1}C3CE^2\varepsilon^2 t^{-2-1}$ $= 6C^2E^3\varepsilon^3/t^4$; the integral from $t = 2M$ is $O(\varepsilon^3)$. We have

$$a_1 = -c'cw_2/r + 3u_t'c'/2r,$$

which can be estimated by $6E\varepsilon C/t^2$; the integral is $O(\varepsilon)$. Thus the number K in (1.4.7) is $O(\varepsilon^3)$. We have

$$a_0(t) = c'c/r = c'(0)(1+O(\varepsilon))/(t+O(1)).$$

When $w_1(2M, r)/\varepsilon$ has a positive lower bound, we obtain from (1.4.8) that

$$c'(0) \log T < (1+O(\varepsilon))/w_1(2M,r).$$

Now $w_1(2M,r) = \varepsilon F''(r-2M) + O(\varepsilon^2)$ with the notation in Theorem 2.2.1, so we obtain with $T = \min (T_\varepsilon, e^{A/\varepsilon})$

$$\overline{\lim_{\varepsilon\to 0}}\, c'(0) \max F'' \varepsilon \min (\log T_\varepsilon, A/\varepsilon) \leqq 1$$

By the definition of A it follows that (2.2.8) holds. The proof is complete.

Remark. As emphasized by John [4] it is sufficient to assume that the equation (2.2.1) is satisfied when $|u'|$ is small; in fact, we verified in (2.2.12) that this will automatically be true in R.

We shall prove in Section 2.4 that the bound for the lifespan of the solution given in Theorem 2.2.1 is in fact asymptotically equal to the precise lifespan. In doing so we shall also consider a rather general class of perturbations of the wave operator, and allow general Cauchy data in C_0^∞. However, we shall show already here that the estimates in the proof of Theorem 2.2.1 suffice to determine the asymptotic behavior of u when $\varepsilon \log t \leqq A$,

$$A < \underline{\lim}_{\varepsilon \to 0} \varepsilon \log T_\varepsilon;$$

by Theorem 2.2.1 this implies

$$A\, c'(0)F'' < 1. \tag{2.2.13}$$

To do so we consider an orbit Γ_λ of $\partial/\partial t + c(u_t')\partial/\partial r$. By (2.2.12) we know that $w_2 = O(\varepsilon^2/r^2)$ on the orbit. For a fixed sufficiently large t we have $\partial v/\partial t = -\varepsilon F'(\lambda) + O(\varepsilon^2)$, and since

$$L_1 v_t' = 2c^2 w_2 = O(\varepsilon^2/r^2)$$

this remains true on the whole orbit. The differential equation of Γ_λ gives, again by (2.2.12),

$$d(r-t)/dt = c-1 = c'(0)u_t' + O(\varepsilon^2/t^2) = c'(0)v_t'/r + O(\varepsilon^2/t^2)$$

which gives a bound for r-t when $\varepsilon \log t \leqq A$. Since

$$1/r - 1/t = (t-r)/(rt) = O(t^{-2}),$$

we conclude that

$$d(r-t)/dt = -c'(0)\varepsilon F'(\lambda)/t + O(\varepsilon/t^2) + O(\varepsilon^2/t).$$

Thus we have on Γ_λ when t is large

$$r-t = \lambda - c'(0)\varepsilon F'(\lambda)\log t + O(\varepsilon),$$

$$u_t' = -\varepsilon F'(\lambda)/t + O(\varepsilon^2/t + \varepsilon/t^2).$$

Now

$$c(w_2-w_1) = \partial^2 v/\partial t\partial r = \partial u/\partial t + r\partial^2 u/\partial t\partial r,$$

so we conclude that $\partial^2 u/\partial t\partial r = O(\varepsilon/r)$, hence that

$$u_t'(t,r) = -\varepsilon F'(\lambda)/t + O(\varepsilon^2/t+\varepsilon/t^2),$$

if

$$r = t+\lambda-c'(0)\varepsilon F'(\lambda)\log t.$$

Since

$$\partial r/\partial\lambda = 1-c'(0)F''(\lambda)\varepsilon\log t \geqq \min(1, 1-Ac'(0)F''(\lambda))$$

has a positive lower bound by (2.2.13), we can solve this equation for λ as a function of r and t. Integrating the equation for u_t' when r is fixed, noting that u = 0 when r-t is large enough, we obtain

$$ru(t,r) = -\varepsilon\int F'(\lambda)dt + O(\varepsilon^2+\varepsilon/t).$$

For fixed r we have

$$\partial t/\partial\lambda(1-c'(0)\varepsilon F'(\lambda)/t) = -(1-c'(0)\varepsilon F''(\lambda)\log t),$$

so it follows that

$$-\int F'(\lambda)dt = \int F'(\lambda)(1-c'(0)F''(\lambda)\varepsilon\log t)(1+O(\varepsilon/t))d\lambda$$

$$= F(\lambda) - \tfrac{1}{2}c'(0)F'(\lambda)^2\varepsilon\log t + O(\varepsilon).$$

Returning to the λ variable we have for $0 < s \leqq A$

$$\varepsilon^{-1}e^{s/\varepsilon}u(e^{s/\varepsilon}, e^{s/\varepsilon}+\lambda-c'(0)sF'(\lambda)) \to F(\lambda)-\tfrac{1}{2}c'(0)F'(\lambda)^2 s.$$

The function defined by

$$U(s,r) = F(\lambda)-c'(0)sF'(\lambda)^2/2 \text{ when } r=\lambda-c'(0)sF'(\lambda)$$

is the solution of the Cauchy problem

$$\partial U/\partial s - \tfrac{1}{2}c'(0)(\partial U/\partial r)^2=0;\ U(0,r) = F(r),$$

for the bicharacteristic starting at $(0,\lambda)$ (with the cotangent vector $(\tfrac{1}{2}c'(0)F'(\lambda)^2, F'(\lambda))$) is defined by $r = \lambda-c'(0)F'(\lambda)s$, and the value of U there is

$$F(\lambda) + \tfrac{1}{2}sc'(0)F'(\lambda)^2-c'(0)F'(\lambda)sF'(\lambda) = F(\lambda)-\tfrac{1}{2}sc'(0)F'(\lambda)^2$$

as claimed. Thus we have

Theorem 2.2.2. With the notation in Theorem 2.2.1 let

$$0 < A < \underline{\lim}_{\varepsilon \to 0} \varepsilon \log T_\varepsilon .$$

For the solution $u_\varepsilon(t,r)$ of (2.2.1), (2.2.2) we have then uniformly on compact subsets of $(0,A] \times \mathbf{R}$

(2.2.14) $\quad \varepsilon^{-1} e^{s/\varepsilon} u_\varepsilon(e^{s/\varepsilon}, e^{s/\varepsilon}+r) \to U(s,r)$

where U is the solution of the Cauchy problem

(2.2.15) $\quad \partial U/\partial s - \frac{1}{2}c'(0)(\partial U/\partial r)^2 = 0,$

(2.2.16) $\quad U(0,r) = F(r),$

where F is defined by (2.1.7).

Note that the solution of (2.2.14) introduces the bicharacteristics $r = \lambda - c'(0)sF'(\lambda)$, as seen above. This equation can be solved for λ as a C^1 function of r precisely when

$$s \max c'(0)F''(\lambda) < 1.$$

This is a bound of the same form as in (2.2.8). For rather general non-linear perturbations of the wave operator we shall prove in Section 2.4 that when an analogue of (2.2.15),(2.2.16) has a smooth solution for $0 \leqq s \leqq A$, then the Cauchy problem can be solved for $0 \leqq t \leqq e^{A/\varepsilon}$, and an analogue of (2.2.14) is valid. In the proof it is essential that one starts by constructing an approximate solution suggested by (2.2.14). We shall in fact present a direct and much more natural motivation for the equation (2.2.15) then. Since we shall obtain another completely independent proof of Theorem 2.2.2 we have left out some details above; in particular one should prove (2.2.12) also if R is replaced by any wider strip between characteristics Γ_λ.

2.3. <u>L^2 estimates and some function spaces.</u> The estimates used so far have been based on L^∞ and L^1 norms. When we pass to several space dimensions it is necessary to switch to L^2 estimates which can be proved by energy integral arguments. From John-Klainerman [5] we quote the following one:

<u>Proposition 2.3.1.</u> Let $v \in C^2$ satisfy a linear differential equation

$$\Box v + \sum_{j,k=0}^{n} \gamma_{jk}(x)\partial_j\partial_k v = f, \quad 0 \leqq x_0 < T,$$

where $\Box = \partial_0^2 - \partial_1^2 - \dots - \partial_n^2$, $\gamma_{00} = 0$ and

$$|\gamma| = \sum |\gamma_{jk}| \leqq \tfrac{1}{2} \text{ for } 0 \leqq x_0 < T.$$

Assume that v vanishes for large $|x|$. Then we have for $0 \leqq x_0 < T$

$$(2.3.1) \qquad \|v'(x_0,.)\| \leqq 3(\|v'(0,.)\| + \int_0^{x_0} \|f(\tau,.)\| d\tau) \exp \int_0^{x_0} |\gamma'(\tau)| d\tau,$$

where the norms are L^2 norms in $\mathbf{R}^n$ and

$$(2.3.2) \qquad |\gamma'(\tau)| = \sum \sup |\partial_i \gamma_{jk}(\tau,.)|.$$

Note that the hypothesis $|\gamma| \leqq \frac{1}{2}$ implies hyperbolicity of the equation. The proof given in Klainerman [7] follows by partial integration in the product of the equation by $\partial u/\partial x_0$.

Energy estimates of the form (2.3.1) suggest that one should work with $L^\infty L^2$ norms in $\mathbf{R}^{1+n}$. Following Klainerman [6] (see also Hörmander [2] and Section 2.1) we shall define such Sobolev spaces associated with the Lie algebra spanned by the vector fields

$$(2.3.3) \qquad \partial/\partial x_j, \; j = 0, \dots, n;$$

$$(2.3.4) \qquad \sum_0^n x_j \partial/\partial x_j; \; \lambda_j x_j \partial/\partial x_k - \lambda_k x_k \partial/\partial x_j, \; j,k = 0,\dots,n.$$

Here $\lambda_0 = 1$ and $\lambda_j = -1$ for $j \neq 0$. These vector fields commute with $\Box$ apart from the fact that

$$(2.3.5) \qquad [\Box, \sum x_j \partial/\partial x_j] = 2\Box.$$

Let s be a non-negative integer and let

$$W_s^{\#} = \{u;\ Z_1 \ldots Z_j u \in L^\infty L^2(\mathbf{R}^{1+n}) \text{ if } 0 \leqq j \leqq s\},$$

where $Z_1, \ldots, Z_j$ are chosen arbitrarily among the vector fields (2.3.3), (2.3.4). This is a Banach space with the norm

(2.3.6) $$\|u\|_s^{\#} = \sum \sup_{x_0} \|Z_1 \ldots Z_j u(x_0, .)\|,$$

where the sum is taken over the same products and $\| \ \|$ denotes the L^2 norm. By Theorem 5.1 in [2] we have

Proposition 2.3.2. If $\frac{1}{2}n < s_0$ then

(2.3.7) $$\sup |\mu^{\frac{1}{2}} u| \leqq C\|u\|_{s_0}^{\#}, \quad u \in W_{s_0}^{\#},$$

where

(2.3.8) $$\mu(x) = (1+|x|)^{n-2}(1+|x|+|x_0{}^2 - x_1{}^2 - \ldots - x_n{}^2|),$$

and for every $h \in C^s$ we have

(2.3.9) $$\|h(u)-h(0)\|_s^{\#} \leqq C(h, s, \|u\|_{s_0}^{\#})\|u\|_s^{\#}, \quad u \in W_s^{\#}.$$

When $Z_1, \ldots, Z_j$ are chosen among the vector fields (2.3.3), (2.3.4) then

(2.3.10) $$\|\mu^{\frac{1}{2}}(Z_1 \ldots Z_i u)(Z_{i+1} \ldots Z_j v)\| \leqq C(\|u\|_s^{\#}\|v\|_{s_0}^{\#} + \|u\|_{s_0}^{\#}\|v\|_s^{\#}),$$

if $u, v \in W_s^{\#}$ and $0 \leqq i \leqq j \leqq s$.

In (2.3.9) we may allow u to take its values in $\mathbf{R}^N$ when $h \in C^s(\mathbf{R}^N)$. We shall need a supplement to (2.3.9) when the components of u are split into two groups v and w:

Proposition 2.3.3. Let $h \in C^{s+2}(\mathbf{R}^{N'+N''})$ and let $v, w \in W_s^{\#}$ for some $s \geqq 0$, with values in $\mathbf{R}^{N'}$ and $\mathbf{R}^{N''}$ respectively. Then

(2.3.11) $$\|h(v,w)-h(v,0)-h(0,w)+h(0,0)\|_s^{\#} \leqq$$
$$\leqq C(h,s,\|v\|_{s_0}^{\#} + \|w\|_{s_0}^{\#})(\|v\|_s^{\#}\|w\|_{s_0}^{\#} + \|v\|_{s_0}^{\#}\|w\|_s^{\#}).$$

Proof. By Taylor's formula we can write

$$h(y,z)-h(y,0)-h(0,z)+h(0,0) = \sum h_{jk}(y,z) y_j z_k$$

where $h_{jk} \in C^s$. For $h_{jk}(0)v_j w_k$ we have the required estimate by (2.3.10); we could even gain a weight factor $\mu^{\frac{1}{2}}$. Using (2.3.10) we also obtain

$$\|(h_{jk}(v,w)-h_{jk}(0))v_jw_k\|_s^\# \leqq C(\|h_{jk}(v,w)-h_{jk}(0)\|_s^\#\|v_jw_k\|_{s_0}^\# + \\ +\|h_{jk}(v,w)-h_{jk}(0)\|_{s_0}^\# \|v_jw_k\|_s^\#) \leqq \\ \leqq C'((\|v\|_s^\#+\|w\|_s^\#)\|v_j\|_{s_0}^\# \|w_k\|_{s_0}^\# +\|v_jw_k\|_s^\#),$$

where the constants depend on h, s, $\|v\|_{s_0}^\# + \|w\|_{s_0}^\#$. Another application of (2.3.10) proves (2.3.11).

In our application of (2.3.11) we shall actually estimate $h(v, w) - h(0, w) = h(v,w) - h(0,w) - h(v,0) + h(0,0) + h(v,0) - h(0, 0)$ by applying (2.3.9) to the last difference and (2.3.11) to the first four terms. In that context we would have gained nothing by including in the estimate (2.3.11) the additional weight which is given in (2.3.10).

In this section we have only considered the space $W_s^\#$ in the whole of $\mathbf{R}^{n+1}$. By standard extension techniques (see Section 4 of [2] for example) the estimates remain valid if we take the norms (2.3.6) only over a half space, defined by $x_0<T$, say. This will be used in Section 2.4.

Later on it will be important that the operators

$$Z_{jk} = x_j\partial/\partial x_k - x_k\partial/\partial x_j;\ j,k = 1, \ldots, n;$$

of the form (2.3.4) are tangents to the unit sphere S^{n-1}, since they annihilate $x_1^2+\ldots+x_n^2$, and that they span all vector fields on S^{n-1}. In particular,

$$\sum_{j,k=1}^{n} Z_{jk}^2 = 2|x|^2\Delta - 2\sum x_jx_k\partial^2/\partial x_j\partial x_k-2(n-1)\sum x_j\partial/\partial x_j = \\ = 2r^2(\Delta -\partial^2/\partial r^2- (n-1)r^{-1}\partial/\partial r),\ r = (x_1^2+\ldots+x_n^2)^{\frac{1}{2}},$$

is twice the Laplacean in the unit sphere.

2.4. An existence and approximation theorem for a general non-linear wave equation. In this section we shall discuss equations of the form

$$(2.4.1) \qquad \sum_{i,j=0}^{3} g_{ij}(u')\partial_i\partial_j u = 0$$

where g_{ij} are C^∞ functions and $(g_{ij}(0))$ is the Lorentz matrix, $g_{ij}(0) = 0$ for $i\neq j$, $g_{00}(0) = 1$ and $g_{jj}(0) = -1$ for $j\neq 0$. (This can of course always be attained by a linear change of variables if the equation (2.4.1) is hyperbolic when $u' = 0$.) We pose the initial conditions

$$(2.4.2) \qquad u = \varepsilon u_0,\ \partial_0 u = \varepsilon u_1 \text{ when } x_0 = 0,$$

where $u_0,\ u_1 \in C_0^\infty$. Sometimes we shall write $t = x_0$, $x = (x_1,x_2,x_3)$ and $\hat{x} = (x_0,\ x_1,\ x_2,x_3)$.

The first step is to construct an approximate solution. When t is so small that the effect of the non-linear terms is not yet substantial, we expect of course that u/ε is well approximated by the solution w_0 of the wave equation with Cauchy data u_0, u_1. According to Friedlander (see Section 2.1) rw_0 is of the form $F(\omega,\ 1/r,\ r-t)$ where $F(\omega,\ z,\ r)$ is a C^∞ function of $(\omega,\ z,\ r) \in S^2\times[0,\ 1]\times\mathbf{R}$, of compact support. For $z = 0$ we obtain the radiation field in (2.1.15); we shall write $F_0(\omega,\ r) = F(\omega,\ 0,\ r)$.

(2.2.14) suggests that we should look for an approximate solution to (2.4.1), (2.4.2) of the form

$$(2.4.3) \qquad u(t,\ r\omega) = \varepsilon r^{-1} U(\omega,\ \varepsilon\log t,\ r-t).$$

Since

$$\Box u = r^{-1}((\partial_t-\partial_r)(\partial_t+\partial_r)-r^{-2}\Delta_\omega)(ru),$$

the main term in $\Box u$ is obtained when $\partial_t+\partial_r$ acts on the argument $s = \varepsilon\log t$ and $\partial_t-\partial_r$ acts on $q = r-t$, which gives

$$-2\varepsilon^2(\mathrm{tr})^{-1}U''_{sq}(\omega, s, q).$$

Writing

(2.4.4) $g_{jk}(U) = g_{jk}(0) + \sum g_{jkl}U_l + O(|U|^2)$

we find that the main non-linear terms are

$$\varepsilon^2 r^{-2}\sum g_{jkl}\hat{\omega}_j\hat{\omega}_k\hat{\omega}_l U'_q U''_{qq};\ \hat{\omega}=(-1,\omega_1,\omega_2,\omega_3).$$

Thus it is natural to require that

$$2U''_{sq}(\omega, s, q) = \sum g_{jkl}\hat{\omega}_j\hat{\omega}_k\hat{\omega}_l U'_q(\omega, s, q)U''_{qq}(\omega, s, q),$$

or equivalently, if U vanishes for large q,

(2.4.5) $\partial U(\omega, s, q)/\partial s = \frac{1}{4}\sum g_{jkl}\hat{\omega}_j\hat{\omega}_k\hat{\omega}_l(\partial U(\omega, s, q)/\partial q)^2.$

When t is large but $\varepsilon \log t$ is small we see that it is natural to require the initial condition

(2.4.6) $U(\omega, 0, q) = F(\omega, 0, q).$

We shall next discuss the solution of (2.4.5), (2.4.6) and then show how to define our approximate solution of (2.4.1), (2.4.2).

<u>Lemma 2.4.1.</u> If

(2.4.7) $A = (\max \frac{1}{2}\sum g_{jkl}\hat{\omega}_j\hat{\omega}_k\hat{\omega}_l\,\partial^2 F_0(\omega,\rho)/\partial\rho^2)^{-1},$

then the Cauchy problem (2.4.5), (2.4.6) has a unique C^∞ solution for $0 \leqq s < A$, but the second order derivatives are unbounded when $s \to A$ if $A < \infty$.

<u>Proof.</u> This is an immediate consequence of the proof of (1.1.4), if we regard ω as a parameter and note that (2.4.5) with $\partial U/\partial q = u$ is equivalent to

$$\partial u/\partial s = \tfrac{1}{2}\sum g_{jkl}\hat{\omega}_j\hat{\omega}_k\hat{\omega}_l u\,\partial u/\partial q.$$

(Note that the condition $\int u\,dq = 0$ is fulfilled for all s if it holds for the initial data, and that $|\rho| \leqq M$ in supp u and in supp U if $|y| \leqq M$ in supp u_0 and in supp u_1. One could equally well prove the lemma using the standard Hamilton-Jacobi integration theory.)

Let $\chi \in C^\infty(\mathbf{R})$ be a decreasing function which is equal to 1 in $(-\infty, 1)$ and 0 in $(2, \infty)$, and set

(2.4.8) $$w(t, x) = \varepsilon\{\chi(\varepsilon t)w_0(t, x) + (1-\chi(\varepsilon t))r^{-1}U(\omega, \varepsilon \log(\varepsilon t), r-t)\}, \quad 0 \leqq \varepsilon t < e^{A/\varepsilon}.$$

Thus w is equal to the solution of the wave equation with Cauchy data (2.4.2) if $t < 1/\varepsilon$. When $t > 2/\varepsilon$ we obtain essentially the approximation (2.4.3) apart from a slight shift of the argument intended to make $\varepsilon\log(\varepsilon t) = O(\varepsilon)$ in the transition zone. Our next lemma examines how well w satisfies the equation (2.4.1); the initial conditions (2.4.2) are fulfilled exactly.

Lemma 2.4.2. With w defined by (2.4.8) and

(2.4.9) $$f = \sum g_{jk}(w')\partial_j\partial_k w,$$

we have $f, w \in C^\infty$ when $\varepsilon t < e^{A/\varepsilon}$, where A is defined by (2.4.7), and $|r-t|$ is bounded in $\operatorname{supp} w \cup \operatorname{supp} f$ by the supremum of $|x|$ in $\operatorname{supp} u_0 \cup \operatorname{supp} u_1$. If $0 < B < A$ we have for small ε and all α if $\varepsilon\log t \leqq B$

(2.4.10) $$|Z^\alpha w(t, x)| \leqq C_{\alpha,B}\varepsilon(1+t)^{-1},$$

(2.4.11) $$|Z^\alpha f(t, x)| \leqq C_{\alpha,B}\varepsilon^2(1+t)^{-2}(1+\varepsilon t)^{-1}.$$

Here Z^α is defined as in Sections 2.1 and 2.3.

Proof. If w is replaced by the solution $\varepsilon w_0(t, x)$ of the wave equation with initial data εu_0, εu_1, then (2.4.10) follows from (2.1.13). As in the proof of Theorem 2.1.2 we can write $Z^\alpha r^{-1}U(\omega, \varepsilon \log(\varepsilon t), r-t)$ as a linear combination of terms of the form

$$\varepsilon^{\beta+1}a(t,x)D_\omega^{\ \alpha}D_z^{\ \beta}\rho^\nu D_\rho^{\ \gamma}U(\omega,z,\rho); \quad z=\varepsilon\log(\varepsilon t), \ \rho=r-t,$$

where $0\leqq\nu\leqq\gamma$ and a is homogeneous of degree $\leqq -1$. In fact, $Z\varepsilon\log(\varepsilon t)$ is equal to ε times a function homogeneous of degree $\leqq 0$. When $\varepsilon t \geqq 1$ such terms are bounded by $C\varepsilon(1+t)^{-1}$. Since $Z^\alpha\chi(\varepsilon t)$ is bounded when $1\leqq\varepsilon t\leqq 2$, for any α, it follows that

(2.4.10) holds.

To prove (2.4.11) we must distinguish three cases:

a) When $\varepsilon t \leqq 1$ we have $w(t,x) = \varepsilon w_0(t, x)$, which is a solution of the wave equation, so

$$f = \sum (g_{jk}(w') - g_{jk}(0))\partial_j\partial_k w.$$

Using (2.4.10) we obtain (2.4.11) at once in this case.

b) Now consider the transition zone where $1 \leqq \varepsilon t \leqq 2$. In addition to the arguments in a) we must examine $\Box w$, which is no longer equal to 0. However, some cancellation occurs since

$$\Box w = \varepsilon r^{-1}((\partial_t-\partial_r)(\partial_t+\partial_r)-r^{-2}\Delta_\omega)((1-\chi(\varepsilon t))(U(\omega,\varepsilon\log\varepsilon t,r-t) - \\ - F(\omega, r^{-1}, r-t))).$$

Here the factor $(\partial_t+\partial_r)$ must act either on $1-\chi(\varepsilon t)$ or on the second argument in $U(\omega,\varepsilon\log(\varepsilon t), r-t) - F(\omega, r^{-1}, r-t)$. In that case we gain a factor ε/t or r^{-2}, both of which are $O(\varepsilon^2)$. In the first case we only gain a factor ε, but

$$U(.,\ \varepsilon\log(\varepsilon t),\ .) - F(.,\ r^{-1},\ .) = U(.,\ \varepsilon\log(\varepsilon t),\ .) - \\ - U(.,\ 0,\ .) + F(.,0,.) - F(.,\ r^{-1},\ .)$$

and its derivatives with respect to the variables indicated by dots are all $O(\varepsilon+r^{-1}) = O(\varepsilon)$. If no derivatives ever fall on the middle argument this gains us a factor ε, and we obtain the bound $O(\varepsilon^4)$ stated in (2.4.11) when $\alpha=0$. (Recall that $|r-t|$ is bounded and that we assume $1 \leqq \varepsilon t \leqq 2$ now.) As in the proof of (2.4.10) we see that (2.4.11) also holds for arbitrary α.

c) Let $2/\varepsilon \leqq t \leqq e^{B/\varepsilon}$. Then

$$\Box w = \varepsilon r^{-1}((\partial_t-\partial_r)(\partial_t+\partial_r)-r^{-2}\Delta_\omega)U(\omega,\varepsilon\log(\varepsilon t),\ r-t).$$

Here $\partial_t+\partial_r$ must act on $\varepsilon\log(\varepsilon t)$, producing a factor ε/t. With $s = \varepsilon\log(\varepsilon t)$ and $q = r-t$ we obtain

$$|\Box w + 2\varepsilon^2 r^{-1} t^{-1} U''_{sq}(\omega, s, q)| \leqq C\varepsilon t^{-3}.$$

Moreover, with $\hat{\omega} = (-1, \omega_1, \omega_2, \omega_3)$ we have

$$|\partial^\alpha w - \varepsilon r^{-1}\hat{\omega}^\alpha \partial_q^{|\alpha|} U(\omega, s, q)| \leqq C\varepsilon r^{-2},$$

which proves that

$$|\sum (g_{jk}(w') - g_{jk}(0))\partial_j\partial_k w - \varepsilon^2 r^{-2} \sum g_{jkl}\hat{\omega}_j\hat{\omega}_k\hat{\omega}_l U'_q U''_{qq}| \leqq$$
$$\leqq C\varepsilon^2 r^{-3}.$$

Recalling that

$$2U''_{sq} = \sum g_{jkl}\hat{\omega}_j\hat{\omega}_k\hat{\omega}_l U'_q U''_{qq}$$

by (2.4.5) we conclude that $|f| \leqq C\varepsilon t^{-3}$, which is equivalent to the estimate (2.4.11) when $\alpha = 0$. By the arguments at the beginning of the proof the estimate follows for general α.

We shall write the solution u of (2.4.1), (2.4.2) in the form $u = v + w$ where w is the approximate solution just discussed. Then the Cauchy problem (2.4.1), (2.4.2) becomes

(2.4.1)' $\quad \sum g_{jk}(v'+w')\partial_j\partial_k v + f + \sum(g_{jk}(v'+w')-g_{jk}(w'))\partial_j\partial_k w = 0,$

(2.4.2)' $\quad v = \partial_0 v = 0$ when $x_0 = 0$

where f is defined by (2.4.9). The following lemma is the main step towards an existence theorem.

Lemma 2.4.3. Assume that (2.4.1),(2.4.2) has a C^∞ solution for $0 \leqq t \leqq T$ where $\varepsilon \log T \leqq B < A$, with A defined by (2.4.7). If $0 < \varepsilon < \delta_B$ it follows that

(2.4.12) $\quad (\int |Z^\alpha(u-w)'(t,x)|^2 dx)^{\frac{1}{2}} \leqq C_{\alpha,B}\varepsilon^2 \log\frac{1}{\varepsilon}, \; 0 \leqq t \leqq T,$

where δ_B and $C_{\alpha,B}$ are independent of T and ε.

Proof. We shall use the notation

(2.4.13) $\quad N_s(v; x_0) = \sum_{j\leqq s} \|Z_1 \ldots Z_j v(x_0, .)\|$

where the norms are L^2 norms in $\mathbf{R}^3$ and $Z_1, \ldots, Z_j$ are chosen arbitrarily among the vector fields (2.3.3),(2.3.4) (with n=3). As observed at the end of Section 2.3 the norms

(2.4.14) $$N_s^{\#}(v;\,t) = \sup_{x_0<t} N_s(v;\,x_0)$$

for functions in the half space defined by $x_0<t$ have all the properties of the norms $\| \ \|_s^{\#}$ listed in Propositions 2.3.2 and 2.3.3, with $s_0=2$. We shall write

$$N_s(v';\,t) = \sum_0^3 N_s(\partial_j v;\,t)$$

and similarly for $N_s^{\#}$.

When proving the estimates (2.4.12) we shall assume that we already know the estimate

(2.4.15) $$N_3^{\#}(v';\,T) \leqq \varepsilon.$$

Since $\varepsilon^2 \log \frac{1}{\varepsilon}$ is much smaller than ε for small ε it will be easy to eliminate this assumption afterwards.

To prove (2.4.12) we shall regard (2.4.1)' as a linear hyperbolic equation for v and apply Proposition 2.3.1 with

$$\Box + \sum \gamma_{jk}(x)\partial_j\partial_k = \sum g_{jk}(u')\partial_j\partial_k.$$

By (2.3.7) it follows from (2.4.15) that

(2.4.16) $$|D^\alpha v'(x)| \leqq C\varepsilon/(1+|x_0|) \text{ if } 0 \leqq x_0 \leqq T \text{ and } |\alpha| \leqq 1,$$

and (2.4.10) gives a similar bound for w'. Hence $|u'/\varepsilon|$ has a fixed bound. Dividing the equation (2.4.1) by g_{00}, which is $> \frac{1}{2}$ for small u', we may therefore assume that $g_{00} = 1$, hence that $\gamma_{00}=0$, and we have $|\gamma| < \frac{1}{2}$ if ε is small enough. Moreover, for some new constant C of course,

(2.4.17) $$|\gamma'(\tau)| \leqq C\varepsilon/(1+\tau),$$

hence

$$\int_0^T |\gamma'(\tau)|\,d\tau \leqq C\varepsilon\log(1+T) \leqq C(B+1),$$

which gives a bound for the exponential in (2.3.1). The inhomogeneous terms in (2.4.1)' consist of the second sum and of f, which can be estimated by $C\varepsilon(1+t)^{-1}|v'|$ and by

$$C\,\varepsilon^2(1+t)^{-2}(1+\varepsilon t)^{-1}.$$

The measure of the support of f for fixed t is $O(1+t)^2$, and

$$(2.4.18)\qquad \int_0^\infty \varepsilon^2(1+t)^{-1}(1+\varepsilon t)^{-1}dt = \varepsilon^2(1-\varepsilon)^{-1}\log\frac{1}{\varepsilon},$$

so (2.3.1) gives

$$(2.4.19)\qquad \|v'(t,.)\| \leqq C(\varepsilon^2\log\frac{1}{\varepsilon} + \varepsilon\int_0^t \|v'(\tau,.)\|(1+\tau)^{-1}d\tau).$$

Now an estimate of the form

$$(2.4.20)\qquad \varphi(t) \leqq E + \int_0^t \varphi(\tau)k(\tau)d\tau\ ,\ 0 \leqq t \leqq T,$$

with non-negative φ and k, implies

$$(2.4.21)\qquad \varphi(t) \leqq E\exp\int_0^t k(\tau)d\tau\ .$$

In fact, if the right-hand side of (2.4.20) is denoted by Φ then $\varphi \leqq \Phi$ and (2.4.20) gives

$$\Phi' = \varphi k \leqq \Phi k,$$

so (2.4.21) is a consequence of Grönwall's lemma. When $E = C\varepsilon^2\log\frac{1}{\varepsilon}$ and $k(t) = C\varepsilon(1+t)^{-1}$ we have

$$\exp\int_0^t k(\tau)d\tau = (1+t)^{C\varepsilon} \leqq 2^{C\varepsilon}e^{CB},\ t \leqq T,$$

so we conclude from (2.4.19) that (2.4.12) holds for $\alpha=0$.

The preceding discussion was just meant to indicate the lines of the argument without technicalities. We start now with the main proof, the estimate of $N_s^{\#}(v';\ t)$ when $s \geqq 3$. Fixing s and applying Z^α to the equation (2.4.1)' for any α with $|\alpha|\leqq s$, we obtain

$$\sum g_{jk}(u')\partial_j\partial_k Z^\alpha v = \Psi = -Z^\alpha f+f_0+f_1+f_2+f_3$$

where

$$f_0 = [\Box,\ Z^\alpha]v,\quad f_1 = \sum[\gamma_{ij},Z^\alpha]\partial_i\partial_j v,$$

$$f_2 = \sum \gamma_{ij}[\partial_i\partial_j,Z^\alpha]v,\ f_3 = -Z^\alpha\sum(g_{jk}(v'+w')-g_{jk}(w'))\partial_j\partial_k w.$$

We shall prove that

$$(2.4.22)\qquad \int_0^t \|\Psi(\tau,.)\|d\tau \leqq C(\varepsilon^2\log\frac{1}{\varepsilon}+\varepsilon\int_0^t N_s^{\#}(v';\tau)d\tau/(1+\tau)).$$

Combined with (2.3.1) this will prove that

(2.4.19) is valid with $\|v'(t,.)\|$ replaced by $N_s^{\#}(v';t)$ (and some other constant). The implication (2.4.20) ⇒ (2.4.21) then yields (2.4.12) when $|\alpha| \leqq s$. (Note that $Z^\alpha \partial_j = \sum_{|\beta|\leqq|\alpha|} c_{\alpha j\beta k}\partial_k Z^\beta$.)

By Lemma 2.4.2 and (2.4.18) we have

$$\int_0^t \|Z^\alpha f(\tau,.)\|\, d\tau \leqq C\varepsilon^2 \log\frac{1}{\varepsilon}. \tag{2.4.23}$$

To estimate the contribution from f_0 in (2.4.22) we note that for some constants $c_{\alpha\beta}$

$$[\Box, Z^\alpha]v = \sum_{|\beta|<s} c_{\alpha\beta} Z^\beta \Box v,$$

$$\Box v = \sum (g_{jk}(0)-g_{jk}(u'))\partial_j\partial_k v - f - \sum (g_{jk}(v'+w')-g_{jk}(w'))\partial_j\partial_k w.$$

By Proposition 2.3.2 we have for $|\beta| < s$

$$(1+t)\|Z^\beta(g_{jk}(0)-g_{jk}(u'))\partial_j\partial_k v(t,.)\| \leqq$$

$$\leqq C(N_{s-1}^{\#}(g_{jk}(0)-g_{jk}(u');t)N_2^{\#}(v'';t)+N_2^{\#}(g_{jk}(0)-g_{jk}(u');\, t)\cdot$$

$$\cdot N_{s-1}(v'';t))\leqq C(N_{s-1}^{\#}(u';t)N_3^{\#}(v';t) + N_2^{\#}(u';t)N_s^{\#}(v';t)).$$

Here

$$N_2^{\#}(u';t) \leqq N_2^{\#}(v';\, t) + N_2^{\#}(w';\, t) \leqq C\,\varepsilon$$

by (2.4.15) and Lemma 2.4.2, and

$$N_{s-1}^{\#}(u';\, t) \leqq N_s^{\#}(v';\, t) + C\varepsilon;\ N_3^{\#}(v';t) \leqq C\varepsilon.$$

Since $s \geqq 3$ it follows that the right-hand side is bounded by

$$C\varepsilon N_s^{\#}(v';\, t),$$

which gives an estimate of the desired form (2.4.22) for the first term in f_0. Next we apply Propositions 2.3.3 and 2.3.2 to

$$g_{jk}(v'+w')-g_{jk}(w')-g_{jk}(v')+g_{jk}(0) \text{ and } g_{jk}(v')-g_{jk}(0),$$

which gives

$$N_s^{\#}(g_{jk}(v'+w')-g_{jk}(w');t) \leqq C(N_s^{\#}(v';t)+N_s^{\#}(w';t)N_2^{\#}(v';t)) \leqq$$

$$\leqq C'\, N_s^{\#}(v';t).$$

Since

$$Z^\beta \partial_j\partial_k w = O(\varepsilon/(1+t)) \text{ for all } \beta$$

an estimate of the form (2.4.22) follows for f_0 and also for f_3.

Since $[\partial_i\partial_j, Z^\alpha]$ is a sum of terms of the form $Z^\beta\partial_k$ with $|\beta|\leqq|\alpha|\leqq s$ and $|\gamma_{ij}|\leqq C\varepsilon/(t+1)$, an estimate of the form (2.4.22) for f_2 is obvious. It remains to examine f_1, which is a sum of terms of the form

$$(Z^\beta(g_{ij}(u')-g_{ij}(0))Z^\gamma\partial_i\partial_j v$$

with $|\beta|+|\gamma|=|\alpha|=s$ and $\beta\neq 0$. Thus one factor Z falls on $g_{ij}(u')-g_{ij}(0)$ and the remaining s-1 factors are distributed arbitrarily on the two factors. By Proposition 2.3.2 the norm can thus be estimated by $(1+t)^{-1}$ times

$$C(N_s^\#(g_{ij}(u')-g_{ij}(0);t)N_2^\#(v'';t)+N_3^\#(g_{ij}(u')-g_{ij}(0);t)N_{s-1}^\#(v'';t))\leqq$$
$$C'((N_s^\#(v';t)+N_s^\#(w';t))N_3^\#(v';t)+N_3^\#(u';t)N_s^\#(v';t))\leqq C''\varepsilon N_s^\#(v';t).$$

This completes the proof of (2.4.22).

What remains is to eliminate the assumption (2.4.15). When ε is small the estimate (2.4.12) implies (2.4.15) with ε replaced by $\frac{1}{2}\varepsilon$ in the right-hand side. Hence the supremum of all $T'\leqq T$ such that (2.4.15) holds with T replaced by T' must be equal to T, which completes the proof.

The proof of the main theorem in this section is now straightforward:

<u>Theorem 2.4.4.</u> The Cauchy problem (2.4.1), (2.4.2) with $u_j\in C_0^\infty(\mathbf{R}^3)$ has a C^∞ solution u_ε for $0\leqq t<T_\varepsilon$ where

(2.4.24) $$\varliminf_{\varepsilon\to 0}\ \varepsilon\log T_\varepsilon\geqq A=(\max\tfrac{1}{2}\sum g_{jkl}\hat\omega_j\hat\omega_k\hat\omega_l\partial^2F_0(\omega,\rho)/\partial\rho^2)^{-1}.$$

Here $\omega\in S^2$ and $\hat\omega=(-1,\omega)\in\mathbf{R}^4$. If U is the solution of (2.4.5), (2.4.6) then

(2.4.25) $$\varepsilon^{-1}e^{s/\varepsilon}u_\varepsilon(e^{s/\varepsilon},(e^{s/\varepsilon}+r)\omega)-U(\omega,s,r)\to 0,\ \varepsilon\to 0,$$

locally uniformly in $S^2\times(0,A)\times\mathbf{R}$; in fact, the difference is locally uniformly $O(\varepsilon\log\varepsilon)$.

Proof. Let $0 < B < A$. For small ε the Cauchy problem (2.4.1), (2.4.2) has a C^∞ solution for $0 \leqq t \leqq e^{B/\varepsilon}$. In fact, it follows from Lemma 2.4.3 and the local existence theorem that the set of all $T \leqq e^{B/\varepsilon}$ such that a C^∞ solution exists for $0 \leqq t \leqq T$ is both closed and open. By (2.4.12) and Lemma 2.3.2 we have

$$|u'(t,x) - w'(t,x)| \leqq C(1+t)^{-1}\varepsilon^2\log\frac{1}{\varepsilon}\ , \quad 0 \leqq t \leqq e^{B/\varepsilon},$$

and since $u = w = 0$ when $|x|>t+M$, for some M, it follows that

$$|\varepsilon^{-1}t(u(t,x) - w(t,x))| \leqq C(1+||x|-t|)\varepsilon\log\frac{1}{\varepsilon},\ 0 \leqq t \leqq e^{B/\varepsilon}.$$

($||x|-t|$ could be replaced by the square root here.) Since

$$\varepsilon^{-1}e^{s/\varepsilon}w(e^{s/\varepsilon},(e^{s/\varepsilon}+r)\omega) = U(\omega,\ s+\varepsilon\log\varepsilon,\ r)/(1+re^{-s/\varepsilon})$$

when $\varepsilon e^{s/\varepsilon}>2$, the theorem is proved.

The fact that $U(\omega, s, r)$ blows up when $s \to A$ suggests that the upper limit of $\varepsilon\log T_\varepsilon$ cannot be larger than A if T_ε is the precise lifespan of the solution. However, we do not have any proof of that except for the results of John [4] in the rotationally symmetric case, discussed in Section 2.2. Then we obtain by combining Theorems 2.2.1 and 2.4.4:

Corollary 2.4.5. If the Cauchy problem (2.2.1), (2.2.2) with radially symmetric f and g, $c = c(\partial u/\partial t)$ and $c(0) = 1$, has a C^∞ solution precisely when $0 \leqq t < T_\varepsilon$, then

$$\lim_{\varepsilon\to 0} \varepsilon\log T_\varepsilon = (\max c'(0)F''(\rho))^{-1} \tag{2.4.24}$$

where F is defined by (2.1.7).

Unless f and g are identically 0 the right-hand side of (2.4.24) is finite when $\sum g_{jkl}\omega_j\omega_k\omega_l$ is not divisible by $\omega_0{}^2-\omega_1{}^2-\omega_2{}^2-\omega_3{}^2$. In that case Klainerman [8] has recently proved global existence for small ε. The result was announced in [9] but came to the author's attention only when this paper was completed. Recent related work by D.Christodoulou is also quoted in [8].

2.5. The case of two space dimensions. The arguments in this section will be parallel to those in Section 2.4 although somewhat more complicated because Huygens' principle is not valid. We shall therefore be brief where no substantial changes are required.

In dimension 1+2 we shall look for an approximate solution to (2.4.1), (2.4.2) of the form

$$u(t, r\omega) = \varepsilon r^{-\frac{1}{2}}U(\omega,\varepsilon t^{\frac{1}{2}},r-t)$$

where $r = |x|$ and $\omega = x/r$. Since

$$\Delta u = r^{-\frac{1}{2}}(\partial^2/\partial r^2+r^{-2}(4^{-1}+\partial^2/\partial\omega^2))(r^{\frac{1}{2}}u),$$

we have

$$\Box u = \varepsilon r^{-\frac{1}{2}}((\partial/\partial t+\partial/\partial r)(\partial/\partial t-\partial/\partial r)-r^{-2}(4^{-1}+\partial^2/\partial\omega^2))U(\omega,\varepsilon t^{\frac{1}{2}},r-t).$$

Here $\partial/\partial t+\partial/\partial r$ must act on $s = \varepsilon t^{\frac{1}{2}}$, while $\partial/\partial t-\partial/\partial r$ may act on $q = r-t$. The main contribution is therefore $-\varepsilon^2(rt)^{-\frac{1}{2}}U''_{sq}$, at least when $r^{-1} = O(\varepsilon)$. With $g_{jkl} = \partial g_{jk}(0)/\partial U_l$ we find that the main non-linear terms are

$$\sum g_{jkl}\partial u/\partial x_l\partial^2u/\partial x_j\partial x_k \sim \varepsilon^2 r^{-1}G(\omega)U'_qU''_{qq};\ G(\omega) = \sum g_{jkl}\hat\omega_j\hat\omega_k\hat\omega_l.$$

To balance them we pose the equation $U''_{sq} = G(\omega)U'_qU''_{qq}$, or equivalently

$$\partial U(\omega,s,q)/\partial s = \tfrac{1}{2}G(\omega)(\partial U(\omega,s,q)/\partial q)^2. \tag{2.5.1}$$

To get agreement with the solution of the linear wave equation for small $\varepsilon t^{\frac{1}{2}}$ we pose the initial condition

$$U(\omega,0,q) = F_0(\omega,q), \tag{2.5.2}$$

where F_0 is the radiation field for the solution of the wave equation with Cauchy data u_0 and u_1, defined by (2.1.14). We shall now prove an analogue of Lemma 2.4.1:

Lemma 2.5.1. If

$$A = (\max \sum g_{jkl}\hat\omega_j\hat\omega_k\hat\omega_l\partial^2F_0(\omega,\rho)/\partial\rho^2)^{-1} \tag{2.5.3}$$

then the Cauchy problem (2.5.1), (2.5.2) has a unique C^∞ solution for $0 \leqq s < A$ but the second order derivatives are not locally bounded when $0 \leqq s \leqq A<\infty$. If $F_0(\omega,q) = 0$ for $q > M$ then $U(\omega,s,q) = 0$ for $q > M$, and U is a polyhomogeneous symbol of degree $-\frac{1}{2}$ with respect to q.

Before the proof we remark that the maximum in (2.5.3) is always attained and that it is strictly positive unless $f = g = 0$ or $G(\omega) \equiv 0$. In fact, if neither condition is fulfilled we can choose ω so that $F_0(\omega, \rho)$ is not 0 for all ρ and $G(\omega)\neq 0$, for $G(\omega)$ has at most six zeros in S^1. Since $F_0(\omega,\rho) = 0$ for $\rho>M$ and F_0 is a symbol of order $-\frac{1}{2}$ we see that $\partial^2 F_0(\omega,\rho)/\partial\rho^2$ is integrable with integral 0 but not identically 0. Hence it takes both positive and negative values. Thus the supremum in (2.5.3) is positive, and since the sum in (2.5.3) tends to 0 as $\rho\to-\infty$ the maximum is attained.

<u>Proof of Lemma 2.5.1.</u> As in the proof of Lemma 2.4.1 we set $\partial U/\partial q = u$ and obtain the equation

$$\partial u/\partial s = G(\omega)u\partial u/\partial q;\ u = \partial F_0/\partial q \text{ when } s = 0.$$

As in Section 1.1 this means that with $F_0' = \partial F_0/\partial q$

$$u(s,q-sG(\omega)F_0'(\omega,q)) = F_0'(\omega,q).$$

The equation $q - sG(\omega)F_0'(\omega,q) = \rho$ has a unique solution $q = Q(\omega,s,\rho)$ when $0 \leqq s < A$ since $1-sG(\omega)\partial^2 F_0(\omega,q)/\partial q^2 > 0$ by (2.5.3) and all derivatives of $F_0(\omega, q)$ tend to 0 at infinity. We can find an asymptotic expansion as $\rho\to\infty$ by defining $Q_0(\omega,s,\rho) = \rho$ and then recursively, suppressing the ω dependence in Q_k,

$$Q_{k+1}(s, \rho) - sG(\omega)F_0'(\omega, Q_k(s,\rho)) = \rho,\ k = 0, 1, \dots$$

Then

$$Q_{k+2}(s,\rho) - Q_{k+1}(s,\rho) = sG(\omega)(F_0'(\omega, Q_{k+1}(s,\rho))-F_0'(\omega,Q_k(s,\rho))).$$

Since $Q_1(s,\rho) - Q_0(s,\rho) = sG(\omega)F_0'(\omega,\rho)$ is polyhomogeneous of degree $-3/2$ we find by induction that $Q_{k+1}(s,\rho)-Q_k(s,\rho)$ is polyhomogeneous of degree $-(3+5k)/2$ (with half integer steps). Thus

$$Q_k(s,\rho) - sG(\omega)F_0'(\omega,Q_k(s,\rho)) = \rho + O(|\rho|^{-(3+5k)/2})$$

and it follows that

$$Q(s,\rho) - Q_k(s,\rho) = O(|\rho|^{-(3+5k)/2}).$$

Hence $Q(s,\rho)$ is a polyhomogeneous symbol of order 1, and

$$u(\omega,s,\rho) = F_0'(\omega, Q(\omega,s,\rho))$$

is a polyhomogeneous symbol of degree $-3/2$. We have $\int u(\omega,s,\rho)d\rho = 0$ for every s since this is true when $s = 0$ and

$$\int \partial u(\omega,s,\rho)/\partial s \, d\rho = G(\omega)\int u\partial u/\partial\rho \, d\rho = 0.$$

Thus

$$U(\omega,s, q) = \int_{-\infty}^{q} u(\omega,s,\rho)d\rho$$

is polyhomogeneous of degree $-\frac{1}{2}$ and satisfies (2.5.1), (2.5.2). Integrating from $+\infty$ instead we find that $U(\omega,s,q) = 0$ for $q > M$, which completes the proof.

Let w_0 be the solution of the linear wave equation with Cauchy data u_0, u_1. Choose $\chi\in C^\infty(\mathbf{R})$ equal to 1 in $(-\infty,1)$ and 0 in $(2,\infty)$, and set with U defined by Lemma 2.5.1

$$w(t,x) = \varepsilon\Big(\chi(\varepsilon t)w_0(t,x) + (1-\chi(\varepsilon t))\chi(3\varepsilon(t-r))r^{-\frac{1}{2}}U(\omega,\varepsilon t^{\frac{1}{2}},r-t)\Big). \tag{2.5.4}$$

Note that $\chi(3\varepsilon(t-r))=0$ when $r-t < -2/(3\varepsilon)$; since $t\geqq 1/\varepsilon$ in the support of the second term it follows that $r\geqq t/3$ there. We shall now prove an analogue of Lemma 2.4.2.

<u>Lemma 2.5.2.</u> With w defined by (2.5.4) and

$$f = \sum g_{jk}(w')\partial_j\partial_k w, \tag{2.5.5}$$

we have $f, w\in C^\infty$ when $\varepsilon^2 t<A^2$, where A is defined by (2.5.3), and $r-t\leqq M$ in $\operatorname{supp} f \cup \operatorname{supp} w$ if $|x|\leqq M$ in $\operatorname{supp} u_0 \cup \operatorname{supp} u_1$. When $t>2/\varepsilon$ we have $r-t\geqq -2/(3\varepsilon)$ in $\operatorname{supp} f\cup\operatorname{supp} w$. If $0 < B < A$ we have for small ε and all α

(2.5.6) $|Z^\alpha w(t,x)| \leqq C_{\alpha,B}\varepsilon(1+t)^{-\frac{1}{2}}(1+|r-t|)^{-\frac{1}{2}},\ 0 \leqq \varepsilon^2 t \leqq B^2,$

(2.5.7) $\|Z^\alpha f(t,.)\| \leqq C_{\alpha,B}\varepsilon^2(1+t)^{-\frac{1}{2}}(1+\varepsilon t)^{-1},\ 0 \leqq \varepsilon^2 t \leqq B^2.$

Proof. By (2.1.13) we have for all α

$$|Z^\alpha w_0(t,x)| \leqq C_\alpha (1+t)^{-\frac{1}{2}}(1+|r-t|)^{-\frac{1}{2}}.$$

Set $U_\varepsilon(\omega,z,\rho) = \chi(-3\varepsilon\rho)U(\omega,z,\rho)$. Then U_ε is bounded as a symbol of order $-\frac{1}{2}$ for $0 < \varepsilon < 1$, and the arguments in the proof of Theorem 2.1.2 show that $Z^\alpha(\varepsilon r^{-\frac{1}{2}}U_\varepsilon(\omega,\varepsilon t^{\frac{1}{2}},r-t))$ is a sum of terms of the form

$$\varepsilon^{\beta+1}a(x,t)D_\omega{}^\alpha D_z{}^\beta \rho^\nu D_\rho{}^\gamma U_\varepsilon(\omega,z,\rho)$$

where $z = \varepsilon t^{\frac{1}{2}}$, $\rho = r-t$, $\nu \leqq \gamma$ and a is homogeneous of degree $\leqq (\beta-1)/2$. Such a term can be estimated by

$$C\varepsilon^{\beta+1}t^{\frac{1}{2}(\beta-1)}(1+|\rho|)^{-\frac{1}{2}} \leqq C'\varepsilon t^{-\frac{1}{2}}(1+|\rho|)^{-\frac{1}{2}}$$

when $\varepsilon t^{\frac{1}{2}} \leqq B$. Since $Z^\alpha\chi(\varepsilon t)$ is uniformly bounded for $0 \leqq \varepsilon \leqq 1$ it follows that (2.5.6) holds.

(2.5.7) is somewhat more laborious to prove than (2.4.11) since we do not have compact support in the ρ variable. We shall study three cases separately:

a) When $0 \leqq \varepsilon t \leqq 1$ we have $w = \varepsilon w_0$ so

$$f = \sum (g_{jk}(w')-g_{jk}(0))\partial_j\partial_k w.$$

By (2.5.6) we obtain

$$|Z^\alpha f(t,x)| \leqq C'_{\alpha,B}\,\varepsilon^2(1+t)^{-1}(1+|r-t|)^{-1}.$$

Hence

$$\|Z^\alpha f(t,.)\| \leqq C_{\alpha,B}\varepsilon^2(1+t)^{-1}\Big(\int_0^{t+M}(1+|r-t|)^{-2}\pi r dr\Big)^{\frac{1}{2}}$$
$$\leqq C'_{\alpha,B}\varepsilon^2(1+t)^{-\frac{1}{2}}$$

since $r \leqq t+M$ and $\int_{-\infty}^{\infty}(1+|\rho|)^{-2}d\rho = 2$.

b) When $1 \leqq \varepsilon t \leqq 2$ the same estimate holds for the non-linear terms so we just have to examine

$$\Box w = \Box\, \varepsilon\Big((1-\chi(\varepsilon t))(\chi(3\varepsilon(t-r))r^{-\frac{1}{2}}U(\omega,\varepsilon t^{\frac{1}{2}},r-t) - w_0(t,x))\Big).$$

In the support of $1-\chi(3\varepsilon(t-r))$ we have $3\varepsilon(t-r) \geqq 1$, hence

$1-r/t \geqq 1/(3\varepsilon t) \geqq 1/6$, so $r/t \leqq 5/6$. We have already seen that $r/t \geqq 1/3$ in the support of $\chi(3\varepsilon(t-r))$. Since w_0 behaves like a symbol of order -1 when $r/t \leqq 5/6$ and $(1-\chi(\varepsilon t))(1-\chi(3\varepsilon(t-r))$ is uniformly bounded as a symbol of order 0 in this set, we conclude that for

$$R_0 = \Box\varepsilon(1-\chi(\varepsilon t))(\chi(3\varepsilon(t-r))-1)w_0(t,x)$$

we have the estimates

$$|Z^\alpha R_0| \leqq C_\alpha \varepsilon t^{-3}, \; 1 \leqq \varepsilon t \leqq 2.$$

Hence

$$\|Z^\alpha R_0(t,.)\| \leqq C_\alpha \varepsilon t^{-2}$$

which is better than required in (2.5.7).

It remains to estimate

$$R_1 = \Box\varepsilon(1-\chi(\varepsilon t))\chi(3\varepsilon(t-r))r^{-\frac{1}{2}}(U(\omega,\varepsilon t^{\frac{1}{2}},r-t)-F(\omega,r^{-1},r-t)).$$

In doing so we write as above

$$\Box = r^{-\frac{1}{2}}(\partial_t-\partial_r)(\partial_t+\partial_r)r^{\frac{1}{2}} - r^{-2}(4^{-1}+\partial^2/\partial\omega^2).$$

Since $\partial/\partial\omega = x_1\partial/\partial x_2 - x_2\partial/\partial x_1$ is one of the operators Z in (2.1.12) we see using Theorem 2.1.2 that for

$$R_2 = -r^{-2}(4^{-1}+\partial^2/\partial\omega^2)\varepsilon(1-\chi(\varepsilon t))\chi(3\varepsilon(t-r))r^{-\frac{1}{2}}F(\omega,r^{-1},r-t)$$

we have

$$|Z^\alpha R_2| \leqq C_\alpha \varepsilon t^{-5/2}(1+|r-t|)^{-\frac{1}{2}}.$$

This implies that

$$\|Z^\alpha R_2(t,.)\| \leqq C_\alpha \varepsilon t^{-2}(\log t)^{\frac{1}{2}}$$

which is better than required in (2.5.7). We can argue similarly for the term R_3 defined as R_2 but with $F(\omega,r^{-1},r-t)$ replaced by $U(\omega,\varepsilon t^{\frac{1}{2}},r-t)$, noting that $Z\varepsilon t^{\frac{1}{2}}$ is equal to ε times a homogeneous function of degree $\leqq \frac{1}{2}$, hence bounded, and that $Z(r-t)$ is homogeneous of degree 0 or divisible by r-t. (See the proof of Theorem 2.1.2.) This gives the same estimate for R_3 as for R_2, and it remains to estimate

$$R_4 = r^{-\frac{1}{2}}\varepsilon(\partial_t-\partial_r)(\partial_t+\partial_r)(1-\chi(\varepsilon t))\chi(3\varepsilon(t-r))(U(\omega,\varepsilon t^{\frac{1}{2}},r-t) - F(\omega,r^{-1},r-t)).$$

If no derivative falls on the middle arguments in U and in F, we get terms which can be estimated by

$$C\varepsilon^2 r^{-\frac{1}{2}}(\varepsilon t^{\frac{1}{2}}+r^{-1}(1+|\rho|))(1+|\rho|)^{-3/2}$$

where $\rho = r-t$. Here we have used that $\partial_t+\partial_r$ must act on the cutoff functions and used the cancellation which follows from the initial condition (2.5.2) when we compare with the same terms with middle argument in U and in F equal to 0. The L^2 norm can be estimated by

$$C\varepsilon^2(\varepsilon t^{\frac{1}{2}}+t^{-1}(\log t)^{\frac{1}{2}}) \leqq C'\varepsilon^2 t^{-\frac{1}{2}}.$$

On the other hand, if a derivative falls on the middle arguments in U or F we get terms estimated by

$$C\varepsilon r^{-\frac{1}{2}}(\varepsilon t^{-\frac{1}{2}}(1+|\rho|)^{-3/2}+t^{-2}(1+|\rho|)^{-\frac{1}{2}});$$

the L^2 norm can be estimated by

$$C\varepsilon(\varepsilon t^{-\frac{1}{2}}+t^{-2}(\log t)^{\frac{1}{2}}) \leqq C'\varepsilon^2 t^{-\frac{1}{2}},$$

which completes the proof of (2.5.7) when $\alpha = 0$. When $\alpha \neq 0$ the estimate of $Z^\alpha R_4$ can be made in the same way by treating separately the terms where no differentiation ever falls on the middle arguments in U and in F; in these terms we get some cancellation using (2.5.2). We leave the repetition of the other details for the reader.

c) Now assume that $\varepsilon t^{\frac{1}{2}} \leq B$ and that $\varepsilon t \geqq 2$, hence

$$w(t,x) = \varepsilon r^{-\frac{1}{2}} U_\varepsilon(\omega,\varepsilon t^{\frac{1}{2}},r-t).$$

Then $\Box w = R_0 - R_1$ where

$$R_0 = \varepsilon r^{-\frac{1}{2}}(\partial_t-\partial_r)(\partial_t+\partial_r)U_\varepsilon(\omega,\varepsilon t^{\frac{1}{2}},r-t),$$

$$R_1 = \varepsilon r^{-5/2}(4^{-1}+ \partial^2/\partial\omega^2)U_\varepsilon(\omega,\varepsilon t^{\frac{1}{2}},r-t).$$

The usual arguments give

$$|Z^\alpha R_1| \leqq C_{\alpha,B}\varepsilon r^{-5/2}(1+|r-t|)^{-\frac{1}{2}},$$

which implies that

$$\|Z^\alpha R_1(t,.)\| \leqq C'_{\alpha,B}\varepsilon t^{-3/2}.$$

With the notation $s=\varepsilon t^{\frac{1}{2}}$, $q=r-t$, we can write

$$R_0 = \varepsilon r^{-\frac{1}{2}}(\partial_t-\partial_r)\tfrac{1}{2}\varepsilon t^{-\frac{1}{2}}\partial U_\varepsilon/\partial s = -\varepsilon^2(rt)^{-\frac{1}{2}}\partial^2 U_\varepsilon/\partial s\partial q + R_2$$

where

$$R_2 = 4^{-1}(\varepsilon^3 r^{-\frac{1}{2}}t^{-1}\partial^2 U_\varepsilon/\partial s^2 - \varepsilon^2 r^{-\frac{1}{2}}t^{-3/2}\partial U_\varepsilon/\partial s).$$

We have

$$|Z^\alpha R_2| \leqq C_{\alpha,B}(\varepsilon^3 t^{-3/2}+\varepsilon^2 t^{-2}),$$

and since $|t| \leqq B^2/\varepsilon^2$ it follows that

$$\|Z^\alpha R_2(t,.)\| \leqq C'_{\alpha,B}\varepsilon t^{-3/2}.$$

We write

$$-\varepsilon^2(rt)^{-\frac{1}{2}}\partial^2 U_\varepsilon/\partial s\partial q = -\varepsilon^2 r^{-1}\partial^2 U_\varepsilon/\partial s\partial q + R_3$$

where

$$R_3 = \varepsilon^2 r^{-1}(1-(1-q/r)^{-\frac{1}{2}})\partial^2 U_\varepsilon/\partial s\partial q.$$

Writing $1 - (1-q/r)^{-\frac{1}{2}} = aq/r$ where a is a smooth homogeneous function of degree 0 when $r \geqq t/3$, we find that

$$|Z^\alpha R_3| \leqq C_{\alpha,B}\varepsilon^2 r^{-2}(1+|q|)^{-\frac{1}{2}},$$

hence that

$$\|Z^\alpha R_3(t,.)\| \leqq C'_{\alpha,B}\varepsilon^2 t^{-3/2}(\log t)^{\frac{1}{2}} \leqq C''_{\alpha,B}\varepsilon t^{-3/2}.$$

Summing up the estimates of R_1, R_2, R_3, we have proved that

$$\|Z^\alpha(\Box w+\varepsilon^2 r^{-1}\partial^2 U_\varepsilon/\partial s\partial q)(t,.)\| \leqq C_{\alpha,B}\varepsilon t^{-3/2}.$$

Finally we must study the non-linear term

$$R_4 = \sum (g_{jk}(w')-g_{jk}(0))\partial_j\partial_k w.$$

To do so we observe that

$$|Z^\alpha(\partial^\beta w - \varepsilon r^{-\frac{1}{2}}\hat{\omega}^\beta\partial_q^{|\beta|}U_\varepsilon| \leqq C_{\alpha,\beta,B}\varepsilon r^{-\frac{1}{2}}t^{-1}$$

since a factor in ∂^β which does not fall on the q variable will give rise to a factor homogeneous of degree -1 or a factor $\varepsilon t^{-\frac{1}{2}}$ which is equally small since $\varepsilon t^{\frac{1}{2}} \leqq B$. It follows that

$$|Z^\alpha(R_4-\varepsilon^2 r^{-1}G(\omega)\partial U_\varepsilon/\partial q\partial^2 U_\varepsilon/\partial q^2)| \leqq C_{\alpha,B}\varepsilon^2 r^{-2}.$$

Hence the L^2 norm is at most

$$C'_{\alpha,B}\varepsilon^2 t^{-1} \leqq C''_{\alpha,B}\varepsilon t^{-3/2}.$$

The last error to discuss is

$$R_5 = \varepsilon^2 r^{-1}(\partial^2 U_\varepsilon/\partial q\partial s - G(\omega)\partial U_\varepsilon/\partial q\partial^2 U_\varepsilon/\partial q^2).$$

Using (2.5.1) we can write

$$R_5 = \tfrac{1}{2}G(\omega)\varepsilon^2 r^{-1}(\partial_q(\chi(-3\varepsilon q)(\partial_q U)^2)-\partial_q(\partial_q\chi(-3\varepsilon q)U)^2)$$

which shows that $-2/(3\varepsilon) \leqq q \leqq -1/(3\varepsilon)$ in the support and that

$$|Z^\alpha R_5| \leqq C\varepsilon^2 r^{-1}\varepsilon^4.$$

Hence

$$\|Z^\alpha R_5(t,\cdot)\| \leqq C_{\alpha,B}\varepsilon^6 \leqq C'_{\alpha,B}\varepsilon^3 t^{-3/2}.$$

The proof of the lemma is now complete.

Remark. All the estimates were somewhat crude in case c). The reason is that we could have improved the estimates (2.5.6), (2.5.7) by patching together w where $t \sim \varepsilon^{-2/3}$ instead. However, the lemma is adequate as it stands for our purposes and the proof of the stronger result would have been even longer.

From this point on we can follow Section 2.4 even more closely. We shall write the solution u of (2.4.1), (2.4.2) (with two space dimensions) in the form $u = v+w$ where w is the approximate solution discussed in Lemma 2.5.2. The Cauchy problem then takes the form (2.4.1)', (2.4.2)'.

Lemma 2.5.3. Assume that (2.4.1), (2.4.2) has a C^∞ solution for $0 \leqq t \leqq T$ where $\varepsilon T^{\frac{1}{2}} \leqq B < A$, with A defined by (2.5.3). If $0 < \varepsilon < \delta_B$ it follows that

(2.5.8) $(\int |Z^\alpha(u-w)'(t,x)|^2 dx)^{\frac{1}{2}} \leqq C_{\alpha,B}\varepsilon^{3/2}, \quad 0 \leqq t \leqq T.$

Here δ_B and $C_{\alpha,B}$ are independent of T and ε.

Proof. We define the norms $N_s^\#$ as in Section 2.4. When proving the estimates (2.5.8) we shall assume that we already know the estimate

$$(2.5.9) \qquad N_3^{\#}(v'; T) \leqq \varepsilon.$$

Since $\varepsilon^{3/2}$ is much smaller than ε when ε is small, this assumption can easily be eliminated afterwards.

With the notation $\gamma_{jk}(x) = g_{jk}(u') - g_{jk}(0)$ we obtain from (2.5.9)

$$(2.5.10) \qquad |\gamma'(\tau)| \leqq C\varepsilon(1+\tau)^{-\frac{1}{2}}, \ 0 \leqq \tau \leqq T,$$

for in view of (2.3.7) it follows from (2.5.9) that

$$(2.5.11) \ |D^{\alpha}v'(x)| \leqq C\varepsilon(1+|x_0|)^{-\frac{1}{2}} \text{ if } 0 \leqq x_0 \leqq T \text{ and } |\alpha| \leqq 1,$$

and w' has a similar bound by (2.5.6). Thus

$$\int_0^T |\gamma'(\tau)|d\tau \leqq 2C\varepsilon(1+T)^{\frac{1}{2}} < 3CB,$$

which gives a bound for the exponential in (2.3.1). The inhomogeneous terms in (2.4.1)' consist of the second sum and of f, and by (2.5.7) we have

$$(2.5.12) \qquad \int_0^T \|Z^{\alpha}f(t,.)\|dt \leqq C\varepsilon^{3/2}.$$

Instead of (2.4.19) we now obtain

$$(2.5.13) \qquad \|v'(t,.)\| \leqq C(\varepsilon^{3/2} + \varepsilon \int_0^t \|v'(\tau,.)\|(1+\tau)^{-\frac{1}{2}}d\tau).$$

Since

$$\int_0^t \varepsilon(1+\tau)^{-\frac{1}{2}} \, d\tau < 2C\varepsilon(1+T)^{\frac{1}{2}} < 3CB,$$

we obtain (2.5.8) when $\alpha=0$ in view of the implication (2.4.20) $\Rightarrow$ (2.4.21). The arguments used to prove (2.4.22) can be applied with no change apart from obvious changes of exponents. It follows that (2.5.13) is valid with $\|v'(t,.)\|$ replaced by $N_s^{\#}(v';t)$ (and some other constant). This yields (2.5.8); we leave for the reader to supply the missing details of the proof.

Repetition of the proof of Theorem 2.4.4 gives now:

<u>Theorem 2.5.4.</u> The Cauchy problem (2.4.1), (2.4.2) with $u_j \in C_0^{\infty}(\mathbf{R}^2)$ has a C^{∞} solution u_{ε} for $0 \leqq t < T_{\varepsilon}$ where

$$(2.5.14) \qquad \underline{\lim}_{\varepsilon\to 0} \varepsilon \, T_{\varepsilon}^{\frac{1}{2}} \geqq A = (\max \ \sum g_{jkl}\hat{\omega}_j\hat{\omega}_k \ \hat{\omega}_l \partial^2 F_0(\omega,\rho)/\partial\rho^2)^{-1}.$$

Here $\omega \in S^1$ and $\hat{\omega} = (-1,\omega) \in \mathbf{R}^3$. If U is the solution of (2.5.1), (2.5.2) then

(2.5.15) $$s\varepsilon^{-2} u_\varepsilon(s^2/\varepsilon^2, (s^2/\varepsilon^2 + r)\omega) - U(\omega, s, r) \to 0, \quad \varepsilon \to 0,$$

locally uniformly in $S^1 \times (0,A) \times \mathbf{R}$; in fact, the difference is locally uniformly $O(\varepsilon^{\frac{1}{2}})$.

References

[1] R. DiPerna and A. Majda, The validity of nonlinear geometric optics for weak solutions of conservation laws. Report PAM-235, June 1984, Center for pure and applied mathematics, University of California, Berkeley.

[2] L. Hörmander, On Sobolev spaces associated with some Lie algebras. Report 4, 1985, Institute Mittag-Leffler.

[3] F. John, Formation of singularities in one-dimensional non-linear wave propagation. Comm.Pure Appl.Math. 27(1974), 377-405.

[4] - , Blowup of radial solutions of $u_{tt} = c^2(u_t)\, u$ in three space dimensions. Preprint 1984.

[4a] - , A lower bound for the life span of solutions of nonlinear wave equations in three space dimensions. Preprint 1986.

[5] F. John and S. Klainerman, Almost global existence to non-linear wave equations in three space dimensions. Comm. Pure Appl. Math. 37(1984), 443-455.

[6] S. Klainerman, Uniform decay estimates and the Lorentz invariance of the classical wave equation. Preprint 1984.

[7] - , Global existence for nonlinear wave equations. Comm. Pre Appl. Math. 33(1980), 43-101.

[8] - , The null condition and global existence to nonlinear wave equations. Preprint 1985.

[9] - , Long time behaviour of solutions to nonlinear wave equations. Proc. Int. Congr. Math., Warszawa 1983, 1209-1215.

[10] P.D. Lax, Hyperbolic systems of conservation laws and the mathematical theory of shock waves. Regional conf. series in applied mathematics 11, SIAM 1973.

[11] - , Hyperbolic systems of conservation laws II. Comm. Pure Appl. Math. 10(1957), 227-241.

Gevrey-hypoellipticity and Pseudo-differential operators on Gevrey class

Chisato Iwasaki
Department of Mathematics
Himeji Institute of Technology
Shosha, Himeji 671-22, Japan

1. Introduction

It is known that elliptic differential operators with analytic coefficients are analytic hypoelliptic. But in general degenerate operators are not analytic hypoelliptic even if they are hypoelliptic in C^∞ sense. In fact Baouendi-Goulaouic[1] gave an example of such an operator. They showed that $P = D_x^2 + x^2 D_y^2 + D_z^2$ is not analytic hypoelliptic at the origin, moreover they showed that P is not G^s-hypoelliptic if s is less than 2.

We study Gevrey-hypoellipticity for degenerate operators of the following form. Let $p(x,\xi)$ be a $k \times k$ matrix of the form

$$p(x,\xi) = p_m(x,\xi) + p_{m-1}(x,\xi) + p_{m-2}(x,\xi), \quad p_m(x,\xi) = q_m(x,\xi)\, I$$

where q_m belongs to $S^m_{1,0}$, I is the identity matrix, p_j is a matrix whose elements belong to $S^j_{1,0}$ $(j=m-1,m-2)$ and p_m, p_{m-1} are homogeneous with respect to ξ.

The main theorem of this paper is the following

Theorem 1. We assume the following assumption $(\mathfrak{A})$ for $p(x,\xi)$

$(\mathfrak{A})$ $(\mathfrak{A}$-1) q_m is a non-negative function.

$(\mathfrak{A}$-2) There exists a positive constant c such that

$$\min_{1\le j\le k} \operatorname{Re} \mu_j(x,\xi) + \frac{1}{2} \operatorname{tr}^+ A(x,\xi) \;\geq\; c|\xi|^{m-1}$$

on the characteristic set $\Sigma=\{X\in T^*(R^n);\ q_m(X)=0\}$, where $\mu_j(x,\xi)$ $(j=1,\cdots,k)$ are the eigenvalues of the matrix p_{m-1}, $A(x,\xi)=A=iJH$ $\left(J=\begin{pmatrix}0, I\\ -I, 0\end{pmatrix}\right.$, $H=\begin{pmatrix}\partial_{xx}q_m, \partial_{x\xi}q_m\\ \partial_{\xi x}q_m, \partial_{\xi\xi}q_m\end{pmatrix}$) and $\operatorname{tr}^+ A$ means the sum of all positive eigenvalues of A. Let the coefficients of $P=p(x,D)$ belong to G^s and $s\geq 1/(m-1)$. Then P is micro-locally G^σ-hypoelliptic if $\sigma\geq\max(s,2)$.

In this paper we use the Weyl symbol for pseudo-differential operator that is

$$p(x,D)u(x) = (2\pi)^{-n} \int_{R^n}\int_{R^n} e^{i(x-y)\xi}\, p(\tfrac{x+y}{2},\xi)\, u(y)\, dyd\xi ,$$

(See Hörmander [5]). We give a more precise statement of Theorem 1 in Section 2 (See Theorem 4).

About the assumption $(\mathfrak{A})$ we will give some remarks. If k=1, by Melin [11], $(\mathfrak{A})$ is equivalent to the fact that P satisfies the following estimate

$$(1.1) \qquad \mathrm{Re}\,(Pu, u) \geq c'\|u\|^2_{(m-1)/2} - C\,\|u\|_0^2$$

for some positive constants c' and C. When $p(x,\xi)$ is a matrix and satisfies $(\mathfrak{A})$, (1.1) is also true. One example of such an operator P is $\Box_b$ on q-forms. It is shown that $(\mathfrak{A}\text{-}2)$ is equivalent to the condition Y(q) in Folland-Kohn[3] (See[8]).

We can show Theorem 1 by the method of Durand [2]. But in this paper the method to prove Theorem 1 is to construct a parametrix instead of using the estimate.

In the C^∞ case, under the assumption $(\mathfrak{A})$, the author constructed the fundamental solution E(t) of the heat equation $L=\frac{d}{dt} + p(x,D)$ as a pseudo-differential operator of class $S^0_{1/2,1/2}$ with parameter t in [8]. Then $Q = \int_0^T e(t)dt$ is a parametrix of P because $e(t)$ belongs to $S^{-\infty}$ for positive t. If we intend to apply the same method in the case the coefficients of P belong to a Gevrey class, we have to review the way of constructing the fundamental solution of the C^∞ case.

The method of construction of E(t) is as follows. We obtain the asymptotic expansion $E_k(t)$ of E(t) such that

$$L\,\widetilde{E}_k \sim 0, \qquad \widetilde{E}_k = \sum_{j=0}^{k} E_j ,$$

where E_j belongs to $S^{-\varepsilon j}_{1/2,1/2}$ for some positive constant ε. The symbol $e_0(t)$ of $E_0(t)$ is given by the following formula near $\Sigma \times \{t=0\}$

$$e_0(t) = [\det\{\cosh(At/2)\}]^{-1/2}$$

$$\times \exp[\{q_m t + \langle \nabla q_m t, F(At/2)J\nabla q_m t\rangle/4\}\, I - p_{m-1}t],$$

where $F(s)=(is)^{-1}(1-s^{-1}\tanh(s))$ $\nabla q_m = {}^t(\partial_{x_1} q_m, \cdots, \partial_{x_n} q_m, \partial_{\xi_1} q_m, \cdots, \partial_{\xi_n} q_m)$

and $J=\begin{pmatrix} 0, & I \\ -I, & 0 \end{pmatrix}$. The symbol e_j $(j\geq 1)$ of E_j is obtained by solving a kind of transport equation approximately. We want to find the fundamental solution $E(t)$ of the form

$$(1.2) \qquad E(t) = \widetilde{E}_k(t) + \int_0^t \widetilde{E}_k(t-s)\Phi(s)ds.$$

Then $\Phi(t)$ must satisfy the equation of Volterra type

$$(1.3) \qquad \Phi(t) = K(t) + \int_0^t K(t-s)\Phi(s)ds = K(t) + \mathcal{B}\Phi(t),$$

where $K(t) = -L\widetilde{E}_k(t)$ and $\mathcal{B}\Phi(t) = \int_0^t K(t-s)\Phi(s)ds$. So we find that $\Phi(t) = \sum_{j=0}^{\infty} \mathcal{B}^j K(t)$ is the solution of (1.2).

2. Symbol classes and results

We introduce subclasses of $S^m_{\rho,\delta}$.

Definition 2.1 For $0\leq\delta\leq\rho\leq 1$, $\delta<1$, $s\geq 1$, let

$$\widetilde{S}^m_{\rho,\delta;s} = \{p\in S^m_{\rho,\delta}(R^n)\text{; There exists a constant } h \text{ such that } |p|_{\ell,h} \text{ is bounded for any } \ell\},$$

where

$$|p|_{\ell,h} = \sup_{|\alpha|\leq\ell}\ \sup_{\beta}\ \sup_{(x,\xi)}[|p^{(\alpha)}_{(\beta)}(x,\xi)|h^{-|\beta|}/\{\beta!^s+\beta!^{s(1-\delta)}\langle\xi\rangle^{\delta|\beta|}\}\langle\xi\rangle^{m-\rho|\alpha|}],$$

$$S^m_{\rho,\delta;s} = \{p\in\widetilde{S}^m_{\rho,\delta;s}\text{; There exists a constant } h \text{ such that } |p|_h \text{ is bounded}\},$$

where

$$|p|_h = \sup_{\alpha,\beta}\ \sup_{(x,\xi)}[|p^{(\alpha)}_{(\beta)}(x,\xi)|h^{-|\alpha|+|\beta|}/\{\alpha!^s\langle\xi\rangle^{-|\alpha|}+\alpha!^{s\rho}\langle\xi\rangle^{-\rho|\alpha|}\} \times\{\beta!^s+\beta!^{s(1-\delta)}\langle\xi\rangle^{\delta|\beta|}\}\langle\xi\rangle^m],$$

$$R_s = \{p\in S^{-\infty}\text{; There exist positive constants } r,h \text{ and } C_\alpha \text{ such that } |p^{(\alpha)}_{(\beta)}(x,\xi)| \leq C_\alpha h^{|\beta|}\beta!^s \exp(-r\langle\xi\rangle^{1/s}) \text{ for any } \alpha \text{ and } \beta\}.$$

R_s is the set of symbols of smoothing operators in the Gevrey sense.

To construct $E(t)$ we must solve the integral equation (1.2) of Volterra type. So we need the following theorem which is the version of Iwasaki [7] in the C^∞ case.

Theorem 2. If P_j belongs to $\tilde{S}^{m(j)}_{\rho,\delta;s}$ $(j=1,\cdots,\nu)$, then the multi product $P=P_1\cdots P_\nu$ is also a pseudo-differential operator whose symbol p belongs to $\tilde{S}^m_{\rho,\delta;s}$ $(m=\sum_{j=1}^{\nu} m(j))$. Moreover if $s(1-\delta)\geq 1$, then for any ℓ there exists ℓ' such that

$$|p|_{\ell,h} \leq C_\ell M^\nu \prod_{j=1}^{\nu} |p_j|_{\ell+\ell',h'},$$

where C_ℓ, M and h' are independent of ν.

Remark. If we replace $\tilde{S}^m_{\rho,\delta;s}$ in the statement of Theorem 2 by $S^m_{\rho,\delta;s}$, the statement will not be true even if $\nu=2$.

If we consider the operators $\tilde{S}^m_{\rho,\delta;s}$ on ultradistributions, we obtain

Theorem 3. There exists γ_0 such that for all γ $(|\gamma|\leq\gamma_0)$

$$e^{\gamma\langle D\rangle^\kappa} p(x,D)\, e^{-\gamma\langle D\rangle^\kappa}$$

belongs to $\tilde{S}^m_{\tilde{\rho},\delta;s}$, if $s\kappa\leq 1$, $\delta+\kappa\leq 1$ and p belongs to $\tilde{S}^m_{\rho,\delta;s}$, where $\tilde{\rho}=\min(\rho,1-\kappa)$.

We can prove that $e_j(t)$ belongs to $S^{-\varepsilon j}_{1/2,1/2;s}$ if p belongs to $S^m_{1,0;s}$. So applying Theorem 2 to solve the equation (1.3) we have

Theorem 4. If $p(x,\xi)$ belongs to $S^m_{1,0;s}$ for $s\geq\max(2,1/(m-1))$ and satisfies the assumption $(\mathfrak{A})$, then $E(t)$ belongs to $\tilde{S}^0_{1/2,1/2;s}$, $E(t)$ belongs to R_s for positive t and Q belongs to $\tilde{S}^{-(m-1)}_{1/2,1/2;s}$. Moreover $WF_\sigma(Pu)=WF_\sigma(u)$ for any $\sigma\geq s$. Here $WF_s(u)$ is $WF_L(u)$ of Definition 8.4.3 in Hörmander [6] when $L_\ell=(\ell+1)^s$ for any ℓ.

3. Remarks

We give some remarks. In Theorem 2 the condition $s(1-\delta)\geq 1$ is not needed if we take h depending on ν. Métivier[11] intruduced the class $S^m_{\rho,\delta;1}$ to prove analytic hypoellipticity for degenerate operators. But in our case $\delta=1/2$, so our method can be used to prove only G^s-hypoellipticity for $s\geq 2$. On the other hand we can apply Theorem 2 to prove Gevrey-hypoellipticity for operators discussed in Hashimoto-

Matsuzawa-Morimoto[4] and obtain better results than theirs.

If p satisfies $(\mathfrak{A})$, p_m has exactly double characteristic Σ and Σ is symplectic, then P is analytic hypoelliptic. This fact is shown by Métivier[11], Tartakoff[12] and Trèves[13].

4. Proof of Theorem 2

For the proof of Theorem 2 we prepare basic lemmas.

Lemma 1. Let W be a meamurable set in $R^{n\nu} = \{ \mathbb{\eta}=(\eta^1,\cdots,\eta^\nu);\ \eta^j \in R^n\}$. Assume $g(\mathbb{y},\mathbb{\eta};\xi)$ is a smooth function defined on $R^{n\nu}\times W\times R^n$ which satisfies for any $\mathbb{b}=(b^1,\cdots,b^\nu)$, $b^j\in N^n$, $\mathbb{y}=(y^1,\cdots,y^\nu)$, $y^j\in R^n$, $\xi\in R^n$

$$\left|\partial_{\mathbb{y}}^{\mathbb{b}} g(\mathbb{y},\mathbb{\eta};\xi)\right| \leq M_{\mathbb{b}}(\xi)\prod_{j=1}^{\nu}\left[(1+\Xi_j^{2\delta\tilde{n}}|y^j-y^{j+1}|^{2\tilde{n}})^{-1}(\Xi_j+\Xi_{j-1})^{\delta|b^j|}\right]$$

where $\Xi_j=\langle\xi+\eta^j\rangle$, $\tilde{n}=[n/2]+1$, $y^{\nu+1}=-y^1$, $\eta^0=-\eta^\nu$.

Set

$$I(\xi) = \text{Os-}\int_W\int_{R^{n\nu}} e^{i\phi(\mathbb{y},\mathbb{\eta})}\, g(\mathbb{y},\mathbb{\eta};\xi)\, d\mathbb{y}d\mathbb{\eta},$$

where

$$\phi(\mathbb{y},\mathbb{\eta}) = \sum_{j=1}^{\nu}(\eta^j-\eta^{j-1})\,y^j. \tag{4.1}$$

Then we have

$$|I(\xi)| \leq C^\nu\Big(\sup_{\substack{\mathbb{b}=(b^1,\cdots b^\nu)\\ |b^j|\leq 2N}} M_{\mathbb{b}}(\xi)\Big),$$

where C is a constant independent of ν and N is an integer satisfying

$$2N(1-\delta) > n$$

The sketch of the proof will be given in section 7.

Lemma 2. Let $f(\mathbb{y},\mathbb{\eta};x,\xi)$ be a smooth function defined on $R^{n\nu}\times R^{n\nu}\times R^n\times R^n$ which satisfies for any $\beta\in N^n$, $\mathbb{a}=(a^1,\cdots,a^\nu)$, $a^j\in N^n$, $|a^j|\leq 2\tilde{n}$, $\mathbb{b}=(b^1,\cdots,b^\nu)$, $b^j\in N^n$, $x\in R^n$

$$\left|\partial_x^\beta\partial_{\mathbb{\eta}}^{\mathbb{a}}\partial_{\mathbb{y}}^{\mathbb{b}} f(\mathbb{y},\mathbb{\eta};x,\xi)\right| \leq K_1 K_2^{|\beta|+|\mathbb{b}|} Q(\mathbb{b})(\beta!^s+\beta!^{s(1-\delta)}\Xi^{\delta|\beta|})\prod_{j=1}^{\nu}\Xi_j^{m^j-\rho|a^j|}$$
$$\times\, e^{-\gamma\tilde{\Xi}^\kappa},$$

where $\Xi_j=\langle\xi+\eta^j\rangle$, $\Xi=\max_{1\leq j\leq\nu}\Xi_j$, $\tilde{\Xi}=\min_{1\leq j\leq\nu}\Xi_j$, $\gamma\geq 0$ and

$Q(\mathbb{b})=\prod_{j=1}^{\nu}\{b^j!^s+b^j!^{s(1-\delta)}(\Xi_j+\Xi_{j-1})^{\delta|b^j|}\}$. Assume $s\kappa\leq 1$ and $\kappa+\delta\leq 1$. Then there exists a constant γ_0 such that $I(x,\xi)$ defined by

$$I(x,\xi) = \text{Os-}\int_{R^{n\nu}}\int_{R^{n\nu}} e^{i\phi(\mathbb{y},\mathbb{\eta})} f(\mathbb{y},\mathbb{\eta};x,\xi)\, d\mathbb{y}d\mathbb{\eta}$$

satisfies

$$\left|\partial_x^\beta I(x,\xi)\right| \leq K_1' K_2' C^\nu (\beta!^s + \beta!^{s(1-\delta)}\langle\xi\rangle^{\delta|\beta|})\langle\xi\rangle^m e^{-\gamma\langle\xi\rangle^\kappa},$$

where $m=\sum_{j=1}^{\nu} m^j$, the constants K_j' are independent of β and also independent of ν if $s(1-\delta)\geq 1$.

<u>Proof</u>. In the following discussion C_j denotes a constant which is independent of β and ν $(j=0,1,2,3,4)$.

We have

$$Y_j L_j e^{i\phi} = e^{i\phi}, \qquad j=1,\cdots,\nu,$$

where

$$Y_j = \{1+ \Xi_j^{2\delta\tilde{n}}|y^j-y^{j+1}|^{2\tilde{n}}\}^{-1}, \quad L_j = 1+ \Xi_j^{2\delta\tilde{n}}(-\Delta_{\eta^j})^{\tilde{n}}.$$

Set

$$r(\mathbb{y},\mathbb{\eta};x,\xi) = {}^tL_1 Y_1 {}^tL_2 Y_2 \cdots {}^tL_\nu Y_\nu \partial_x^\beta f.$$

Then we have

$$\partial_x^\beta I(x,\xi) = \text{Os-}\int_{R^{n\nu}}\int_{R^{n\nu}} e^{i\phi(\mathbb{y},\mathbb{\eta})} r(\mathbb{y},\mathbb{\eta};x,\xi)\, d\mathbb{y}d\mathbb{\eta}$$

and $r(\mathbb{y},\mathbb{\eta};x,\xi)$ satisfies

$$(4.2)\qquad \left|\partial_{\mathbb{y}}^{\mathbb{b}} r\right| \leq C_1^\nu K_1 K_2^{|\beta|+|\mathbb{b}|}(\beta!^s + \beta!^{s(1-\delta)}\Xi^{\delta|\beta|})\, Q(\mathbb{b})\, e^{-\gamma\tilde{\Xi}^\kappa}\prod_{j=1}^{\nu} Y_j \Xi_j^{m^j}.$$

At first we assume $\gamma=0$. We divide the integral domain $R^{n\nu}=\{\mathbb{\eta}=(\eta^1,\cdot\cdots,\eta^\nu);\ \eta^j\in R^n\}$ into $(\nu+1)$ parts $\Gamma_0, \Gamma_1, \cdots, \Gamma_\nu$, where

$$\Gamma_0 = \{\mathbb{\eta}\in R^{n\nu};\quad |\eta^j|\leq\langle\xi\rangle/2, \quad 1\leq j\leq\nu\},$$

$$\underset{(\ell\geq 1)}{\Gamma_\ell} = \{\mathbb{\eta}\in R^{n\nu};\quad |\eta^\ell| = \max_{1\leq j\leq\nu}|\eta^j| > \langle\xi\rangle/2,\ |\eta^j|<|\eta^\ell|\ (j<\ell)\}.$$

It is clear that

<u>Proposition 1.</u> (i) If $\mathbb{\eta}\in\Gamma_0$, then $\langle\xi\rangle/2\leq\Xi_j\leq 2\langle\xi\rangle$ for any j.

(ii) If $\mathbb{\eta}\in\Gamma_\ell$ $(\ell\geq 1)$, then $\Xi_j\leq\langle\xi\rangle+|\eta^\ell|\leq 3|\eta^\ell|$ for any j.

For $\ell\geq 1$, we divide Γ_ℓ into $\{W_{\ell,N}\}_{N=0}^{\infty}$, where

$$W_{\ell,N} = \{\mathbb{\eta}\in\Gamma_\ell;\quad \frac{\eta^\ell}{B}\in J_N\}.$$

Here

$$J_N = \{ \eta \in R^n ; \quad N^s \leq |\eta| < (N+1)^s \}$$

and B is a constant which will be chosen later. Set T^ℓ be a vector of differential operator of first order with respect to y such that

$$\frac{\eta^\ell \cdot T^\ell}{|\eta^\ell|^2} \quad e^{i\phi(y,\eta)} = e^{i\phi(y,\eta)}.$$

Then we have

$$\begin{aligned} \partial_x^\beta I(x,\xi) &= Os-\int_{\Gamma_0}\int_{R^{n\nu}} e^{i\phi} r(y,\eta;x,\xi)\, dy d\eta \\ (4.3) \qquad &+ \sum_{\ell=1}^{\nu}\sum_{N=0}^{\infty} Os-\int_{W_{\ell,N}}\int_{R^{n\nu}} e^{i\phi} \; g_{\ell,N}(y,\eta;x,\xi)\, dy d\eta \quad , \end{aligned}$$

where

$$g_{\ell,N} = \left(- \frac{\eta^\ell \cdot T^\ell}{|\eta^\ell|^2}\right)^N \; r(y,\eta;x,\xi) \; .$$

By Proposition 1 (i) and (4.2) we have

$$(4.4) \qquad \left|\partial_y^b r\right| \leq C_1^\nu\, 2^{|m|}\, K_1 (2K_2)^{|\beta|+|b|} (\beta!^s + \beta!^{s(1-\delta)} \langle\xi\rangle^{\delta|\beta|}) \langle\xi\rangle^m\, Q(b) \prod_{j=1}^{\nu} Y_j$$

$$\text{for} \quad \eta \in \Gamma_0,$$

and

$$\begin{aligned} &\left|\partial_y^b g_{\ell,N}\right| \leq C_1^{\nu}\, K_1\, K_2^{|\beta|+|b|} (\beta!^s + \beta!^{s(1-\delta)}\, \Xi^{\delta|\beta|}) \langle\xi\rangle^m\, C_2^{|b|}\, \tilde{K}_2^N\, Q(b) \\ (4.5) \qquad & \\ &\times \left(\frac{N!^s + N!^{s(1-\delta)}\;\Xi^{\delta N}}{|\eta^\ell|^N}\right) (3|\eta^\ell|)^{\tilde{M}} \prod_{j=1}^{\nu} Y_j \quad \text{for} \quad \eta \in \Gamma_\ell , \end{aligned}$$

where $\tilde{K}_2 = C_0 K_2 \nu^*$, $* = \max\,(\,1-s(1-\delta),\,0)$ and $\tilde{M} = \sum_{j=1}^{\nu} |m^j|$.

If $\eta \in W_{\ell,N}$, there exists C_3 such that

$$N!^s + N!^{s(1-\delta)} \quad \Xi^{\delta N} \;\leq\; (C_3 B^{\delta-1})^N \; |\eta^\ell|^N$$

and

$$\beta!^{s(1-\delta)}\; \Xi^{\delta|\beta|} \leq \beta!^s\; \varepsilon^{-\delta s|\beta|} \;(\;\exp(3B^{1/s}\delta s\varepsilon))^{N+1} \quad \text{for any } \varepsilon>0$$

by Proposition 1 (ii). If we choose ε and B such that $\tilde{B} < 1$, where

$$\tilde{B} = \tilde{K}_2\; C_3\; B^{\delta-1} \quad \exp(3B^{1/s}\; \delta s \varepsilon),$$

we can apply Lemma 1 for $g=r$, $g_{\ell,N}$, $W = \Gamma_0$, $W_{\ell,N}$ by (4.4) and (4.5). Then we get the result for $\gamma=0$. Now let us consider the case γ is positive. For the sake of $\kappa \leq 1$, we have

$$\left|\tilde{\Xi}^{\kappa} - \langle\xi\rangle^{\kappa}\right| \leq |\eta^\ell|^{\kappa} \quad \eta \in \Gamma_\ell \qquad (\ell \geq 1).$$

So we have

$$(4.6) \qquad e^{-\gamma\tilde{\Xi}^{\kappa}} \leq e^{-\gamma\langle\xi\rangle^{\kappa}}(e^{\gamma B^{\kappa}})^{N+1}, \qquad \eta \in W_{\ell,N},$$

where we use $s\kappa \leq 1$. We divide Γ_0 as follows. $\Gamma_0 = \sum_{\ell=1}^{\nu}\sum_{N=0}^{\infty}\tilde{W}_{\ell,N}$, where

$$\tilde{W}_{\ell,N} = \{\eta\in\Gamma_0;\ \sup_{1\leq j\leq\nu}|\eta^j| = |\eta^{\ell}|,\ \frac{\eta^{\ell}}{B\langle\xi\rangle^{1-\kappa}} \in J_N\}.$$

It is clear that

$$\left|\langle\xi + \eta^j\rangle^{\kappa} - \langle\xi\rangle^{\kappa}\right| \leq 2\langle\xi\rangle^{\kappa-1}\,|\eta^j|.$$

So we have

$$(4.7) \qquad e^{-\gamma\tilde{\Xi}^{\kappa}} \leq e^{-\gamma\langle\xi\rangle^{\kappa}}(e^{2\gamma B})^{N+1} \qquad \eta \in W_{\ell,N}.$$

We can apply the same method to $\tilde{W}_{\ell,N}$ and $W_{\ell,N}$ by (4.6), (4.7), used for the estimates of $W_{\ell,N}$ of $\gamma=0$. We get the result if we choose γ_0 and B satisfying $\tilde{B}e^{2\gamma_0 B} < 1$ in this case.

Proof of Theorem 2. We have

$$p(x,\xi) = 2^n(2\pi)^{-n\nu}\ \text{Os-}\int_{R^{n\nu}}\int_{R^{n\nu}} e^{i\phi(\mathbf{y},\eta)}\prod_{j=1}^{\nu} p_j(x+\frac{y^j+y^{j+1}}{2}, \xi+\eta^j)\,d\mathbf{y}d\eta.$$
$$(y^{\nu+1}=-y^1).$$

So we get Theorem 2 by taking $f(\mathbf{y},\eta;x,\xi) = \sum_{\alpha^1+\cdots+\alpha^{\nu}=\alpha}\frac{\alpha!}{\alpha^1!\cdots\alpha^{\nu}!}\times$ $\prod_{j=1}^{\nu} p_j^{(\alpha^j)}(x+\frac{y^j+y^{j+1}}{2}, \xi+\eta^j)$, $m^j = m(j)-\rho|\alpha^j|$, $\gamma=0$ in Lemma 2. In this case we have $K_1 \leq C_4{}^{\nu}C_{\alpha}\prod_{j=1}^{\nu}|p_j|_{|\alpha^j|+4n,h}$.

We get the following theorem by the same method if $s>1$, $s\rho \geq 1$ and $s(1-\delta) \geq 1$.

Theorem 2'. If P_j belongs to $S^{m(j)}_{\rho,\delta;s}$ $(j=1,\cdots,\nu)$, then $P=P_1\cdots P_{\nu}$ has a aymbol $p=q_1+q_2$, where q_1 belongs to $S^{m}_{\rho,\delta;s}$ $(m=\sum_{j=1}^{\nu} m(j))$ and q_2 belongs to R_s.

Proof. Fix a smooth function $\psi \in C_0^{\infty}(R)$ such that $\psi(x)=1$ $(|x|\leq 1/4)$, $\psi(x)=0$ $(|x|\geq 1/2)$. Let q_1 be defind by

$$q_1(x,\xi) = 2^n(2\pi)^{-n}\ \text{Os-}\int_{R^{n\nu}}\int_{R^{n\nu}} e^{i\phi(\mathbf{y},\eta)}\prod_{j=1}^{\nu}\psi(\eta^j/\xi)$$
$$\times\prod_{j=1}^{\nu} p_j(x+\frac{y^j+y^{j+1}}{2}, \xi+\eta^j)\,d\mathbf{y}d\eta.$$

Then we obtain the result.

5. Proof of Theorem 3 and some properties.

Proposition 2.

$$\text{Os-}\int_{R^n}\int_{R^n} e^{i(x-y)\cdot\xi} s_1(\xi,y)u(y)dyd\xi = r_1(x,D)u(x),$$

$$\text{Os-}\int_{R^n}\int_{R^n} e^{i(x-y)\cdot\xi} s_2(x,\xi)u(y)dyd\xi = r_2(x,D)u(x),$$

where

$$r_1(x,\xi) = (2\pi)^{-n}\ \text{Os-}\int_{R^n}\int_{R^n} e^{iy\cdot\eta} s_1(\xi+\eta, x-\frac{y}{2})\ dyd\eta,$$

$$r_2(x,\xi) = (2\pi)^{-n}\ \text{Os-}\int_{R^n}\int_{R^n} e^{iy\cdot\eta} s_2(x+\frac{y}{2},\ \xi+\eta)\ dyd\eta\ .$$

Applying Lemma 2 to the above proposition, we can prove that r_j belongs to $\tilde{S}^m_{\rho,\delta;s}$ if s_j belongs to $\tilde{S}^m_{\rho,\delta;s}$ $(j=1,2)$.

Proof of Theorem 3.

$$e^{\gamma\langle D\rangle^\kappa}\ p(x,D)\ e^{-\gamma\langle D\rangle^\kappa}u(x)$$

$$= (2\pi)^{-n}\cdot\text{Os-}\int_{R^n}\int_{R^n} e^{i(x-y)\cdot\xi} e^{\gamma\langle\xi\rangle^\kappa} q_\kappa(\xi,y)u(y)dyd\xi \quad (\gamma \geq 0),$$

$$= (2\pi)^{-n}\ \text{Os-}\int_{R^n}\int_{R^n} e^{i(x-y)\cdot\xi}\ e^{-\gamma\langle\xi\rangle^\kappa}\tilde{q}_\kappa(x,\xi)u(y)dyd\xi \quad (\gamma \leq 0),$$

where

$$q_\kappa(\xi,x) = 2^n(2\pi)^{-2n}\ \text{Os-}\int_{R^{2n}}\int_{R^{2n}} e^{i\phi(y,\eta)-\gamma\langle\xi+\eta^2\rangle^\kappa} p(x+\tfrac{3}{2}y^1+\tfrac{1}{2}y^2, \xi+\eta^1)dV,$$

$$q_\kappa(x,\xi) = 2^n(2\pi)^{-2n}\ \text{Os-}\int_{R^{2n}}\int_{R^{2n}} e^{i\phi(y,\eta)+\gamma\langle\xi+\eta^1\rangle^\kappa} p(x-\tfrac{3}{2}y^1-\tfrac{1}{2}y^2, \xi+\eta^2)dV,$$

$$\Big(\ \phi(y,\eta) = \sum_{j=1}^{2} (\eta^j-\eta^{j-1})y^j,\ \eta^0=-\eta^2,\ dV = dy^1dy^2d\eta^1d\eta^2\ \Big).$$

By $\kappa \leq \min(1/s, 1-\delta)$ and Lemma 2 we can prove both $q_\kappa(\xi,x)e^{\gamma\langle\xi\rangle^\kappa}$ $(\gamma \geq 0)$ and $q_\kappa(x,\xi)e^{-\gamma\langle\xi\rangle^\kappa}$ $(\gamma \leq 0)$ belong to $\tilde{S}^m_{\tilde{\rho},\delta;s}$ if γ is sufficiently small. Then from Proposition 2 we get the result.

Lemma 3. (i) It is equivalent that p belongs to R_s and there exists a positive constant γ such that $\exp(\gamma\langle D\rangle^{1/s})p(x,D)$ and $p(x,D)\exp(\gamma\langle D\rangle^{1/s})$ belongs to $\tilde{S}^m_{\rho,\delta;s}$ for any m. (ii) If $p(x,\xi)$ belongs to R_s, then both $p(x,D)q(x,D)$ and $q(x,D)p(x,D)$ belong to R_s for any $q(x,\xi)\in R_s$.

Let ϕ and ψ be functions which belong to G^s $(s > 1)$ and suppϕ and suppψ are disjoint. Then we have

Lemma 4. $\phi(x)p(x,D)\psi(x)$ belongs to R_s if $p(x,\xi)$ belongs to $S^m_{\rho,\delta;s}$.

Proof. $\sigma(\phi(x)p(x,D)\psi(x))(x,\xi)$

$$=(2\pi)^{-n}\ \mathrm{Os}\text{-}\int_{R^n}\int_{R^n} e^{iy\cdot\eta}|y|^{-N}\ \phi(x+\tfrac{y}{2})(\Delta_\eta)^N p(x,\xi+\eta)\psi(x-\tfrac{y}{2})dyd\eta\ .$$

So if we apply Lemma 2 to $f(y,\eta;x,\xi)=|y|^{-N}\ \phi(x+\frac{y}{2})(\Delta_\eta)^N p(x,\xi+\eta)\psi(x-\frac{y}{2})$, we obtain Lemma 4, noting

$$\inf_{N\geq 0}\left\{ N!^s\langle\xi+\eta\rangle^{-N} + N!^{s\rho}\langle\xi+\eta\rangle^{-\rho N}\right\} \leq C\, e^{-\gamma\langle\xi+\eta\rangle^{1/s}}$$

for some positive constants γ and C.

6. Proof of Theorem 4.

Under the assumption $(\mathfrak{M})$ we have the following proposition for the symbol $e_j(t;x,\xi)$ $(j\geq 0)$.

Proposition 3. There exists a positive constant ε such that $e_j(t)$ belongs to $S^{-\varepsilon j}_{1/2,1/2;s}$ for any j. Moreover $\exp(ct\langle\xi\rangle^{m-1})e_j(t)$ belongs to $S^{-\varepsilon j}_{1/2,1/2;s}$ for some positive constant c and $\int_0^T e_j(t)dt$ belongs to $S^{-(m-1)-\varepsilon j}_{1/2,1/2;s}$ for any positive T.

If k is sufficiently large, $K(t) = -L\tilde{E}_k(t)$ belongs to $\tilde{S}^0_{1/2,1/2;s}$ by the above proposition and there exists a constant h such that

$$|k(t)|_{\ell,h} \leq B_\ell \quad \text{for any } \ell ,$$

where $k(t)$ is the symbol of $K(t)$. By (1.3) we have

$$\Phi(t) = K(t) + \sum_{j=1}^{\infty}\int_0^t\int_0^{t_1}\cdots\int_0^{t_{j-1}} K(t-t_1)K(t_1-t_2)\cdots K(t_{j-1}-t_j)$$
$$\times K(t_j)dt_j\cdots dt_2dt_1 \quad (t_0=t).$$

By Theorem 2 $\Phi(t)$ belongs to $\tilde{S}^0_{1/2,1/2;s}$ and satisfies

$$\left|\sigma(\Phi(t))\right|_{\ell,h'} \leq B_\ell + \sum_{j=1}^{\infty} t^jM^j(B_{\ell+\ell_0})^{j+1}/j! \leq B_{\ell+\ell_0}\, e^{tMB_{\ell+\ell_0}} .$$

Then $E(t)$ belongs to $\tilde{S}^0_{1/2,1/2;s}$ by (1.2).

Let a positive constant T fix. From Proposition 3 and $m-1\geq 1/s$ we have

$$(6.1)\qquad e^{\gamma T\langle D\rangle^{1/s}}\, e_0(T-\sigma)\, e^{-\gamma\sigma\langle D\rangle^{1/s}} \in \tilde{S}^0_{1/2,1/2;s}.$$

By the way of construction of $e_j(t)$, we have

$$(6.2)\qquad \tilde{K}(t_1,t_2) = e^{\gamma t_1\langle D\rangle^{1/s}}\, K(t_1-t_2)\, e^{-\gamma t_2\langle D\rangle^{1/s}} \in \tilde{S}^0_{1/2,1/2;s} \quad (T\geq t_1 \geq t_2 \geq 0),$$

for some positive constant γ. By (1.3) and (6.2) and applying Theorem 2 to $P_j=\tilde{K}(t_{j-1},t_j)$, we have

$$(6.3)\qquad e^{\gamma\sigma\langle D\rangle^{1/s}}\, \Phi(\sigma) \in \tilde{S}^0_{1/2,1/2;s} \qquad (\sigma \leq T).$$

By (6.1)(6.3) and (1.2) we get also

$$e^{\gamma T\langle D\rangle^{1/s}}\, E(T) \in \tilde{S}^0_{1/2,1/2;s}.$$

Then by Lemma 3 it is proved that $E(T)$ belongs to R_s. Put $Q' = \int_0^T E(t)dt$. Then Q' belongs to $\tilde{S}^{-(m-1)}_{1/2,1/2;s}$ and $PQ' = I-E(T)$. If we replace P by P^*, we can construct $Q\in \tilde{S}^{-(m-1)}_{1/2,1/2;s}$ such that $QP=I-V$, with $V \in R_s$. Fix functions ϕ,ψ which satisfy $\phi, \psi \in G^s(U)$ and $\phi=1$ on suppψ . Put $Pu=f$. Then we have

$$(6.4)\qquad e^{\gamma\langle D\rangle^\kappa}\psi u=e^{\gamma\langle D\rangle^\kappa}\psi Q e^{-\gamma\langle D\rangle^\kappa}\, e^{\gamma\langle D\rangle^\kappa}\phi f + e^{\gamma\langle D\rangle^\kappa}\psi Q(1-\phi) f + e^{\gamma\langle D\rangle^\kappa}\psi V u .$$

If $s\kappa\leq 1$ and γ is sufficiently small, then $e^{\gamma\langle D\rangle^\kappa}\psi Q e^{-\gamma\langle D\rangle^\kappa}$ belongs to $\tilde{S}^{-(m-1)}_{1/2,1/2;s}$ by Theorem 3. Noting $p\in S^m_{1,0;s}$ and $e_j(t) \in S^0_{1/2,1/2;s}$, we have $\sigma(Q) = q_1 + q_2$, where $q_1\in S^{-(m-1)}_{1/2,1/2;s}$ and $q_2\in R_s$ by Theorem 2'.

Owing to Lemma 4 we have $\psi Q(1-\phi)\in R_s$. So both $e^{\gamma\langle D\rangle^\kappa}\psi Q(1-\phi)$ and $e^{\gamma\langle D\rangle^\kappa}\psi V$ belong to R_s. By (6.4) we can prove $WF_\sigma(u) = WF_\sigma(f)$ if $\sigma\geq s$.

7. <u>Sketch of the proof of Lemma 1.</u>

In the proof C_j ($j=1,2,3,4$) denotes a constant independent of ν. If we define

$$X_j = \{1+(\Xi_j + \Xi_{j-1})^{-2\delta}\, |\eta^j - \eta^{j-1}|^2\}^{-1}, \qquad \eta^0 = -\eta^\nu ,$$

$$R_j = 1+(\Xi_j + \Xi_{j-1})^{-2\delta}(-\Delta_{y^j})$$

for j=1, ⋯ ,ν , it follows that

$$I(\xi) = \int_W \int_{R^{n\nu}} e^{i\phi(\mathbb{y},\mathbb{m})} \prod_{j=1}^{\nu} (R_j X_j)^N g(\mathbb{y},\mathbb{m};\xi)\, d\mathbb{y}d\mathbb{m} .$$

Then by the assumption we have

$$\left|I(\xi)\right| \leq C_1^{\nu}\{ \sup_{\substack{\mathbb{b}=(b^1,\cdots,b^{\nu}) \\ |b^j|\leq 2N}} M_{\mathbb{b}}(\xi)\} \int_W \int_{R^{n\nu}} \prod_{j=1}^{\nu} (X_j^N Y_j)\, d\mathbb{y}d\mathbb{m} ,$$

where

$$Y_j = \{ 1 + \Xi_j^{2\delta\tilde{n}} |y^j - y^{j+1}|^{2\tilde{n}} \}^{-1} .$$

Set $G(\xi,\eta) = 1 + |\xi-\eta| (\langle\xi\rangle + |\xi-\eta|)^{-\delta}$. Then the following inequalities hold .

$$\int_{R^{n\nu}} \prod_{j=1}^{\nu} Y_j \, d\mathbb{y} \leq C_2^{\nu} \prod_{j=1}^{\nu} \Xi_j^{-\delta n} ,$$

$$X_j \leq C_3\, G(\xi+\eta^{j-1}, \xi+\eta^j)^{-2} , \quad j=1, \cdots ,\nu .$$

We get Lemma 1 according to the following fact which was proved in Lemma A.2.3 of [9].

$$\int_W \prod_{j=1}^{\nu} G(\xi+\eta^{j-1}, \xi+\eta^j)^{-2N} \langle \xi+\eta^j\rangle^{-\delta n} \, d\mathbb{m} \leq C_4^{\nu}$$

if N satisfies $2N(1-\delta) > n$.

References

[1] M.S.Baouendi and C.Goulaouic: Nonanalytic hypoellipticity for some degenerate operators, Bull.Amer.Math. Soc.78(1972),483-486.

[2] M.Durand: Régularité Gevrey d'une classe d'opérateurs hypoelliptiques J.Math.Pures Appl.57(1978),323-350.

[3] G.B.Folland and J.J.Kohn: The Neumann problem for the Cauchy-Riemann complex, Ann.of Math.Studies 75(1972).

[4] S.Hashimoto, T.Matsuzawa et Y.Morimoto: Opératuers pseudodifférentiels et classes de Gevrey, Comm.Partial Differential Equations 8 (1983),1277-1289.

[5] L.Hörmander: The Weyl calculus of pseudo-differential operators, Comm.Pura Appl.Math.32(1979),359-443.

[6] L.Hörmander: The Analysis of Linear Partial Differential Operators I, Springer-Verlag,1983.

[7] C.Iwasaki: The fundamental solution for pseudo-differential operators of parabolic type, Osaka J.Math.14(1977),569-592.

[8] C.Iwasaki: Construction of the fundamental solution for degenerate parabolic systems and its application to construction of a parametrix of $\Box_b$, Osaka J.Math.21(1984),931-954.
[9] C.Iwasaki and N.Iwasaki: Parametrix for a degenerate parabolic equation and its application to the asymptotic behavior of spectral function for ststionary problems, Publ.Res.Inst. Math.Soc. 17(1981), 557-655.
[10] A.Melin: Lower bounds for pseudo-differential operators, Ark.Mat. 9(1971),117-140.
[11] G.Métivier: Analytic hypoellipticity for operators with multiple characteristics, Comm.Partial Differential Equations 6(1981),1-90.
[12] D.S.Tartakoff: Elementary proofs of analytic hypoellipticity for $\Box_b$ and $\bar{\partial}$-Neumann problem, Asterisque 89-90(1981),85-116.
[13] F.Trèves: Analytic hypoellipticity of a class of pseudo-differential operators, Comm.Partial Differential Equations 3(1978),475-642.

PROPAGATION OF THE SECOND ANALYTIC WAVE FRONT SET ALONG DIFFRACTIVE RAYS

Pascal Laubin
Department of Mathematics, University of Liège
Avenue des Tilleuls, 15, B-4000 LIEGE

1. Introduction

Let M be an analytic n-dimensional manifold with boundary ∂M and let P be a second order differential operator with analytic coefficients on M. We assume that the boundary is non-characteristic for P and that P has a real principal symbol p. Let $x_0 \in \partial M$. It is well-known, [10], that, after multiplication and conjugation of P with non-vanishing functions, we can choose real analytic coordinates near x_0 such that M is given by $x_n \geqslant 0$ and

$$P = D^2_{x_n} + R(x, D_{x'}).$$

We assume that the tangential operator $R(x', 0, D_{x'})$ is of real principal type i.e.

$$\partial_{\xi'} r_0 \neq 0 \text{ if } r_0 = 0$$

where $r_0(x',\xi') = r(x',0,\xi')$ and r is the principal symbol of R.

We consider solutions $u \in D'(M)$ to the Dirichlet problem

$$Pu = 0 \text{ in } M \tag{1.1}$$

$$u|_{\partial M} = 0. \tag{1.2}$$

The elliptic, hyperbolic and glancing regions are

$$\begin{Bmatrix} E \\ G \\ H \end{Bmatrix} = \{(x',\xi') \in \dot{T}^{\star}\partial M : r_0(x',\xi') \begin{Bmatrix} > \\ = \\ < \end{Bmatrix} 0\}.$$

In this paper we are interested in the so-called diffractive region

$$G_+ = \{(x',\xi') \in \dot{T}^{\star}\partial M : r_0(x',\xi') = 0,\ \partial_{x_n} r(x',0,\xi') < 0\}.$$

J. Sjöstrand has obtained a complete description of the propagation of the analytic wave front set in $\dot{T}^\star(\partial M) \cup \dot{T}^\star\mathring{M}$ near G_+, [12],[13], see also [14]. In [8], G. Lebeau constructed an analytic parametrix near a point of G_+ and proved refined results on the propagation of Gevrey singularities. This parametrix also provides a result on the partial holomorphy along the glancing hypersurface.

Nevertheless the situation remains intricate. It seems to the author that the constructions of Lebeau indicate that the study of diffraction could be more transparent if we look at the second analytic wave front set.

Our purpose here is to study the 2-microlocal structure of the parametrix of (1.1)-(1.2). We construct two 2-microlocal parametrices and, using them, we prove propagation theorems for the second analytic wave front set.

Let

$$V_o = \{(x',\xi') \in \dot{T}^\star(\partial M) : r_o(x',\xi') = 0\}.$$

This is an involutive submanifold of $\dot{T}^\star\partial M$ of codimension 1, and, for a solution u to (1.1), we define

$$WF^{(2)}_{a,V_o}(u) = WF^{(2)}_{a,V_o}(u|_{x_n=0}) \cup WF^{(2)}_{a,V_o}(D_{x_n}u|_{x_n=0}) \subset T_{V_o}(T^\star(\partial M)).$$

See [6],[7] and [15] for the definition of the second analytic wave front set. Since $\partial_{x_n} r < 0$ we can introduce

$$V = \{\exp(sH_p)(x',0,\xi',0) : (x',\xi') \in V_o,\ s > 0\} \subset T^\star M.$$

The hamiltonian flow defines a symplectic map for each s, hence V is an involutive submanifold of $T^\star M$ of codimension 2. So we can also define the second analytic wave front set

$$WF^{(2)}_{a,V}\, u \subset \dot{T}_V(\dot{T}^\star M)$$

of u along V. Of course

$$V \subset \Sigma = \{(x,\xi) \in \dot{T}^\star M : p(x,\xi) = 0\}.$$

Since Pu = 0 we have

$$WF^{(2)}_{a,V}\, u \subset \dot{T}_V(\Sigma).$$

Therefore, at each point of V, there are only two normal directions that may belong to $WF^{(2)}_{a,V}u$. Moreover, $WF^{(2)}_{a,V}u$ is invariant under the flow defined by the hamiltonian vector field H_p.

If $\rho'_o = (x'_o, \xi'_o) \in V_o$, we can define the hyperbolic (resp. elliptic) normal direction ν at ρ'_o by the condition

$$dr_o(\rho'_o).\nu < 0 \text{ (resp. } > 0).$$

This definition extends by continuity to $T_V(\Sigma)$ in a neighbourhood of ∂M.

Our main result on the propagation of singularities is the following.

Theorem 1.1. *Let* u *be a solution to* (1.1) - (1.2) *and let* ρ'_o *be a point of* V_o. *The hyperbolic (resp. elliptic) normal direction to* V_o *at* ρ'_o *belongs to* $WF^{(2)}_{a,V_o}u$ *if and only if the hyperbolic (resp. elliptic) normal direction to* V *at* $\exp(sH_p)(x'_o,0,\xi'_o,0)$ *in* Σ, $s > 0$ *belongs to* $WF^{(2)}_{a,V}(u)$.

In particular the second analytic wave front do not propagate in the shadow. Theorem 1.1. contains the result of Lebeau on partial holomorphy. Moreover the results proved by Sjöstrand in [12] follow easily from theorem 1.1. and the microlocal version of the analytic extension theorem.

In this paper we shall emphasize on the construction of the phase functions that describes the propagation process and on the definition of the asymptotic solutions that are involved in the proof. The details of the proofs will appear elsewhere.

2. F.B.I. - transform of second kind

We use the Fourier-Bros-Iagolnitzer-transforms in the spirit of Sjöstrand to characterize the analytic wave front set. In [6] we introduced a class of phase functions adapted to the study of conical refraction. The class that we use here is a little bit more general.

These phase functions appear naturally as solutions of 2-microlocal eiconal equations.

Let V be an involutive submanifold of $\dot{T}^\star\mathbb{R}^n$ with codimension k and let $\rho_o = (y_o,\eta_o)$ be a point of V. Denote by F_o the bicharacteristic leaf of V containing ρ_o. We assume that the rank of the projection π_{F_o} from F_o to $\mathbb{R}^n$ is k at ρ_o. In the applications this means that we do not have propagation of singularities occuring in ξ-variables only. We use the splitting $z = (z',z'') \in \mathbb{C}^k \times \mathbb{C}^{n-k}$.

Definition 2.1. An holomorphic function $\phi_\mu(z,y)$ in a neighbourhood of a point $(z_o,y_o,0) \in \mathbb{C}^n \times \mathbb{R}^n \times \mathbb{R}$ is a F.B.I.-*phase function of the second kind adapted to* V near $\tau_o \in \dot{T}_V(T^\star\mathbb{R}^n)_{\rho_o}$ if

(i) $\phi_\mu(z,y) = \phi_o(z'',y) + \mu^2\phi_2(z,y) + \mathcal{O}(\mu^3)$;

(ii) $\partial_y\phi_o(z''_o,y_o) = -\eta_o$ and for z" close to z''_o

$$N_{z''} = \{y \in \mathbb{R}^n : \mathcal{I}(\partial_y\phi_o)(z'',y) = 0\}$$

is a submanifold of $\mathbb{R}^n$ near y_o with dimension k, $\mathcal{I}\phi_o$ is transversely positive on $N_{z''}$;

(iii) $V = \{(y,-\partial_y\phi_o(z'',y)) : y \in N_{z''},\ z'' \in \mathbb{C}^{n-k}\}$ near ρ_o;

(iv) the function

$$N_{z''_o} \ni y \to \mathcal{I}\phi_2(z_o,y)$$

has a non-degenerated critical point at y_o with signature (k,0) and

$$(0,-\mathcal{R}(\partial_y\phi_2)(z_o,y_o))$$

is equal to τ_o in $T_V(T^\star\mathbb{R}^n)_{\rho_o}$;

(v) $\det(\partial_z,\partial_y\phi_2,\partial_{z''}\partial_y\phi_o) \neq 0$.

If

$$rg(\pi_{F_o}) = k$$

there are F.B.I.-phase functions of second type near each point of $\dot{T}_V(T^\star\mathbb{R}^n)_{\rho_0}$, see [6]. To analyse the consequences of (i)-(v) the following lemma is usefull.

Lemma 2.2. *Let* f *be a real analytic function near* $(x_0,y_0,0) \in \mathbb{R}^p \times \mathbb{R}^n \times \mathbb{R}$ *with the expansion*

$$f(x,y,\mu) = f_0(x,y) + \mu f_1(x,y) + \mu^2 f_2(x,y) + O(\mu^3).$$

Assume that

(a) *in a neighbourhood of* (x_0,y_0),

$$V_x = \{y \in \mathbb{R}^n : \partial_y f_0(x,y) = 0\}$$

is a submanifold of $\mathbb{R}^n$ *with dimension* k *and* f_0 *is transversely non degenerated with signature* (α_+,α_-);

(b) $f_1(x,y) = 0$ *if* $y \in V_x$;

(c) *using* (a) *we see that* $\partial_y^2 f_0$ *defines a non degenerated bilinear map on* $N_y(V_x)$, *by duality this gives a bilinear map* $Q(x,y)$ *on* $N_y^\star(V_x)$ *; from* (b) *it follows that* $\partial_y f_1(x,y) \in N_y^\star(V_x)$ *if* $y \in V_x$ *; we assume that*

$$V_x \ni y \to f_2(x,y) - \frac{1}{2} Q(x,y)(\partial_y f_1(x,y),\ \partial_y f_2(x,y))$$

has a non degenerated critical point with signature (β_+,β_-) *at* y_0 *when* $x = x_0$. *Then one can find* $\mu_0 > 0$ *such that, in a neighbourhood of* (x_0,y_0) *and when* $0 < \mu < \mu_0$, *the function* $y \to f(x,y,\mu)$ *has a unique non-degenerated critical point with signature* $(\alpha_+ + \beta_+,\alpha_- + \beta_-)$. *The critical point* $Y(x,\mu)$ *is an analytic function and* $Y(x_0,0) = y_0$.

For the proof we refer to [7] . Moreover the point $Y(x,0)$ of V_x and the class of $\partial_\mu Y(x,0)$ in $N(V_x)$ depend only on f_0,f_1,f_2. If

$$g_\mu(x) = g_0(x) + \mu g_1(x) + \mu^2 g_2(x) + O(\mu^3)$$

then g_0,g_1,g_2 depend only on f_0,f_1,f_2. If the function f_1 is identically equal to 0 then $\partial_\mu Y(x,0)$ is tangent to V_x. There is also an holomorphic version of Lemma 2.2. The statement is the same apart from the fact that they are no signature conditions.

If ϕ_μ is a F.B.I.-phase function of second type, it follows from Lemma 2.2 that

$$\mathbb{R}^n \ni y \to \mathcal{I}\partial_y\phi_\mu(z,y)$$

has a non-degenerated critical point $y_\mu(z)$ if z is close to z_o and $0 < \mu < \mu_o$. On the analogy of the classical F.B.I.-phase functions we introduce

$$\eta_\mu(z) = -\partial_y\phi_\mu(z,y_\mu(z)) \ , \ \rho_\mu(z) = (y_\mu(z),\eta_\mu(z))$$

and the weight function

$$\Phi_\mu(z) = -\mathcal{I}\phi_\mu(z,y_\mu(z)).$$

It follows from (ii) and (iv) that

$$-\mathcal{I}\phi_\mu(z,y) \leqslant \Phi_\mu(z) - c\mu^2|y-y_\mu(z)|^2 \tag{2.1}$$

when y is real.

Since $y_o(z) \in N_{z''}$ we have $\rho_o(z) \in V$ for every z. The leaves of V are given locally by

$$F_{z''} = \{(y,-\partial_y\phi_o(z'',y)) : y \in N_{z''}\}.$$

To prove it, we have to show that

$$T_\rho(F_{z''}) = T_\rho^\sigma(V) \text{ if } \rho \in F_{z''}$$

where σ is the symplectic form. On one hand, it follows from (ii) that

$$T_\rho(V) = \{\begin{pmatrix} h \\ -\partial_{z''}\partial_y\phi_o.k''-\partial_y^2\phi_o.h \end{pmatrix} : h \in \mathbb{R}^n, k \in \mathbb{C}^{n-k},$$
$$\mathcal{I}(\partial_{z''}\partial_y\phi_o.k''+\partial_y^2\phi_o.h) = 0\} .$$

On the other hand we have

$$\mathcal{I}(\partial_y\phi_o)(z'',y_o(z)) = 0$$

hence

$$\partial_{z''}\partial_y\phi_o(z'',y_o(z)) = \frac{2}{i}\,\mathcal{I}(\partial_y^2\phi_o)(z'',y_o(z)).\partial_{z''}y_o(z).$$

Therefore

$$\sigma\left(\begin{pmatrix} h \\ -\partial_{z''}\partial_y\phi_o.k''-\partial_y^2\phi_o.h \end{pmatrix}, \begin{pmatrix} h_1 \\ -\partial_y^2\phi_o.h_1 \end{pmatrix}\right)$$

$$= - \langle h_1, \partial_{z''}\partial_y\phi_o.k''\rangle = 2i \langle I(\partial_y^2\phi_o).k_1, \partial_{z''}y_o(z).k''\rangle = 0$$

if

$$h_1 \in T_{y_o(z)}(N_{z''}) .$$

This proves the assertion.

We know that

$$\partial_\mu y_\mu(z)|_{\mu=o} \in T_{y_o(z)} N_{z''}$$

hence

$$\partial_\mu \rho_\mu(z)|_{\mu=o} = (\partial_\mu y_\mu(z)|_{\mu=o}, - \partial_y^2\phi_o(z))\partial_\mu y_\mu(z)|_{\mu=o}) \in T_{\rho_o(z)}(F_{z''}). \tag{2.2}$$

Using the previous phase functions we can define F.B.I.-transforms of second kind. A classical analytic 2-symbol is an holomorphic function $a(z,y,\mu,\lambda)$ near (z_o,y_o), defined when $0 < \mu < \mu_o$ and $\lambda > f(\mu)$ for some $\mu_o > 0$ and decreasing function f, which has an expansion

$$a(z,y,\mu,\lambda) = \sum_{0\leq k<\lambda\mu^2/eC} (\lambda\mu^2)^{-k} a_k(z,y,\mu) + O(e^{-\varepsilon\lambda\mu^2}).$$

The functions a_k are holomorphic and have to satisfy the growth condition

$$|a_k(z,y,\mu)| \leq C_\mu C^k k!$$

for some $C > 0$ and decreasing function C_μ, $0 < \mu < \mu_o$. If $\chi \in D(\mathbb{R}^n)$ is equal to 1 in a neighbourhood of y_o and u is a distribution in an open set containing the support of χ, define

$$(T^{(2)}u)(z,\mu,\lambda) = u_{(y)}(e^{i\lambda\phi_\mu(z,y)} a(z,y,\mu,\lambda)\chi(y)).$$

The inequality (2.1) shows that there are m,r,μ_o,f such that

$$|(T^{(2)}u)(y,\mu,\lambda)| \leq \lambda^m e^{\lambda\Phi_\mu(z)} \quad \text{if } |z-z_o| < r,\ 0 < \mu < \mu_o, \lambda > f(\mu).$$

The 2-symbol a is elliptic at (z_o,y_o) if there exists an increasing function $c_\mu > 0$ in $0 < \mu < \mu_o$ such that

$$|a_o(z,y,\mu)| \geq c_\mu \quad \text{near } (z_o,y_o).$$

Using 2-microdifferential operators, we can prove that, when z_1 is close to z_o and a is elliptic at $(z_1,y_o(z_1))$, the class τ_1 of

$$(0,-R(\partial_y\phi_2)(z_1,y_o(z_1))) \text{ in } T_V(\dot{T}^\star\mathbb{R}^n)_{\rho_o(z_1)}$$

does not belong to $WF^{(2)}_{a,V}u$ if and only if there are constants $s,\varepsilon,\mu_o > 0$ and a decreasing function f satisfying

$$|(T^{(2)}u)(z,\mu,\lambda)| \leq e^{\lambda\Phi_\mu(z)-\varepsilon\lambda\mu^2} \quad \text{if } |z-z_1| < s,\ 0 < \mu < \mu_o, \lambda > f(\mu).$$

3. Critical points and phase functions

Let U be an open neighbourhood of 0 in $\mathbb{R}^{n-1}$, $T > 0$ and let

$$P(x,D) = D^2_{x_n} + R(x,D_{x'})$$

be a second order differential operator with analytic coefficients on $U \times]-T,T[$. We assume that the principal symbol of P

$$p(x,\xi) = \xi_n^2 + r(x,\xi')$$

is real. Let $(0,\xi_o')$ be a point of $\dot{T}^\star U$ satisfying

$$r(0,\xi_o') = 0,\ \partial_{x_n} r(0,\xi_o') < 0 \text{ and } \partial_{\xi'} r(0,\xi_o') \neq 0.$$

Using the implicit functions theorem write

$$r(x,\xi') = -(x_n + h(x',\xi'))S(x,\xi')$$

with $S(0,\xi_o') > 0$, $h(0,\xi_o') = 0$ and $\partial_{\xi'}h(0,\xi_o') \neq 0$. To fix the ideas we shall also assume

$\partial_{\xi_1} h(0,\xi'_o) > 0.$

We use the splitting $x'=(x_1,x'') \in \mathbb{C} \times \mathbb{C}^{n-2}$. Consider the holomorphic function $\psi_\mu(x',\eta')$ in a neighbourhood of $(x',\eta') = (0,\sigma'_o) = (0,0,\xi''_o)$ and $\mu = 0$ that satisfies

$$\begin{cases} h(x',\partial_{x'}\psi_\mu(x',\eta')) = \mu^2\eta_1 \\ \psi_\mu(0,x'',\eta'') = x''.\eta'' \ , \ \partial_{x'}\psi_o(0,\sigma'_o) = \xi'_o \ . \end{cases} \tag{3.1}$$

It is clear from the definition that $\psi_\mu(x',\eta')$ is in fact an holomorphic function of $(x',\mu^2\eta_1,\eta'')$. In the same way we can define an holomorphic function $\tilde{\phi}_\mu(z',y')$ in a neighbourhood of $z' = -i\sigma'_o$, $y' = 0$ and $\mu = 0$ by the equation

$$\begin{cases} h(y',-\partial_{y'}\tilde{\phi}_\mu(z',y')) = i\mu^2(z_1-y_1) \\ \tilde{\phi}_\mu(z',0,y'') = \frac{i}{2}(z''-y'')^2 \ , \ -\partial_{y'}\tilde{\phi}_\mu(-i\sigma'_o,0) = \xi'_o. \end{cases} \tag{3.2}$$

Here again $\tilde{\phi}_\mu(z',y')$ is in fact an holomorphic function of $(\mu^2 z_1,z'',y')$.

Let us show that $\tilde{\phi}_\mu$ is a F.B.I.-phase function of second kind adapted to V_o at any point $(z'_o,y'_o) = (-i\eta'_o,0)$ such that $\eta'_o = (\eta_{o1},\sigma''_o)$, $\eta_{o1} \neq 0$. Denote by $\eta_1(z'')$ the solution of

$$h(0,\mathcal{R}z'',\eta_1(z''),-\mathcal{I}z'') = 0.$$

Applying the Hamilton-Jacobi theory to (3.2) at $\mu = 0$ we get a function $y'(s,z'')$ satisfying

$$\exp(sH_h)(0,\mathcal{R}z'',\eta_1(z''),-\mathcal{I}z'') = (y'(s,z''),-\partial_y\tilde{\phi}_o(z'',y'(s,z''))). \tag{3.3}$$

Let

$$N_{z''} = \{y'(s,z'') : s \text{ is real}\} \ .$$

Since h is real we obtain

$$\mathcal{I}(\partial_{y'}\tilde{\phi}_o)(z'',y') = 0 \text{ if } y' \in N_{z''}.$$

By definition $N_{z''}$ is a submanifold of $\mathbb{R}^{n-1}$ with dimension 1. Moreover it follows from the initial data in (3.2) that

$$y' \to I\tilde{\phi}_o(z'',y')$$

is transversely positif on $N_{z''}$. The curves in (3.3) are null-bicharacteristics of h, hence

$$V_o = \{(y',-\partial_{y'}\tilde{\phi}_o(z'',y')) \; : \; y' \in N_{z''}\}$$

in a neighbourhood of $(0,\xi_o')$. The first derivative with respect to μ^2 at $\mu = 0$ of (3.2) gives

$$\begin{cases} \partial_{\xi'}h(y',-\partial_{y'}\tilde{\phi}_o(z'',y'))\partial_{y'}\tilde{\phi}_2(z',y') = i(y_1-z_1) \\ \tilde{\phi}_2(z',0,y'') = 0 \; , \; -\partial_{y'}\tilde{\phi}_2(z_o',0) = (\eta_{o1}/\partial_{\xi_1}h(0,\xi_o'),0). \end{cases}$$

Therefore

$$I(\partial_{y'}\tilde{\phi}_2)(z_o',0) = 0$$

and

$$\partial_{\xi'}h(y'(s,z''),\eta'(s,z''))\,.\,\partial_{y'}\tilde{\phi}_2(z',y(s,z'')) = i(y_1(s,z'')-z_1), \quad (3.4)$$

if

$$\eta'(s,z'') = -\partial_{y'}\tilde{\phi}_o(z'',y'(s,z'')).$$

The derivative of (3.4) with respect to s at s=0 is

$$\partial_{\xi'}h(0,\xi_o')I(\partial^2_{y'}\tilde{\phi}_o)(z_o',0)\partial_{\xi'}h(0,\xi_o') = \partial_{\xi_1}h(0,\xi_o') > 0.$$

This proves that the function

$$N_{z_o''} \ni y' \to I\tilde{\phi}_2(z_o',y')$$

has a non degenerated critical point with signature (1,0) at 0. Finally we have

$$(\partial_{z_1}\partial_{y'}\tilde{\phi}_2, \partial_{z''}\partial_{y'}\tilde{\phi}_o)_{(z_o',0)} = \begin{pmatrix} -i/\partial_{\xi_1}h(0,\xi_o') & \star \\ & \\ \cdot & -iE'' \end{pmatrix}.$$

So $\tilde{\phi}_\mu$ is a F.B.I.-phase function of second kind adapted to V_o. The normal direction ν which corresponds to z_o' is the class of $(0,-\mathcal{R}\partial_{y'}\tilde{\phi}_2(z_o',0))$, hence it satisfies

$$dr(\rho_o).\nu = \eta_{o1}. \tag{3.5}$$

Now consider the solution $G_\mu'(x',\xi_n,\eta')$ of

$$\begin{cases} \xi_n^2 + r(x',-\partial_{\xi_n}G_\mu,\partial_{x'}G_\mu) = 0 \\ G_\mu(x',0,\eta') = \psi_\mu(x',\eta'). \end{cases} \tag{3.6}$$

Since $\partial_{x_n} r \neq 0$, this is again a well-posed Cauchy-problem near $x' = 0$, $\xi_n = 0$, $\eta' = \sigma_o'$ and $\mu = 0$. As above, G_μ depends on (η_1,μ) as a function of $\mu^2\eta_1$. The Taylor expansion of G_μ with respect to ξ_n is

$$G_\mu(x',\xi_n,,\eta') = \psi_\mu(x',\eta') + \mu^2\eta_1\xi_n - \frac{1}{3}\,a(x',\eta',\mu)\xi_n^3 + \mathcal{O}(\xi_n^4)$$

where

$$a(x',\eta',\mu) = -1/\partial_{x_n} r(x',-\mu^2\eta_1,\partial_{x'}\psi_\mu(x',\eta')).$$

Using the Weierstrass preparation theorem as in [5], we can study the critical points of

$$\xi_n \to x_n\xi_n + G_\mu(x',\xi,\eta').$$

It turns out that there are two critical points near 0 which are holomorphic functions of x', the square root of $x_n + \mu^2\eta_1$ and η',μ. They are given by

$$\xi_n^{\pm}(x',\sqrt{x_n+\mu^2\eta_1},\eta',\mu) = \pm\sqrt{x_n+\mu^2\eta_1}\,\sqrt{Y(x,\eta',\mu)} + X(x,\eta',\mu)(x_n+\mu^2\eta_1)$$

where

$$Y(x',-\mu^2\eta_1,\eta',\mu) = -\partial_{x_n} r(x',-\mu^2\eta_1,\partial_{x'}\psi_\mu(x',\eta')).$$

The critical values are

$$\phi^{\pm}(x',\sqrt{x_n+\mu^2\eta_1},\eta') = x_n\xi_n^{+}(x',\sqrt{x_n+\mu^2\eta_1},\eta',\mu) +$$

$$+ G_\mu(x',\xi_n^{\pm}(x',\sqrt{x_n+\mu^2\eta_1},\eta',\mu),\eta',\mu)$$

$$= \psi_\mu(x',\eta') \pm \frac{2}{3} e(x,\eta',\mu)(x_n+\mu^2\eta_1)^{3/2} + f(x,\eta',\mu)(x_n+\mu^2\eta_1)^2$$

where e and f are holomorphic functions and

$$e(x',-\mu^2\eta_1,\eta',\mu) = \sqrt{-\partial_{x_n} r(x', -\mu^2\eta_1,\partial_{x'}\psi_\mu(x',\eta'))}.$$

Of course there is a problem to choose the sign of the square root of $x_n + \mu^2\eta_1$. However there are situations where we can fix this sign. If $\eta_{o1} > 0$, the argument of $x_n + \mu^2\eta_1$ is always close to 0 since $x_n \geqslant 0$ and $0 < \mu < \mu_o$. In this case we choose the square root whose real part is positive. The same choice is possible if x_n is near some positive value and μ^2 is small with respect to x_n. If $x_n = 0$ and $\eta_{o1} < 0$ we choose the square root whose imaginary part is positive. We have to be carefull when $x_n > 0$, $\eta_{o1} < 0$ and μ^2 cannot be taken small with respect to x_n.

We write ϕ_μ instead of ϕ_μ^{+}. Denote by $\tilde{\phi}_\mu$ the holomorphic conjugate function of $-\tilde{\phi}_\mu$. If $\eta_o' = (\eta_{o1},\sigma_o'')$, $\eta_{o1} \neq 0$, then ϕ_μ is a F.B.I.-phase function of second type adapted to $W_o = \{(x',\xi') : (x',-\xi') \in V\}$ near $(z_o',y_o') = (i\eta_o',0)$. The normal direction ν to W_o which corresponds to z_o' satisfies

$$dr.\nu = -\eta_{o1}$$

instead of (3.5). Therefore $\eta_{o1} > 0$ (resp. < 0) defines the hyperbolic (resp. elliptic) normal direction.

Using these functions we have the following basic result.

Proposition 3.1. *The holomorphic function*

$$(y',\eta') \to G_\mu(x',\xi_n,\eta') - \phi_\mu(y',\mu\sqrt{\eta_1},\eta') + \phi_\mu(z',y') \tag{3.7}$$

satisfies the hypothesis of Lemma 2.2 *at* $(y',\eta') = (0,\eta_o')$ *and* $(z',\xi_n,\eta') = (z_o',0,\eta_o')$. *The critical value* $H_\mu(z',\xi_n,x')$ *has the expansion*

$$H_\mu(z',\xi_n,x') = \phi_\mu(z',x') - \frac{2}{3}\,\frac{\mu^3(i(x_1-z_1))^{3/2}}{\sqrt{a_o(z'',x')}} + i\mu^2\eta_n(x_1-z_1)$$

$$-\frac{1}{3}\,a_o(z'',x')\,\xi_n^3 + O(\mu^4 + |\xi_n|^4),$$

with

$$a_o(z'',x') = -\frac{1}{\partial_{x_n} r(x',0,\partial_{x'}\phi_o(z'',x'))}\,.$$

If $\eta_{o1} > 0$ (resp. < 0) *we choose the square roots whose real (resp. imaginary part) is positive.*

The solution of the eiconal equation is obtained as the critical value of

$$\xi_n \to x_n\xi_n + H_\mu(z',\xi_n,x'). \tag{3.8}$$

We can study the critical points of (3.8) using the Weierstrass preparation theorem as in [5]. There are two critical points near 0 which have the form

$$\xi_n^{\pm}(z',x',\sqrt{x_n+\mu^2 b(z',x',\mu)},\mu)$$

$$= \tilde{\xi}_n(z',x',\mu) \pm \sqrt{S(z',x,\mu)(x_n+\mu^2 b(z',x',\mu)}$$

$$+ T(z',x,\mu)(x_n+\mu^2 b(z',x',\mu))$$

where

$$b(z',x',\mu) = i(x_1-z_1) + O(\mu)$$

$$S(z',x',0,0) = -\partial_{x_n} r(x',0,\partial_{x'}\phi_o(z'',x'))$$

and

$$\tilde{\xi}_n(z',x',\mu) = O(\mu^2).$$

Denote by

$$\theta_\mu(z',x',\sqrt{x_n+\mu^2 b(z',x',\mu)}) = x_n\xi_n^+ + H_\mu(z',\xi_n^+,x')$$

the critical value which corresponds to ξ_n^+. As indicated above we can fix the choice of the square roots if $x_n = 0$ or $x_n > 0$ and μ^2 is small with respect to x_n.

Let

$$W = \{\exp(sH_p)(x',0,\xi',0) : (x',\xi') \in W_o,\ s > 0\}.$$

Since the hamiltonian flow of H_p is a symplectic map for every s, W is an involutive submanifold with codimension two. We also introduce

$$W_{x_n} = \{(x',\xi') \in \dot{T}^\star U : \exists\xi_n \text{ s.t. } (x',x_n,\xi',\xi_n) \in W\},\ x_n > 0.$$

If x_n is small enough, W_{x_n} is an involutive submanifold of $\dot{T}^\star U$ with codimension 1. Denote by ${}^n y'_\mu(z')$ the solution of

$$I(\partial_{y'}\phi_\mu)(z',y'_\mu(z')) = 0$$

and let

$$\eta'_\mu(z') = \partial_{y'}\phi_\mu(z',y'_\mu(z')),\ \rho'_\mu(z') = (y'_\mu(z'),\eta'_\mu(z'))$$

$$\Phi_\mu(z') = -\,I\phi_\mu(z',y'_\mu(z')).$$

Proposition 3.2. *The function* θ_μ *satisfies the eiconal equation*

$$\begin{cases} (\partial_{x_n}\theta_\mu)^2 + r(x,\partial_x,\theta_\mu) = 0 \\ \theta_\mu(z',x',\mu\sqrt{b(z',x',\mu)}) = \phi_\mu(z',x'). \end{cases}$$

If $x_n > 0$ *and* μ^2 *is small with respect to* x_n , *we have*

$$\theta_\mu(z',x',\sqrt{x_n+\mu^2 b(z',x',\mu)})$$

$$= \phi_\mu(z',x') + \frac{1}{\sqrt{a_o(z'',x')}}\left(\tfrac{2}{3}x_n^{3/2}+i\mu^2(x_1-z_1)x_n^{1/2}-\tfrac{2}{3}\mu^3(i(x_1-z_1))^{3/2}\right)$$

$$+\ O(x_n^2+x_n\mu^2+x_n^{1/2}\mu^3+\mu^4).$$

Moreover

$$(z',x') \to \theta_\mu(z',x', \sqrt{x_n+\mu^2 b(z',x',\mu)})$$

is a F.B.I.*-phase function of second type adapted to* W_{x_n} *for every small* $x_n > 0$. *If* $\Phi_u(z')$ *is the weight function of* ϕ_μ *then the weight function of* θ_μ *is*

$$\psi_\mu(z') = \Phi_\mu(z')$$

for every x_n *when* $\eta_{o1} > 0$ *and is*

$$\psi_\mu(z') = \Phi_\mu(z') - \frac{2}{3}\mu^3 \frac{(-Iz_1)^{3/2}}{\sqrt{-\partial_{x_n} r_o(\rho_o'(z'))}} + O(\mu^4)$$

if $\eta_{o1} < 0$.

The fact that ψ_μ is smaller than Φ_μ when η_{o1} is negative corresponds to a gain of regularity in the elliptic region. Note that this gain is weaker than the exponential decay required in the definition of the second analytic wave front set.

One can also look at the real point x' such that

$$\partial_{x'}\theta_\mu(z',x', \sqrt{x_n+\mu^2 b(z',x',\mu)})$$

is real. The analytic expression of this point is not the same when $\eta_{o1} > 0$ and when $\eta_{o1} < 0$ but in any cases, its value when $\mu = 0$ is the x'-component of

$$\exp(sH_p)(y_o'(z'),0,\eta_o'(z'),0)$$

when $s > 0$ is chosen such that the x_n-component is equal to the given x_n. We denote this point by

$$x_o'(z',\sqrt{x_n}).$$

So θ_μ characterizes the analytic wave front set of points that lie on the bicharacteristic of p starting at $\rho_o'(z')$.

The function θ_μ blows up as a F.B.I.-phase function when x_n

converges to 0. However there is a good bound for the imaginary part of θ_μ when x is real.

Proposition 3.3. *Let* $c_o > 0$ *be such that*

$$|b(z',x',\mu)| < c_o$$

in a complex neighbourhood of $(z_o',0,0)$. *Then there are constants* c,C, $r > 0$ *such that*

$$-I\theta_\mu(z',x',\sqrt{x_n+\mu^2 b(z',x',\mu)}) \leqslant \Phi_\mu(z')-c\mu^2|x'-x_o'(z',\sqrt{x_n})|^2 + C\mu^3$$

if $0 \leqslant x_n < r$, $|z'-z_o'| < r$, x' *is real* , $|x'| < r$ *and* $0 < \mu < \mu_o$. *This is true for every choice of the square root if* $0 \leqslant x_n \leqslant c_o\mu^2$ *and for the square root whose real part is positive otherwise.*

If $x_n > 0$ the function

$$(z,x) \to \theta_\mu(z',x',\sqrt{x_n+\mu^2 b(z',x',\mu)}) + \frac{i\mu^2}{2}(z_n-x_n)^2$$

is an F.B.I.-phase function of second kind. Its weight function is

$$\psi_\mu(z') + \frac{\mu^2}{2}(Iz_n)^2.$$

4. Asymtotic solutions

Let us show how to get asymtotic solutions to the Dirichlet problem using the previous phase functions. We use the spaces $H_\phi^{(2)}$ of Sjöstrand, see [15] for the definitions.

It is wellknown that, solving transport equations, one can find a classical analytic symbol satisfying

$$\begin{cases} (\xi_n^2 + r(x',-\tilde{D}_{\xi_n},\tilde{D}_{x'},\lambda))\,[e^{i\lambda G_\mu(x',\xi_n,\eta')}a(x',\xi_n,\eta',\mu,\lambda)] = 0 \\ a(x',0,\eta',\mu,\lambda) = 1. \end{cases}$$

If

$$u \in H^{(2)}_{-I\phi_\mu,(z_o',0)}$$

let

$$Lu(z',\xi_n,x',\mu,\lambda)$$

$$= \mu^2(\frac{\lambda}{2\pi})^{n-1}\iint_{\Gamma_\mu(z',\xi_n,x')} e^{i\lambda(G_\mu(x',\xi_n,\eta') - \phi_\mu(y',\mu\sqrt{\eta_1},\eta'))}$$

$$a(x',\xi_n,\eta',\mu,\lambda)u(z',y',\mu,\lambda)dy'd\eta'$$

where Γ_μ is a family of good contours for $G_\mu - \phi_\mu + \phi_\mu$. It follows from Proposition 3.1 that

$$Lu \in H^{(2)}_{-IH_{\mu,(z'_o,0,0)}}.$$

Furthermore we put

$$Ku(z',x,\mu,\lambda) = \frac{(\lambda\mu)^{1/2}}{2i\pi}\int_{Re^{i\pi/2}}^{Re^{-i\pi/6}} e^{i\lambda x_n\xi_n} Lu(z',\xi_n,x',\mu,\lambda)d\xi_n.$$

From Proposition 3.3 and the study of paths of steepest descend in [5], it follows that

$$|Ku(z',x,\mu,\lambda)| \leqslant C_\varepsilon e^{\lambda\Phi_\mu(z')-c\lambda\mu^2|x'-x'_o(z',\sqrt{x_n})|^2+\varepsilon\lambda\mu^2}$$

if x' is real, $x_n \geqslant 0$ and $0 < \mu < \mu_o(\varepsilon)$, $\lambda > f(\mu,\varepsilon)$ for some functions μ_o, f.

In the level of formal analytic symbol we have

$$e^{-i\lambda(x_n\xi_n+G_\mu(x',\xi_n,\eta'))}P(x,\tilde{D},\lambda)(e^{i\lambda(x_n\xi_n+G_\mu(x',\xi_n,\eta'))}a)$$

$$= e^{-i\lambda G}(\xi_n^2+r(x',-\tilde{D}_{\xi_n},\tilde{D}_{x'},\lambda))(e^{i\lambda G}a)+e^{-i\lambda(x_n\xi_n+G_\mu)}\tilde{D}_{\xi_n}(e^{i\lambda(x_n\xi_n+G_\mu)}c)$$

where c is a classical analytic symbol. Therefore we have

$$P(x,D)(Ku)(z',x,\mu,\lambda) = \mathcal{O}(e^{\lambda\Phi_\mu(z')-\varepsilon\lambda})$$

where $\varepsilon > 0$. Moreover it turns out that the operator

$$u \to Ku(z',x',0,\mu,\lambda)$$

is an elliptic 2-microdifferential operator. Hence we can choose u such that

$$Ku(z',x',0,\mu,\lambda) = e^{i\lambda\phi_\mu(z',x')} + O(e^{-\lambda I\phi_\mu(z',x')-\varepsilon\lambda\mu^2}).$$

This shows that Ku is an asymptotic solution to the Dirichlet problem (1.1)-(1.2).

REFERENCES

1. F.G. Friedlander, R.B. Melrose, The wave front set of the solution of a simple initial boundary value problem with glancing rays, II, Math. Proc. Comb. Phil. Soc., 81, 1977, 97-120.
2. L. Hörmander, The analysis of linear partial differential operators, I-IV, Springer Verlag, 1983-1985.
3. K. Kataoka, Microlocal theory of boundary value problems, I-II, J. Fac. Sci. Univ. Tokyo 27(2), 1980, 355-399, and preprint.
4. P. Laubin, Analyse microlocale des singularités analytiques, Bull. Soc. Roy. Sc. Liège, 2, 1983, 103-212.
5. P. Laubin, Asymptotic solutions of hyperbolic boundary value problems with diffraction, Proceedings of the Nato ASI on Advances in microlocal analysis, D. Reidel, 165-202.
6. P. Laubin, Propagation of the second analytic wave front set in conical refraction, to appear
7. G. Lebeau, Deuxième microlocalisation sur les sous-variétés isotropes, Thèse, Orsay, 1983.
8. G. Lebeau, Régularité Gevrey 3 pour la diffraction, Comm. in P.D.E., 9(15), 1984, 1437-1494.
9. N. Levinson, Transformation of an analytic function of several variables to a canonical form, Duke Math. J. 28, 345-353, 1961.
10. R.B. Melrose, J. Sjöstrand, Singularities of boundary value problems I, Comm. Pure Appl. Math., 31, 1978, 593-617.
11. P. Schapira, Propagation at the boundary and reflexion of analytic singularities of solutions of linear partial differential equations, I and II, Publ. RIMS, Kyoto Univ., 12, 1977, 441-453 and Sem. Goulaouic-Schwartz, IX, 1976-77.
12. J. Sjöstrand, Propagation of analytic singularities for second order Dirichlet problems, I and II, Comm in P.D.E., 5, 1980, 41-94 and 187-207.
13. J. Sjöstrand, Analytic singularities and microhyperbolic boundary value problems, Math. Qnn., 254, 1980, 211-256.
14. J. Sjöstrand, Analytic singularities of solutions of boundary value problems, Proceeding of the Nato ASI on Singularities in boundary value problems, D. Reidel, 1980, 235-269.
15. J. Sjöstrand, Singularités analytiques microlocales, Astérisque 95, 1982.

BOUNDARY REGULARITY FOR ONE-SIDED SOLUTIONS OF LINEAR PARTIAL DIFFERENTIAL EQUATIONS WITH ANALYTIC COEFFICIENTS

Otto Liess

§ 1. Statement of the main results.

1. In this paper we study boundary regularity in normal directions for one-sided solutions of linear partial differential equations with analytic coefficients at points where the boundary is noncharacteristic. Our main result is theorem 1.8, and its generalization from theorem 1.10, below. It is modelled on the Hörmander-Sato regularity theorem on interior regularity at noncharacteristic points, but, due to the fact that we are close to the boundary, it is weaker, in that it only gives regularity at the level of second microlocalization (also called two-microlocalization later on). Nevertheless, it is a natural complement to results on tangential boundary regularity from Liess [5] for solutions of Cauchy problems with regular Cauchy data (also cf. Schapira [1] for results related to those from Liess [5]) and in a forthcoming paper we shall show that when one combines theorem 1.10 from this paper with the tangential boundary regularity results from Liess, loc. cit., then one can extend the results from Lebeau [1] to the case of Gevrey regularity (Of course this is only possible if suitable additional assumptions are added to those from Lebeau [1].) Although we shall not give any details here to make the last statement more precise, we would like to mention that it is not possible to extend the method of proof from Lebeau, loc. cit. to the Gevrey category.

2. Let us now assume that $p = p(x,t,D_x,D_t)$ is a linear partial differential operator of form

$$(1) \qquad p(x,t,D_x,D_t) = D_t^m + \Sigma\, a_{\alpha j}(x,t)\, D_x^\alpha D_t^j ,$$

where the sum is for $|\alpha| + j \le m$, $j < m$, and where the coefficients are germs of real-analytic functions defined in a neighborhood of $0 \in R^{n+1} = R^n_x \times R_t$. It follows that the surface $t = 0$ is noncharacteristic for p at 0 , but no further assumption on p will be made until theorem 1.10 . Let us now also consider some germ u of a distribution defined in a full neighborhood of 0 in R^{n+1} and assume that , in the sense of germs,

(2) $p\,u = 0$, for $t > 0$.

Thus u solves $p\,u = 0$ only on one side of $t = 0$, but u itself is defined on both sides of 0 . (Likewise,we could have assumed that u is defined for $t > 0$, that it solves (2) there and that it is extendible across $t = 0$.) It is classical (cf. e.g. Hörmander [1]) that u is then for small $t \ge 0$ a C^∞ function with values which are distributions in x . (The "=" in "$t \ge 0$" means : up to the boundary.) Of course,to obtain this kind of regularity in t ,it would have sufficed to assume that the coefficients $a_{\alpha j}$ were germs of C^∞ functions near 0 .

Here now we have assumed that the coefficients are real-analytic,so one may ask if more is true in this case . This is indeed so,in that, essentially, u is then a real-analytic function in t for $0 \le t < d$ (for some d) with values which are hyperfunctions in x . We shall give a more precise statement,due to K.Kataoka, of what we mean by this in a moment,but before we do so,we want to mention that it is not true in general that any extendible distribution solution of (2) is a real-analytic function in t ,$0 \le t < d$ (for suitable d) ,with distributional values in x . (To give an example , let $n = 1$,and let p be the Cauchy-Riemann operator $(1/2)(\partial/\partial t + i\partial/\partial x)$. If $u = 1/(t + ix)$, then u solves $pu = 0$ for $t > 0$, but it is not in general true that $f : R_+ = \{ t \in R; t > 0 \} \to C$ defined for $g \in C^\infty_o(R)$ by $f(t) = \int g(x)u(x,t)\,dx$ is real-analytic in t up to $t=0$.)

3. Before we can state the result of K.Kataoka to which we aluded a moment ago, we must introduce the following definition:

Definition 1.1. Let f be a germ of a hyperfunction defined in a neighborhood of

0 in R^{n+1}. We shall then say that f is mild from the positive side of $t=0$ if we can find $\varepsilon > 0$, $\xi^1,\dots,\xi^k \in R^n$, $|\xi^i| = 1$, and holomorphic functions f^j, $j=1,\dots,k$, defined on

$$D_{j,\varepsilon} = \{(x,t) \in C^{n+1};\ |t| < \varepsilon,\ |x| < \varepsilon, <\mathrm{Im}\, x,\xi^j> > \varepsilon|\mathrm{Im} x|+(1/\varepsilon)[|\mathrm{Im} t|+(\mathrm{Re}-t)_+]\}$$

such that

$$(3)\quad f(x,t) = \sum_{j=1}^{k} b(f_j)(x,t) \qquad \text{for } x \in R^n,\ |x| < \varepsilon \text{ and } 0 < t < \varepsilon.$$

Here $b(f_j)$ is the (cohomological) boundary value of f_j, computed from the relevant wedges, and a_+ is, for $a \in R$, the positive part of a.
It is a remarkable thing about mild hyperfunctions that, although (3) refers only to the part $t > 0$, the f_j have to exist also in a region where $\mathrm{Re}\, t < 0$. Furthermore, note that for fixed $t^o > 0$, ($t^o \in R$), the intersection of $D_{j,\varepsilon}$ with $t = t^o$ is just a standard wedge in C^n with edge in R^n. The functions $x \to f_j(x,t^o)$ then define for each t^o a hyperfunction in R^n, so we have a natural trace of f to $t = t^o$ defined by

$$(4)\quad f(x,t^o) = \sum b(f_j(\cdot\,,t^o))(x),$$

if f is given by (3). (The boundary values are now computed of course in C^n.) The same is true of course for $(\partial/\partial t)^k f$ for any k, since, obviously, $(\partial/\partial t)^k f$ is also mild form the positive side of $t = 0$. More generally, if $q = q(D_x,D_t)$ is an infinite order differential operator of infraexponential type (i.e., if q has the form $\sum_{\alpha,j} b_{\alpha j} D_x^\alpha D_t^j$, where the $b_{\alpha j}$ satisfy the following condition: $\forall\, \varepsilon > 0, \exists\ c_\varepsilon$ such that $|b_{\alpha j}| < c_\varepsilon \varepsilon^{|\alpha|+j}/(\alpha!\, j!)$) then q acts in a natural way on holomorphic functions, respectively on hyperfunctions, and we have that

$$q f = \sum b(q f_j)$$

if (3) was valid. Thus, qf is also mild from the positive side of $t=0$, if f was, and, once more, the traces of qf to $t=t^o$ exist up to $t^o=0$. It is in this sense that we shall say that f is then a real analytic function up to the boundary $t=0$ in t with values in hyperfunctions. (Here we alude to the fact that a function g which is defined on an open set in R is real-analytic there if and only if $q(D_t)g$ is continuous for any infinite order differential operator $q(D_t)$ of infra-exponential type. Of course, $q(D_t)$ is then even real-analytic.)

The result of K.Kataoka which we wanted to mention is now this :

Theorem 1.2. *Let U be a germ of a hyperfunction defined in a neighborhood of* 0 *in* R^{n+1} *and which satisfies* (2) *for* $t > 0$. *Then* u *is mild from the positive side of* $t = 0$.

Remark 1.3. It follows in particular that u has a natural trace to $t = 0$. This fact had been observed first (and without any reference to mildness) by H.Komatsu [1] and P.Schapira [1] .

Remark 1.4. In theorem 1.2 there is of course no need to assume explicitly that u is defined in a full neighborhood of 0 , since the sheaf of hyperfunctions is flabby.

4. The result of K.Kataoka is very beautiful, but it is not directly useful for the applications which we have in mind, since it does not contain any quantitative information (in terms of inequalities) on what happens when $t \to 0$, $t > 0$. This is due of course to the fact that hyperfunctions defined for $t > 0$ may "grow" arbitrarily wild for $t \to 0$, even though they can be extended across $t = 0$. In this paper we look however at distribution solutions of (2) , so we may still try to obtain such a quantitative information. Before we state our main result, it is useful to recall one of the main definitions from microlocal analysis, namely that of the analytic wave front set. (We do this since in our discussion we shall explicitly refer to three different definitions for this concept.) In doing so, we shall denote (x,t) by z , the Fourier-dual variables of x by ξ or ζ , those of t by τ and (ζ,τ) by λ .

Definition 1.5. Consider $v \in \mathcal{D}'(U)$ where U is open in R^{n+1} . Further consider z^o in U and λ^o in $R^{n+1} \setminus \{0\}$. We shall then say that v is microanalytic at (z^o,λ^o) , and write that $(z^o,\lambda^o) \notin WF_A v$, if one of the following three equivalent conditions is verified:

a) There is $\varepsilon > 0$, some natural number k ,some open cones G^j in R^{n+1},$j=1,\dots,k$, and analytic functions h_j,$j=1,\dots,k$, defined on $\{z \in C^{n+1};\ |z - z^o| < \varepsilon, \mathrm{Im}\, z \in G^j\}$ with the following properties :

$\forall\, j$, $\exists \eta \in G^j$ such that $<\eta\, ,\lambda^o > < 0$,

$v = \Sigma\, b(h_j)$ for $|z - z^o| < \varepsilon$.

b) There are $\varepsilon > 0$, $c > 0$, an open cone $\Gamma \subset R^{n+1}$ which contains λ^o and a bounded sequence v_j of distributions with compact support such that

$v = v_j$ for $|z - z^o| < \varepsilon$, $\forall\, j$,

(5) $|\hat{v}_j(\lambda)| < c(cj/\,|\lambda|)^j$ for $\lambda \in \Gamma$.

c) There are $b \in R$, $d > 0$, $c > 0$ and an open cone $\Gamma \subset R^{n+1}$ which contains λ^o such that $|v(g)| < c$ for any $g \in C_o^\infty(R^{n+1})$ such that

(6) $|\hat{g}(\lambda)| \leq \exp\,(d(|\mathrm{Re}\,\zeta| + |\mathrm{Re}\tau|) + d(|\mathrm{Im}\,\zeta| + |\mathrm{Im}\tau|) + <z^o, \mathrm{Im}\,\lambda> +$
$+\, b\ln(1 + |\lambda|))$, if $-\mathrm{Re}\,\lambda \in \Gamma$,

(7) $|\hat{g}(\lambda)| \leq \exp\,(d(|\mathrm{Im}\,\zeta| + |\mathrm{Im}\tau|) + <z^o, \mathrm{Im}\lambda> + b\ln(1 + |\lambda|))$,
if $-\mathrm{Re}\,\lambda \notin \Gamma$.

Here $\hat{v}_j$, $\hat{g}$, is the Fourier-Borel transform of v_j,respectively g .

(The set of points (z,λ) for which v is not microanalytic at (z,λ) is called the analytic wave front set,or the analytic singular spectrum,of v.)

Remark 1.6. The conditions from a),respectively b) here are precisely the initial definitions for microanalyticity of M.Sato [1] ,respectively L.Hörmander [3] . The fact that the two definitions are equivalent (for distributions.The definition of M.Sato was in fact stated directly for hyperfunctions.) has been proved by J.M.Bony [1] . The fact that c) is equivalent with b) is a consequence of results from Liess [1] .This has been observed in Liess [2] .(For more details and for a direct proof of the equivalence a) <=> b)<=> c) see Liess [4] .)

5. It follows in particular from definition 1.5,using condition a), that if (3) is valid ,then $(x^o,t^o, \pm N) \notin WF_A f$, where $N = (0,\ldots,0,1)$, provided that (x^o,t^o) is small enough and $t^o > 0$. (We tacitely assume now that f is a distribution. This is not strictly speaking necessary , but in definition 1.5 we have only considered distributions.) In fact,locally,this is all information which one can obtain from mildness from the positive side of t= 0, for one can prove:

Lemma 1.7. Let v be a distribution defined in a neighborhood of some point $z^o \in R^{n+1}$. Then the following two statements are equivalent :

α) $(z^o, N) \notin WF_A v$ and $(z^o,-N) \notin WF_A v$.

ß) There are $\xi^j \in R^n$, $|\xi^j| = 1$,j=1,...,k, $\varepsilon > 0$ and analytic functions g_j defined on $\{z \in C^{n+1} ; |z - z^o|< \varepsilon, < \mathrm{Im} x , \xi^j> > \varepsilon |\mathrm{Im}\, x|+ (1/\varepsilon) | \mathrm{Im}\, t| \}$ such that $v = \Sigma\, b(g_j)$ in a neighborhood of z^o .

The implication ß) => α) is here obvious (if we use condition a) from definition 1.5 as a definition for the analytic wave front set) and to prove that α) implies ß) it suffices to write down the assumptions from α) in terms of condition a) from definition 1.5 and to apply the edge-of-the-wedge theorem.

6. We can now state the main result from this paper for the case of analytic microlocalization. (We shall state a related result for Gevrey-microlocalization later on.)

Theorem 1.8. Let u be a germ of a distribution defined in a full neighborhood of 0 in R^{n+1} and assume that u satisfies (2) . Then there are constants c_1,c_2,d,b , such that $|u(v)| \le c_1$ for any $v \in C_o^\infty(R^{n+1})$ which satisfies

(8) $|\hat{v}(\lambda)| \le \exp (d|\mathrm{Re}\, \zeta| + d|\mathrm{Im}\, \zeta| + d\, \mathrm{Im}\, \tau_+ + b \ln(1+ |\lambda|))$, if $|\tau| \ge c_2|\zeta|$,

(9) $|\hat{v}(\lambda)| \le \exp (d|\mathrm{Im}\, \zeta| + d\, \mathrm{Im}\, \tau_+ + b \ln(1+ |\lambda|))$, if $|\tau| < c_2|\zeta|$.

7. The pair of inequalities (8) , (9) is very similar to the pair (6), (7) , so a comparision of the two pairs is in order . Let us then note at first that any v which satisfies (8) and (9) must be concentrated in $t \ge 0$. Of course, this is a

natural condition in the present context, for in theorem 1.8 we cannot hope for more than one-sided regularity. Also note that theorem 1.8 is a regularity result at $x^o = 0$, $t^o = 0$. If we set $x^o = 0$, $t^o = 0$, in (6) and (7) we see that now we have explained all differences in the two pairs of inequalities, as far as the imaginary part of λ is concerned. Moreover, we note that $|\tau| > c_2|\zeta|$ is for real λ a conic neighborhood of $\{N, -N\}$ in R^{n+1}, which shows that the conditions from (8) and (9) roughly correspond to simultaneous localization around N and -N. The main difference between (6) and (8) (when we take $\lambda^o = \pm N$ in definition 1.5) is then that in (6) $\hat{g}$ is permitted to grow of order $\exp(d|\lambda|)$ (for real λ close to $\pm N$), whereas in (8) we only admit a growth of order $\exp(d|\zeta| + b \ln(1+|\lambda|))$ for $\hat{v}$. Ideally, what one wants to model in the two cases is a decay of order $\exp(-d|\lambda|)$, respectively $\exp(-d|\zeta| -b \ln(1 + |\lambda|))$ for the "Fourier transform" $\hat{u}$ of u in a real conic neighborhood of $\pm$ N. (The trouble is of course that before we can consider the Fourier transform of u we have to localize in the (x,t)-space and that no suitable cut-off function for such a localization can exist.) Since a decay of order $\exp(-d|\xi|)$ on the Fourier side corresponds to partial analyticity in the x-variables, we can then conclude that the conclusion of theorem 1.8 roughly says the following: if u solves (2), then u is, microlocally near $(0, \pm N)$, and from the positive side of $t=0$, partially analytic in x. This makes it clear that theorem 1.8 is a result on two-microlocalization.

8. Practically, theorem 1.8 means that we have microlocally near $(0, \pm N)$ a good control on high-order x-derivatives of u. In the applications of theorem 1.8 which we have in mind, and perhaps in others, one also needs some control on high-order t-derivatives of u near $(0, \pm N)$. It is therefore interesting to recall that if u is a solution of u, then one can compute (due to the special form of p from (1)) high-order t-derivatives of u in terms of high-order x-derivatives of $(\partial/\partial t)^j u$, for $j = 0,\ldots,m-1$. In particular it is then no surprize that the two-sided variant of theorem 1.8 (where we assume that u solves $pu = 0$ on both sides of $t = 0$, and where, in the conclusion of

theorem 1.8 we replace $\mathrm{Im}\,\tau_+$ by $|\mathrm{Im}\,\tau|$) immediately gives $(0,\pm N) \notin WF_A\, u$, which is the Hörmander-Sato regularity theorem.In this sense,theorem 1.8 is an extension of the Hörmander-Sato regularity theorem to the case of boundary regularity.

9. The conclusion in theorem 1.8 is stated in terms of a duality , similar to that from condition c) in definition 1.5. One may wonder if it is not possible to give a statement in terms closer to condition b) from that definition . (Our interest in condition a) there is more limited,since we are interested in estimates.) The problem is here with the one-sidedness. In fact,two-sided regularity of the type from theorem 1.8 can be characterized easily in terms related to Hörmander's definition of the analytic wave front set. We state a result to this effect , which is a consequence of propsoition 1.4.3 from Liess-Rodino [1] (or rather of its proof,since,strictly speaking , $1 + |\xi|$ is not an admissible weight function on R^{n+1} in Liess-Rodino [1] .This is of no importance for proposition 1.4.3 in that paper however.) .

Proposition 1.9. Assume u is C^∞ in a neighborhood of the origin. Then the following two conditions are equivalent :

i) There are $\varepsilon > 0$, $c > 0$, $c' > 0$ and a bounded sequence of distributions u_j with compact support with the following properties :

$$u = u_j \quad \text{for} \quad |(x,t)| < \varepsilon,$$

$$|\hat{u}_j(\xi,\tau)| \le c(cj/(1 + |\xi|))^j \quad \text{if} \quad \lambda \in R^{n+1} \quad \text{and} \quad |\tau| \ge c'|\xi| .$$

ii) There are $d > 0$, $c > 0$ and for every b some c' such that $|v(u)| \le c'$ for any distribution with compact support v which satisfies

$$|\hat{v}(\lambda)| \le \exp(d\,|\mathrm{Re}\zeta| + d|\mathrm{Im}\,\zeta| + d|\mathrm{Im}\tau| + b\,\ln(1+|\lambda|)) \text{ ,if } |\mathrm{Re}\tau| \ge c|\mathrm{Re}\,\zeta| ,$$

$$|\hat{v}(\lambda)| \le \exp(d\,|\mathrm{Im}\,\zeta| + d\,|\mathrm{Im}\,\tau| + b\,\ln(1+|\lambda|)) \text{ , if } |\mathrm{Re}\,\tau| < c|\mathrm{Re}\,\zeta| .$$

(The reason why we have assumed that u is C^∞ here is that this is also assumed in proposition 1.4.3 from Liess-Rodino [1] . One can prove a related result for u in $\mathcal{D}'$,but we have no use for a result of this type at this moment.)

Until now, we have always measured (micro-) regularity in the analytic class. If however the operator under consideration has a natural quasihomogeneous structure, then it is more natural to measure regularity in the associated anisotropic Gevrey class . Here we recall that when $M = (M_1,\dots,M_n,M_{n+1})$, $M_j \geq 1$, then the Gevrey class G^M associated with M is introduced as follows :

f is said to be of class G^M at (x^o,t^o) if f is C^∞ near (x^o,t^o) and if there are constants c , c', such that

$$|D_x^\alpha D_t^j \; f(x,t)| \leq c^{|\alpha|+j+1} (\alpha_1!)^{M_1} \cdots (\alpha_n!)^{M_n} (j!)^{M_{n+1}} \quad \text{for} \quad |x-x^o|+|t-t^o| < c' .$$

Thus for example , for the Schrödinger operator $\Delta + i\,\partial/\partial t$, $\Delta = -\Sigma(\partial/\partial x_j)^2$ (which is not of form (1) in the initial variables x and t . Actually in theorem 1.10 below, we may think of one of the variables x to play the role of t and of t being ,after a renotation , one of the x - variables .), it is more natural to measure regularity in the class G^M ,where $M = (1,\dots,1,2)$.

It is then also natural to study microregularity with the aid of notions derived from G^M and it was in fact an attempt to prove a variant of the result from Lebeau [1] for this type of microlocalization which was the starting point for this paper.

Due to lack of space,we shall not comment on the opportunity of this type of microlocalization any further (cf. e.g. Liess-Rodino [1] and [2] and,as far as the "geometry" of this kind of microlocalization is concerned,also Parenti-Segala [1] and references quoted there,) but rather state the main result from this paper in its general form. Assume then that some weight-vector $M = (M_1,\dots,M_n,M_{n+1})$ is given with $M_i \geq 1$ and $M_{n+1} = 1$. Assume further that the sum in (1) is now only for $\Sigma\, \alpha_k M_k + j \leq m$, $j < m$.(This is in fact also all what we need for the result of the type from Lebeau [1] to which we have referred in the above.) Further denote by $\psi(\xi) = \Sigma\,(+|\xi_j|)^{1/M_j}$. We can then prove the following result:

Theorem 1.10 . Let M and p be as before and consider an extendible solution u of (2) . Then we can find c_1, c_2 , d and b such that $|u(v)| \leq c_1$ for any $v \in C_o^\infty(R^{n+1})$ for which

$$|\hat{v}(\lambda)| \leq \exp\ (\ d\psi(\mathrm{Re}\ \zeta) + d\ |\mathrm{Im}\ \zeta| + d\ \mathrm{Im}\ \tau_+ + b\ \ln(1+|\lambda|))\ ,\text{if } |\tau| \geq c_2\ \psi(\zeta)\ ,$$

$$|\hat{v}(\lambda)| \leq \exp\ (\ d|\mathrm{Im}\ \zeta| + d\ \mathrm{Im}\ \tau_+ + b\ \ln(1+|\lambda|))\ ,\ \text{if}\ |\tau| < c_2\ \psi(\zeta)\ .$$

Thus,when passing from theorem 1.8 to theorem 1.10 , two changes did occur : we have replaced $|\mathrm{Re}\ \zeta|$ by $\psi(\mathrm{Re}\ \zeta)$ in the exponent from (8) (which we did since now we want to measure Gevrey regularity instead of analytic regularity), and, furthermore, localization is now on $|\tau| > c_2\ \psi(\zeta)$ instead of on $|\tau| > c_2|\zeta|$. Here we note that $T = \{\ \lambda\ ;\ |\tau| > c_2\ \psi(\zeta)\ \}$ is essentially a union of "parabolic" rays and that T is larger (if at least one of the M_j is strictly larger than 1) than $|\tau| > c_2|\zeta|$. This corresponds of course to the quasihomogeneous structure of the operator under consideration.Also note that when all $M_j = 1$,then theorem 1.10 reduces to theorem 1.8.

§ 2. Sketch of the proof of theorem 1.8.

1. As we have seen in § 1 ,the theorems 1.8 and 1.10 are related to the Hörmander-Sato regularity theorem. The statements being by duality, it is no surprize that also in the proofs we shall have to argue by duality . Since the inequalities from the statement refer to C^{n+1} ,it is moreover natural that at some moment we shall have to use complex methods.

Due to lack of space , we shall only describe the general line of argument and leave a number of details in the proofs for a future publication. Moreover ,for notational reasons, it is convenient to restrict our attention to the case of theorem 1.8 , the proof of theorem 1.10 being, appart from obvious modifications, very similar . Finally we shall assume , to simplify the situation (from a technical point of view) even more , that u is a germ of a C^∞ function in a neighborhood of the origin. (Here we recall that u is known to be a C^∞ function in t for $t \geq 0$ small, if $pu = 0$, and this makes the case of distribution solutions to be very close to the case of C^∞ solutions.)

2. The first step in the proof of theorem 1.8 is the following result, which shows that we can localize (8), §1, near some previously fixed $\alpha \in Z^n$ (Z denotes here the integers) :

Proposition 2.1. Let d_1, b, c_1, c_2, be given. Then we can find d, b_1, c_3 and c_4 such that any $v \in E'(R^{n+1})$ (the space of distributions with compact support) which satisfies

$$(1) \quad |\hat{v}(\lambda)| \leq \exp\,(d|\mathrm{Re}\,\zeta| + d|\mathrm{Im}\zeta| + d\,\mathrm{Im}\,\tau_+ + b\,\ln(1+|\lambda|)), \text{ if } |\tau| \geq c_3|\zeta|\,,$$

$$(2) \quad |\hat{v}(\lambda)| \leq \exp\,(d|\mathrm{Im}\,\zeta| + d\,\mathrm{Im}\,\tau_+ + b\,\ln(1+|\lambda|))\,, \text{ if } |\tau| < c_3|\zeta|\,,$$

can be decomposed in the form

$$(3) \quad v = \sum_{\alpha \in Z^n} v_\alpha\,,$$

where the v_α, $\alpha \in Z^n$ are in $E'(R^{n+1})$ and satisfy the following estimates :

$$(4) \quad |\hat{v}_\alpha(\lambda)| \leq c_4(1+|\alpha|)^{-n-2} \exp(d_1|\mathrm{Im}\zeta| + d_1\mathrm{Im}\,\tau_+ + b_1\,\ln(1+|\lambda|))\,,$$

$$\text{if } |\tau| < c_1|\zeta| \text{ or } |\mathrm{Re}\zeta - \alpha| \geq c_2|\alpha|\,,$$

respectively , for all other λ,

$$(5) \quad |\hat{v}_\alpha(\lambda)| \leq c_4(1+|\alpha|)^{-n-2} \exp(d_1|\mathrm{Re}\zeta| + d_1|\mathrm{Im}\,\zeta| + d_1\mathrm{Im}\,\tau_+ + b_1\,\ln(1+|\lambda|))\,.$$

(Here b_1 depends only on b.)

Once we have proved proposition 2.1 , the following result is very natural :

Proposition 2.2 Let u be a germ of a C^∞ function defined in a neighborhood of the origin and assume that we can find d_1, c_1 and c_2 with the following property :

for every b_1 there is c_5 such that $|v_\alpha(u)| \leq c_4c_5(1+|\alpha|)^{-n-1}$ whenever $v_\alpha \in E'(R^{n+1})$ satisfies (4) and (5).

Then we can find d and c_3 such that for every b there is c_6 such that $|v(u)| \le c_6$, whenever $v \in \mathcal{E}'(R^{n+1})$ satisfies (1) and (2).

The proof of proposition 2.2 is based on an approximation procedure very similar to the one used in the proof of proposition 3.2 from Liess [1]. Similarily,one can prove proposition 2.1 with arguments similar to those used in the proof of lemma 3.6 a) in the same paper. (The arguments from Liess ,loc.cit., use some auxiliary plurisubharmonic function. One additional difficulty in the present situation comes from the fact that,apparently,there is no similar universal plurisubharmonic function which one can use for all v from the proposition.) We shall not give the details of these proofs.

3. One of the reasons why we prefer to work with the estimates (4) and (5) rather than with (1) and (2) is that the following result is very easy to prove :

Proposition 2.3. Let d_2, c_7 and c_8 be given. Then we can find d_1, c_1 and c_2 such that for every b_1 there are b_2 and c_9 with the following property : if v_α satisfies (4) and (5) and if f is an analytic function defined on the set $\{ z \in C^{n+1} ; |z| < 1 \}$ which satisfies $|f(z)| \le 1$ on its domain of definition, then it follows that $w_\alpha = f v_\alpha$ satisfies (4) and (5) with d_1,b_1,c_1,c_2,c_4 replaced by d_2,b_2,c_7,c_8 and $c_4 \cdot c_9$.

The proof of proposition 2.3 is not difficult. A similar situation appears in lemma 2.8 from Liess [3] and the proof from there can be adapted to the present situation. We omit the details of the argument.

4. The main step in the proof of theorem 1.8 is now

Proposition 2.4. Let d_2, c_{10}, c_{11} be given. Then we can find d_1, c_1, c_2 and for every b_1 some b_2 and c_{12} such that the following property holds for every $\alpha \in Z^n$:

if v_α satisfies (4) and (5), then we can find v'_α and v''_α in $\mathcal{E}'(R^{n+1})$ such that

(6) $\quad v_\alpha = {}^t p(x,t,D_x,D_t)\, v'_\alpha + v''_\alpha ,$

where

(7) $\quad |\hat{v}'_\alpha(\lambda)| \le c_{12}\, c_4 (1+|\alpha|)^{-n-1} \exp(d_2|\mathrm{Im}\,\zeta| + d_2 \mathrm{Im}\,\tau_+ + b_2 \ln(1+|\lambda|)) ,$

$$\text{if} \quad |\tau| \le c_{10}|\zeta| \ \text{or} \ |\mathrm{Re}\zeta - \alpha| \ge c_{11}|\alpha| ,$$

(8) $\quad |\hat{v}'_\alpha(\lambda)| \le c_{12}\, c_4 (1+|\alpha|)^{-n-1} \exp(d_2|\mathrm{Re}\zeta| + d_2|\mathrm{Im}\zeta| + d_2 \mathrm{Im}\,\tau_+ + b_2 \ln(1+|\lambda|)),$

$$\text{for all other } \lambda,$$

respectively

(9) $\quad |\hat{v}''_\alpha(\lambda)| \le c_{12}\, c_4 (1+|\alpha|)^{-n-1} \exp(d_2|\mathrm{Im}\zeta| + d_2 \mathrm{Im}\,\tau_+ + b_2 \ln(1+|\lambda|)),$

$$\text{for all } \lambda \in C^{n+1} .$$

Here ${}^t p$ is the formal adjoint of p .

Note here that theorem 1.8 is now a consequence of the propositions 2.2 and 2.4. In fact in view of proposition 2.2 it suffices to check that

(10) $\quad |v_\alpha(u)| \le c'(1+|\alpha|)^{-n-1}$

for all α and all v_α which satisfy the inequalities (4) and (5) for suitable d_1, b, c_1, c_2 .(We are here and from now on somewhat sloppy about the "b" . This is not really important,since we have assumed that u is C^∞ .)
To check this,we can apply proposition 2.4. In fact , if d_1, b, c_1, c_2 have been suitable , then we may assume that we can write v_α in the form (6) with v'_α and v''_α satisfying (7),(8) and (9) . For small d_2 we can conclude that

(11) $\quad v_\alpha(u) = v''_\alpha(u) \quad .$

Indeed,for small d_2 the dualities $v_\alpha(u)$, $v''_\alpha(u)$ and ${}^t p(x,t,D_x,D_t) v'_\alpha(u)$ all make sense and we have $({}^t p\, v'_\alpha)(u) = v'_\alpha(p\, u)$.(Here we use that u is C^∞ . If u had not been C^∞ ,than more care would have been needed with the b's .) It remains to recall that $pu = 0$ for $t \ge 0$ and that supp v'_α lies in $\{(x,t);\ t \ge 0\}$, so that $v'_\alpha(pu) = 0$.

We have thus proved (11) and (10) is then a consequence .

5. In order to conclude the sketch of the proof of theorem 1.8 , it remains to give a sketch of the proof of proposition 2.4. To do so , we shall fix α and drop the index α in v_α , v'_α , v''_α henceforth . Let us moreover, fix some constants d_2 and c_{13} and denote by B the set

$$B = \{ \; z \in C^{n+1} \; ; \; |z| < 2\, d_2 \, , \; |\tau| > c_{13}|\zeta| \; \} \; .$$

If d_2 is small enough and c_{13} is sufficiently large (note for example that by the nature of the statement in proposition 2.4 we may always assume that d_2 is as small as we please,) then we can find a formal analytic symbol $\sum_{j \geq 0} q_j$ of order $-m$ such that

$$(12) \qquad (\Sigma\, q_j) \circ \tilde{p} \sim 1 \quad \text{on } B \; ,$$

where $\tilde{p} = \tau^m + \Sigma\, a_{\alpha j}(x,t)\zeta^\alpha \tau^j$ is the symbol of p . (We drop the ~ henceforth and denote both the operator and its symbol by p .)

Here the terminology and the notations are from the theory of analytic pseudo-differential operators, but since we are working in a complex region,let us observe that what we effectively mean by this is the following :

a) all q_j are defined on B , are positively homogeneous in λ of degree $-m-j$ (this requirement,which makes all our symbols classical,is not strictly speaking necessary,but it can be achieved here and it simplifies the exposition) and satisfy estimates of form

$$|D_z^\alpha D_\lambda^\beta \; q_j(z,\lambda)| \leq c_{14}^{|\alpha| + \beta + j + 1} \alpha!\, \beta!\, j!\, (1+|\lambda|)^{-m-j-|\beta|} \quad \text{on } B,$$

b) if $\sum_{j \geq 0} a_j$, $\sum_{s \geq 0} b_s$ are formal sums of functions which are positively homogeneous in λ of degrees $\mu-j$ respectively $\mu'-s$,then $(\Sigma\, a_j) \circ (\Sigma\, b_s)$ is by definition the formal sum $\Sigma\, r_k$ where r_k is defined by

$$r_k = \Sigma\, a_j^{(\alpha)} \, b_{s(\alpha)} \,/\, \alpha! \; ,$$

where the sum is for $j+s+ \quad = k$.

Here $a_j^{(\alpha)} = (\partial/\partial\lambda)^\alpha a_j$ and $b_{s(\alpha)} = (-i\partial/\partial z)^\alpha b_s$.

c) In b) we have of course assumad that the a_j and b_s are C^∞. We shall assume in fact even more,namely that they are analytic in z and λ .

d) The sysmbol p is regarded as $p = \sum_{s \geq 0} p_s$ where $p_o = \tau^m + \sum_{|\alpha|+j=m} -$ $a_{\alpha j}(z)\ \zeta^\alpha \tau^j$ is the principal symbol of the operator and where p_s for $s \geq 1$ is defined by $p_s = \sum_{|\alpha|+j=m-s} a_{\alpha j}(z)\ \zeta^\alpha\ \tau^j$ when $s \leq m$ and by $p_s \equiv 0$ when $s > m$.

e) If $\Sigma\, a_j$ and $\Sigma\, b_j$ are given,we shall say that $\Sigma\, a_j \sim \Sigma\, b_j$ if all terms of the same degree of homogeneity in $\Sigma\, a_j$ and $\Sigma\, b_j$ coincide .

Thus $(\Sigma\, a_j) \circ (\Sigma\, b_s)$ is here the standard composition rule for formal symbols of pseudodifferential operators and the equality $(\Sigma\, q_j) \circ p \sim 1$ (which should be read as $(\Sigma\, q_j) \circ (\Sigma\, p_s) \sim 1$) says that $\Sigma\, q_j$ is an inverse to $\Sigma\, p_s$.

Note that for d_2 small and for c_{13} large B is an ellipticity region for p , so the standard proofs to obtain a parametrix for elliptic pseudodifferential operators in the analytic category will work .(Cf. e.g. Treves [1]) .

We have now explained (12) and can describe our first guess for v' (or rather for v'_α) . It is :

$$(13) \qquad h(\lambda) = v(\exp(-i \langle z,\lambda\rangle)\ \sum_{j \leq \chi|\alpha|} q_j(z\ ,\ -\lambda)) ,$$

which is defined for λ in D ,

$$D = \{ \lambda \in C^{n+1}\ ;\ |\tau| > c_{13}|\zeta| \} .$$

Here $\chi \leq 1$ is some positive constant to be chosen according to later needs. None of the constants c in later estimates must however depend on χ ,once χ is small enough . Note that we can apply proposition 2.3 and obtain if d_3 , c_{15} and c_{16} are given and if we we did start with suitable d_1, c_1, c_2 ,that

$$|h(\lambda)| \leq c_{17}\ \chi\ |\alpha|(1+|\alpha|)^{-n-2} \exp(d_3\ |\mathrm{Re}\ \zeta| + d_3|\mathrm{Im}\ \zeta| + d_3\ \mathrm{Im}\ \tau_+ + b_3 \ln(1+|\lambda|)),$$

$$\text{if } |\mathrm{Re}\zeta - \alpha| \leq c_{15}|\alpha| \text{ and } |\tau| \geq c_{16}|\zeta| ,$$

respectively

$$|h(\lambda)| \le c_{17}\chi|\alpha| \ (1+|\alpha|)^{-n-2} \exp\ (d_3|\operatorname{Im}\zeta| + d_3 \operatorname{Im}\tau_+ + b_3 \ln(1+|\lambda|)) \ ,$$

for the remaining λ .

Of course, in both of the preceding inequalities we have assumed that λ is in D .

We have called h a "first guess" for $\hat{v}'$, but it is clear that we are not allowed to set $\hat{v}' = h$. In fact, h is not even defined on all of C^{n+1} . It is then natural to cut down h in a suitable neighborhood of $\operatorname{Re}\zeta = \alpha$ and to use a $\bar{\partial}$- argument to find an entire function which does not differ from h very much near $\operatorname{Re}\zeta = \alpha$.

Let us then choose $\psi \in C^\infty\ (C^{n+1})$ such that

$$\psi(\lambda) = 1 \ , \text{ if } \ |\tau| > 2\, c_{13}|\zeta| + 1 \ \text{ and } |\operatorname{Re}\zeta - \alpha| \le 4\, c_{15}|\alpha| \ ,$$

$$\psi(\lambda) = 0 \ , \text{ if } \ |\tau| \le c_{13}|\zeta| \ \text{ or } \ |\operatorname{Re}\zeta - \alpha| \ge 5\, c_{15}|\alpha| \ ,$$

$$|D^\gamma_{\operatorname{Re}\lambda}\ \psi(\lambda)| \le c_{18}^{|\gamma|+1}\ \chi^{|\gamma|}\ (|\alpha|)^{|\gamma|} \qquad \text{if } \ |\gamma| \le \chi|\alpha| \ .$$

Once again the existence of such functions follows from Hörmander [3] .

If c_{15} is small enough, then we will have $|\operatorname{Re}\zeta| \sim |\alpha|$ on supp ψ .

We then define an auxiliary function h' on C^{n+1} by

$$h'(\lambda) = \begin{cases} \psi(\lambda)h(\lambda) \ \text{ if } \ |\tau| > c_{13}|\zeta| \ , \\ 0 \ , \quad \text{for all other } \lambda \ . \end{cases}$$

It follows that

$$|\bar{\partial}\ h'(\lambda)| \le c_{19}(1+|\alpha|)^{-n-1} \exp(\ d_3|\operatorname{Im}\zeta| + d_3\ \operatorname{Im}\tau_+ + b_3\ln(1+|\lambda|)) \ ,$$

for all $\lambda \in C^{n+1}$, where $\bar{\partial}$ is the Cauchy-Riemann operator on C^{n+1} .

We can now apply classical results on the theory of the $\bar{\partial}$ operator from Hörmander [2] and find c_{20} , b_4 and a C^∞ function f on C^{n+1} such that $\bar{\partial} f = \bar{\partial}\ h'$ and for which

(14) $\quad |f(\lambda)| \le c_{20}(1+|\alpha|)^{-n-1} \exp\ (d_3\ |\operatorname{Im}\zeta| + d\operatorname{Im}\tau_+ + b_4\ \ln(1+|\tau|))$, for all λ .

(When u is a distribution solution, this point is a little bit more complicated: then we must maintain some cpntrol over the part "$b_4\ln(1+|\zeta|)$"in $b_4\ln(1+|\lambda|))$.

On the other hand, no control on the part "$b_4 \ln(1+|\tau|)$" is needed, due to the fact that u is C^∞ in the t-variables. The idea is then to treat $\ln(1+|\zeta|)$ and $\ln(1+|\tau|)$ separately. A similar situation appears in the proof of proposition 10.3 from Liess [5] .)

We finally define $\hat{v}'$ by

$$\hat{v}' = h' - f ,$$

such that $\hat{v}'$ will indeed satisfy estimates of the desired type.

We have now completed the construction of v' and are then forced to define v'' by

$$(15) \qquad v'' = v - {}^t p(x,t,D_x,D_t)\, v'$$

if we want (6) to be satisfied. The proof will then come to an end if we can estimate the v'' defined by (15) as in (9). That this is possible is not completely obvious at this moment, but the main idea of the proof is easy to describe. Assume in fact that it were possible to find an analytic pseudodifferential operator $q: C^\infty(U) \to C^\infty(U)$, U some neighborhood of the origin, such that $p \circ q$ were the identity. In this case, we would simply set $v'_\alpha = {}^t q\, v_\alpha$, $v''_\alpha = 0$, where ${}^t q$ is the adjoint of q, and were done. If we would use the representation of Kohn-Nirenberg [1] of a pseudodifferential operator with the symbol of its adjoint, this would lead to

$$(16) \qquad {}^t q \hat{v}\,(\lambda) = v(\exp(-\langle z, \lambda\rangle)\ \tilde{q}(z, -\lambda)),$$

where $\tilde{q}$ denotes the symbol of q.

Even though it is not possible to find q as described before, we now see that (13) is very close to (16), in the region where we were able to invert p. After this discusion it is now not surprizing that the arguments which are needed to prove (9) are parallel to those which one uses when one studies composition of analytic pseudo-differential operators. (A similar situation appears in Liess [6] .) Of course, in the present context it is more natural to work on the Fourier side.

We omit further details.

References

J.M.Bony [1] : Propagation des singularités differentiables pour une classe d'oper. differentiels à coefficients analytiques, Asterisque, 34-35, 1976, 43-91

L.Hörmander [1] : Linear partial differential operators,Springer Verlag,1963,

[2] : An introduction to complex analysis in several complex variables D.Van Nostrand Comp.,1966,

[3] : Uniqueness theorems and wave front sets for solutions of linear diff.equ.with analytic coefficients.C.P.A.M.,34,1971,671-704.

K.Kataoka [1] : Microlocal analysis of boundary value problems with applications to diffraction.NATO ASI Series,vol.C65 (ed.by Garnir),121-133.

J.J.Kohn-L.Nirenberg [1] : On the algebra of pseudodifferential operators, C.P.A.M.,18,1965,269-305.

H.Komatsu [1] :Boundary values for solutions of elliptic equations,Proc.Int.Conf. Funct.Anal.Rel.Topics,Univ.of Tokyo Press,Tokyo 1970,107-121.

G.Lebeau [1] : Une proprieté d'invariance pour le spectre des traces de solutions d'op.diff.,C.R.Acad.Sci.Paris,Ser.I Math.,294:22 (1982),723-725.

O.Liess [1] : Intersection properties of weak analytically uniform classes of functions,Ark.Mat.,14:1,1976,93-111,

[2] : The microlocal Cauchy problem for constant coefficient linear partial diff. op.,Rev.Roum.,21:9,1976,1221-1239,

[3] : Uniqueness for the characteristic Cauchy problem ...,Astérisque, 89-90,1981,163-203,

[4] : Boundary behaviour of analytic functions,analytic wave front sets and the intersection problem of Ehrenpreis,Rend.Sem.Mat.Univers. Politecn.Torino,42:3,1984,103-152,

[5] : Microlocality of the Cauchy problem in inhom.Gevrey classes to appear in the C.P.D.E.,

[6] : Microlocal regularity for global solutions of partial diff.equ. with polynomial coeff.,to appear in the Annals of Global Analysis and Geometry,

O.Liess-L.Rodino [1] : Inhomogeneous Gevrey classes and related pseudodifferential operators,Boll.U.M.I.,Serie VI,III-C,1984,233-323,

[2] : Fourier integral operators in inhomogeneous Gevrey classes, to appear,

C.Parenti-F.Segala [1]: Propagation and reflection of singularities for a class of evolution equations,C.P.D.E.,6,1981,741-742,

P.Schapira [1] : Propagation at the boundary and reflection of analytic singularities ...,Publ.R.I.M.S.,Kyoto Univ.,12 Suppl.,1977,441-453,

F.Treves [1] : Introduction to pseudodifferential and Fourier integral operators, Plenum Press,New York and London,1980,

M.Sato [1] : Hyperfunctions and partial differential equations,Proc.Int.Conf. Funct.Anal.Rel.Topics,Univ.of Tokyo Press,Tokyo 1970,91-94.

Institut für Angewandte Mathematik
Universität Bonn
5300 Bonn,W.Germany.

Estimates for the Norm of Pseudo-differential Operators by Means of Besov Spaces

Tosinobu MURAMATU
Dept. Math., University of Tsukuba
Sakura-mura Niihari-gun Ibaraki, 305 Japan

Introduction.

In this paper we shall explain that our theory of Besov spaces, in particular characterization by "regularization of distribution", is one of very useful tools to study boundedness of pseudo-differential operators. Adopting this method, we can improve the regularity assumption on symbols to the extreme.

For example, consider Cordes' results ([4], Theorem B_1');

"If the derivatives $\partial_x^\alpha \partial_\xi^\beta a(x,\xi)$ of a symbol $a(x,\xi)$ exist and bounded for $|\alpha| \leq [n/2] + 1$, $|\beta| \leq [n/2] + 1$, then the operator $a(X,D_x)$ corresponding to $a(x,\xi)$ is L_2-bounded."

Here $[\lambda]$ means the integral part of a real number λ. As pointed out by Miyachi [11], from Theorem D in [4] we obtain the more precise results; that is, the operator $a(X,D_x)$ is L_2-bounded if

$$\begin{aligned} &|\partial_x^\alpha \partial_\xi^\beta a(x,\xi)| \leq C \text{ for } |\alpha| \leq [n/2],\ |\beta| \leq [n/2], \\ &|\Delta_y \partial_x^\alpha \partial_\xi^\beta a(x,\xi)| \leq C|y|^\theta \text{ for } |\alpha| = [n/2],\ |\beta| \leq [n/2], \\ &|\Delta_\eta \partial_x^\alpha \partial_\xi^\beta a(x,\xi)| \leq C|\eta|^\theta \text{ for } |\alpha| \leq [n/2],\ |\beta| = [n/2], \\ &|\Delta_y \Delta_\eta \partial_x^\alpha \partial_\xi^\beta a(x,\xi)| \leq C|y|^\theta|\eta|^\theta \text{ for } |\alpha| = [n/2],\ |\beta| = [n/2], \end{aligned} \tag{1}$$

with $\theta > n/2 - [n/2]$, where Δ_y and Δ_η denote the difference operators with respect to x and ξ, respectively. But the condition $\theta > n/2 - [n/2]$ can not be replaced by $\theta = n/2 - [n/2]$ in view of an example $a(x,\xi) = e^{ix\xi-|x|^2}(1 + |\xi|^2)^{-n/4}$ due to Coifman-Meyer [3]. In connection with these results a question arise; Does the regularity of order n/2 in some sense on symbols guarantee boundedness of the operators ? The answer is "yes" if we measure regularity of symbols by Besov type norms, that is,

" If $a(x,\xi) \in B^{(n/2,n/2)}_{(\infty,\infty),(1,1)}(\mathbb{R}^n_x \cdot \mathbb{R}^n_\xi)$, then $a(X,D_x)$ is L_2-bounded."

(Theorem 3.1). In the above notation, the superscript figures indicate the order of regularity in x and ξ, the subscript ∞ indicates the maximum norm, and the subscript 1 is the subexponent which is proper to Besov spaces. Notice that the space of symbols satisfying the condition (1) is expressed as $B^{(\kappa,\kappa)}_{(\infty,\infty),(\infty,\infty)}$ with $\kappa = [n/2] + \theta$ in our terminology, and generally $B^{(\sigma,\sigma)}_{(\infty,\infty),(\infty,\infty)} \subsetneq B^{(\tau,\tau)}_{(\infty,\infty),(1,1)} \subsetneq B^{(\tau,\tau)}_{(\infty,\infty),(\infty,\infty)}$ if $\sigma > \tau$ (see Definition 1.2 and § 1.6).

In § 4.1 we introduce the symbol space $S^{\mu}_{\rho,\delta}B^{(\sigma,\tau)}_{\mathbb{p},\mathbb{q}}(\mathbb{R}^n_x\cdot\mathbb{R}^n_\xi)$ which is a Besov space version of symbol class $S^m_{\rho,\delta}$ in the sense of Hörmander [6]. By making use of this notion we can state our main result;

"If $a(x,\xi) \in S^0_{\rho,\delta}B^{(\lambda,n/2)}_{(\infty,\infty),(1,1)}(\mathbb{R}^n_x\cdot\mathbb{R}^n_\xi)$, $0 \le \delta \le \rho < 1$, $\lambda = \frac{1-\rho}{1-\delta}\cdot\frac{n}{2}$, then $a(X,D_x)$ is L_2-bounded." (Theorem 4.7)

This result is a precision of the theorem in Kato [8] which states (with some improvement by Coifman-Meyer [3]) that

"If $|\partial^\alpha_x\partial^\beta_\xi a(x,\xi)| \le C(1 + |\xi|)^{\rho|\alpha|-\rho|\beta|}$ for $|\alpha| \le k$, $|\beta| \le k$, (2)
$k = [n/2] + 1$, $0 \le \rho < 1$, then $a(X,D_x)$ is L_2-bounded."

In our terminology condition (2) is identical with $a(x,\xi) \in S^0_{\rho,\rho}W^{(k,k)}_{(\infty,\infty)}$. We shall also show a precision of a result of Wang-Li [28] for classical symbols (Theorem 5.3).

There are several methods to treat Besov spaces. Well-known one is that of Fourier transform (see Peetre [22], Triebel [26], [27]), but our method is different from that. In stead of Fourier transform we adopt regularization of distributions, and instead of dyadic partition of unity in the space of variables for Fourier transform we make use of the integral representation formulas (1.4.5) and (1.4.7). We shall report in this paper that these formulas, which are essential in our theory for Besov spaces (see [13] ∿ [15]), are very much fit for studying boundedness of pseudo-differential operators. As a result, our theory is able to include the case where values of symbols are operators in Hilbert spaces.

Recently Sugimoto [24] has proved a theorem including Theorem 3.1 for scalar-valued symbols. He used theory of Besov spaces similar to Peetre [22] and Triebel [26], [27]. His result also include an improvement of Childs [2] and Coifman-Meyer [3] p.18 Théorèm 3 which is not included Theorem 3.1, but we are able to get the same result by using our method. Sugimoto also has proved an L_p-boundedness result by means of Besov spaces which is an improvement of Miyachi [11].

We shall deal only L_2-boundedness in the present paper.

Notations. For typographical convenience we shall freqently use the notation $\|x:X\| = \|x\|_X$, the (quasi-)norm of x in the space X.

For a quasi-normed space X, $0 < p \leq \infty$, and a measure space (Ω,μ), by $L_p(\Omega,\mu;X)$ we denote the space of X-valued strongly measurable functions f(x) such that

$$\|f:L_p(\Omega,\mu;X)\| = \begin{cases} \{\int_\Omega \|f(x)\|_X^p d\mu(x)\}^{1/p} < \infty & , \text{ if } p < \infty, \\ \text{ess.sup}\|f(x)\|_X < \infty & , \text{ if } p = \infty. \end{cases} \qquad (3)$$

And for a measurable set Ω in the real n-dimensional space $\mathbb{R}^n$ we adopt the abbreviations $L_p(\Omega;X) = L_p(\Omega,\mu;X)$ with μ = the Lebesgue measure, and $L_p^*(\Omega;X) = L_p(\Omega,\mu;X)$ with $d\mu(x) = |x|^{-n}dx$. We shall abbreviate as $\|f\|_p = \|f:L_p(\mathbb{R}^n)\|$, and use the notation $đ\xi = (2\pi)^{-n}d\xi$ for $\xi \in \mathbb{R}^n$.

$\mathcal{L}(X,Y)$ = the space of all bounded linear operators from X to Y. For multi-index $\alpha = (\alpha_1,\ldots,\alpha_n)$ and $x = (x_1,\ldots,x_n) \in \mathbb{R}^n$, $\partial_{x_j} = \partial/\partial x_j$, $x^\alpha = x_1^{\alpha_1}\cdots x_n^{\alpha_n}$, $\partial_x^\alpha = \partial_{x_1}^{\alpha_1}\cdots\partial_{x_n}^{\alpha_n}$, $\alpha! = \alpha_1!\cdots\alpha_n!$, $|\alpha| = \alpha_1 + \cdots + \alpha_n$.

$C_0^\infty(\mathbb{R}^n;X)$ denotes the space of all infinitely differentiable X-valued functions with compact support, $\mathcal{D}(\mathbb{R}^n) = C_0^\infty(\mathbb{R}^n)$, $\mathcal{D}'(\mathbb{R}^n;X) = \mathcal{L}(\mathcal{D}(\mathbb{R}^n),X)$ the space of all X-valued distributions, and $\mathcal{S}(\mathbb{R}^n;X)$ denotes the space of all rapidly decreasing, infinitely differentiable X-valued functions.

§ 1. Besov spaces of multiple orders.

1.1. In this paper j, k, ℓ and m denotes integer, σ and τ real numbers, Δ_y the difference operator with increment y for functions of x, and $\Delta_{y'}$ denotes that with increment y' for functions of x'. Thus for a function f(x,x') of (x,x') and positive integers ℓ and m

$$\Delta_y^\ell \Delta_{y'}^m f(x,x') = \sum_{j=0}^{\ell}\sum_{k=0}^{m}\binom{\ell}{j}\binom{m}{k}(-1)^{\ell+m-j-k}f(x+jy,x'+ky').$$

Definition. Let X be a quasi-normed space, Ω an open set in $\mathbb{R}^n$, and let $0 < p, q \leq \infty$.

For $k \geq 0$ $W_p^k(\Omega;X)$ consists of all functions whose all derivatives of order up to k belong to $L_p(\Omega;X)$, and its quasi-norm is defined by

$$\|f:W_p^k(\Omega;X)\| = \sum_{|\alpha|\leq k} \|\partial_x^\alpha f:L_p(\Omega;X)\|. \qquad (1)$$

For $k < 0$ $W_p^k(\Omega;X)$ consists of all distributions f of the form

$$f(x) = \sum_{|\alpha|\leq -k}\partial_x^\alpha f_\alpha(x), \qquad f_\alpha \in L_p(\Omega;X), \qquad (2)$$

and its quasi-norm is defined by

$$\|f:W_p^k(\Omega;X)\| = \inf \sum_{|\alpha|\leq -k}\|f_\alpha:L_p(\Omega;X)\|, \qquad (3)$$

where the infimum is taken over all possible representations of the form (2). We write as $W_p^{-\infty}(\Omega;X) = \bigcup_k W_p^k(\Omega;X)$.

The Besov space $B^{\sigma}_{p,q}(\Omega;X)$ is defined as follows. Let k be the largest integer and m the smallest integer such that $k < \sigma < k+m$.

When $k \geq 0$, the space $B^{\sigma}_{p,q}(\Omega;X)$ consists of all functions f in $W^k_p(\Omega;X)$ for which the quasi-seminorm

$$|f:B^{\sigma}_{p,q}(\Omega;X)| = \sum_{|\alpha|=k}\| |y|^{k-\sigma}\{\|\Delta^m_y\partial^\alpha_x f(x):L_p(\Omega_{m,y};X)\|\}:L^*_q(\mathbb{R}^n)\| \quad (4)$$

is finite, and its quasi-norm is defined by

$$\|f:B^{\sigma}_{p,q}(\Omega;X)\| = \|f:W^k_p(\Omega;X)\| + |f:B^{\sigma}_{p,q}(\Omega;X)|, \quad (5)$$

where $\Omega_{m,y} = \Omega \cap \ldots \cap (\Omega - my) = \{x;\ x+jy \in \Omega \text{ for } j=0,\ldots,m\}$.

When $k < 0$, $B^{\sigma}_{p,q}(\Omega;X)$ consists of all distributions f of the form

$$f(x) = \sum_{|\alpha|\leq -k}\partial^\alpha_x f_\alpha(x), \qquad f_\alpha \in B^{\sigma-k}_{p,q}(\Omega;X), \quad (6)$$

and its quasi-norm is defined by

$$\|f:B^{\sigma}_{p,q}(\Omega;X)\| = \inf \sum_{|\alpha|\leq -k}\|f_\alpha:B^{\sigma-k}_{p,q}(\Omega;X)\|, \quad (7)$$

where the infimum is taken over all possible representations of the form (6).

<u>1.2</u>. We define here the spaces of "multiple order".

<u>Definition</u>. Let X be a quasi-normed space, $0 < p_1,p_2,q,q_1,q_2 \leq \infty$, and let Ω and Ω' be open sets in $\mathbb{R}^n$ and in $\mathbb{R}^{n'}$, respectively.

We write as $W^{(j,k)}_{(p_1,p_2)}(\Omega\cdot\Omega';X) = W^j_{p_1}(\Omega;W^k_{p_2}(\Omega';X))$, in particular, $L_{(p_1,p_2)}(\Omega\cdot\Omega';X) = W^{(0,0)}_{(p_1,p_2)}(\Omega\cdot\Omega';X) = L_{p_1}(\Omega;L_{p_2}(\Omega';X)$, and also write as $B^{\sigma}_{p_1,q}W^k_{p_2}(\Omega\cdot\Omega';X) = B^{\sigma}_{p_1,q}(\Omega;W^k_{p_2}(\Omega';X))$, $W^j_{p_1}B^{\tau}_{p_2,q}(\Omega\cdot\Omega';X) = W^j_{p_1}(\Omega;B^{\tau}_{p_2,q}(\Omega';X))$, $B^{\sigma}_{p_1,q_1}B^{\tau}_{p_2,q_2}(\Omega\cdot\Omega';X) = B^{\sigma}_{p_1,q_1}(\Omega;B^{\tau}_{p_2,q_2}(\Omega';X))$.

Let $j \geq 0$, and let $\tau > 0$. Take the largest integer k and smallest positive integer m such that $k < \tau < k+m$. The Sobolev-Besov space $WB^{(j,\tau)}_{(p_1,p_2),q}(\Omega\cdot\Omega';X)$ consists of all functions f in $W^{(j,k)}_{(p_1,p_2)}(\Omega\cdot\Omega';X)$ for which the quasi-seminorm

$$|f:WB^{(j,\tau)}_{(p_1,p_2),q}(\Omega\cdot\Omega';X)| = \sum_{|\beta|=k}\| |y'|^{k-\tau}\{\|\Delta^m_{y'}\partial^\beta_{x'}f:W^{(j,0)}_{(p_1,p_2)}(\Omega\cdot\Omega'_{m,y'};X)\|\}:L^*_q(\mathbb{R}^n)\| \quad (1)$$

is finite, where $\Omega'_{m,y'} = \{x';\ x'+hy' \in \Omega',\ h = 0,\ldots,m\}$, and its quasi-norm is defined by

$$\|f:W^{(j,k)}_{(p_1,p_2)}(\Omega\cdot\Omega';X)\| + |f:WB^{(j,\tau)}_{(p_1,p_2),q}(\Omega\cdot\Omega';X)|. \quad (2)$$

Let $\sigma > 0$ and take the largest integer j and the smallest integer ℓ such that $j < \sigma < j+\ell$. The Besov space $B^{(\sigma,\tau)}_{(p_1,p_2),(q_1,q_2)}(\Omega\cdot\Omega';X)$

consists of all $f \in WB^{(j,\tau)}_{(p_1,p_2),q_2}(\Omega\cdot\Omega';X)$ for which the quasi-seminorm

$$|f:B^{(\sigma,\tau)}_{(p_1,p_2),(q_1,q_2)}(\Omega\cdot\Omega';X)| \tag{3}$$
$$= \sum_{|\alpha|=j}\| |y|^{j-\sigma}\{\|\Delta^{\ell}_{y}\partial^{\alpha}_{x}f(x,x'):WB^{(0,\tau)}_{(p_1,p_2),q_2}(\Omega_{\ell,y}\cdot\Omega';X)\|\}:L^{*}_{q_1}(\mathbb{R}^n_y)\|$$

is finite, and its quasi-norm is defined by

$$\|f:WB^{(j,\tau)}_{(p_1,p_2),q_2}(\Omega\cdot\Omega';X)\| + |f:B^{(\sigma,\tau)}_{(p_1,p_2),(q_1,q_2)}(\Omega\cdot\Omega';X)|. \tag{4}$$

For the case where j is negative, $\sigma \le 0$, or $\tau \le 0$ the spaces $WB^{(j,\tau)}_{(p_1,p_2),q}$ and $B^{(\sigma,\tau)}_{(p_1,p_2),(q_1,q_2)}$ are defined in the same way as the Sobolev space of negative order and Besov spaces of non-positive order.

1.3. Analogously we can define the space $WB^{(j_1,\ldots,j_\ell,\sigma_{\ell+1},\ldots,\sigma_m)}_{(p_1,\ldots,p_m),(q_{\ell+1},\ldots,q_m)}(\Omega_1\cdot\ldots\cdot\Omega_m;X)$ and the space $B^{(\sigma_1,\ldots,\sigma_m)}_{(p_1,\ldots,p_m),(q_1,\ldots,q_m)}(\Omega_1\cdot\ldots\cdot\Omega_m;X)$, where $j_1,\ldots,j_\ell$ are integers, $\sigma_1,\ldots,\sigma_m$ are real numbers, $0 < p_1,\ldots,p_m,q_1,\ldots,q_m \le \infty$, and $\Omega_1,\ldots,\Omega_m$ are open sets in $\mathbb{R}^{n_1},\ldots,\mathbb{R}^{n_m}$, respectively.

1.4. Integral representations. In the following by X we always denote a Banach space, and use the notations $\phi_t(x) = t^{-n}\phi(x/t)$ and

$$\phi*f(x) = \langle\phi(x-y),f(y)\rangle_y, \tag{1}$$

where the right-hand side is the duality on $\mathcal{D}(\mathbb{R}^n) \times \mathcal{D}'(\mathbb{R}^n;X)$.

Let ϕ be a C_0^∞-function, and set

$$\psi(x) = \sum_{j=1}^{n}\partial_{x_j}\{x_j\phi(x)\}. \tag{2}$$

Then we have for any f in $\mathcal{D}'(\mathbb{R}^n;X)$ and any positive number s

$$cf(x) = \int_0^s \psi_t*f(x)\frac{dt}{t} + \phi_s*f(x), \quad c = \int\phi(x)dx, \tag{3}$$

which is a special case of the formula in [14] Lemma 3.1. Here the integral with respect t converges in the topology of $\mathcal{D}'$.

Now, let $e_1 \in C_0^\infty$ such that $\int e_1(x)dx = 1$, and $m > 0$. Putting $e_\alpha(x) = x^\alpha e_1(x)/\alpha!$ and

$$e_m(x) = \sum_{|\alpha|<m}\partial_x^\alpha e_\alpha(x), \quad M(x) = \sum_{|\alpha|=m} m\partial_x^\alpha e_\alpha(x), \tag{4}$$

by (3) we have the first integral representation formula:

$$f(x) = \int_0^s M_t*f(x)\frac{dt}{t} + e_{m,s}*f(x). \tag{5}$$

Making use of this formula twice, partial integration and the fact that for non-negative integer k less than m, we can express as

$$M(x) = \sum_{|\beta|=k}\partial_x^\beta M_\beta(x), \tag{6}$$

we have the second integral representation formula;

$$f(x) = \sum_{|\alpha|=k}\int_0^s M_{\alpha,t}*u^{\alpha}(t,x)\frac{dt}{t} + \sum_{|\beta|=h}\int_0^s M_t^{(\beta)}*u_{\beta}(t,x)\frac{dt}{t} + \{e_m + e_{\ell} - e_m*e_{\ell}\}_s*f(x), \tag{7}$$

where ℓ is a positive integer, h a non-negative integer less than ℓ,

$$L(x) = \sum_{|\beta|=\ell}\partial_x^{\beta}e_{\beta}(x) = \sum_{|\beta|=h}\partial_x^{\beta}L_{\beta}(x), \tag{8}$$

$$u^{\alpha}(t,x) = \int_t^s\left(\frac{t}{t'}\right)^k L_{t'}^{(\alpha)}*f(x)\frac{dt'}{t'}, \tag{9}$$

and

$$u_{\beta}(t,x) = \int_0^t\left(\frac{t'}{t}\right)^h L_{\beta,t'}*f(x)\frac{dt'}{t'}. \tag{10}$$

Here $L^{(\alpha)}(x) = \partial_x^{\alpha}L$. (see pp.329-350 in [14] or pp.219-231 in [15]).

<u>1.5</u>. In the folowing we always assume that $1 \le p,p_1,p_2,q,q_1,q_2 \le \infty$, and $s > 0$.

<u>Theorem</u> (Characterization of Besov spaces). Let ℓ and m be non-negative integers such that $\sigma < \ell$, $\tau < m$, and set $I = (0,s]$.

(a) A distribution $f \in B^{\sigma}_{p,q}(\mathbb{R}^n;X)$ if and only if $\phi*f \in L_p(\mathbb{R}^n;X)$ for any $\phi \in C_0^{\infty}$, and $t^{-\sigma}\phi_t*f \in L_q^*(I;L_p(\mathbb{R}^n;X))$ for any ϕ of the form

$$\phi(x) = \sum_{|\alpha|=\ell}\partial_x^{\alpha}\phi_{\alpha}(x), \qquad \phi_{\alpha} \in C_0^{\infty}. \tag{1}$$

(b) Let $\tau > 0$. Then a distribution $f \in WB^{(j,\tau)}_{(p_1,p_2),q}(\mathbb{R}^n_x\cdot\mathbb{R}^{n'}_{x'};X)$ if and only if $\phi\overset{x'}{*}f \in W^{(j,0)}_{(p_1,p_2)}(\mathbb{R}^n_x\cdot\mathbb{R}^{n'}_{x'};X)$ and $t^{-\tau}\psi_t\overset{x'}{*}f \in L_q^*(I;W^{(j,0)}_{(p_1,p_2)}(\mathbb{R}^n_x\cdot\mathbb{R}^{n'}_{x'};X))$ for any $\psi = \sum_{|\alpha|=m}\partial_{x'}^{\alpha}\psi_{\alpha}$, $\psi_{\alpha} \in C_0^{\infty}(\mathbb{R}^{n'})$.

Here $\phi\overset{x'}{*}f(x,x') = \langle\phi(x'-y'),f(x,y')\rangle_{y'}$.

(c) Let $\sigma, \tau > 0$. Then a distribution $f \in B^{(\sigma,\tau)}_{(p_1,p_2),(q_1,q_2)}(\mathbb{R}^n_x\cdot\mathbb{R}^{n'}_{x'};X)$ if and only if $\phi\overset{x}{*}f \in WB^{(0,\tau)}_{(p_1,p_2),q_2}(\mathbb{R}^n_x\cdot\mathbb{R}^{n'}_{x'};X)$ for any $\phi \in C_0^{\infty}(\mathbb{R}^n)$, and $t^{-\sigma}\phi_t\overset{x}{*}f \in L^*_{q_1}(I;WB^{(0,\tau)}_{(p_1,p_2),q_2}(\mathbb{R}^n_x\cdot\mathbb{R}^{n'}_{x'};X))$ for any ϕ of the form (1).

Here $\phi\overset{x}{*}f(x,x') = \langle\phi(x-y),f(y,x')\rangle_y$.

<u>Proof</u>. See [14] or [15] Chapter 10.

<u>1.6</u>. From Theorem 1.5 and its proof we also get the following facts:

(a) $B^{\sigma}_{p,q}(\mathbb{R}^n;X) \subset B^{\tau}_{p,r}(\mathbb{R}^n;X)$, if $q \le r$ and $\sigma \le \tau$ or if $\sigma < \tau$.

(b) Let $0 < \sigma < m$. Then $f \in B^{\sigma}_{p,q}(\mathbb{R}^n;X)$ if and only if

$$f \in W_p^{-\infty}(\mathbb{R}^n;X), \quad\text{and}\quad |y|^{-\sigma}\Delta_y^m f(x) \in L_q^*(\mathbb{R}^n;L_p(\mathbb{R}^n;X)).$$

(c) Let $m \ge 0$. Then, $f \in B^{\sigma}_{p,q}(\mathbb{R}^n;X))$ if and only if

$$f(x) = \sum_{|\alpha|\le m}\partial_x^{\alpha}f_{\alpha}(x), \quad f_{\alpha} \in B^{\sigma+m}_{p,q}(\mathbb{R}^n;X).$$

(d) Let $m \geq 0$. Then, $f \in B^{\sigma}_{p,q}(\mathbb{R}^n;X))$ if f and only if

$$f \in W^{-\infty}_p(\mathbb{R}^n;X) \text{ and } \partial^\alpha f \in B^{\sigma-m}_{p,q}(\mathbb{R}^n;X)) \text{ for any } \alpha \text{ with } |\alpha| = m.$$

(e) Take a non-negative integer h and non-negative integer k so that $-h < \sigma < k$, and let $\ell = k+h$, $I = (0,s]$. Then the norm of the space $B^{\sigma}_{p,q}(\mathbb{R}^n;X)$ is equivalent with

$$\sum_{|\alpha|=k}\|t^{-\sigma}L^{(\alpha)}_t * f : L^*_q(I;L_p(\mathbb{R}^n;X))\|$$
$$+ \sum_{|\beta|=h}\|t^{-\sigma}L_{\beta,t} * f : L^*_q(I;L_p(\mathbb{R}^n;X))\| + \|\psi_s * f : W^m_p(\mathbb{R}^n;X)\|,$$

where e_ℓ, L and L_β are as in § 1.5, and $\psi = 2e_\ell - e_\ell * e_\ell$.

(f) Let $m > \sigma > 0$. Then the norm of $B^{(\sigma,\tau)}_{(p_1,p_2),(q_1,q_2)}(\mathbb{R}^n_x \cdot \mathbb{R}^{n'}_{x'};X)$ is equivalent with the norm

$$(1+s^\sigma)\{\sum_{|\alpha|=0,m}\| \| t^{-\sigma}M^{(\alpha)}_t \overset{x}{*} f(x,x')\|_\tau : L^*_{q_1}(I)\| + s^{-\sigma}\|f\|_\tau\},$$

where $\|f\|_\tau = \|f : WB^{(0,\tau)}_{(p_1,p_2),q_2}(\mathbb{R}^n_x \cdot \mathbb{R}^{n'}_{x'} : X)\|$, and M is given by (1.4.4).

(g) $B^j_{p,1} \subsetneq W^j_p \subsetneq B^j_{p,\infty}$, $B^{(j,k)}_{(p_1,p_2),(1,1)} \subsetneq W^{(j,k)}_{(p_1,p_2)} \subsetneq B^{(j,k)}_{(p_1,p_2),(\infty,\infty)}$.

<u>1.7</u>. Here we discuss multiplier in Besov spaces. The following results are not best one, but they are sufficient to our purpose.

<u>Theorem</u>. Assume that $\ell \geq |j|$ or $\ell > |\sigma|$, and that $m \geq |k|$ or $m > |\tau|$.

(a) If $1 \leq \tilde{p} \leq p \leq \infty$, and if $1/p + 1/r = 1/\tilde{p}$, then multiplication by a function $g \in W^\ell_r(\mathbb{R}^n)$ is a bounded operator from $W^j_p(\mathbb{R}^n;X)$ into $W^j_{\tilde{p}}(\mathbb{R}^n;X)$, and from $B^\sigma_{p,q}(\mathbb{R}^n;X)$ into $B^\sigma_{\tilde{p},q}(\mathbb{R}^n;X)$, respectively, whose norm does not exceed $C\|g;W^\ell_r\|$.

(b) If $1 \leq \tilde{p}_i \leq p_i \leq \infty$, and if $1/p_i + 1/r_i = 1/\tilde{p}_i$ for $i = 1,2$, then multiplication by a function $g \in W^{(\ell,m)}_{(r_1,r_2)}(\mathbb{R}^n \cdot \mathbb{R}^{n'})$ is a bounded operator

from $W^{(j,k)}_{(p_1,p_2)}(\mathbb{R}^n \cdot \mathbb{R}^{n'};X)$ into $W^{(j,k)}_{(\tilde{p}_1,\tilde{p}_2)}(\mathbb{R}^n \cdot \mathbb{R}^{n'};X)$,

from $WB^{(j,\tau)}_{(p_1,p_2),q}(\mathbb{R}^n \cdot \mathbb{R}^{n'};X)$ into $WB^{(j,\tau)}_{(\tilde{p}_1,\tilde{p}_2),q}(\mathbb{R}^n \cdot \mathbb{R}^{n'};X)$,

and from $B^{(\sigma,\tau)}_{(p_1,p_2),(q_1,q_2)}(\mathbb{R}^n \cdot \mathbb{R}^{n'};X)$ into $B^{(\sigma,\tau)}_{(\tilde{p}_1,\tilde{p}_2),(q_1,q_2)}(\mathbb{R}^n \cdot \mathbb{R}^{n'};X)$,

respectively, whose norm does not exceed $C\|g : W^{(\ell,m)}_{(r_1,r_2)}\|$. Here, C is a constant independent of ϕ.

<u>Proof</u>. Case W^j_p. If $j = 0$, then Hölder's inequality gives the result. For the case $j > 0$ the result follows from the Leibniz formula and that for the case $j = 0$. For the case $j = -h < 0$ the result follows from the adjoint Leibniz formula

$$g\sum_{|\alpha|\le h}\partial_x^\alpha f_\alpha = \sum_{|\beta|\le h}\partial_x^\beta\{\sum_{|\alpha|\le h,\ \alpha\ge\beta}\binom{\alpha}{\beta}(-1)^{|\alpha-\beta|}(\partial_x^{\alpha-\beta}g)f_\alpha\}. \tag{1}$$

Case $B^\sigma_{p,q}$. Let j be the largest integer and h the least integer such that $j < \sigma < j + h$. First consider the case $j = 0$. From the Leibniz formula for the difference operator, i.e.

$$\Delta_y^{2h}\{g(x)f(x)\} = \sum_{i=0}^{2h}\binom{2h}{i}\Delta_y^{2h-i}g(x+hy)\Delta_y^i f(x), \tag{2}$$

Hölder's inequality and the inequality

$$\|\Delta_y^k f(x):L_p(\mathbb{R}^n_x;X)\| \le 2^{k-h}\|\Delta_y^h f(x):L_p(\mathbb{R}^n_x;X)\| \text{ for } k \ge h, \tag{3}$$

we have

$$\begin{aligned}&\|\Delta_y^{2h}\{g(x)f(x)\}:L_{\tilde p}(\mathbb{R}^n_x;X)\|\\ &\le 2^{3h-1}\{\|g\|_r\|\Delta_y^h f(x):L_p(\mathbb{R}^n_x;X)\| + \|\Delta_y^h g(x):L_r(\mathbb{R}^n_x)\|\cdot\|f(x):L_p(\mathbb{R}^n_x;X)\|\}.\end{aligned} \tag{4}$$

Taking $L^*_q(\mathbb{R}^n_y)$-norm after multiplying the both sides by $|y|^{-\sigma}$, we get

$$\|gf:B^\sigma_{\tilde p,q}(\mathbb{R}^n;X)\| \le C\|g:W^h_r\|\cdot\|f:B^\sigma_{p,q}(\mathbb{R}^n;X)\|. \tag{5}$$

The case $j > 0$ and the case $j < 0$ can be reduced to the case $j = 0$ by making use of the Leibniz formula and (1), respectively.

Thus Part (a) is proved, and Part (b) can be proved analogously.

§ 2. Estimate for operators with symbols decreasing at infinity.

We begin by discussing symbols which decrease at infinity with respect to ξ. In this section we always assume that X and Y are Hilbert spaces.

2.1. Our starting results is the following

Theorem. Let $1 \le r \le 2 \le p \le \infty$, and assume that $a(x,\xi)$ belongs to the space $\Sigma_r = WB^{(0,n/r)}_{(\infty,r),1}(\mathbb{R}^n_x\cdot\mathbb{R}^n_\xi;\mathcal{L}(X,Y))$.

Then, the operator A with symbol $a(x,\xi)$ is bounded from $L_p(\mathbb{R}^n;X)$ to $L_p(\mathbb{R}^n;Y)$, and the correspondence $a \to A$ is bounded from the space Σ_r into $\mathcal{L}(L_p(\mathbb{R}^n;X),L_p(\mathbb{R}^n;Y))$.

Proof. The identity (1.4.7) and Theorem 1.5 imply that $a(x,\xi)$ can be expressed as a finite sum of symbols of the following types;

First type.

$$a_1(x,\xi) = \phi \overset{\xi}{*} b(x,\xi),\quad \phi\in C_0^\infty,\ b(x,\xi)\in L_{(\infty,r)}(\mathbb{R}^n_x\cdot\mathbb{R}^n_\xi;\mathcal{L}(X,Y)), \tag{1}$$

$$\|b:L_{(\infty,r)}(\mathbb{R}^n_x\cdot\mathbb{R}^n_\xi;\mathcal{L}(X,Y))\| \le C\|a:L_{(\infty,r)}(\mathbb{R}^n_x\cdot\mathbb{R}^n_\xi;\mathcal{L}(X,Y)\|. \tag{2}$$

Second type.

$$\begin{aligned}&a_2(x,\xi) = \int_0^1 t^{n/r}\phi_t \overset{\xi}{*} b(t,x,\xi)\frac{dt}{t},\qquad \phi\in C_0^\infty,\\ &\qquad b(t,x,\xi)\in L_1^*(I;L_{(\infty,r)}(\mathbb{R}^n_x\cdot\mathbb{R}^n_\xi;\mathcal{L}(X,Y)),\ I=(0,1],\end{aligned} \tag{3}$$

$$\| b(t,x,\xi):L_1^*(I;L_{(\infty,r)}(\mathbb{R}_x^n\cdot\mathbb{R}_\xi^n;\mathcal{L}(X,Y)))\| \le C\|a\|_{\Sigma_r}. \tag{4}$$

Therefore, to prove the theorem it suffices to consider the operators with symbols of these types.

2.2. For convenience of later use we formulate here the result on symbols of the first type.

Lemma. If the symbol of A_1 is of the form (2.1.1), then

$$\| A_1:\mathcal{L}(L_p(\mathbb{R}^n;X),L_p(\mathbb{R}^n;Y))\| \le c_{n,r}\|g\|_r\|b:L_{(\infty,r)}(\mathbb{R}_x^n\cdot\mathbb{R}_\xi^n;\mathcal{L}(X,Y))\|, \tag{1}$$

where g is the inverse Fourier transform of ϕ.

Proof. For any function u in $\mathcal{S}(\mathbb{R}^n;X)$ we have

$$A_1u(x) = \int a_1(x,\xi)d\xi\int e^{i(x-y)\xi}u(y)dy = \int b(x,\eta)d\eta\int e^{iy\eta}g(y)u(x-y)dy. \tag{2}$$

Hence, by Hölder's inequality and the Hausdorff-Young theorem we get

$$\| A_1u(x)\|_Y \le c_{n,r}\|b\|\cdot\|g(y)u(x-y):L_r(\mathbb{R}_y^n;X)\|, \tag{3}$$

where $\|b\| = \|b(x,\xi):L_{(\infty,r)}(\mathbb{R}_x^n\cdot\mathbb{R}_\xi^n;\mathcal{L}(X,Y))\|$, which implies that

$$\| A_1u:L_p(\mathbb{R}^n;Y)\| \le c_{n,r}\|b\|\cdot\|g\|_r\|u:L_p(\mathbb{R}^n;X)\|.$$

2.3. Proof of Theorem 2.1. Part (ii). Let A_2 be the operator whose symbol is of the form (2.1.3), and g is the inverse Fourier transform of ϕ. Since the inverse Fourier transform of ϕ_t is equal to g(tx), it follows from (2.2.3) that

$$\| A_2u(x)\|_Y \le c_{n,r}\int_0^1 t^{n/r}v(t)\| g(ty)u(x-y):L_r(\mathbb{R}_y^n;X)\|\frac{dt}{t}, \tag{1}$$

where $v(t) = \| b(t,x,\xi):L_{(\infty,r)}(\mathbb{R}_x^n\cdot\mathbb{R}_\xi^n;\mathcal{L}(X,Y))\|$, and this shows that

$$\| A_2u:L_p(\mathbb{R}^n;Y)\| \le c_{n,r}\|b\|\cdot\| g\|_r\|u;L_p(\mathbb{R}^n;X)\|, \tag{2}$$

where $\|b\| = \| b(t,x,\xi):L_1^*(I;L_{(\infty,r)}(\mathbb{R}^{2n};\mathcal{L}(X,Y))\|$.

2.4. Next we shall discuss the continuity as operators from L_p into Besov spaces.

Theorem. Let $1 \le r \le 2 \le p \le \infty$, $a(x,\xi)$ the symbol of an operator A , h the least non-negative integer such that $\sigma > -h$, and let k be a non-negative integer such that $\sigma + 2h < k$. Assume that $a(x,\xi)\xi^\alpha \in B^{(\sigma,n/r)}_{(\infty,r),(q,1)}(\mathbb{R}_x^n\cdot\mathbb{R}_\xi^n;\mathcal{L}(X,Y))$ for any α with $|\alpha| \le k$.

Then, A is bounded from $L_p(\mathbb{R}^n;Y)$ to $B^\sigma_{p,q}(\mathbb{R}^n;Y)$, and its norm is not greater than $C\sum_{|\alpha|\le k}\| a(x,\xi)\xi^\alpha:B^{(\sigma,n/r)}_{(\infty,r),(q,1)}\|$. In particular, if σ is positive, then

$$\begin{aligned}\|\phi_t*Au:L_p(\mathbb{R}^n;Y)\| \le C\{&\textstyle\sum_{|\alpha|<k}t^{|\alpha|}\|\phi_{\alpha,t}\overset{x}{*}a(x,\xi)\xi^\alpha:WB^{(0,n/r)}_{(\infty,r),1}\| \\ &+ \textstyle\sum_{|\alpha|=k}t^k\|\phi_\alpha\|_1\|a(x,\xi)\xi^\alpha:WB^{(0,n/r)}_{(\infty,r),1}\|\}\|u:L_p(\mathbb{R}^n;X)\|\end{aligned} \tag{1}$$

for any $\phi \in C_0^\infty$, where $\phi_\alpha(x) = x^\alpha\phi(x)/(\alpha!)$, and C is a constant independent of a, u, t and ϕ.

Proof. Case h = 0, i.e. $k > \sigma > 0$. With the aid of the Taylor formula we get

$$\begin{aligned}\phi_t{*}Au(x) &= \int\phi(x')dx'\int e^{ix\xi}e^{-itx'\xi}a(x-tx',\xi)\hat{u}(\xi)d\xi \\ &= \sum_{|\alpha|<k}(-t)^{|\alpha|}\int e^{ix\xi}a_\alpha(t,x,\xi)\hat{u}(\xi)d\xi \\ &+ \sum_{|\alpha|=k}k\int_0^t(-1)^k t'^{k-1}dt'\int\phi_\alpha(x')V(t,t',x,x')dx',\end{aligned}$$

where

$$a_\alpha(t,x,\xi) = \int\phi_\alpha(x')a(x-tx',\xi)\xi^\alpha dx' = \phi_{\alpha,t}\overset{x}{*}a(x,\xi)\xi^\alpha, \tag{2}$$

$$V(t,t',x,x') = \int e^{i\{x-(t-t')x'\}\xi}a(x-tx',\xi)\xi^\alpha\hat{u}(\xi)d\xi. \tag{3}$$

On the other hand, it follows from Theorem 2.1 that

$$\left\|\int e^{ix\xi}a_\alpha(t,x,\xi)\hat{u}(\xi)d\xi : L_p(\mathbb{R}^n;Y)\right\| \leq C\|a_\alpha(t,x,\xi)\|\cdot\|u:L_p(\mathbb{R}^n;X)\|,$$

and that, with the aid of the change of variables $x - (t-t')x' \to x$,

$$\begin{aligned}\|V(t,t',x,x'):L_p(\mathbb{R}_x^n;Y)\| &= \left\|\int e^{ix\xi}a(x-t'x',\xi)\xi^\alpha\hat{u}(\xi)d\xi:L_p(\mathbb{R}^n;Y)\right\| \\ &\leq C\|a\|\cdot\|u:L_p(R^n;X)\|,\end{aligned}$$

which gives (1). Here $\|a\| = \|a:WB^{(0,n/r)}_{(\infty,r),1}(R^n\cdot R^n;\mathcal{L}(X,Y))\|$.

Now, multiply by $t^{-\sigma}$, and take the $L^*_q(I)$-norm of (1), where I = (0,s]. Noting that $\|t^{-\sigma}t^k:L^*_q(I)\| \leq Cs^ks^{-\sigma}$, we obtain that

$$\|t^{-\sigma}\phi_t{*}Au:L^*_q(I;L_p)\| \leq C\sum_{|\alpha|\leq k}\|a(x,\xi)\xi^\alpha:B^{(\sigma,n/r)}_{(\infty,r),(q,1)}\|\cdot\|u:L_p(R^n;X)\|.$$

Thus, in view of Theorem 1.5 (a), this and the fact that $Au \in L_p$, which is also a consequence of Theorem 2.1, give the conclusion.

Case h > 0. In this case $a(x,\xi)$ has an expression $a(x,\xi) = \sum_{|\alpha|\leq h}\partial_x^\alpha a_\alpha(x,\xi)$, $a_\alpha(x,\xi)\xi^\alpha \in B^{(\sigma+h,n/r)}_{(\infty,r),(q,1)}$ for any $|\alpha| \leq k$ (see Lemma 3.1 Corollary 2 in [14]). Let $\tilde{A}_\beta$ be the operator with symbol

$$\tilde{a}_\beta(x,\xi) = \sum_{|\alpha|\leq h,\ \alpha\geq\beta}\binom{\alpha}{\beta}a_\alpha(x,\xi)(-i\xi)^{\alpha-\beta}. \tag{4}$$

Then $Au = \sum_{|\beta|\leq h}\partial_x^\beta\{\tilde{A}_\beta u\}$, by which we can reduce the case h > 0 to the case h = 0.

§ 3 L_2 boundedness of operators with $S^0_{0,0}$ symbols.

In this section we denote by $\hat{u}$ the Fourier transform of u.

3.1. First we state our main results in this section:

Theorem. Let X and Y be Hilbert spaces, and assume that the symbol $a(x,\xi)$ of an operator A belongs to

$$B^{(n/2,n/2)}_{(\infty,\infty),(1,1)}(\mathbb{R}^n_x\cdot\mathbb{R}^n_\xi;\mathcal{L}(X,Y)). \tag{1}$$

Then A is bounded from $L_2(\mathbb{R}^n;X)$ into $L_2(\mathbb{R}^n;Y)$ and its norm is estimated by $C\|a\|$, where $\|a\|$ is the norm of a in the space (1).

Outline of the proof of the theorem. Take $g \in C_0^\infty$ such that $\int g(\xi)d\xi = 1$. Then we see that $Au(x)$ is equal to

$$\iiint e^{i(x-y)\xi}a(x,\xi)g(\xi-\eta)u(y)dyd\xi d\eta = \int e^{ix\eta}d\eta\int e^{ix\xi}a(\eta,x,\xi)\hat{v}(\eta,\xi)d\xi,$$

for $u \in \mathcal{S}(\mathbb{R}^n;X)$, where $a(\eta,x,\xi) = a(x,\xi+\eta)g(\xi)$, $v(\eta,x) = e^{-ix\eta}u(x)$, and $\hat{v}(\eta,\xi) = \int e^{-ix\xi}v(\eta,x)dx = \hat{u}(\xi+\eta)$. Hence the theorem follows from the following three statements:

(i) $u \to e^{i\eta x}u(x)$ is bounded from $L_2(\mathbb{R}^n;X)$ into $W_2^{-m}(\mathbb{R}^n_x;L_2(\mathbb{R}^n_\eta;X))$.

(ii) The operator $v \to \int e^{ix\xi}a(\eta,x,\xi)\hat{v}(\eta,x)d\xi$ is bounded from $W_2^{-m}(\mathbb{R}^n_x;L_2(\mathbb{R}^n_\eta;X))$ into $B^{n/2}_{2,1}(\mathbb{R}^n_x;L_2(\mathbb{R}^n_\eta;Y))$.

(iii) The operator $w \to \int e^{ix\eta}w(\eta,x)d\eta$ is bounded from $B^{n/2}_{2,1}(\mathbb{R}^n_x;L_2(\mathbb{R}^n_\eta;Y))$ into $L_2(\mathbb{R}^n;Y)$.

3.2. Part (i) and (iii) can be proved as follows:

Lemma. (a) The operator $u \to v(\eta,x) = e^{-ix\eta}u(x)$ is bounded from $L_2(\mathbb{R}^n;X)$ into $B^{-n/2}_{2,\infty}(\mathbb{R}^n_x;L_2(\mathbb{R}^n_\eta;X))$.

(b) The operator $w \to \int e^{ix\eta}w(\eta,x)d\eta$ is bounded from $B^{n/2}_{2,1}(\mathbb{R}^n_x;L_2(\mathbb{R}^n_\eta;X))$ into $L_2(\mathbb{R}^n;X)$.

Proof. Let $u \in \mathcal{S}(\mathbb{R}^n;X)$, and $\phi \in \mathcal{S}(\mathbb{R}^n)$. Since

$$\int e^{-ix\xi}dx\int \phi_t(x-y)e^{-iy\eta}u(y)dy = \iint e^{-ix\xi}\phi_t(x)e^{-iy(\xi+\eta)}u(y)dxdy = \hat{\phi}(t\xi)\hat{u}(\xi+\eta),$$

by the Parseval equality we have

$$\|t^{n/2}\phi_t\overset{x}{*}v(\eta,x):L_2(\mathbb{R}^n_x\cdot\mathbb{R}^n_\eta;X)\| = c_n\|\phi\|_2\|u:L_2(\mathbb{R}^n;X)\|.$$

With the aid of Theorem 1.5 (a) this gives Part (a).

It follows from the Parseval equality that

$$\begin{aligned}\left\|\int e^{ix\eta}\phi\overset{x}{*}v(\eta,x)d\eta:L_2(\mathbb{R}^n_x;X)\right\| &= c_n\left\|\int e^{-ix\xi}dx\int e^{ix\eta}\phi\overset{x}{*}v(\eta,x)d\eta:L_2(\mathbb{R}^n_\xi;X)\right\| \\ &= c_n\left\|\int\hat{\phi}(\xi-\eta)\hat{v}(\eta,\xi-\eta)d\eta:L_2(\mathbb{R}^n_\xi;X)\right\| \\ &\le C\|\phi\|_2\|v:L_2(\mathbb{R}^{2n};X)\|,\end{aligned}$$

which gives that

$$\begin{aligned}\left\|\int e^{ix\eta}d\eta\int_0^s\phi_t\overset{x}{*}u(t,\eta,x)\frac{dt}{t}:L_2(\mathbb{R}^n;X)\right\| &\le c_n\int_0^s\|\phi_t\|_2\|u(t,\eta,x):L_2(\mathbb{R}^{2n};X)\|\frac{dt}{t} \\ &= c_n\|\phi\|_2\|t^{-n/2}u(t,\eta,x):L_1^*(I;L_2(\mathbb{R}^n;X)\|.\end{aligned}$$

With the aid of Theorem 1.5 these two inequalities imply Part (b).

3.3. Proof of Part (ii) in § 3.1. Let $v \in W_2^{-m}(\mathbb{R}^n_x;L_2(\mathbb{R}^n_\eta;X))$, that is, $v(\eta,x) = \sum_{|\alpha|\le m}\partial_x^\alpha v_\alpha(\eta,x)$, $v_\alpha(\eta,x) \in L_2(\mathbb{R}^{2n};X)$. Then

$$\int e^{ix\xi}a(\eta,x,\xi)\hat{v}(x,\xi)d\xi = \sum_{|\alpha|\le m}\int e^{ix\xi}a(\eta,x,\xi)(i\xi)^\alpha\hat{v}_\alpha(\eta,\xi)d\xi.$$

Therefore, to prove Part (ii) it suffices to prove the boundedness of the operator with symbol $a(\eta,x,\xi)(i\xi^\alpha)$ from $L_2(\mathbb{R}^n;L_2(\mathbb{R}^n_\eta;X)$ into $B^{n/2}_{2,1}(\mathbb{R}^n_x;L_2(\mathbb{R}^n_\eta;Y))$, which follows from the fact that

$$a(\eta,x,\xi)(i\xi^\alpha) \in B^{(n/2,n/2)}_{(\infty,2),1}(\mathbb{R}^n_x\cdot\mathbb{R}^n_\xi;L_\infty(\mathbb{R}^n_\eta;\mathcal{L}(X,Y))), \qquad (1)$$

in view of Theorem 2.4, since $L_\infty(\mathbb{R}^n_\eta;\mathcal{L}(X,Y)) \subset \mathcal{L}(L_2(\mathbb{R}^n_\eta;X),L_2(\mathbb{R}^n_\eta;Y))$.

Now, the fact (1) follows from the assumptions and the following

Lemma. For a Banach space X and $g \in C_0^\infty$ the operator $a(x,\xi) \to a(x,\xi+\zeta)g(\xi)$ is bounded from

$$B^{(\sigma,\tau)}_{(\infty,\infty),(q_1,q_2)}(\mathbb{R}^n_x\cdot\mathbb{R}^n_\xi;X) \text{ into } B^{(\sigma,\tau)}_{(\infty,p),(q_1,q_2)}(\mathbb{R}^n_x\cdot\mathbb{R}^n_\xi;L_\infty(\mathbb{R}^n_\zeta;X)).$$

Proof. The lemma follows from Theorem 1.7 and the fact that the operator $a(x,\xi) \to a(x,\xi+\zeta)$ is bounded from

$$B^{(\sigma,\tau)}_{(\infty,\infty),(q_1,q_2)}(\mathbb{R}^n_x\cdot\mathbb{R}^n_\xi;X) \text{ into } B^{(\sigma,\tau)}_{(\infty,\infty),(q_1,q_2)}(\mathbb{R}^n_x\cdot\mathbb{R}^n_\xi;L_\infty(\mathbb{R}^n_\zeta;X)).$$

§ 4. Symbols of class $S^\mu_{\delta,\rho}$.

In the following we always assume that $0 \le \delta,\rho \le 1$, μ a real number, and by B we denote the unit ball in R^n.

4.1. First we introduce the class of symbols which is a Besov space version of symbol class $S^m_{\rho,\delta}$ in the sense of Hörmander [8].

Definition. Let $F^{(j)}$ be one of the spaces $W^{(j,k)}_{(p_1,p_2)}(\mathbb{R}^n_x\cdot\mathbb{R}^n_\xi;X)$, $W^j_{p_1}B^\tau_{p_2,q}(\mathbb{R}^n_x\cdot\mathbb{R}^n_\xi;X)$ and $WB^{(j,\tau)}_{(p_1,p_2),q}(\mathbb{R}^n_x\cdot\mathbb{R}^n_\xi;X)$, $F^\sigma_{q_1}$ one of the spaces $B^\sigma_{p_1,q_1}W^k_{p_2}(\mathbb{R}^n_x\cdot\mathbb{R}^n_\xi;X)$, $B^\sigma_{p_1,q_1}B^\tau_{p_2,q_2}(\mathbb{R}^n_x\cdot\mathbb{R}^n_\xi;X)$ and $B^{(\sigma,\tau)}_{(p_1,p_2),(q_1,q_2)}(\mathbb{R}^n_x\cdot\mathbb{R}^n_\xi;X)$, $F = F^{(0)}$, ϕ a fixed C^∞-function which is identically equal to 1 on B and vanishes outside 2B, $\psi(\xi) = \phi(\xi/2) - \phi(\xi)$, and let $I = (0,1]$.

For $j \ge 0$ $S^\mu_{\rho,\delta}F^{(j)}$ is the space of all symbols $a(x,\xi)$ satisfying the conditions that $\phi(\xi)a(x,\xi) \in F^{(j)}$ and $t^\mu\psi(t^{1-\rho}\xi)a(t^\delta x,t^{-\rho}\xi) \in L_\infty(I;F^{(j)})$, and its norm is defned by

$$\|a(x,\xi)\phi(\xi):F^{(j)}\| + \|t^\mu a(t^\delta x,t^{-\rho}\xi)\psi(t^{1-\rho})\xi):L_\infty(I;F^{(j)})\|. \qquad (1)$$

For $j < 0$ $S^\mu_{\rho,\delta}F^{(j)}$ is the space of all symbols $a(x,\xi)$ such that

$$a(x,\xi) = \sum_{|\alpha|\le -j}\partial_x^\alpha a_\alpha(x,\xi),\ a_\alpha(x,\xi) \in S^{\mu-\delta|\alpha|}_{\rho,\delta}F, \qquad (2)$$

and its norm is defined by

$$\|a:S^{\mu}_{\rho,\delta}F^{(j)}\| = \inf\textstyle\sum_{|\alpha|\leq -j}\|a_\alpha:S^{\mu-\delta|\alpha|}_{\rho,\delta}F\|. \tag{3}$$

For $\sigma > 0$ $S^{\mu}_{\delta,\rho}F^{\sigma}_{q_1}$ is the space of all symbols $a(x,\xi)$ satisfing the conditions that $t^{\mu}a(t^{\delta}x,t^{-\rho}\xi)\psi(t^{1-\rho}\xi) \in L_\infty(I;F^{(j)})$, $a(x,\xi)\phi(\xi) \in F^{\sigma}_{q_1}$ and $t^{\mu}|y|^{j-\sigma}\Delta^{\ell}_{y}\partial^{\alpha}_{x}a(t^{\delta}x,t^{-\rho}\xi)\psi(t^{1-\rho}\xi) \in L^{*}_{q_1}(\mathbb{R}^n;L_\infty(I;F))$ for all $|\alpha| = j$, and its norm is defined by

$$\begin{aligned}\|a:S^{\mu}_{\rho,\delta}F^{\sigma}_{q_1}\| &= \|a(x,\xi)\phi(\xi):F^{\sigma}_{q_1}\| + \|t^{\mu}a(t^{\delta}x,t^{-\rho}\xi)\psi(t^{1-\rho}\xi):L_\infty(I;F^{(j)})\| \\ &+ \textstyle\sum_{|\alpha|=j}\|t^{\mu}|y|^{j-\sigma}\Delta^{\ell}_{y}\partial^{\alpha}_{x}a(t^{\delta}x,t^{-\rho}\xi)\psi(t^{1-\rho}\xi):L^{*}_{q_1}(\mathbb{R}^n;L_\infty(I;F)\|,\end{aligned} \tag{4}$$

where j is the largest integer and ℓ is the smallest integer such that $j < \sigma < j+\ell$.

For $\sigma \leq 0$ $S^{\mu}_{\rho,\delta}F^{\sigma}_{q_1}$ is the space of all symbols of the form (2) with $a_\alpha \in S^{\mu-\delta|\alpha|}_{\rho,\delta}F^{\sigma-j}_{q_1}$ and its norm is defined by

$$\|a:S^{\mu}_{\rho,\delta}F^{\sigma}_{q_1}\| = \inf\textstyle\sum_{|\alpha|\leq -j}\|a_\alpha:S^{\mu-\delta|\alpha|}_{\rho,\delta}F^{\sigma-j}_{q_1}\|, \tag{5}$$

where j is the largest integer such that $j < \sigma$.

<u>4.2</u>. Let ϕ_0 be another function satisfying the condition in Definition 4.1. Then there is a C^{∞}_{0}-function f and g such that

$$\phi_0 = f(\phi + \psi), \quad \psi_0(\xi) = g(\xi)\{\psi(2\xi) + \psi(\xi) + \psi(\xi/2)\}, \tag{1}$$

since $\phi(\xi) + \psi(\xi) = 1$ on the support of ϕ_0 and $\psi(2\xi) + \psi(\xi) + \psi(\xi/2) = 1$ on the support of $\psi_0(\xi) = \phi_0(\xi/2) - \phi_0(\xi)$. With the aid of Theorem 1.7 we easily see that the spaces $S^{\mu}_{\delta,\rho}F^{(j)}$ and $S^{\mu}_{\delta,\rho}F^{\sigma}_{q_1}$ are independent of the choice of ϕ.

In the same way we can prove the following

<u>Lemma</u>. Let $F^{(j)}$ and $F^{\sigma}_{q_1}$ be as in Definition 4.1, $j \geq 0$, $\sigma > 0$, $g \in C^{\infty}_{0}$, and let f be C^{∞}_{0}-function vanishing near the origin.

(a) If $a(x,\xi) \in S^{\mu}_{\rho,\delta}F^{(j)}$, then $t^{\mu}a(t^{\delta}x,t^{-\rho}\xi)f(t^{1-\rho}\xi) \in L_\infty(I;F^{(j)})$, $a(x,\xi)g(\xi) \in F^{(j)}$, and their norms are estimated by $C\|a:S^{\mu}_{\rho,\delta}F^{(j)}\|$.

(b) If $a(x,\xi) \in S^{\mu}_{\rho,\delta}F^{\sigma}_{q_1}$, then $t^{\mu}|y|^{j-\sigma}\Delta^{\ell}_{y}\partial^{\alpha}_{x}a(t^{\delta}x,t^{-\rho}\xi)f(t^{1-\rho}\xi) \in L^{*}_{q_1}(\mathbb{R}^n;L_\infty(I;F))$, $a(x,\xi)g(\xi) \in F^{\sigma}_{q_1}$, and their norms are estimated by $C\|a:S^{\mu}_{\rho,\delta}F^{\sigma}_{q_1}\|$, where $j < \sigma < j + \ell$.

<u>4.3</u>. From Definition 4.1, Lemma 4.2 and the adjoint Leibniz formula (1.7.1) we get

<u>Corollary</u>. If $a(x,\xi) \in S^{\mu}_{\delta,\rho}W^{(j,k)}_{(p_1,p_2)}$, $S^{\mu}_{\delta,\rho}W^{j}_{p_1}B^{\tau}_{p_2,q}$,

$S^{\mu}_{\delta,\rho}WB^{(j,\tau)}_{(p_1,p_2),q}$, $S^{\mu}_{\delta,\rho}B^{\sigma}_{p_1,q}W^{k}_{p_2}$, $S^{\mu}_{\delta,\rho}B^{\sigma}_{p_1,q_1}B^{\tau}_{p_2,q_2}$, or $S^{\mu}_{\delta,\rho}B^{(\sigma,\tau)}_{(p_1,p_2),(q_1,q_2)}$, then $\partial_x^{\alpha}\partial_{\xi}^{\beta}a(x,\xi) \in S^{\mu+\delta|\alpha|-\rho|\beta|}_{\delta,\rho}W^{(j-|\alpha|,k-|\beta|)}_{(p_1,p_2)}$ $S^{\mu+\delta|\alpha|-\rho|\beta|}_{\delta,\rho}W^{j-|\alpha|}_{p_1}B^{\tau-|\beta|}_{p_2,q}$, $S^{\mu+\delta|\alpha|-\rho|\beta|}_{\delta,\rho}WB^{(j-|\alpha|,k-|\beta|)}_{(p_1,p_2),q}$ $S^{\mu+\delta|\alpha|-\rho|\beta|}_{\delta,\rho}B^{\sigma-|\alpha|}_{p_1,q}W^{k-|\beta|}_{p_2}$, $S^{\mu+\delta|\alpha|-\rho|\beta|}_{\delta,\rho}B^{\sigma-|\alpha|}_{p_1,q_1}B^{\tau-|\beta|}_{p_2,q_2}$, or $S^{\mu+\delta|\alpha|-\rho|\beta|}_{\delta,\rho}B^{(\sigma-|\alpha|,\tau-|\beta|)}_{(p_1,p_2),(q_1,q_2)}$, respectively.

4.4. Lemma. Let $k \geq 0$, $\ell \geq 0$, $0 \leq \delta \leq \rho \leq 1$, $1 \leq r \leq \infty$, f a C_0^{∞}-function vanishing near the origin, $a(x,\xi)$ a symbol, and let $\phi \in C_0^{\infty}$. For s, t > 0 put

$$a(s,x,\xi) = \int \phi_s(x-x')a(x',\xi)dx', \tag{1}$$

$$a_{s,t}(x,\xi) = a(st^{\delta},t^{\rho}x,t^{-\rho}\xi)f(t^{1-\rho}\xi). \tag{2}$$

By $F_p^{(j)}$ we denote one of the spaces

$$W^{(j,k)}_{(\infty,p)}(\mathbb{R}^n_x\cdot\mathbb{R}^n_\xi;X),\quad WB^{(j,\tau)}_{(\infty,p),q}(\mathbb{R}^n_x\cdot\mathbb{R}^n_\xi;X) \text{ or } W^j_\infty B^{\tau}_{p,q}(\mathbb{R}^n_x\cdot\mathbb{R}^n_\xi;X),$$

$F_p = F_p^{(0)}$, and by $F^{\sigma}_{p,r}$ we denote one of the spaces

$$B^{\sigma}_{\infty,r}W^{k}_{p}(\mathbb{R}^n_x\cdot\mathbb{R}^n_\xi;X),\ B^{(\sigma,\tau)}_{(\infty,p),(r,q)}(\mathbb{R}^n_x\cdot\mathbb{R}^n_\xi;X) \text{ or } B^{\sigma}_{\infty,r}B^{\tau}_{p,q}(\mathbb{R}^n_x\cdot\mathbb{R}^n_\xi;X).$$

(a) Assume that $a(x,\xi) \in S^{\mu}_{\rho,\delta}F^{(\ell)}_{\infty}$, and ϕ is of the form (1.5.1). Then

$$\|a_{s,t}:F_p\| \leq C(1+s^h)t^{-\mu-(1-\rho)n/p}s^{\ell}\|a:S^{\mu}_{\rho,\delta}F^{(\ell)}_{\infty}\|. \tag{3}$$

(b) Assume that $a(x,\xi) \in S^{\mu}_{\rho,\delta}F^{\sigma}_{\infty,r}$, $\kappa > 0$, ϕ of the form (1.5.1) with $\ell > 0$,and h is the smallest non-negative integer such that $-h < \sigma$. Then

$$\|a_{s,t}:F_p\| \leq (1+s^h)t^{-\mu-(1-\rho)n/p}s^{\sigma}\Phi(s), \tag{4}$$

$$\|a_{s,t}:F^{\kappa}_{\infty,1}\| \leq t^{-\mu}(1+s^h)\{1+(t^{\rho-\delta}/s)^{\kappa}\}s^{\sigma}\Phi(s), \tag{5}$$

$$\|\Phi:L^*_{r_1}(\mathbb{R}_+)\| \leq C\|a:S^{\mu}_{\rho,\delta}F^{\sigma}_{\infty,r}\| \text{ for any } r \leq r_1 \leq \infty. \tag{6}$$

(c) Assume that $a(x,\xi)$ $S^{\mu}_{\rho,\delta}F^{\sigma}_{\infty,1}$, $\sigma > 0$, and $\kappa > 0$. Then

$$\|a_{s,t}:F^{\kappa}_{\infty,1}\| \leq Ct^{-\mu}\{1+(t^{\rho-\delta}/s)^{\kappa}\}(1+s^{\sigma})\|a:S^{\mu}_{\rho,\delta}F^{\sigma}_{\infty,r}\|. \tag{7}$$

Here C is a constant independent of s, t and the symbol a.

Proof. It suffices to consider only the case $\ell = 0$ in (a) and $0 < \sigma \leq 1$ in (b) and (c), since the other cases can be reduced to these cases by making use of partial integretion and Corollary 4.3.

On the other hand there is a C_0^{∞}-function g identically equal to 1 on the support of f, we get by Lemma 1.7

$$\|a_{s,t}:F_p\| = \|a(st^{\delta},t^{\rho}x,t^{-\rho}\xi)f(t^{1-\rho}\xi)g(t^{1-\rho}\xi):F_p\| \leq Ct^{-(1-\rho)n/p}\|a_{s,t}:F_\infty\|.$$

Thus, it suffices to consider only the case $p = \infty$.

Proof of (a). By Lemma 4.2 we have

$$\|a_{s,t}:F_\infty\| \le \int |\phi_s(y)|\|a(t^\rho x - t^\delta y, t^{-\rho}\xi)f(t^{1-\rho}\xi):F_\infty\| dy \le Ct^{-\mu}\|a:S^\mu_{\rho,\delta}F_\infty\|.$$

Proof of (4). There is a C^∞-function H_ℓ with support contained in $bB\times B$, $b > 0$, such that

$$a(s,x,\xi) = \iint s^{-2n}H_\ell(\frac{x-x'}{s}, \frac{x-x'-y}{s})\Delta^\ell_y a(x',\xi)dx'dy. \tag{8}$$

(see [13] Lemma 2.4). Put

$$\Psi(y) = \|t^\mu_-|y|^{-\sigma}\Delta^\ell_y a(t^\delta x, t^{-\rho}\xi)f(t^{1-\rho}\xi):L_\infty(I;F_\infty)\|. \tag{9}$$

Then, by Lemma 4.2 we get

$$\|\Psi(y):L^*_r(\mathbb{R}^n)\| \le C\|a:S^\mu_{\rho,\delta}F^\sigma_{\infty,r}|. \tag{10}$$

Now, (8) and (9) imply that

$$\|a_{s,t}:F_\infty\| = \|a(st^\delta, t^\rho x, t^{-\rho}\xi)f(t^{1-\rho}\xi):F_\infty\|$$

$$\le \|\iint s^{-2n}H_\ell(\frac{t^{\rho-\delta}x-x'}{s}, \frac{t^{\rho-\delta}x-x'-y}{s})\Delta^\ell_y a(t^\delta x', t^{-\rho}\xi)f(t^{1-\rho}\xi)dx'dy:F_\infty\|$$

$$\le C_1\int_{csB} s^{-n}|y|^\sigma\Psi(y)dy = s^\sigma\Phi(s),$$

where $c = 1 + b$, and $\Phi(s) = C_1\int_{csB}(|y|/s)^{n+\sigma}\Psi(y)|y|^{-n}dy$. Furthermore, by the theorem of boundedness of integral operators we get (6).

Proof of (5) and (7). Let $\phi \in C_0^\infty$, $\psi(x) = \sum_{|\alpha|=m}\partial^\alpha\psi_\alpha(x)$, $\psi_\alpha \in C_0^\infty$, where $m > \kappa$, σ. Then

$$\psi_{s'} \overset{x}{*} a(st^\delta, t^\rho x, \xi) = (t^{\rho-\delta}s'/s)^m\sum_{|\alpha|=m}\int\psi_{\alpha,s'}(y)a_\alpha(st^\delta, t^\rho(x-y),\xi)dy,$$

where $a_\alpha(s,x,\xi) = \int\phi^{(\alpha)}_s(x-y)a(y,\xi)dy$. This formula and (4) give

$$\|\psi_{s'} \overset{x}{*} a_{s,t}(x,\xi):F_\infty\| \le (s't^{\rho-\delta}/s)^m t^{-\mu}(1 + s^h)s^\sigma\Phi_1(s),$$

$$\|\Phi_1(s);L^*_{r_1}\| \le \sum_{|\alpha|=m}\|\psi_\alpha\|_1\|\Phi_\alpha:L^*_{r_1}\| \le C\|a:S^\mu_{\rho,\delta}F^\sigma_{\infty,r}\|.$$

Hence, by § 1.7 (f), taking $I = (0, st^{\delta-\rho}]$, we get

$$\|a_{s,t}:F^\kappa_{\infty,1}\| \le C\{1 + (t^{\rho-\delta}/s)^\kappa\}\{t^{-\mu}(1 + s^h)s^\sigma\Phi_1(s) + \|a_{s,t}:F_\infty\|\}. \tag{11}$$

If ϕ is of the form (1.5.1), (4) and (11) imply (5), where $\Phi(s) = C_1\Phi_1(s) + C_0\Phi_0(s)$.

Moreover, if $\sigma > 0$, then by (3), (6) and (11) we have (7).

<u>4.5</u>. In the following of this section X and Y denote Hilbert spaces and $\|T\|$ means the norm of T in $\mathcal{L}(L_2(\mathbb{R}^n;X), L_2(\mathbb{R}^n;Y))$.

<u>Lemma</u> Let $F^\sigma_{\infty,1} = B^{(\sigma,n/2)}_{(\infty,\infty),(1,1)}(\mathbb{R}^n_x\cdot\mathbb{R}^n_\xi;\mathcal{L}(X,Y))$, $0 \le \delta \le \rho \le 1$, and assume that $\phi \in C_0^\infty(\mathbb{R}^n_x)$, $a(x,\xi) \in S^\mu_{\delta,\rho}\ F^\sigma_{\infty,1}$. Let $A(s,t)$ be the pseudo-differential operator with symbol $a(st^\delta, x, \xi)f(t\xi)$, where $a(s,x,\xi)$ and

f are functions as in Lemma 4.4.

(a) Let $\ell \geq 0$, $\ell > \sigma$, and assume that ϕ is of the form (1.5.1). Then

$$\|A(s,t)\| \leq t^{-\mu-(1-\rho)n/2}(1+s^h)s^\sigma\Phi(s^\delta), \tag{1}$$

$$\|A(s,t)\| \leq t^{-\mu}(1+s^h)\{1+(t^{\rho-\delta}/s)^{n/2}\}s^\sigma\Phi(s), \tag{2}$$

$$\|\Phi:L_r^*(\mathbb{R}_+)\| \leq C\|a:S^\mu_{\rho,\delta}F^\sigma_{\infty,1}\| \quad \text{for any } 1 \leq r \leq \infty, \tag{3}$$

where h is the smallest non-ngative integer larger than $-\sigma$.

(b) Assme that $\sigma > 0$. Then

$$\|A(s,t)\| \leq Ct^{-\mu}\{1+(t^{\rho-\delta}/s)^{n/2}\}(1+s^\sigma)\|a:S^\mu_{\rho,\delta}F^\sigma_{\infty,1}\|. \tag{4}$$

Here C is a constant independent of s, t and the symbol a.

Proof. Let $\tilde{A}(s,t)$ be the operator with symbol $a_{s,t}(x,\xi) = a(st^\delta, t^\rho x, t^{-\rho}\xi)f(t^{1-\rho}\xi)$, and let $\Lambda(t)u(x) = t^{n/2}u(tx)$. Then, by a simple calculation we have $A(s,t)u = \Lambda(t^{-\rho})\tilde{A}(s,t)\Lambda(t^\rho)u$ and $\|\Lambda(t)u\|_2 = \|u\|_2$ for $u \in \mathcal{S}(\mathbb{R}^n;X)$, so we get $\|A(s,t)\| = \|\tilde{A}(s,t)\|$. Therefore, the desired results follow from Theorm 2.1, Theorem 3.1 and Lemma 4.4.

4.6. The following lemma is the key to our main theorem.

Lemma. Let $a(t,x,\xi)$ be $\mathcal{L}(X,Y)$-valued symbols with parameter t (0,b], b >0, f a C_0^∞-function vanishing near the origin, and let $A(t)$, $A_j(t)$, $B_j(t)$, $A_j^0(t)$ and $B_j^0(t)$ be the operators whose symbols are $a(t,x,\xi)f(t\xi)$, $a(t,x,\xi)f(t\xi)t\xi_j|t\xi|^{-2}$, $a(t,x,\xi)f(t\xi)t\xi_j$, $\partial_{x_j}a(t,x,\xi)f(t\xi)t\xi_j|t\xi|^{-2}$ and $\partial_{x_j}a(t,x,\xi)f(t\xi)$, respectively. Assume that $0 \leq \delta < 1$, $\|A(t)\| \leq K$, $\|A_j(t)\| \leq K$, $\|B_j(t)\| \leq K$, $\|A_j^0(t)\| \leq Kb^{\delta-1}t^{-\delta}$, and $\|B_j^0(t)\| \leq Kb^{\delta-1}t^{-\delta}$, where K is a constant independent of t, b and δ. Then

$$\left\|\int_0^b A(t)\frac{dt}{t}\right\| \leq 4\{4+2/(1-\delta)\}\sqrt{n}K. \tag{1}$$

Proof. Putting $D_j = -i\partial_{x_j}$, by a simple calculation we have

$$D_jA(t) + B_j^0(t) = A(t)D_j = t^{-1}B_j(t), \tag{2}$$

$$t\textstyle\sum_{j=1}^n D_jA_j(t) + it\sum_{j=1}^n A_j^0(t) = t\sum_{j=1}^n A_j(t)D_j = A(t). \tag{3}$$

Therefore, we get

$$\begin{aligned} A(t)^*A(s) &= t\textstyle\sum_j\{A_j(t)^*D_jA(s) - iA_j^0(t)^*A(s)\} \\ &= t\textstyle\sum_j\{s^{-1}A_j(t)^*B_j(s) - iA_j(t)^*B_j^0(s) - iA_j^0(t)^*A(s)\}, \end{aligned} \tag{4}$$

and

$$A(t)^*A(s) = s\textstyle\sum_j\{t^{-1}B_j(t)^*A_j(s) + iB_j^0(t)^*A_j(s) + iA(t)^*A_j^0(s)\}. \tag{5}$$

It follows from (4), (5) and the assumption that

$$\|A(t)^*A(s)\| \leq nK^2h_1(t,s)^2, \tag{6}$$

where $h_1(t,s) = \min\{\sqrt{t/s} + b^{-\kappa}t^{\kappa}, \sqrt{s/t} + b^{-\kappa}s^{\kappa}\}$, $\kappa = (1-\delta)/2$.

Likewise, from the identities

$$A(t)A(s)^* = t/s\sum_{j=1}^{n} A_j(t)B_j(s)^* = s/t\sum_{j=1}^{n} B_j(t)A_j(s)^*, \tag{7}$$

it follows that

$$\|A(t)A(s)^*\| \leq nK^2 h_2(t,s)^2, \text{ with } h_2(t,s) = \min\{\sqrt{t/s}, \sqrt{s/t}\}. \tag{8}$$

Since the function $h(t,s) = \int_0^b h_1(t,\tau)h_2(\tau,s)\frac{d\tau}{\tau}$ has the properties that

$$\int_0^b h(t,s)\frac{dt}{t} \leq C_\kappa = 4(4 + 1/\kappa), \quad \int_0^b h(t,s)\frac{ds}{s} \leq C_\kappa,$$

the norm of the integral operator with kernel $h(t,s)$ as an operator in $L_2^*((0,1])$ does not exceed C_κ, hence the desired result follows from this with the aid of Cotlar-Stein-Calderón-Vaillancourt's lemma, noticing that by (3) we have

$$\int_0^b \|A(t)u:L_2(\mathbb{R}^n;Y)\|\frac{dt}{t} \leq \sum_{j=1}^{n}\int_0^b \|A_j(t)\partial_{x_j} u:L_2(\mathbb{R}^n;Y)\|dt$$
$$\leq bK\sum_{j=1}^{n}\|\partial_{x_j} u:L_2(\mathbb{R}^n;X)\|.$$

<u>4.7</u>. Now we are in a position to state our main theorem.

<u>Theorem</u>. Let X and Y be Hilbert spaces, $0 \leq \delta \leq \rho < 1$, and assume that the symbol $a(x,\xi)$ of an operator A belongs to

$$S^0_{\rho,\delta}B^{(\lambda,n/2)}_{(\infty,\infty),(1,1)}(\mathbb{R}^n_x\cdot\mathbb{R}^n_\xi;\mathcal{L}(X,Y)), \ \lambda = \frac{1-\rho}{1-\delta}\cdot\frac{n}{2}. \tag{1}$$

Then, A is a bounded operator from $L_2(\mathbb{R}^n;X)$ to $L_2(\mathbb{R}^n;Y)$, and $|A| \leq C\|a\|$, where $\|a\|$ denotes the norm of $a(x,\xi)$ in the space (1).

<u>Proof</u>. Take a C_0^∞-function f_0 with support contained in [1/2,2] such that $\int_0^\infty f_0(t)\frac{dt}{t} = 1$. Then $g(\xi) = 1 - \int_0^1 f_0(t|\xi|)\frac{dt}{t}$ is a C_0^∞-function with support contained in 2B, and we get, putting $f(\xi) = f_0(|\xi|)$,

$$a(x,\xi) = a(x,\xi)g(\xi) + \int_0^1 a(x,\xi)f(t\xi)\frac{dt}{t}. \tag{2}$$

On the other hand by the formula (1.4.5) we have

$$a(x,\xi) = \int_0^{t^\rho} a(s,x,\xi)\frac{ds}{s} + e(t^\rho,x,\xi), \tag{3}$$

$$a(s,x,\xi) = \int M_s(x-y)a(y,\xi)dy, \ e(s,x,\xi) = \int e_{m,s}(x-y)a(y,\xi)dy, \tag{4}$$

where $m > n/2$, M and e_m are the functions defined by (1.4.4). Thus, let A_0, $A(s,t)$ and $E(t)$ be the operators with the symbols $a(x,\xi)g(\xi)$, $a(st^\delta,x,\xi)f(t\xi)$ and $e(t^\rho,x,\xi)f(t\xi)$, respectively. Then we have

$$\begin{aligned} Au &= A_0u + \int_0^1\frac{dt}{t}\{\int_0^{t^{1-\delta}} + \int_{t^{1-\delta}}^{t^{\rho-\delta}}\}A(s,t)u\frac{ds}{s} + \int_0^1 E(t)u\frac{dt}{t} \\ &= A_0u + A_1u + A_2u + Eu. \end{aligned} \tag{5}$$

Estimate of A_0 and A_1. Since it follows from Lemma 4.2 that $a(x,\xi)g(\xi) \in WB^{(0,n/2)}_{(\infty,2),1}$, the boundednes of A_0 follows from Theorem 2.1.

It follows from Lemma 4.5 (a) that

$$\|A_1\| \leq \int_0^1 \frac{dt}{t}\int_0^{t^{1-\delta}} \Phi(s)s^\lambda t^{-(1-\rho)n/2}\frac{ds}{s} \leq \frac{2}{n(1-\rho)}\int_0^1 \Phi(s)\frac{ds}{s} \leq C_1\|a\|.$$

Estimate of E. Let $E_j(t)$, $F_j(t)$, $E_j^0(t)$ and $F_j^0(t)$ be the operators with symbols $e(t^\rho,x,\xi)f(t\xi)t\xi_j|t\xi|^{-2}$, $e(t^\rho,x,\xi)f(t\xi)t\xi_j$, $\partial_{x_j}e(t^\rho,x,\xi)f(t\xi)t\xi_j|t\xi|^{-2}$ and $\partial_{x_j}e(t^\rho,x,\xi)f(t\xi)$, respectively. Then, it follows from Lemma 4.5 (b) that $\|E(t)\| \leq M$, $\|E_j(t)\| \leq M$, $\|F_j(t)\| \leq M$, $\|E_j^0(t)\| \leq Mt^{-\rho}$ and $\|F_j^0(t)\| \leq Mt^{-\rho}$, where M is a constant such that $M \leq C\|a\|$. In view of Lemma 4.6 this implies $\|E\| \leq C_2\|a\|$.

Estimate of A_2. If $\delta < \rho$, it follows from Lemma 4.5 (a) that

$$\|A_2\| \leq \int_0^1 \Phi(s)s^{\lambda-n/2}\frac{ds}{s}\int_{s^\varepsilon}^{s^\kappa} t^{(\rho-\delta)n/2}\frac{dt}{t} \leq \frac{2}{(\rho-\delta)n}\int_0^1 \Phi(s)\frac{ds}{s} \leq C\|a\|.$$

where $\kappa = 1/(1-\delta)$ and $\varepsilon = 1/(\rho-\delta)$.

Finally consider the case $\rho = \delta$. Let $A_j(s,t)$, $B_j(s,t)$, $A_j^0(s,t)$ and $B_j^0(s,t)$ be the operators with symbols $a(st^\delta,x,\xi)f(t\xi)t\xi_j|t\xi|^{-2}$, $a(st^\delta,x,\xi)f(t\xi)t\xi_j$, $\partial_{x_j}a(st^\delta,x,\xi)f(t\xi)t\xi_j|t\xi|^{-2}$ and $\partial_{x_j}a(st^\delta,x,\xi)f(t\xi)$ respectively. Then, it follows from Lemma 4.5 (a) that $\|A(s,t)\| \leq \Phi(s)$, $\|A_j(s,t)\| \leq \Phi(s)$, $\|B_j(s,t)\| \leq \Phi(s)$, $\|A_j^0(s,t)\| \leq \Phi(s)s^{-1}t^{-\rho}$ and $\|B_j^0(s,t)\| \leq \Phi(s)s^{-1}t^{-\rho}$, which, in view of Lemma 4.6, gives

$$\|A_2\| \leq \int_0^1 \frac{ds}{s}\left\|\int_0^{s^\kappa} A(s,t)\frac{dt}{t}\right\| \leq C\int_0^1 \Phi(s)\frac{ds}{s} \leq C'\|a\|.$$

The proof of the theorem is complete.

§ 5. Further results.

5.1. For a weight functon ω (that is, a positive measurable function) on R_+ the generalized Besov space $B^\omega_{p,q}$ is defined by replacing $|y|^{-\sigma}$ with $\omega(|y|)^{-1}$ in Definition 1.1. The space $WB^{(j,\omega)}_{(p_1,p_2),q}$ and $B^{(\omega_1,\omega_2)}_{(p_1,p_2),(q_1,q_2)}$ are defined analogously. Also we can introduce the symbol classes corresponding these type of spaces.

5.2. Theorem 2.1 with $WB^{(0,n/r)}_{(\infty,r),1}$ replaced by $L_\infty B^\omega_{r,p'}$, where $1/p + 1/p' = 1$ and ω is a weight on $\mathbb{R}_+$ such that $\omega(t)t^{-n/r} \in L^*_p((0,1])$, is valid. Theorem 3.1 with $B^{(n/2,n/2)}_{(\infty,\infty),(1,1)}$ replaced by $B^{n/2}_{\infty,1}B^\omega_{\infty,2}$ and Theorem 4.7 with

$B^{(\lambda,n/2)}_{(\infty,\infty),(1,1)}$ replaced by $B^{\lambda}_{\infty,1}B^{\omega}_{\infty,2}$, where $\omega(t)t^{-n/2} \in L^*_2((0,1])$, are also valid. The proofs of these facts are similar to that of the corresponding theorems. Notice that $B^{\kappa}_{\infty,\infty}B^{\kappa}_{\infty,\infty} = B^{(\kappa,\kappa)}_{(\infty,\infty),(\infty,\infty)} \subset B^{n/2}_{\infty,1}B^{\omega}_{\infty,2}$, so these are also improvement of results in Cordes [4] and Kato [7].

5.3. For symbols of class $S^0_{1,\delta}$ we can prove the following

Theorem. Let X and Y be Hilbert spaces, $0 \leq \delta < 1$, $\zeta(t) = (1 + \log_+ t^{-1})^{-1}$, and assume that $a(x,\xi)$ belongs to one of the spaces $S^0_{1,\delta}B^{(\zeta,n/2)}_{\infty,1}(\mathbb{R}^n\cdot\mathbb{R}^n;\mathcal{L}(X,Y))$, and $S^0_{1,\delta}B^{\zeta}_{\infty,1}B^{\omega}_{\infty,2}(\mathbb{R}^n\cdot\mathbb{R}^n;\mathcal{L}(X,Y))$ with $\omega(t)t^{-n/2} \in L^*_2((0,1])$. Then the operator $a(x,D_x)$ is a bounded operator from $L_2(\mathbb{R}^n;X)$ into $L_2(\mathbb{R}^n;Y)$. Here $f_+(t) = \max\{0,f(t)\}$.

REFERENCES

[1] Calderón, A. P. and Vaillancourt, R.; On the boundedness of pseudo differential operators, J. Math. Soc. Japan 23 (1971), 374-378.

[2] Childs, A. G.; On the L^2-boundedness of pseudo-differential operators, Proc. A.M.S., 61 (1976), 252-254.

[3] Coifman, R. R. et Meyer, Y.; Au delà opérateurs pseudo-differentiels, Asterisque, 57 (1978), 1-185.

[4] Cordes, H. O.; On compactness of commutators of multiplications and convolutions, and boundedness of pseudo-differential operators, J. Functional Analysis, 18 (1975), 115-131.

[5] Fefferman, C.; L^p-bounds for pseudo-differential operators, Israel J. Math., 14 (1973), 413-417.

[6] Hörmander, L.; Pseudo-differential operators and hypo-elliptic equations, Proc. Symp. Pure Math. 10 (1967), 138-183, A.M.S. Providence, R.I., U. S. A.

[7] Illner, R.; A class of L^p-bounded pseudo-differential operators, Proc. Amer. Math. Soc., 51 (1975), 347-355.

[8] Kato, T.; Boundedness of some pseudo-differential operators, Osaka J. Math., 13 (1976), 1-9.

[9] Kumano-go, H.; Pseudo-Differential Operators, MIT Press, Cambridge, Massachusettes, U.S.A. and London, England, 1982.

[10] Kumano-go, H., and Nagase, M.; Pseudo-differential operators with non-regular symbols and applications, Funkcial. Ekvac., 22 (1978), 151 -192.

[11] Miyachi, A.; Estimates for pseudo-differential operators of class $S^0_{0,0}$, Research Reports Dep. Math. Hitotsubashi Univ., 1985.

[12] Mossaheb, S. et Okada, M.; Une class d'operateurs pseudo-diffe-

rentiels bornes sur $L^r(R^n)$, $1 < r < \infty$, C. R. Acad. Sc. Paris, 285 (1977), 1045-1061.

[13] Muramatu, T.; On Besov spaces of functions defined in general regions, Publ. R. I. M. S. Kyoto Univ., 6 (1970/1971), 515-543.

[14] Muramatu, T.; Besov spaces and Sobolev spaces of generalized functions defined on a general region, Publ. R. I. M. S. Kyoto Univ., 9 (1974), 325-396.

[15] Muramatu, T.; Interpolation Spaces and Linear Operators, Kinokuniya-Shoten, Tokyo, Japan, 1985. (in Japanese)

[16] Muramatu, T. and Nagase, M.; On sufficient conditions for the boundeness of pseudo-differential operators, Proc. Japan Acad., 55 Ser. A (1979), 613-616.

[17] Muramatu, T. and Nagase, M.; L^2-boundedness of pseudo-differential operators with non-regular symbols, Canadian Math. Soc. Conference Proc., 1 (1981), 135-144.

[18] Nagase, M.; The L^p-boundedness of pseudo-differential operators with non-regular symbols, Comm. P. D. E., 2 (1977), 1045-1061.

[19] Nagase, M.; On the boundedness of pseudo-differential operators in L^p-spaces, Sci. Rep. College Gen. Ed. Osaka Univ., 32 (1983), 9-19.

[20] Nagase, M.; On a class of L^p-bounded pseudo-differential operators, Sci. Rep. College General Ed. Osaka Univ., 33 (1984), 1-7.

[21] Nagase, M.; On some classes of L^p-bounded pseudo-differential operators, Osaka Math. J. to appear.

[22] Peetre, J.; New Thoughts on Besov Spaces, Duke Univ. Math. Ser. 1 (1974). Duruham, N.C., U. S. A.

[23] Stein, E. M.; Singular Integrels and Differentiability Properties of Functions, Princton Univ. Press, Princeton, New Jersey, U.S.A. 1970

[24] Sugimoto, M.; L^p-boundedness of pseudo-differential operators satisfying Besov estimates I, in preparation.

[25] Taibleson, M. H.; On the theory of Lipschitz spaces of distributions on Eucledean n-space I, J. Math. Mech. 13 (1964), 407-479.

[26] Triebel, H.; Fourier Analysis and Function Spaces, Teubner-Texte Math., Teubner, Leipzig, 1977.

[27] Triebel, H.; Spaces of Besov-Hardy-Sobolev Type, Teubner-Texte Math., Teubner, Leipzig, 1978.

[28] Wang Rouhuai and Li Chengzhang; On the L^p-boundedness of several classes of pseudo-differential operators, Chinese Ann. Math., 5 B (1984), 193-213.

[29] Yabuta, K.; Calderón-Zygmund operators and pseudo-differential operators, Comm. P. D. E., to appear.

ON SUFFICIENT CONDITIONS FOR PSEUDO-DIFFERENTIAL OPERATORS TO BE L^p-BOUNDED

M. Nagase
Department of Mathematics
College of General Education
Osaka University
Toyonaka, Osaka, 560
Japan

1. Introduction.

In [12], the author gives several L^p-boundedness theorems for pseudo-differential operators with non-smooth symbols for $2 \leqq p \leqq \infty$. The symbol classes in [12] are originated from Hörmander's class $S^m_{\rho,0}$, $0 < \rho \leqq 1$. In the present paper we give sufficient conditions for L^p-boundedness of pseudo-differential operators with non-smooth symbols which are originated from Hörmander's class $S^m_{\rho,\delta}$, $0 \leqq \delta < \rho \leqq 1$. We also give results for $1 < p \leqq 2$.

For $2 \leqq p < \infty$, we obtain the L^p-boundedness (Theorem A in Section 5) under $\kappa = [n/2]+1$ differentiability in ξ of symbols $p(x,\xi)$. This fact corresponds to the result of Hörmander in [3] on the Fourier multipliers. We note also that, if $\delta = 0$, the result is just the same as Theorem 4.5 in [12]. On the other hand, for $1 < p < 2$, it is known that the κ-differentiability in ξ of symbols $p(x,\xi)$ does not always imply the L^p-boundedness of operators $p(X,D_x)$ (see, for example, [7], [8] and [13]). For $1 < p \leqq 2$, we give an L^p-boundedness theorem (Theorem B in Section 5) under $(n+1)$-differentiability in ξ of symbols $p(x,\xi)$.

In Section 2 we give fundamental lemmas as preliminaries and in Section 3 we state a regularization of non-smooth symbols. The regularization of symbols in Section 3 is used in [12] and we also use it in the proof of our main theorems. In Section 4, we give boundedness theorems for operators of symbols which have high decreasing order in ξ as $|\xi| \to \infty$. The main theorems of the present paper are Theorem A and Theorem B in Section 5. Finally in Section 6 we give some remarks on the main theorems.

2. Preliminaries.

We use a standard notation which is used in the theory of pseudo-differential operators (see, for example, [6] or [14]). Let $0 \leqq \delta \leqq \rho \leqq 1$ and $\delta < 1$. Then the Hörmander class $S^m_{\rho,\delta}$ of symbols is defined by

$$S^m_{\rho,\delta} = \{p(x,\xi) \in C^\infty(\mathbb{R}^n\times\mathbb{R}^n): |p^{(\alpha)}_{(\beta)}(x,\xi)| \leqq C_{\alpha,\beta}\langle\xi\rangle^{m-\rho|\alpha|+\delta|\beta|} \text{ for any } \alpha \text{ and } \beta\},$$

where $p^{(\alpha)}_{(\beta)}(x,\xi) = \partial^\alpha_\xi D^\beta_x p(x,\xi)$ and $\langle\xi\rangle = (1+|\xi|^2)^{1/2}$.
For $p(x,\xi)$ in $S^m_{\rho,\delta}$ we define the pseudo-differential operator $p(X,D_x)$ by

$$p(X,D_x)u(x) = (2\pi)^{-n} \int e^{ix\cdot\xi}p(x,\xi)\hat{u}(\xi)d\xi,$$

where $\hat{u}(\xi)$ denotes the Fourier transform of $u(x)$, that is, $\hat{u}(\xi) = \int e^{-ix\cdot\xi}u(x)dx$. We define the operator $p(X,D_x)$ for non-smooth symbol $p(x,\xi)$ in a similar way.

It is known that, if all $p(X,D_x)$ with symbols $p(x,\xi)$ in $S^m_{\rho,\delta}$ are L^p-bounded, then $m \leqq -n(1-\rho)|1/2 - 1/p|$ (see [4]). So we write $m_p = n(1-\rho)|1/2 - 1/p|$ and say that m_p is the critical decreasing order for the L^p-boundedness. In the present paper L^p means $L^p(\mathbb{R}^n)$. In the following we denote $\kappa = [n/2] + 1$ and $\kappa_p = [n/p] + 1$ for $1 \leqq p \leqq 2$. We give some known boundedness results as lemmas.

Lemma 2.1 ([1], [2] and [5]). Let $0 \leqq \rho < 1$. If a symbol $p(x,\xi)$ satisfies

$$(2.1)\qquad |p^{(\alpha)}_{(\beta)}(x,\xi)| \leqq C_{\alpha,\beta}\langle\xi\rangle^{\rho(|\beta|-|\alpha|)} \quad \text{for } |\alpha| \leqq \kappa \text{ and } |\beta| \leqq \kappa,$$

then the operator $p(X,D_x)$ is L^2-bounded, that is,

$$(2.2)\qquad \|p(X,D_x)u\|_{L^2} \leqq C\,\|u\|_{L^2}\,.$$

Lemma 2.2 ([7] and [13]). (i) Let $0 \leqq \rho < 1$ and $2 \leqq p < \infty$. If a symbol $p(x,\xi)$ satisfies

$$(2.3)\qquad |p^{(\alpha)}_{(\beta)}(x,\xi)| \leqq C_{\alpha,\beta}\langle\xi\rangle^{-m_p-\rho(|\alpha|-|\beta|)} \quad \text{for } |\alpha|\leqq\kappa \text{ and } |\beta|\leqq\kappa,$$

then the operator $p(X,D_x)$ is L^p-bounded.
(ii) Let $1 < p \leqq 2$. If the inequality (2.3) holds for $|\alpha| \leqq n+1$ and $|\beta| \leqq \kappa$, then the operator $p(X,D_x)$ is L^p-bounded.

When $\rho = 1$, we have the following.

Lemma 2.3 ([10] and [11]). Let $0 \leqq \delta < 1$ and $0 < \mu \leqq 1$. If a symbol $p(x,\xi)$ satisfies

$$(2.4)\qquad \begin{cases} |p^{(\alpha)}(x,\xi)| \leqq C_\alpha \langle\xi\rangle^{-|\alpha|} \\ |p^{(\alpha)}(x,\xi) - p^{(\alpha)}(y,\xi)| \leqq C_\alpha |x-y|^\mu \langle\xi\rangle^{-|\alpha|+\mu\delta} \end{cases}$$

for $|\alpha| \leqq \kappa$, then $p(X,D_x)$ is L^p-bounded for $2 \leqq p < \infty$.

3. Regularization of non-smooth symbols.

In this section we give a regularization of non-regular symbols, which is useful to investigate the boundedness in $L^p(\mathbb{R}^n)$ of pseudo-differential operators.

Let $0 \leqq \delta < \rho \leqq 1$, $\mu > 0$ and let ν be a positive integer. In this section we assume that a symbol $p(x,\xi)$ satisfies

$$(3.1)\qquad |p^{(\alpha)}_{(\beta)}(x,\xi)| \leqq C_{\alpha,\beta} \langle\xi\rangle^{m-\rho|\alpha|+\delta|\beta|}$$

for $|\alpha| \leqq \nu$ and $|\beta| \leqq [\mu]$, and

$$(3.2)\qquad |p^{(\alpha)}_{(\beta)}(x,\xi) - p^{(\alpha)}_{(\beta)}(x+y,\xi)| \leqq C_{\alpha,\beta} |y|^{\mu-[\mu]} \langle\xi\rangle^{m-\rho|\alpha|+\delta\mu}$$

for $|\alpha| \leqq \nu$ and $|\beta| = [\mu]$.

Let $\psi(z)$ be a Schwartz rapidly decreasing function satisfying

$$(3.3)\qquad \int \psi(z)dz = 1, \quad \int z^\alpha \psi(z)dz = 0 \text{ for any } \alpha \neq 0.$$

Then we define new symbols $\tilde{p}(x,\xi)$ and $q(x,\xi)$ by

$$(3.4)\qquad \tilde{p}(x,\xi) = \int \psi(y) p(x - \langle\xi\rangle^{-\rho'} y, \xi) dy$$
$$= \int \psi(\langle\xi\rangle^{\rho'}(x-y)) p(y,\xi) \langle\xi\rangle^{\rho' n} dy,$$

$$(3.5)\qquad q(x,\xi) = \tilde{p}(x,\xi) - p(x,\xi),$$

where $\delta < \rho' \leqq 1$. Then we have the following theorem.

Theorem 3.1. The symbols $\tilde{p}(x,\xi)$ and $q(x,\xi)$ satisfies;

$$(3.6)\qquad |\tilde{p}^{(\alpha)}_{(\beta)}(x,\xi)| \leqq C_{\alpha,\beta} \langle\xi\rangle^{m-\rho|\alpha|+\delta|\beta|} \quad \text{for } |\alpha| \leqq \nu,\ |\beta| \leqq [\mu],$$

$$(3.7)\qquad |\tilde{p}^{(\alpha)}_{(\beta)}(x,\xi)| \leqq C_{\alpha,\beta}\langle\xi\rangle^{m-\rho|\alpha|+\delta\mu+\rho'(|\beta|-\mu)}$$

for $|\alpha| \leqq \nu$, $|\beta| \geqq \mu$ and

$$(3.8)\qquad |q^{(\alpha)}(x,\xi)| \leqq C_\alpha\langle\xi\rangle^{m-(\rho'-\delta)\mu-\rho|\alpha|} \qquad \text{for } |\alpha| \leqq \nu.$$

Proof. We give only a sketch of the proof.

For $|\beta| \leqq [\mu]$, writing

$$(3.9)\qquad \tilde{p}^{(\alpha)}_{(\beta)}(x,\xi) = \partial^\alpha_\xi\{ \int \psi(\langle\xi\rangle^{\rho'}(x-y))p_{(\beta)}(y,\xi)\langle\xi\rangle^{\rho' n}dy\}$$

we can prove (3.6).

When $|\beta| \geqq [\mu]$, we take β^1 and β^2 such that $\beta = \beta^1 + \beta^2$, $|\beta^1| = [\mu]$ and $\beta^2 \neq 0$. Then we can write

$$(3.10)\qquad \tilde{p}^{(\alpha)}_{(\beta)}(x,\xi) = \partial^\alpha_\xi\{ \int (\partial^{\beta^2}_y \psi)(\langle\xi\rangle^{\rho'}(x-y))\{p_{(\beta^1)}(y,\xi) - p_{(\beta^2)}(x,\xi)\} \times \langle\xi\rangle^{\rho'(n+|\beta^2|)}dy\}.$$

Using (3.10), we can prove the inequality (3.7).

When $\mu > 1$, by using (3.3) and the Taylor expansion, we can write

$$(3.11)\qquad \tilde{p}(x,\xi) = p(x,\xi) + \sum_{|\beta|=[\mu]} \frac{(-1)^{[\mu]}[\mu]}{\beta\,!}\int_0^1 (1-t)^{[\mu]-1} \times \{ \int \psi(y)y^\beta\{p_{(\beta)}(x-ty,\xi) - p_{(\beta)}(x,\xi)\}\langle\xi\rangle^{\rho'(n-[\mu])}dy\}dt.$$

From this equality we can prove the inequality (3.8). When $\mu \leqq 1$, we can prove (3.8) in a similar way. Q.E.D.

4. Boundedness for operators of high decreasing order symbols.

Lemma 4.1. Let $1 \leqq p \leqq 2$ and ε be a positive number. We assume that a symbol $p(x,\xi)$ satisfies

$$(4.1)\qquad |p^{(\alpha)}(x,\xi)| \leqq C_\alpha\langle\xi\rangle^{-n/p-\varepsilon}$$

for $|\alpha| \leqq \kappa_p$, then we have

$$(4.2)\qquad \|p(X,D_x)u\|_{L^r} \leqq C\,\|u\|_{L^r}$$

for any $p \leqq r \leqq \infty$, where the constant C is independent of r.

Proof. We note that, when $p = 2$, the inequality (4.2) has been proved in [12]. Therefore, when $p < 2$, we have only to show the boundedness for $r = p$ by the Riesz-Thorin interpolation theorem.

We can write

$$(4.3)\qquad p(X,D_x)u(x) = \int K(x,x-y)u(y)dy,$$

where

$$(4.4)\qquad K(x,z) = (2\pi)^{-n} \int e^{iz\cdot\xi} p(x,\xi)d\xi.$$

Moreover we can see that when $1 < p < 2$, for $|\alpha| \leqq \kappa_p$

$$(4.5)\qquad \{ \int |z^\alpha K(x,z)|^q dz\}^{1/q} \leqq c_n \{\int |p^{(\alpha)}(x,\xi)|^p d\xi\}^{1/p},$$

because of the Hausdorff-Young inequality, where $1/q + 1/p = 1$. Therefore we have

$$(4.6)\qquad \|p(X,D_x)u\|_{L^p}^p = \int | \int K(x,x-y)u(y)dy|^p dx$$

$$\leqq \int \{ \int \langle x-y\rangle^{-p\kappa_p} |u(y)|^p dy\}\{ \int (\langle x-y\rangle^{\kappa_p} |K(x,x-y)|)^q dy\}^{p/q} dx$$

$$\leqq C \|u\|_{L^p}^p \{ \int \langle\xi\rangle^{-n-p}\, d\xi\}^{p/q}.$$

When $p = 1$ we can obtain the estimate (4.6) in a similar way.

Thus we obtain (4.2) for $r = p$. Q.E.D.

We note that in Lemma 4.1 we do not assume the differentiability in the space variable x.

Theorem 4.2. Let $0 \leqq \rho \leqq 1$, $1 \leqq p \leqq 2$ and $\varepsilon > 0$. We assume that a symbol $p(x,\xi)$ satisfies

$$(4.7)\qquad |p^{(\alpha)}(x,\xi)| \leqq C_\alpha \langle\xi\rangle^{-n(1-\rho)/p-\varepsilon-\rho|\alpha|} \quad \text{for } |\alpha| \leqq \kappa_p.$$

Then the operator $p(X,D_x)$ is L^r-bounded for $p \leqq r \leqq \infty$, and we have

$$(4.8)\qquad \|p(X,D_x)u\|_{L^r} \leqq C \|u\|_{L^r},$$

where the constant C is independent of $p \leqq r \leqq \infty$.

Proof. Since the boundedness for $2 \leqq r \leqq \infty$ is proved in [12], as in the proof of Lemma 4.1, we have only to prove the boundedness for $r = p$. We take a smooth function $f(t)$ on $\mathbb{R}^1$ such that the support is contained in the interval $[1/2, 1]$ and $\int_0^\infty \frac{f(t)}{t}\, dt = 1$.

By Lemma 4.1, we may assume that the support of $p(x,\xi)$ is contained in $\{\xi ; |\xi| \geqq 4\}$ and

(4.7)' $$|p^{(\alpha)}(x,\xi)| \leqq C_\alpha |\xi|^{-n(1-\rho)/p-\varepsilon-\rho|\alpha|} \quad \text{for} \quad |\alpha| \leqq \kappa_p.$$

Then we have

(4.9) $$p(X,D_x)u(x) = \int_0^{1/2} \frac{dt}{t} \int K(t,x,z)u(x-tz)dz,$$

where

(4.10) $$K(t,x,z) = (2\pi)^{-n} \int e^{iz\cdot\xi} p(x,\tfrac{\xi}{t}) f(|\xi|)d\xi.$$

When $1 < p < 2$, writing

(4.11) $$\int K(t,x,z)u(x-tz)dz = \int_{|z|\leqq t^{\rho-1}} K(t,x,z)u(x-tz)dz$$

$$+ \int_{|z|\geqq t^{\rho-1}} K(t,x,z)u(x-tz)dz = I_t + II_t,$$

by the Hausdorff-Young inequality we have

(4.12) $$\int |I_t|^p dx \leqq \int\{\int_{|z|\leqq t^{\rho-1}} |u(x-tz)|^p dz\}\{\int |K(t,x,z)|^q dz\}^{p/q} dx$$

$$\leqq c_n \int\{\int_{|z|\leqq t^{\rho-1}} |u(x-tz)|^p dz\}\{\int |p(x,\tfrac{\xi}{t}) f(|\xi|)|^p d\xi\} dx.$$

By (4.12) and assumption for $p(x,\xi)$ and $f(t)$, we have

(4.13) $$\|I_t\|_{L^p} \leqq C\, t^\varepsilon \|u\|_{L^p}.$$

Similarly we have

(4.14) $$\int |II_t|^p dx \leqq \int \{\int_{|z|\geqq t^{\rho-1}} |z|^{-p\kappa_p} |u(x-tz)|^p dz\}$$

$$\times \{\int \{|z|^{\kappa_p} |K(t,x,z)|\}^q dz\}^{p/q} dx.$$

Since for $|\alpha| = \kappa_p$

$$|\partial_\xi^\alpha\{p(x,\tfrac{\xi}{t}) f(|\xi|)\}| = \sum_{\alpha^1+\alpha^2=\alpha} \frac{\alpha!}{\alpha^1!\,\alpha^2!} t^{-|\alpha^1|} p^{(\alpha^1)}(x,\tfrac{\xi}{t}) \partial_\xi^{\alpha^2} f(|\xi|)$$

$$\leqq C_\alpha \sum_{|\gamma|\leqq|\alpha|} t^{-|\gamma|} (|\xi|/t)^{-n(1-\rho)/p-\varepsilon-\rho|\gamma|}$$

$$\leqq C_\alpha\, t^{(n/p-\kappa_p)(1-\rho)+\varepsilon}$$

in the support of $f(|\xi|)$, it follows from (4.14) and the Hausdorff-

Young inequality that

$$(4.15)\qquad \|II_t\|_{L^p} \leq C\ t^{(n/p-\kappa_p)(1-\rho)+\varepsilon}\|u\|_{L^p}\{\int_{|z|\geq t^{\rho-1}}|z|^{-p\kappa_p}dz\}^{1/p}$$

$$\leq C\ t^{\varepsilon}\|u\|_{L^p}.$$

Therefore by (4.9), (4.11), (4.13) and (4.15) we have

$$(4.16)\qquad \|p(X,D_x)u\|_{L^p} \leq \int_0^{1/2}\frac{dt}{t}\|I_t + II_t\|_{L^p}$$

$$\leq C\|u\|_{L^p}.$$

When $p = 1$, we can prove the boundedness in a similsr way.

Hence we obtain the theorem. Q.E.D.

5. Main theorems.

First we consider the case $2 \leq p < \infty$. Then we note that $m_p = n(1-\rho)(1/2 - 1/p)$. Let $0 \leq \delta < \rho \leq 1$ and let $\mu_p = \kappa n(1-\rho)/\{p\kappa(\rho-\delta)+n(1-\rho)\}$.

Theorem A. Suppose that a symbol $p(x,\xi)$ satisfies the following; there exists a positive constant $\mu > \mu_p$ such that

$$(5.1)\qquad |p_{(\beta)}^{(\alpha)}(x,\xi)| \leq C_{\alpha,\beta}\langle\xi\rangle^{-m_p-\rho|\alpha|+\delta|\beta|} \quad \text{for } |\alpha| \leq \kappa,\ |\beta| \leq \mu,$$

$$(5.2)\qquad |p_{(\beta)}^{(\alpha)}(x,\xi) - p_{(\beta)}^{(\alpha)}(y,\xi)| \leq C_{\alpha,\beta}|x-y|^{\mu-[\mu]}\langle\xi\rangle^{-m_p-\rho|\alpha|+\mu\delta}$$

for $|\alpha| \leq \kappa$ and $|\beta| = [\mu]$. Then the operator $p(X,D_x)$ is L^p-bounded.

Proof. When $\rho = 1$, the boundedness has already been obtained by Lemma 2.3, and when $\mu \geq \kappa$, theorem has already been given by Lemma 2.2 even for the case $\rho = \delta < 1$. So we may assume $\mu < \kappa$ and $\rho < 1$. We take a Schwartz rapidly decreasing function $\psi(x)$ which satisfies (3.3), and define new symbols $\tilde{p}(x,\xi)$ and $q(x,\xi)$ by (3.4) and (3.5) by taking $\rho' = \rho + n(1-\rho)/(p\kappa)$. We note that $\rho < \rho' < 1$. Then by Theorem 3.1 we have

$$(5.3)\qquad |\tilde{p}_{(\beta)}^{(\alpha)}(x,\xi)| \leq C_{\alpha,\beta}\langle\xi\rangle^{-m_p-\rho|\alpha|+\delta|\beta|} \quad \text{for } |\alpha| \leq \kappa,\ |\beta| \leq \mu,$$

$$(5.4)\qquad |\tilde{p}_{(\beta)}^{(\alpha)}(x,\xi)| \leq C_{\alpha,\beta}\langle\xi\rangle^{-m_p-\rho|\alpha|+\mu\delta+\rho'(|\beta|-\mu)}$$

for $|\alpha| \leqq \kappa$, $|\beta| = [\mu]$, and

$$(5.5) \qquad |q^{(\alpha)}(x,\xi)| \leqq C_\alpha \langle\xi\rangle^{-m_p-\mu\rho'+\mu\delta-\rho|\alpha|} \qquad \text{for } |\alpha| \leqq \kappa.$$

By the definition we see easily that

$$- m_p - \mu\rho' + \mu\delta < - m_p - \mu_p(\rho'-\delta) = -n(1-\rho)/2.$$

Therefore the inequality (5.5) means that $q(x,\xi)$ satisfies the inequality (4.7) in Theorem 4.2 for $p = 2$. Thus we have

$$(5.6) \qquad \|q(X,D_x)u\|_{L^r} \leqq C\|u\|_{L^r} \qquad \text{for } 2 \leqq r \leqq \infty.$$

Moreover for $\mu < |\beta| \leqq \kappa$, we can see that

$$(5.7) \qquad \mu\delta + \rho'(|\beta| - \mu) < - \mu_p(\rho'-\delta) + \rho'|\beta| \leqq \rho|\beta|.$$

Hence it follows from (5.3) and (5.4) that $\tilde{p}(x,\xi)$ satisfies the conditions of Lemma 2.2 (i). Thus $\tilde{p}(X,D_x)$ is L^p-bounded.

Since $p(x,\xi) = \tilde{p}(x,\xi) - q(x,\xi)$, we get the Theorem. Q.E.D.

Next we consider the case $1 < p \leqq 2$. In this case, for $0 \leqq \delta < \rho \leqq 1$, we note that $m_p = n(1-\rho)/(1/p - 1/2)$.

<u>Theorem B.</u> We assume that a symbol $p(x,\xi)$ satisfies the following; there exists a positive constant $\mu > \mu_2 = n(1-\rho)/\{2\kappa(\rho-\delta)+n(1-\rho)\}$ such that

$$(5.8) \qquad |p^{(\alpha)}_{(\beta)}(x,\xi)| \leqq C_{\alpha,\beta} \langle\xi\rangle^{-m_p-\rho|\alpha|+\delta|\beta|} \qquad \text{for } |\alpha| \leqq n+1, \ |\beta| \leqq \mu,$$

$$(5.9) \qquad |p^{(\alpha)}_{(\beta)}(x,\xi) |- p^{(\alpha)}_{(\beta)}(y,\xi)| \leqq C_{\alpha,\beta}|x-y|^{\mu-[\mu]}\langle\xi\rangle^{-m_p-\rho|\alpha|+\delta\mu}$$

for $|\alpha| \leqq n+1$ and $|\beta| = [\mu]$. Then $p(X,D_x)$ is L^p-bounded.

<u>Proof.</u> As in the proof of Theorem A, we may assume that $\mu < \kappa$ and $\rho < 1$. We take $\rho' = \rho + n(1-\rho)/(2\kappa)$ and we define new symbols $\tilde{p}(x,\xi)$ and $q(x,\xi)$ by (3.4) and (3.5) as before. Then by Theorem 3.1 we obtain the estimates (5.3), (5.4) and (5.5).

Therefore by assumptions we have

$$(5.10) \qquad |\tilde{p}^{(\alpha)}_{(\beta)}(x,\xi)| \leqq C_{\alpha,\beta} \langle\xi\rangle^{-m_p-\rho|\alpha|+\delta|\beta|} \qquad \text{for } |\alpha| \leqq n+1, \ |\beta| \leqq \kappa,$$

$$(5.11) \qquad |q^{(\alpha)}(x,\xi)| \leqq C_\alpha \langle\xi\rangle^{-n(1-\rho)/p-\varepsilon-\rho|\alpha|} \qquad \text{for } |\alpha| \leqq n+1,$$

where $\varepsilon = m_p + \mu(\rho'-\delta) - n(1-\rho)/p > -n(1-\rho)/2 + \mu_2(\rho'-\delta) = 0$.
Hence by Theorem 4.2, the operator $q(X,D_x)$ is L^p-bounded, and by

Lemma 2.2 (ii), the operator $\tilde{p}(X,D_x)$ is L^p-bounded.

Thus we get the theorem. Q.E.D.

6. Remarks.

We first note that if $\delta < \rho = 1$ then Theorem A coincides with Lemma 2.3 and if $\delta = \rho < 1$ then Lemma 2.2 (i) and (ii) are a little stronger results than Theorem A and Theorem B respectively.

When $p = 2$, Theorem A and B state the same conclusion, and if $\delta = \rho < 1$, then Lemma 2.1 is a little stronger than our result. We note, in this case, $m_2 = 0$ and $\mu_2 = \kappa n(1-\rho)/\{2\kappa(\rho-\delta)+n(1-\rho)\}$. If we replace κ by $n/2$ then μ_2 changes to $\tilde{\mu}_2 = n(1-\rho)/\{2(1-\delta)\}$. In [9], it is proved that the boundedness in L^2 can be shown under $\tilde{\mu}_2 + \varepsilon$ differentiability in ξ of symbols $p(x,\xi)$.

Moreover when $p > 2$, we can expect that the L^p-boundedness can be obtained under the conditions, in Theorem A, which are replaced by $n/2 + \varepsilon(\varepsilon > 0)$ and μ_p by $\tilde{\mu}_p = n(1-\rho)/\{p(\rho-\delta)+2(1-\rho)\}$, and under some appropriate conditions to the fractional derivatives of symbols.

When $1 < p < 2$, and when $\delta = \rho = 0$ or $\delta < \rho = 1$, we can see the L^p-boundedness under $n/p + \varepsilon$ differentiability in ξ of symbols $p(x,\xi)$ (see [7] and [8]). Therefore for $1 < p < 2$ we can expect that the L^p-boundedness can be shown under the conditions, in Theorem B, which are replaced $n+1$ by $n/p + \varepsilon$.

References

[1] R. R. Coifman and Y. Meyer, Au delà des opérateurs pseudo-differentiels, Asterisque, 57 (1978), 1 – 85.

[2] H. O. Cordes, On compactness of commutators of multiplications and convolutions, and boundedness of pseudo-differential operators, J. Functional Analysis, 18 (1975), 115 – 131.

[3] L. Hörmander, Estimates for translation invariant operators in L^p-spaces, Acta Math., 104 (1960), 93 – 140.

[4] L. Hörmander, Pseudo-differential operators and hypo-elliptic equations, Proc. Symposium on Singular Integrals, A.M.S., 10 (1967), 138 – 183.

[5] T. Kato, Boundedness of some pseudo-differential operators, Osaka J. Math., 13 (1976), 1 – 9.

[6] H. Kumano-go, PSEUDO-DIFFERENTIAL OPERATORS, MIT Press, Cambridge, Massachusetts and London, England, 1982.

[7] A. Miyachi, Estimates for pseudo-differential operators of class $S^0_{0,0}$, to appear.

[8] A. Miyachi and K. Yabuta, L^p-boundedness of pseudo-differential operators with non-regular symbols, Bull. Fac. Sci. Ibaraki Univ. Ser. A., 17 (1985), 1 – 20.

[9] T. Muramatu, Estimates for the norm of pseudo-differential operators by means of Besov spaces I, L^2-theory, to appear.

[10] T. Muramatu and M. Nagase, L^2-boundedness of pseudo-differential operators with non-regular symbols, Canadian Math. Soc. Conference Proc. 1 (1981), 135 – 144.

[11] M. Nagase, On a class of L^p-bounded pseudo-differential operators, Sci. Rep. College of General Education, Osaka Univ., 33 (1984), 1 – 7.

[12] M. Nagase, On some classes of L^p-bounded pseudo-differential operators, Osaka J. Math., 23 (1986), to appear.

[13] Wang Rouhuai and Li Chengzhang, On the L^p-boundedness of several classes of pseudo-differential operators, Chinese Ann. Math., 5 B (1984), 193 – 213.

[14] M. Taylor, PSEUDO-DIFFERENTIAL OPERATORS, Princeton Univ. Press, Princeton, NJ, 1981.

SPACES OF WEIGHTED SYMBOLS AND

WEIGHTED SOBOLEV SPACES ON MANIFOLDS

Elmar Schrohe, Fachbereich Mathematik
Johannes Gutenberg-Universität
Saarstr. 21, D-6500 MAINZ

This paper gives an approach to pseudodifferential operators on noncompact manifolds using a suitable class of weighted symbols and Sobolev spaces introduced by H.O. Cordes on $\mathbb{R}^n$. Here, these spaces are shown to be invariant under certain changes of coordinates. It is therefore possible to transfer them to manifolds with a compatible structure.

INTRODUCTION. The aim of this paper is to make noncompact manifolds and manifolds with singularities accessible to the highly developed methods of pseudodifferential operators on compact manifolds. The concept is based on a class of weighted symbols on $\mathbb{R}^n$, analyzed by H.O. CORDES in [COG]. He introduced 'double order' symbol spaces $SG_1^m(\mathbb{R}^n)$, $m = (m_1,m_2) \in \mathbb{R}^2$, consisting of smooth functions on $\mathbb{R}^n \times \mathbb{R}^n$ such that

(0.1) $$D_\xi^\alpha D_x^\beta a(x,\xi) = O(\langle\xi\rangle^{m_1-|\alpha|}\langle x\rangle^{m_2-|\beta|}).$$

The correspondig weighted Sobolev spaces $H_s(\mathbb{R}^n)$, $s = (s_1,s_2) \in \mathbb{R}^2$, are defined in the canonical way. One then recovers some of the important features of pseudodifferential operator theory on compact manifolds: an operator with a symbol of order m for example maps each $H_s(\mathbb{R}^n)$ continuously into $H_{s-m}(\mathbb{R}^n)$; the imbedding $H_s(\mathbb{R}^n) \hookrightarrow H_t(\mathbb{R}^n)$ is compact, if $s_1>t_1$ and $s_2>t_2$; operators of order $m = (-\infty,-\infty)$ can be written as integral operators with kernels $k(x,y)$ in the Schwartz space $\mathcal{S}(\mathbb{R}^n\times\mathbb{R}^n)$ of rapidly decreasing functions, etc. A short summary is given in section 1.

In order to transfer these spaces to manifolds, one has to show their invariance with respect to changes of coordinates. It is obvious that the diffeomorphisms will have to meet certain requirements, since the weight associated with the symbol is to be preserved under the transformation. It turns out that the corresponding condition on the coordinate changes φ is

COMP 1 $$D_x^\alpha\varphi(x) = O(\langle x\rangle^{1-|\alpha|}).$$

The details are stated in theorem 2.2. It is interesting that the proof requires an additional condition (COMP 2): the diffeomorphism φ is assumed to have an extension to a larger domain, a set so large that, for each x in the original domain, it contains a ball around x of radius $\epsilon\langle x\rangle$ with a fixed $\epsilon > 0$. Considering things from the opposite point of view, this means that the theorem should not be applied to the diffeomorphism on the full domain but only on a 'shrinking'. In the context of manifolds this interpretation seems to be more appropriate.

In section 3, the transition to manifolds is carried out. On SG-compatible manifolds we want all the changes of coordinates to satisfy the conditions COMP 1 and COMP 2. This is achieved by asking not only for a cover by finitely many charts, but also for a 'shrinking', a cover by smaller open sets, so that the larger ones contain -in local coordinates- an $\epsilon\langle x\rangle$ ball around each x in the smaller ones, cf. definition 3.1 and lemma 3.3. Example 3.4 shows that everything holds for manifolds of the 'compact center with finitely many cylindrical ends' type. This includes manifolds that are 'euclidean at infinity', introduced by Y. CHOQUET-BRUHAT and D. CHRISTODOULOU [CHO]; it also covers manifolds with singularities. These are of major interest in recent publications (cf. e.g. J. BRÜNING and R. SEELEY [BRS], R. LOCKHART and R. McOWEN [LOM], W.MÜLLER [MST], [MPS] or S. REMPEL and B.-W. SCHULZE [RES]).

The transition from operators on $\mathbb{R}^n$ to operators on the manifold requires a partition of unity and cut-off functions. Of course, multiplication with these functions has to be part of the algebra, i.e. it must be possible to choose these functions so that they satisfy equation (0.1) (independent of ξ). The construction is carried out in 3.6 - 3.10. Again, the shrinking condition plays an important role.

Section 4 finally contains the actual definitions. They are modeled after the standard case.

The main reason for using the classes $SG_1^m(\mathbb{R}^n)$ is their convenience in applications to complex powers of operators. This will be shown in a forthcoming paper, cf. also [SCD]. Basically, I think that the whole concept is fairly general, and it should be possible to extend it to larger symbol classes, e.g. to some of the classes R. BEALS considered in [BEA]. Actually, this present version does not even use the full generality of H.O. CORDES' original concept. He allowed for two additional parameters ρ,δ, in analogy to L. HÖRMANDER'S classes $S^m_{\rho,\delta}(\mathbb{R}^n)$, cf. [HPO],

[HOE]. The classes $SG_1^m(\mathbb{R}^n)$ represent the most convienient choice $\rho = 1$, $\delta = 0$.

I am also confident that it should be possible to show that the operators with symbols in these classes form a Ψ^*-algebra in the sense of B. GRAMSCH [GRI]. This would be of interest with respect to perturbation theory and the operational calculus in several variables by L. WAELBROEK (cf. [WAE]).

ACKNOWLEDGEMENT. The main part of this concept has been developed during my dissertation project with Prof.B.GRAMSCH as advisor. I would like to thank him for the encouragement and many helpful conversations. I am also grateful to Prof.H.O.CORDES for very valuable suggestions and discussions during the academic year 1984/85 at Berkeley.

1.DEFINITIONS AND PRELIMINARY RESULTS.

1.1 DEFINITION. (a) For $x \in \mathbb{R}^n$ let $\langle x\rangle = (1+|x|^2)^{1/2}$ and $B(x,r) = \{y \in \mathbb{R}^n: |y-x|<r\}$, $r>0$; the ball around x with radius r.

(b) $\mathcal{S}(\mathbb{R}^n)$ is the Schwartz space of rapidly decreasing functions on $\mathbb{R}^n$ equipped with the usual Frechet topology.

(c) For $m = (m_1,m_2,m_3) \in \mathbb{R}^3$ let $SG^m(\mathbb{R}^n)$ denote the class of all [q×q matrix valued] functions $a \in C^\infty(\mathbb{R}^n \times \mathbb{R}^n \times \mathbb{R}^n)$ satisfying the estimate

$$D_\xi^\alpha D_x^\beta D_y^\gamma\, a(x,y,\xi) = O(\langle\xi\rangle^{m_1-|\alpha|} \langle x\rangle^{m_2-|\beta|} \langle y\rangle^{m_3-|\gamma|}) \tag{1.1}$$

for all multiindices α, β, $\gamma \in \mathbb{N}_0^n$. It also carries the usual Frechet topology.

The case of $q\times q$ matrices will later on correspond to the case of a finite (q-) dimensional vector bundle over a manifold. In all the definitions and statements of this paper, a may be both, a function or a matrix of functions.

(d) $SG_1^m(\mathbb{R}^n)$ denotes the subspace of $SG^m(\mathbb{R}^n)$ consisting of all functions a independent of y.

For simplicity we will then write $a(x,\xi)$ instead of $a(x,y,\xi)$ and identify $m^* = (m_1,m_2)$ and $m = (m_1,m_2,0)$. One has the obvious extensions

$$SG^\infty(\mathbb{R}^n) = \bigcup_m SG^m(\mathbb{R}^n) \qquad SG^{-\infty}(\mathbb{R}^n) = \bigcap_m SG^m(\mathbb{R}^n)$$

$$SG_1^\infty(\mathbb{R}^n) = \bigcup_m SG_1^m(\mathbb{R}^n) \qquad SG_1^{-\infty}(\mathbb{R}^n) = \bigcap_m SG_1^m(\mathbb{R}^n).$$

All these functions will be referred to as symbols.

(e) For $m = (m_1, m_2)$, $\mu = (\mu_1, \mu_2) \in \mathbb{R}^2$ write $m \geq \mu$, if $m_1 \geq \mu_1$ and $m_2 \geq \mu_2$ and similarly write $m > \mu$, if $m_1 > \mu_1$ and $m_2 > \mu_2$.

(f) For $a \in SG^m(\mathbb{R}^n)$ define the operator $Op(a)$ on $\mathcal{S}(\mathbb{R}^n)$ by

$$(1.2) \qquad (Op(a))f(x) = (2\pi)^{-n} \int_{\mathbb{R}^n}\int_{\mathbb{R}^n} e^{i(x-y)\xi} a(x,y,\xi)\, f(y)\, dyd\xi, \qquad f \in \mathcal{S}(\mathbb{R}^n),$$

cf. proposition 1.2.

(g) As usual, we will say that the operator A is a pseudodifferential operator with symbol a and order $\leq m$, if $A = Op(a)$ with some $a \in SG^m(\mathbb{R}^n)$.

(h) $\mathcal{K}$ denotes the space of integral operators with kernels in $\mathcal{S}(\mathbb{R}^{2n})$, i.e. all operators K such that for some $k \in \mathcal{S}(\mathbb{R}^{2n})$, we have

$$Kf(x) = \int_{\mathbb{R}^n} k(x,y)\, f(y)\, dy, \qquad f \in \mathcal{S}(\mathbb{R}^n).$$

The following proposition states a few basic properties of the operators with symbols in $SG^m(\mathbb{R}^n)$. Proofs can be found in [COG].

1.2 PROPOSITION. (a) The double integral in equation (1.2) exists in the following sense: the inner integral $\int \ldots dy$ is a Lebesgue integral for all (x,ξ) and defines a function $I(x,\xi) \in C^\infty(\mathbb{R}^n)$. For each $x \in \mathbb{R}^n$, $I(x,\cdot)$ is a Lebesgue integrable function, and the function $g(x) = \int I(x,\xi)\, d\xi$ is in $\mathcal{S}(\mathbb{R}^n)$. Moreover, $Op(a)\colon \mathcal{S}(\mathbb{R}^n) \to \mathcal{S}(\mathbb{R}^n)$ is continuous with respect to the Frechet topology on $\mathcal{S}(\mathbb{R}^n)$, cf. [COG], Thm. 1.1, and remark 1.3 of this paper.

(b) In case $a \in SG_1^m(\mathbb{R}^n)$, definition 1.2(f) coincides with the usual definition:

$$Op(a)f(x) = (2\pi)^{-n/2} \int_{\mathbb{R}^n} e^{ix\xi} a(x,\xi)\, \hat{f}(\xi)\, d\xi, \qquad f \in \mathcal{S}(\mathbb{R}^n),$$

where $\hat{f}(\xi) = (2\pi)^{-n/2} \int e^{-ix\xi} f(y)\, dy$ denotes the Fourier transform of f.

(c) The space of operators with symbols in $SG^{-\infty}(\mathbb{R}^n)$ coincides with $\mathcal{K}$, the space of integral operators with rapidly decreasing kernels, cf. [COG], Thm. 6.6.

(d) Given $a \in SG^m(\mathbb{R}^n)$, there is a symbol $b \in SG_1^\mu(\mathbb{R}^n)$, $\mu = (m_1, m_2+m_3)$, such that $Op(a) = Op(b) + K$ with $K \in \mathcal{K}$, cf. [COG], Thm. 3.2. This function b will also be called a symbol of A.

(e) If $a \in SG_1^m(\mathbb{R}^n)$ and $b \in SG_1^\mu(\mathbb{R}^n)$ then $Op(a)Op(b) = Op(c) + K$ for some $c \in SG_1^{m+\mu}(\mathbb{R}^n)$ and $k \in \mathcal{K}$.

1.3 REMARK. (a) Due to proposition 1.2(d) we can in most cases confine ourselves to operators with symbols in $SG_1^{\infty}(\mathbb{R}^n)$.

(b) In order to prove proposition 1.2(a), one does not need the strong condition (1.1). In [COG], thm. 1.1, it is actually shown that it is sufficient to require

(1.3.a) $$D_\xi^\alpha D_x^\beta D_y^\gamma\, a(x,y,\xi) = O(\langle\xi\rangle^{\kappa(|\beta+\gamma|)}\langle|x|+|y|\rangle^{\lambda(|\alpha|)}),$$

where $\kappa,\lambda\colon \mathbb{N}_0 \to \mathbb{R}$ are monotone functions with

$$\lim_{j\to\infty}(\kappa(j)-j) = \lim_{j\to\infty}(\lambda(j)-j) = -\infty.$$

Using the same methods, this result can even be shown to extend to the case

(1.3.b) $$D_\xi^\alpha D_x^\beta D_y^\gamma\, a(x,y,\xi) = O(\langle\xi\rangle^{\kappa(|\beta+\gamma|)}\langle|x|+|y|\rangle^{\lambda(|\alpha|)+|\beta+\gamma|}),$$

with κ, λ as before.

(c) Using the notation of definition 1.1(e), we can define asymptotic expansions for symbols in $SG_1^m(\mathbb{R}^n)$ similarly as in the standard case, cf. [TAY], ch. II §3. We will then have the classical results. We use the notation $a \sim b$, if $a-b \in SG^{-\infty}(\mathbb{R}^n)$. In particular, the asymptotic expansion of the symbol c of a product of two operators $Op(a)Op(b)$, $a \in SG_1^m(\mathbb{R}^n)$, $b \in SG_1^{\mu}(\mathbb{R}^n)$, has the asymptotic expansion

(1.4) $$c \sim \Sigma_\alpha\, (i^{|\alpha|}/\alpha!)\, D_\xi^\alpha a(x,\xi) D_x^\alpha b(x,\xi),$$ cf. [COG], Thm. 7.3,

and for the symbol b of proposition 1.2(d) we have, cf. [COG], thm. 3.2,

$$b(x,\xi) \sim \Sigma_\alpha\, (i^{|\alpha|}/\alpha!) D_y^\alpha D_\xi^\alpha a(x,y,\xi)\,|_{y=x},$$

this expansion being unique up to a symbol in $SG^{-\infty}(\mathbb{R}^n)$.

1.4 DEFINITION. For $s = (s_1,s_2) \in \mathbb{R}^2$, define the weighted Sobolev space

$$H_s(\mathbb{R}^n) = \{u \in \mathcal{S}'(\mathbb{R}^n)\colon \langle x\rangle^{s_2} Op(\langle\xi\rangle^{s_1})u \in L^2\}.$$

It is equipped with the norm $\|u\|_s = \|\langle x\rangle^{s_2} Op(\langle\xi\rangle^{s_1})u\|_{L^2}$.

We will also use the notations "$>$" and "$\geq$" of definition 1.1(e) with Sobolev space subscripts.

1.5 PROPOSITION. (a) If $a \in SG^m(\mathbb{R}^n)$, then $Op(a)$ is a bounded linear operator from $H_s(\mathbb{R}^n)$ into $H_{s-m}(\mathbb{R}^n)$ for each $s \in \mathbb{R}^2$.

(b) In particular, $H_s(\mathbb{R}^n)$ is continuously imbedded in $H_t(\mathbb{R}^n)$ if $s \geq t$.

(c) If $s > t$, then the imbedding $H_s(\mathbb{R}^n) \hookrightarrow H_t(\mathbb{R}^n)$ is compact, cf. [COG], Cor. 12.3.

1.6 DEFINITION. Let $m \in \mathbb{R}^2$.

(a) An operator $A = Op(a)$, $a \in SG_1^m(\mathbb{R}^n)$, is called md-elliptic of order m, if, for all sufficiently large $|x|+|\xi|$, $[a(x,\xi)]^{-1}$ exists and $a(x,\xi)^{-1} = O(\langle\xi\rangle^{-m_1}\langle x\rangle^{-m_2})$. In that case, also the symbol a is called md-elliptic.

(b) Similarly, an operator $A = Op(a)$, $a \in SG^{\mu}(\mathbb{R}^n)$, is called md-elliptic of order m, if there is an md-elliptic $b \in SG_1^m(\mathbb{R}^n)$, $m = (\mu_1, \mu_2+\mu_3)$, with $Op(a)-Op(b) \in \mathcal{K}$.

(c) A $\mathcal{K}$-parametrix of a pseudodifferential operator A is a pseudodifferential operator B such that $AB - Id$, $BA - Id \in \mathcal{K}$.

1.7 PROPOSITION. A pseudodifferential operator $A \sim Op(a)$ with an md-elliptic symbol $a \in SG_1^m(\mathbb{R}^n)$ has a $\mathcal{K}$-parametrix and is a Fredholm operator from $H_r(\mathbb{R}^n)$ to $H_{r-m}(\mathbb{R}^n)$ for every $r \in \mathbb{R}^2$, cf. [COG], thms.9.1 and 12.4.

2. COORDINATE TRANSFORMS

2.1 DEFINITION. (a) Suppose $\varphi : V \to U$ is a (C^∞) diffeomorphism of open sets $U,V \subseteq \mathbb{R}^n$. For given $f \in C^\infty(U)$, define $f^* \in C^\infty(V)$ by $f^*(x) = f(\varphi(x))$.

(b) If A is a linear operator acting on functions defined in U, then A induces an operator A^* acting on functions defined in V via φ:

$$(A^*f^*)(x) = (Af)(\varphi(x)).$$

The following theorem shows the preservation of the symbol classes $SG_1^m(\mathbb{R}^n)$ under certain changes of coordinates. It is the main result of this section and the rest of § 2 is devoted to its proof.

2.2 THEOREM. Suppose $\varphi: V^\# \to U^\#$ is a diffeomorphism of open sets in $\mathbb{R}^n$ with inverse φ^-, and assume that the derivatives satisfy

COMP 1: $\varphi^{(\alpha)}(x) = O(\langle x\rangle^{1-|\alpha|})$, $x \in V^\#$ and $[\varphi^-]^{(\alpha)}(y) = O(\langle y\rangle^{1-|\alpha|})$, $y \in U^\#$

Assume further

COMP 2: There are subsets U, V of $U^\#$, $V^\#$, resp., such that $\varphi|_V: V \to U$ diffeomorphically, and there is a fixed constant δ_X such that $B(x,\delta_X\langle x\rangle) \subseteq V^\#$ for all $x \in V$ and $B(y,\delta_X\langle y\rangle) \subseteq U^\#$ for all $y \in U$.

Then we will have coordinate invariance of $SG_1^m(\mathbb{R}^n)$ with respect to $\varphi|_V$: If $A = Op(a)$ with $a \in SG_1^m(\mathbb{R}^n)$, supp $a \subseteq U\times\mathbb{R}^n$, then there is a symbol $b \in SG_1^m(\mathbb{R}^n)$ with supp $b \subseteq V\times\mathbb{R}^n$ such that $A^* = Op(b) + K$ with some $K \in \mathcal{K}$.

2.3 REMARK. Condition COMP 2 is needed for technical reasons. Definition 3.1 gives an interpretation for this condition in connection with manifolds.

The proof of theorem 2.2 is based on Kuranishi's method (cf. GRAMSCH and KALB [GRK], 2.2) for showing the coordinate invariance of the standard symbol classes: one chooses a function $\chi(x,y)$ supported near the diagonal $\{x=y\}$ and writes

$$a(x,\xi) = \chi(x,y)a(x,\xi) + (1-\chi(x,y))a(x,\xi) =: a_1(x,y,\xi) + a_2(x,y,\xi),$$

where $a_1, a_2 \in SG^m(\mathbb{R}^n)$. Since it will turn out that the effect of a_2 is negligeable, we check how a_1 behaves. Suppose $u \in \mathcal{S}(\mathbb{R}^n)$, supp$(u) \subseteq U$.

Set $f = (Op\ a_1)u$, $\varphi(x) = z$, $\varphi(y) = w$. Then

$$\text{(2.1.a)}\quad (Op\ a_1)^* u^*(x) = f(z) = (2\pi)^{-n} \iint e^{i(z-w)\eta}\, a_1(z,w,\eta)\, u(w)\, dwd\eta$$

$$= (2\pi)^{-n} \iint e^{i[\varphi(x)-\varphi(y)]\eta}\, a_1(\varphi(x),\varphi(y),\eta)\, u(\varphi(y))\, |\det(\varphi'(y))|\, dyd\eta,$$

with the transformation $w = \varphi(y)$. If the (x,y)-component of the support of a_1 is contained in a sufficiently small neighborhood of the diagonal $\{x=y\}$, $|x-y|$ will be small and we can write

$$\varphi(x) - \varphi(y) = M(x,y)(x-y), \qquad \text{with } M \text{ invertible.}$$

What "small" means in this context and what M precisely is, will be clarified in lemma 2.4. In that case, we may substitute: $\xi = M^T(x,y)\eta$, where $(\cdot)^T$ stands for transposition, and rewrite the last integral as

$$\text{(2.1.b)}\quad = (2\pi)^{-n} \iint e^{i(x-y)\xi}\, a_1(\varphi(x),\varphi(y),M^{-T}(x,y)\xi)\ \times$$

$$\times\ |\det \varphi'(y)|\,|\det M^{-T}(x,y)|\, u^*(y)\, dyd\eta.$$

We now set

$$\text{(2.2)}\quad c(x,y,\xi) = a_1(\varphi(x),\varphi(y),M^{-T}(x,y)\xi)\cdot|\det \varphi'(y)|\,|\det M^{-T}(x,y)|$$

and show that this is a symbol in $SG^m(\mathbb{R}^n)$ by computing an asymptotic expansion in terms of $SG_1^m(\mathbb{R}^n)$-symbols.

2.4 LEMMA. If $x,y \in U$, $|x-y| \le \delta_X\langle x\rangle$, define $M(x,y) = \int_0^1 \varphi'(y + t(x-y))dt$. The prime (') denotes the total differential (here, we use condition COMP 2). Then

(a) $$\varphi(x) - \varphi(y) = M(x,y)(x-y).$$

(b) $M(x,y)$ is invertible for $|x-y| \le k\langle x\rangle$ with a suitable constant k, $0 < k \le \delta_X$.

(c) There is a function $\chi(x,y) \in SG^0(\mathbb{R}^n)$ such that

$$\chi(x,y) = 1, \quad |x-y| \le \frac{k}{2}\langle x\rangle \quad \text{and} \quad \chi(x,y) = 0, \quad |x-y| > k\langle x\rangle$$

with the constant k of (b).

PROOF. (a) is the mean value theorem; (b) follows from the fact that $M(x,x) = \varphi'(x)$ is invertible and φ satisfies the estimates in condition COMP 1.

(c) Choose a function $g \in C^\infty(\mathbb{R})$ such that $g(t) = 1$ for $t \le 1/2$ and $g(t) = 0$ for $t > 1$. Set $\chi(x,y) = g(|x-y|/(k\langle x\rangle))$.

2.5 LEMMA. Write $a_1(x,y,\xi) = \chi(x,y)a(x,\xi)$, $a_2(x,y,\xi) = (1-\chi(x,y))a(x,\xi)$ with the function $\chi(x,y)$ of lemma 2.4(c). Then

(a) $a_1 \in SG^m(\mathbb{R}^n)$, $a_2 \in SG^{-\infty}(\mathbb{R}^n)$.

(b) $(\mathrm{Op}\ a)^* = (\mathrm{Op}\ a_1)^* + (\mathrm{Op}\ a_2)^*$ with the $(\cdot)^*$-notation of definition 2.1.

(c) $(\mathrm{Op}\ a_2)^* \in \mathcal{K}$.

(d) $(\mathrm{Op}\ a_1)^*$ can be written as an operator with the 'symbol' $c(x,y,\xi)$ of equation (2.2). The use of 'symbol' is formal here, it will be justified in remark 2.11.

PROOF. (a) By 1.2(e), a_1 is in $SG^m(\mathbb{R}^n)$, since $\chi \in SG^0(\mathbb{R}^n)$. Computing the asymptotic expansion given in 1.3(c), we find that $a_2 \sim 0 \in SG^{-\infty}(\mathbb{R}^n)$. (b) is due to linearity of $(\cdot)^*$. (c): By (a), $\mathrm{Op}(a_2)$ has a kernel $k \in \mathcal{S}(\mathbb{R}^{2n})$. Then $(\mathrm{Op}\ a_2)^*$ has the kernel l defined by $l(x,y) = k(\varphi(x),\varphi(y))$, which is also in $\mathcal{S}(\mathbb{R}^{2n})$ by the chain rule. Now (d) follows from the calculations in equations (2.1.a,b), since all the transformations are justified by lemma 2.4.

A Taylor expansion of c with respect to the variable y gives the following result.

2.6 LEMMA. For each $N \in \mathbb{N}_0$ we have

$$c(x,y,\xi) = \Sigma_{|\alpha|\le N} (-i)^{|\alpha|}/\alpha!\ D_y^\alpha c(x,y,\xi)|_{y=x} (x-y)^\alpha + r_N(x,y,\xi), \quad \text{where}$$

$$r_N(x,y,\xi) = \Sigma_{|\alpha|=N+1} (N+1)(-i)^{N+1}/\alpha!\ r_\alpha(x,y,\xi)(x-y)^\alpha, \quad \text{and}$$

$$r_\alpha(x,y,\xi) = \int_0^1 D_y^\alpha c(x,x+t(y-x),\xi)(1-t)^N dt.$$

2.7 DEFINITION. For $\theta \in \mathbb{N}_0^n$ let

$$c_\theta(x,y,\xi) = i^{|\theta|}/\theta!\ D_\xi^\theta D_y^\theta c(x,y,\xi),$$

$$c_{\theta,t}(x,y,\xi) = D_\xi^\theta D_y^\theta c(x,x+t(y-x),\xi), \quad 0 \le t \le 1,$$

$$c_N(x,y,\xi) = \Sigma_{|\theta|=N+1} (N+1)i^{N+1}/\theta! \int_0^1 c_{\theta,t}(x,y,\xi)(1-t)^N dt$$

We want to show that $c_\theta(x,x,\xi) \in SG_1^{m^\theta}(\mathbb{R}^n)$, where $m^\theta = (m_1-|\theta|,m_2-|\theta|)$, and $c \sim \Sigma_\theta\ c_\theta(x,x,\xi)$. Since the expression for c in equation (2.2) is rather complicated and we need estimates for its derivatives, it is helpful to define

2.8 DEFINITION. For $0 \le t \le 1$ let

$$F_t(x,y,\xi) = a_1(\varphi(x),\ \varphi(x+t(y-x)),M^{-T}(x,x+t(y-x))\xi),$$

$$G_t(x,y) = |\det \varphi'(x+t(y-x))|,$$

$$H_t(x,y) = |\det M^{-T}(x,x+t(y-x))|.$$

2.9 LEMMA. For all multi-indices $\alpha,\beta,\gamma,\theta \in \mathbb{N}_0^n$,

(a) $c = F_1G_1H_1$; $\quad c_\theta = i^{|\theta|}/\theta!\ D_\xi^\theta D_y^\theta(F_1G_1H_1)$; $\quad c_{\theta,t} = D_\xi^\theta D_y^\theta(F_tG_tH_t)$.

(b) $\langle x+t(y-x)\rangle^{-j} = O(\langle x\rangle^{-j}\langle x-y\rangle^{j})$, independent of t, for $0 \le t \le 1$, $\quad j \ge 0$.

(c) $D_x^\alpha D_y^\beta G_t(x,y) = O(\langle x\rangle^{-|\alpha+\beta|}\langle x-y\rangle^{|\alpha+\beta|})$,

$D_x^\alpha D_y^\beta H_t(x,y) = O(\langle x\rangle^{-|\alpha+\beta|}\langle x-y\rangle^{|\alpha+\beta|})$,

$D_x^\alpha D_y^\beta D_\xi^\gamma F_t(x,y,\xi) = O(\langle\xi\rangle^{m_1-|\gamma|}\langle x\rangle^{m_2-|\alpha+\beta|}\langle x-y\rangle^{|\alpha+\beta|})$.

PROOF. (a) is obvious from the definition, (b) follows from Peetre's inequality. For (c), one uses (b), property COMP 1 of φ, Leibniz's rule and induction.

2.10 COROLLARY. For all multi-indices $\alpha,\beta,\gamma,\theta \in \mathbb{N}_0^n$,

(a) $D_x^\alpha D_y^\beta D_\xi^\gamma c(x,y,\xi) = O(\langle\xi\rangle^{m_1-|\gamma|}\langle x\rangle^{m_2-|\alpha+\beta|}\langle x-y\rangle^{|\alpha+\beta|})$

(b) $D_x^\alpha D_y^\beta D_\xi^\gamma c_{\theta,t}(x,y,\xi) = O(\langle\xi\rangle^{m_1-|\theta+\gamma|}\langle x\rangle^{m_2-|\theta+\alpha+\beta|}\langle x-y\rangle^{|\theta+\alpha+\beta|})$

(c) $c_\theta(x,x,\xi) \in SG_1^{m^\theta}(\mathbb{R}^n)$ where $m^\theta = (m_1-|\theta|,m_2-|\theta|)$.

(d) $D_x^\alpha D_y^\beta D_\xi^\gamma c_N(x,y,\xi) = O(\langle\xi\rangle^{m_1-N-1-|\gamma|}\langle x\rangle^{m_2-N-1-|\alpha+\beta|}\langle x-y\rangle^{N+1+|\alpha+\beta|})$

2.11 REMARK. By condition COMP 1, the derivative of φ^-, $(\varphi^-)'$, is bounded. If C is some constant, then $|\varphi(x)-\varphi(y)| \le C\ \langle\varphi(x)\rangle$ implies $|x-y| \le C'\langle x\rangle$ for a sufficiently large constant C'. Inserting for C the constant k used in lemma 2.4, we conlude that $|x-y| > C'\langle x\rangle$ implies that $|\varphi(x)-\varphi(y)| > k\,\langle\varphi(x)\rangle$. Now $a_1(x,y,\xi)$ is zero, if $|x-y| > k\langle x\rangle$, hence

$$c(x,y,\xi) = a_1(\varphi(x),\varphi(y),M^{-T}(x,y)\xi)\cdot\ldots = 0 \quad \text{for} \quad |x-y| > C'\langle x\rangle.$$

Therefore either $c(x,y,\xi)$ is zero or $\langle x-y\rangle = O(\langle x\rangle)$. This gives us the estimate

$$\text{(2.3)} \qquad D_x^{\alpha}D_y^{\beta}D_{\xi}^{\gamma}c(x,y,\xi) = O(\langle\xi\rangle^{m_1-|\gamma|}\langle x\rangle^{m_2}).$$

So condition (1.3.a) holds and c can be used as a symbol.

2.12 LEMMA. For $f \in \mathcal{S}(\mathbb{R}^n)$ the following integrals exist and we have

$$\iint e^{i(x-y)\xi}\,[(-i)^{|\theta|}/\theta!\; D_y^{\theta}c(x,y,\xi)|_{y=x}(x-y)^{\theta}]\; f(y)\; dyd\xi =$$
$$\iint e^{i(x-y)\xi}\; c_{\theta}(x,x,\xi)\; f(y)\; dyd\xi$$

and, provided $N \geq m_1+n$,

$$\iint e^{i(x-y)\xi}\; r_N(x,y,\xi)\; f(y)\; dyd\xi \;=\; \iint e^{i(x-y)\xi}\; c_N(x,y,\xi)\; f(y)\; dyd\xi.$$

PROOF. The first two integrals exist as iterated integrals by remark 1.3(b). Equality is obtained by using the fact that $D_{\xi}^{\alpha}e^{i(x-y)\xi} = (x-y)^{\alpha}e^{i(x-y)\xi}$ together with partial integration. The last integral is a Lebesgue integral for $N \geq m_1+n$, by corollary 2.10(d). Again, partial integration is applied.

2.13 LEMMA. Choose a symbol $d \in SG_1^m(\mathbb{R}^n)$ with $d(x,\xi) \sim \Sigma_{\theta}\, c_{\theta}(x,x,\xi)$, let $d_N(x,\xi) = \Sigma_{|\theta|\leq N}\, c_{\theta}(x,x,\xi)$. Then the difference $Op(c) - Op(d)$ can be written as an operator $Op(e)$, where $e = (d-d_N) + c_N$ satisfies the estimate

$$\text{(2.4)} \qquad D_x^{\alpha}D_y^{\beta}D_{\xi}^{\gamma}e(x,y,\xi) = O(\langle\xi\rangle^{m_1-N-|\gamma|}\langle x\rangle^{m_2-N-|\alpha+\beta|}\langle x-y\rangle^{N+|\alpha+\beta|}).$$

PROOF. Write $Op(c) - Op(d) = Op(c-d_N) - Op(d-d_N)$. Now $d-d_N \in SG_1^{m-(N,N)}(\mathbb{R}^n)$ and thus satisfies the estimate. Furthermore, for $f \in \mathcal{S}(\mathbb{R}^n)$, we have by lemmas 2.6 and 2.12

$$\iint e^{i(x-y)\xi}\,[c(x,y,\xi)-d_N(x,\xi)]\; f(y)\; dyd\xi \;=\; \iint e^{i(x-y)\xi}\; c_N(x,y,\xi)\; f(y)\; dyd\xi.$$

By corollary 2.10(d), c_N satisfies (2.4), so the lemma is proven. Note: c_N may be used as a symbol, because the estimates of corollary 2.10(d) imply those of (1.3.b).

2.14 LEMMA. Denote by e the operator symbol defined in lemma 2.13.

(a) Op (e), considered as an operator on $\mathcal{S}(\mathbb{R}^n)$, can be written as an operator with the kernel $k(x,y) = \int e^{i(x-y)\xi}\; e(x,y,\xi)\; d\xi$.

(b) $k \in C^{\infty}(\mathbb{R}^n)$ and all the derivatives can be taken under the integral.

(c) For arbitrary M, $\langle x\rangle^M\langle x-y\rangle^M D_x^\alpha D_y^\beta k(x,y) = O(1)$.

PROOF. (a) In lemma 2.13 choose $\alpha = \beta = \gamma = 0$ and N so large that $m_1-N < -n$. Then use Fubini's theorem to interchange $\int \ldots dy$ and $\int \ldots d\xi$ in the integral for $Op(e)f(x)$, $f \in \mathcal{S}(\mathbb{R}^n)$. (b): The estimates (2.4) imply that $D_x^\alpha D_y^\beta(e^{i(x-y)\xi}e(x,y,\xi)) = O(\langle\xi\rangle^{-n-1})$ uniformly in x, y for $|x|,|y| \le$ const. This is sufficient to differentiate under the integral for k in (a). For (c) start with the equality $\langle x-y\rangle^{-2R}(1-\Delta_\xi)^R e^{i(x-y)\xi} = e^{i(x-y)\xi}$. Then apply a partial integration in the integral for $k(x,y)$ and use the estimates (2.4).

2.15 COROLLARY. (a) The kernel k of lemma 2.14 is in $\mathcal{S}(\mathbb{R}^{2n})$.
(b) $e \in SG^{-\infty}(\mathbb{R}^n)$. Moreover, $c \in SG^m(\mathbb{R}^n)$ and $c(x,y,\xi) \sim \Sigma_\theta\, c_\theta(x,x,\xi)$.

PROOF. For (a) use that $1+|x|+|y| = O(\langle x\rangle\langle x-y\rangle)$ and the estimate of lemma 2.14(c). The first statement in (b) follows from proposition 1.2(c). For the second and third use the fact that $Op(c) = Op(d) + Op(e)$, hence $Op(c-d) \sim 0$ by (a). Therefore $c-d \in SG^{-\infty}(\mathbb{R}^n)$ and $c(x,y,\xi) \sim d(x,\xi) \sim \Sigma_\theta\, c_\theta(x,x,\xi) \in SG^{-\infty}(\mathbb{R}^n)$.

3. SG-COMPATIBLE MANIFOLDS

3.1 DEFINITION. Let X be an n-dimensional manifold. We will say that X is SG-compatible if

SG 1 $\quad X$ has a finite atlas,

SG 2 $\quad$ the atlas has a 'good' shrinking,

SG 3 $\quad$ The corresponding changes of coordinates satisfy condition COMP 1.

In order to state this more precisely, we need some notation. Let $\{X_j^\#: j = 1,\ldots,J\}$ be a cover of X by open sets and $\varphi_j^\#: X_j^\# \to U_j^\# \subseteq \mathbb{R}^n$ the chart diffeomorphisms. By asking for a 'good' shrinking, we mean that there is a cover $\{X_j: j = 1,\ldots,J\}$ of X by open sets $X_j \subseteq X_j^\#$ with coordinate maps $\varphi_j = \varphi_j^\#|_{X_j} : X_j \to U_j \subseteq \mathbb{R}^n$ (diffeomorphically) and there is a fixed constant ϵ_X such that for every $y \in U_j$,

$$B(y,\epsilon_X\langle y\rangle) \subseteq U_j^\# \tag{3.1}$$

We then require the changes of coordinates $\varphi^{\#}_{jk} = \varphi^{\#}_k(\varphi^{\#}_j)^- : \varphi^{\#}_j(X^{\#}_j \cap X^{\#}_k) \to \varphi^{\#}_k(X^{\#}_j \cap X^{\#}_k)$ to satisfy the compatibility condition COMP 1.

3.2 REMARK. (a) Once condition SG 2 holds, one may without loss of generality assume to have an additional cover, say $\{\hat{X}_j : j = 1,\dots,J\}$ such that $\{\hat{X}_j\}$ is a 'good' shrinking with respect to $\{X^{\#}_j\}$ and $\{X_j\}$ is a 'good' shrinking with respect to $\{\hat{X}_j\}$: one can define $V_j = \bigcup_{y \in U_j} B(y, \delta\langle y\rangle) \subseteq U^{\#}_j$ with $\delta = \epsilon_X/2$ and $\hat{X}_j = (\varphi^{\#}_j)^-(V_j)$. In particular, we may assume that the coordinate charts and maps extend to an open neighborhood of the $X^{\#}_j$.
(b) For convenience of notation, we shall write φ_j for both, $\varphi^{\#}_j$ and φ_j.

3.3 LEMMA. If conditions SG 1 - SG 3 of definition 3.1 hold, then the changes of coordinates will automatically have the compatibility property COMP 2, i.e. there is a constant $\delta > 0$ such that for $x \in \varphi_j(X_j \cap X_k)$ we have

$$B(x, \delta\langle x\rangle) \subseteq \varphi_j(X^{\#}_j \cap X^{\#}_k).$$

PROOF. By remark 3.2 we may assume that the changes of coordinates extend to a neighborhood of $X^{\#}_j$. Invoking the estimates of COMP 1, we see that for some constant $C \geq 1$, $\langle\varphi_j(x)\rangle/\langle\varphi_k(x)\rangle \leq C$ and $|\varphi_{kj}(x) - \varphi_{kj}(y)| \leq C|x-y|$, whenever the expressions make sense. Now let $x \in \varphi_k(X_j \cap X_k)$ and $\delta = \epsilon_X/C^2 \leq \epsilon_X$. Then $B(x,\delta\langle x\rangle) \subseteq \varphi_k(X^{\#}_j \cap X^{\#}_k)$: if this were not true, we could find $y \in \varphi_k((\hat{X}_j \setminus X^{\#}_j) \cap X^{\#}_k) \cap B(x,\delta\langle x\rangle)$. In that case, $|x-y| < \delta\langle x\rangle$ and $|\varphi_{kj}(x) - \varphi_{kj}(y)| \leq \epsilon_X\langle\varphi_{kj}(x)\rangle$. But by assumption SG 2,

$$B(\varphi_{kj}(x), \epsilon_X\langle\varphi_{kj}(x)\rangle) \subseteq \varphi_j(X^{\#}_j),$$

and therefore $\varphi_{kj}(y) \in \varphi_j(X^{\#}_j)$ or, equivalently, $y \in \varphi_k(X^{\#}_j)$, a contradiction.

3.4 EXAMPLE. Suppose X is an n-dimensional manifold of the following form:

$$X = X_0 \cup X_1 \cup \dots \cup X_N \cup \partial X_1 \cup \dots \cup \partial X_N \qquad \text{disjoint,}$$

where $X_0,\dots,X_N$ are n-dimensional submanifolds and $\partial X_1,\dots,\partial X_N$ are connected (n-1)-dimensional submanifolds. X_0 is assumed to be relatively compact, and its boundary ∂X_0 to satisfy $\partial X_0 = \partial X_1 \cup \dots \cup \partial X_N$. Moreover, for $j = 1,\dots,N$, let X_j be diffeomorphic to $\partial X_j \times (1,\infty)$.
Then X is SG-compatible. In particular, every compact manifold is SG-compatible.

3.5 REMARK. These manifolds have also been considered by H.O. CORDES in

[CST], furthermore in many publications by e.g. J. BRÜNING and R. SEELEY, R. LOCKHART and R. McOWEN, W. MÜLLER, S. REMPEL and B.-W. SCHULZE - cf. [BRS], [LOM], [MST], [MPS], [RES] - in connection with 'manifolds with singularities'. This includes manifolds of the type "euclidean at infinity" introduced by Y. CHOQUET-BRUHAT and D. CHRISTODOULOU, cf. [CHO], because in polar coordinates $\mathbb{R}^n \setminus \overline{B(0,R)}$ is canonically isomorphic to $S^{n-1} \times (R,\infty) \cong S^{n-1} \times (1,\infty)$.

PROOF of Example 3.4. For simplicity we may assume N, the number of 'ends', to be 1. Choose finitely many open sets that cover $X_0 \cup \{\partial X_0 \times (1,3)\}$ and whose images are relatively compact. On $\partial X_0 \times (1,\infty)$ we introduce coordinates in the following way: let $\{M_j^*: j = 1,\dots,J\}$ be an open cover of ∂X_0, $\alpha_j: M_j^* \to V_j^* \subseteq \mathbb{R}^{n-1}$ the corresponding coordinate maps; V_j^* relatively compact. By a compactness argument, there is a cover $\{M_j: j = 1,\dots,J\}$ such that $M_j \subseteq M_j^*$ and $V_j^* = \alpha_j(M_j^*)$ contains an ϵ-neighborhood of $V_j = \alpha_j(M_j)$. As a cover of $\partial X_0 \times (1,\infty)$ choose $Y_j^\# = M_j^\# \times (1,\infty)$ with the coordinate maps $\varphi_j: Y_j^\# \to U_j^\# \subseteq \mathbb{R}^n$ given by $\varphi_j(v,t) = (t\alpha_j(v),t)$.

This is an atlas of finitely many charts. As for condition SG 2, first note that it is sufficient to concentrate on the 'end' $\partial X_0 \times (2,\infty)$, and t large: for bounded parts of the manifold, condition (3.1) is easily seen to be satisfied using compactness arguments. On $\partial X_0 \times (2,\infty)$, $Y_j = M_j \times (2,\infty)$ is a shrinking of $Y_j^\# = M_j^\# \times (1,\infty)$, the coordinate maps $\varphi_j: Y_j \to U_j \subseteq \mathbb{R}^n$ given by restriction. Let $x = (t\alpha_j(v),t) \in U_j$. Condition (3.1) requires that, for some $\delta > 0$,

$$B(x,\delta\langle x\rangle) \subseteq U_j^\# = \{z = (r\alpha_j(u),r): u \in M_j^\#,\ r \in (1,\infty)\}.$$

Since $\langle x\rangle \leq 2|x| = 2t\langle\alpha_j(v)\rangle$, and $\langle\alpha_j(v)\rangle$ is bounded, say by C, we only have to show that, for some $\delta > 0$, $\{z \in \mathbb{R}^n: |z-x| < \delta t\} \subseteq U_j^\#$. Setting $\delta = \epsilon/(3C+3\epsilon)$ with the ϵ and C from above this is seen to be true: writing $z = (r\alpha_j(u),r)$, $|z-x| < \delta t$ implies $|\alpha_j(u)-\alpha_j(v)| < \epsilon$, so that $u \in M_j^\#$, hence $z \in U_j^\#$.

We still have to check SG 3. The case of interest is $|x|$ being large. Fortunately, the changes of coordinates on $\partial X_0 \times (1,\infty)$ are simple: If $\chi = \alpha_j\alpha_k^-$ is a coordinate change on ∂X_0, the corresponding one in $\partial X_0 \times (1,\infty)$ is $\varphi: (yt,t) \mapsto (\chi(y)t,t)$. Write $u = yt$ and note that

$$\varphi(u,t) = (\chi(u/t)t,t) = O(\langle(u,t)\rangle).$$

The first derivative is $D\varphi(u,t) = \begin{bmatrix} a & b \\ 0 & 1 \end{bmatrix}$, where $a(u,t) = D\chi(u/t)$, $b = (b_1,\dots,b_n)^T$

with $b_j = \Sigma_k \partial_k \chi_j(u/t)(-u_k/t) + \chi_j(u/t)$. The boundedness of the derivatives of χ proves that $D\varphi = O(1)$, i.e. $\varphi^{(\alpha)}(x) = O(\langle x\rangle^{1-|\alpha|})$ for $|\alpha| = 1$. In the general case, induction shows that $\varphi^{(\alpha)}(u,t)$ is a linear combination of expressions of the form

$$t^{1-|\alpha|}\cdot\chi_j^{(\beta)}(u/t)(u/t)^{\gamma}, \quad \text{with} \quad |\beta|, |\gamma| \leq |\alpha|.$$

Again, the boundedness of both, $\chi^{(\beta)}$ and (u/t), implies the desired estimate.

We are now going to construct a partition of unity compatible with the symbol classes and subordinate to our cover $\{X_j^{\#}\}$. We do some preliminary work first.

3.6 DEFINITION. (a) Let j denote the real-valued function on $\mathbb{R}^n$ given by

$$j(x) = \begin{cases} a\cdot\exp\,[(|x|^2-1)^{-1}] & |x| < 1, \\ 0 & |x| \geq 1, \end{cases}$$

where a is such that $\int j(x)\,dx = 1$.

(b) For $\epsilon>0$ let $j_\epsilon(x) = \epsilon^{-n}\cdot j(x/\epsilon)$.

NOTE: $j_\epsilon(x) \geq 0$, $j_\epsilon(x) = 0$ for $|x|\geq\epsilon$, $\int j_\epsilon(x)\,dx = 1$.

3.7 LEMMA. Let Ω be an open set in $\mathbb{R}^n$ and χ_Ω its characteristic function. Define $\varphi_\Omega: \mathbb{R}_+ \times \mathbb{R}^n \to \mathbb{R}$ by

$$\varphi_\Omega(\epsilon,x) = \int \chi_\Omega(y)\, j_{\epsilon\langle x\rangle}(x-y)\,dy.$$

Then the following properties of φ_Ω are easily established.

(a) $\varphi_\Omega \in C^\infty(\mathbb{R}^n)$, $0 \leq \varphi_\Omega \leq 1$.

(b) $\varphi_\Omega(\epsilon,x) = 0$, if $\operatorname{dist}(x,\Omega) \geq \epsilon\langle x\rangle$.

(c) $\varphi_\Omega(\epsilon,x) = 1$, if $B(x,\epsilon\langle x\rangle) \subseteq \Omega$.

(d) For ϵ fixed, $D_x^\alpha\varphi_\Omega(\epsilon,x) = O(\langle x\rangle^{-|\alpha|})$.

This has useful consequences. For an open set $\Omega \subseteq \mathbb{R}^n$ and $\epsilon>0$ denote by Ω_ϵ the open set $\bigcup_{x\in\Omega} B(x,\epsilon\langle x\rangle)$. We will then obtain the corollary, below.

3.8 COROLLARY. Let Φ be the function $\varphi_{\Omega_\epsilon}(\epsilon,\cdot)$ as defined in lemma 3.7.

(a) $\Phi \in SG_1^0(\mathbb{R}^n)$, $0 \leq \Phi \leq 1$.

(b) $\Phi \equiv 1$ on Ω; $\Phi \equiv 0$ outside $\Omega_{2\epsilon}$.

3.9 THEOREM. There is a partition of unity on X subordinate to the cover $\{X_j^{\#}: j = 1,\dots,J\}$, i.e. there are functions $\{\Phi_j: j = 1,\dots,J\}$ such that

(a) $\Phi_j: X \to \mathbb{R}^n$, $\operatorname{supp}\Phi_j \subseteq X_j^{\#}$, $0 \leq \Phi_j \leq 1$, $\Sigma_{j=1}^J \Phi_j = 1$.

(b) In local coordinates we have $(\Phi_j)_*^{(\alpha)}(y) = O(\langle y\rangle^{-|\alpha|})$, i.e. $(\Phi_j)_* \in SG_1^0(\mathbb{R}^n)$. Here $(\Phi_j)_*$ is defined on $\mathbb{R}^n$ via the inverse coordinate maps φ_j^- by

$$(\Phi_j)_*(x) = \begin{cases} \Phi_j(\varphi_j^-(x)), & x \in U_j^\# \\ 0 & x \notin U_j^\#. \end{cases}$$

PROOF. Define the function Ψ_j on $U_j^\#$ by $\Psi_j = \Psi_{(U_j)_\delta}(\delta,\cdot)$, where $\delta = \epsilon_X/3$ with the constant ϵ_X of definition 3.1. Write $\Psi_j^*(x) = \begin{cases} \Psi_j(\varphi_j(x)), & x \in X_j^\# \\ 0 & x \notin X_j^\#. \end{cases}$

Now let $\Phi(x) = \Psi_j^*(x)/(\Sigma_{k=1}^J \Psi_k^*(x))$. The denominator is always ≥ 1, since $\Psi_j \equiv 1$ on X_j and the X_j cover X. Now (a) is obvious. For (b), note that, given $y \in U_j^\#$,

$$(\Phi_j)_*(y) = \Psi_j(y)/[\Sigma\, \Psi_k(\varphi_k\varphi_j^-(y))],$$

where the sum is over all k such that $\varphi_j^-(y) \in X_k^\#$. If we differentiate the quotient, we will get the desired estimate by using condition COMP 1 and corollary 3.8(a).

The following corollary furnishes an imporant tool when dealing with pseudodifferential operators on SG-compatible manifolds.

3.10 LEMMA. Let $\{\Phi_j : j = 1,\dots,J\}$ be the partition of unity constructed in theorem 3.9. Then there are functions Θ_j such that

(a) $0 \leq \Theta_j \leq 1$, $\Theta_j \equiv 1$ on $\operatorname{supp} \Phi_j$, $\Theta_j \equiv 0$ outside $X_j^\#$.

(b) In local coordinates, $(\Theta_j)_*^{(\alpha)}(y) = O(\langle y\rangle^{-|\alpha|})$, i.e. $(\Theta_j)_* \in SG_1^0(\mathbb{R}^n)$.

In particular, we have $\Theta_j\Phi_j = \Phi_j$. Moreover, the proof shows that we can iterate this method and obtain a sequence of functions $\{\Theta_j^{(k)} : k = 1,2,\dots\}$ such that $0 \leq \Theta_j^{(k)} \leq 1$, $\Theta_j^{(k+1)} \equiv 1$ on $\operatorname{supp} \Theta_j^{(k)}$ and $\Theta_j^{(k)} \equiv 0$ outside $X_j^\#$.

PROOF. By theorem 3.9 we have $(\Phi_j)_*(x) = 0$ outside $(U_j)_\delta$, where $\delta = 2/3\cdot\epsilon_X$ with ϵ_X from definition 3.1. Let $\Omega_j = (U_j)_\delta$ and define the function θ_j on $\mathbb{R}^n$ by $\theta_j = \Psi_{(\Omega_j)_\mu}(\mu,\cdot)$, where $\mu = \epsilon_X/9$, with the function $\Psi_\Omega(\epsilon,\cdot)$ constructed in lemma 3.7. Obviously, $0 \leq \theta_j \leq 1$, $\operatorname{supp} \theta_j \subseteq U_j^\#$, $\theta_j \equiv 1$ on $\operatorname{supp} \Phi_j$. Then let

$$\Theta_j(x) = \begin{cases} \theta_j(\varphi_j(x)), & x \in X_j^\# \\ 0, & x \notin X_j^\#. \end{cases}$$

By an argument as in theorem 3.9 we will obtain the estimates on the derivatives.

4. THE SPACES $SG_1^m(X)$, $H_s(X)$, $s \in \mathbb{R}^2$.

4.1 NOTATION. (a) In this section let X denote an SG-compatible manifold with covers $\{X_j^{\#}: j = 1,\ldots,J\}$ and $\{X_j: j = 1,\ldots,J\}$ by open sets as in definition 3.1, $\varphi_j: X_j^{\#} \to U_j^{\#} \subseteq \mathbb{R}^n$, $X_j \to U_j \subseteq \mathbb{R}^n$, resp., the coordinate maps, $\{\Phi_j: j = 1,\ldots,J\}$ the partition of unity subordinate to $\{X_j^{\#}\}$ constructed in theorem 3.9.

(b) As in theorem 3.9 and lemma 3.10 we shall use the transfer operators to local coordinates $(\cdot)_*: C^\infty(X_j^{\#}) \to C^\infty(U_j^{\#})$ given via the inverse φ_j^- by

$$(f)_*(x) = f(\varphi_j^-(x)).$$

4.2 DEFINITION. (a) Let $a(\cdot,\cdot) \in C^\infty(U_j^{\#}\times\mathbb{R}^n)$. We say that $a \in SG_1^m(U_j)$, if $a\theta \in SG_1^m(\mathbb{R}^n)$ for every function $\theta = \theta(x) \in SG_1^0(\mathbb{R}^n)$ supported in $\{x \in U_j^{\#}\}$. Here we think of the functions $a\theta$, supported in $U_j^{\#} \times \mathbb{R}^n$, as extended to $\mathbb{R}^n \times \mathbb{R}^n$ by zero.

(b) By $SG_1^m(X)$ we denote the set $\{a = (a_1,\ldots,a_J): a_j \in SG_1^m(U_j)\}$.

(c) For $f \in C^\infty(X)$ we shall write $f|_{X_j} \in \mathcal{S}(X_j)$, if - in local coordinates - it satisfies the estimates required for $\mathcal{S}(\mathbb{R}^n)$-functions on its domain, i.e. if

$$\|f\|_N^{\mathcal{S}} = \sup_{|\alpha|\le N} \{\langle x\rangle^{|\alpha|} |D_x^\alpha f(\varphi_j(x))|: x \in X_j\} < \infty, \quad N = 1,2,\ldots . \tag{4.1}$$

NOTE: (i) by lemma 3.10, $(f|_{X_j})_*$ can be extended to a function in $\mathcal{S}(\mathbb{R}^n)$.

(ii) If $f|_{X_j\cap X_k}$ satisfies the estimates (4.1) with respect to the coordinate map φ_j, then it also will satisfy (4.1) with respect to the coordinate φ_k, because of condition COMP 1 on the changes of coordinates.

We now want to introduce the concept of pseudodifferential operators on SG-compatible manifolds with symbols corresponding to those in the space $SG_1^m(\mathbb{R}^n)$. We proceed as in the standard case. Consider a linear operator $A: \mathcal{S}(X) \to \mathcal{S}(X)$. A induces an operator on $\mathbb{R}^n$ via the coordinate maps in a canonical way: Pick a function of the partition of unity, Φ_j, and choose another function Θ_j, which has the properties stated in lemma 3.10(a),(b). If f is a function in $\mathcal{S}(X)$, denote by f_j the function $(\Phi_j f)_*$, by g_j the function $(\Theta_j A(\Phi_j f))_*$. Then the operator A_j taking f_j to g_j is a linear operator on $\mathcal{S}(\mathbb{R}^n)$ corresponding to the operator $M_{\Theta_j} A M_{\Phi_j}$ on $\mathcal{S}(X)$, where $M_{(\cdot)}$ stands for the multiplication operator with the function in the subscript.

4.3 DEFINITION: In this notation, A is a pseudodifferential operator with symbol $a \in SG_1^m(X)$, if, in local coordinates, $(1-\Theta_j)A\Phi_j$ has a $\mathcal{S}(\mathbb{R}^{2n})$ kernel and $A_j \sim M_{\theta_j} Op(a_j) M_{\varphi_j}$, where $\theta_j = (\Theta_j)_*$, $\varphi_j = (\Phi_j)_*$, i.e. if after the 'cut-off' from the left and right, A can be written in local coordinates on X_j with symbol a_j.

Note that the symbol a_j does not depend on the choice of the cut-off function Θ_j. If we replace Θ_j by a function $\tilde{\Theta}_j$ having also the properties of lemma 3.10(a),(b), then $M_{\theta_j} Op(a_j) M_{\varphi_j} - M_{(\tilde{\theta}_j)} Op(a_j) M_{\varphi_j} = M_{(\theta_j - \tilde{\theta}_j)} Op(a_j) M_{\varphi_j}$ is regularizing, since $\theta_j - \tilde{\theta}_j$ vanishes on supp φ_j.

4.4 DEFINITION: (a) $H_r(X) = \{f: X \to \mathbb{C}: (\Phi_j f)_* \in H_r(\mathbb{R}^n)\}$ is the weighted Sobolev space of order $r \in \mathbb{R}^2$. Its norm is given by $\|f\|_r = (\Sigma_{j=1}^J \|(\Phi_j f)_*\|_r^2)^{1/2}$, where the norms on the right hand side are those on $\mathbb{R}^n$ introduced in definition 1.4.

This definition is independent of the choice of coordinates. By theorem 2.2, a change of coordinates leaves the order of a symbol invariant. Moreover, the transition is continuous with respect to the symbol topology (this follows from the proof of theorem 2.2) and therefore, the resulting norms are equivalent.

4.5 LEMMA: By construction, the pseudodifferential operators defined in 4.3 have the usual properties, i.e.

(a) If A, B are pseudodifferential operators with symbols $a \in SG_1^m(\mathbb{R}^n)$, $b \in SG_1^{\mu}(\mathbb{R}^n)$, then $A+B$, λA are again pseudodifferential operators with symbols $a+b$, λa, resp., for $\lambda \in \mathbb{C}$. The product AB has a symbol $c \in SG_1^{m+\mu}(\mathbb{R}^n)$. The asymptotic expansion of c in local coordinates is given by formula (1.4).

(b) If A has an md-elliptic symbol $a \in SG_1^m(\mathbb{R}^n)$, then A has a $\mathcal{K}$-parametrix and extends to a bounded Fredholm operator from $H_r(\mathbb{R}^n)$ to $H_{r-m}(\mathbb{R}^n)$ for every $r \in \mathbb{R}^2$.

4.6 REMARK. All the definitions of this section extend to the case of a finite dimensional vector bundle over X in an obvious way: we only have to require the manifold X to be SG-compatible and the changes of coordinates to satisfy condition COMP 1 also for the vector bundle coordinates. In local coordinates, the symbols will then be $q \times q$ matrices of functions instead of functions.

5. REFERENCES.

[BEA] BEALS, R.: General Calculus of Pseudodifferential Operators, Duke Math. J. 42, 1 - 42, (1975)

[BEC] BEALS, R.: Characterization of Pseudodifferential Operators and Applications, Duke Math. J. 44, 45 - 57 (1977) and vol. 46, 215 (1979)

[BRS] BRÜNING, J. and SEELEY, R.: The Resolvent Expansion for Second Order Regular Singular Operators, Technical Report No. 2/85

[BSE] BRÜNING, J and SEELEY, R.: Regular Singular Asymptotics, preprint

[CHO] CHOQUET-BRUHAT,Y. and CHRISTODOULOU,D.: Elliptic Systems in $H_{s,\delta}$ Spaces on Manifolds Which are Euclidean at Infinity, Acta Math. 146, 129 - 150 (1981)

[COG] CORDES, H.O.: A Global Parametrix for Pseudodifferential Operators over $\mathbb{R}^n$ with Applications, SFB 72 Preprints, Bonn 1976

[CST] CORDES, H.O.: Spectral Theory of Linear Partial Differential Operators and Comparison Algebras, to appear

[DRD] DROSTE,B.: Fortsetzung des holomorphen Funktionalkalküls in mehreren Variablen auf Algebren mit Zerlegung der Eins, Dissertation, Mainz 1980

[DRO] DROSTE,B.:Holomorphic Approximation of Ultradifferentiable Functions, Math. Ann. 257, 293 - 316 (1981)

[GRI] GRAMSCH,B.: Relative Inversion in der Störungstheorie von Operatoren und Ψ-Algebren, Math. Ann. 269, 27 - 71 (1984)

[GRK] GRAMSCH, B. and KALB, K.G.: Pseudo-locality and Hypoellipticity in Operator Algebras, Semesterberichte Funktionalanalysis, 51 - 61, Tübingen 1985

[HOE] HÖRMANDER, L.: The Analysis of Linear Partial Differential Operators, vols. 1 - 4, Springer, Berlin, Heidelberg, New York 1983 - 1985

[HPO] HÖRMANDER, L.: Pseudo-Differential Operators and Hypoelliptic Equations, AMS Proc. Symp. Pure Math. X, 138 - 183 (1967)

[LOM] LOCKHART, R. und McOWEN, R.: Elliptic Differential Operators on Noncompact Manifolds, Ann.Sc.Norm.Sup. Pisa 12, 409 - 447 (1986)

[MEM] MELROSE, R. and MENDOZA, G.: Elliptic Boundary Problems on spaces with Conic Points, preprint

[MEY] MEYER, Y.: Les opérateurs pseudo-différentiels classiques et leurs conjugués par changement de variables, Sem. Goulaouic-Meyer-Schwartz, 1980-1981

[MLE] MÜLLER, W.: L^2-Index of Elliptic Operators on Manifolds with Cusps of Rank One, Report R-Math-06/85, Akad. d. Wiss. der DDR, Berlin 1985

[MST] MÜLLER, W.: Spectral Theory for Riemannian Manifolds with Cusps and a Related Trace Formula, Math. Nachr. 111, 197 - 288 (1983)

[MPS] MÜLLER, W.: The Point Spectrum and Spectral Geometry for Riemannian Manifolds with Cusps, preprint Akad. d. Wiss. der DDR, Berlin 1984

[RES] REMPEL, S. and SCHULZE, B.-W.: Complete Mellin and Green Symbolic Calculus in Spaces with Conormal Asymptotics, Ann.Glob.An. & Geo.4, 137 - 224 (1986)

[SCH] SCHROHE, E.: Complex Powers of Elliptic Pseudodifferential Operators, Int. Eq. Op. Th. 9, 337 - 354 (1986)

[SCD] SCHROHE, E.: Komplexe Potenzen elliptischer Pseudodifferentialoperatoren, Dissertation, Mainz 1986

[SCE] SCHULZE, B.-W.: Ellipticity and Continuous Conormal Asymptotics on Manifolds with Conical Singularities, Prépublications Univ. de Paris-Sud 1985

[TAY] TAYLOR, M.: Pseudodifferential Operators, Princeton University Press, Princeton, 1981

[WAE] WAELBROEK,L.: Topological Vector Spaces and Algebras, Springer LN 230, Berlin, Heidelberg, New York 1971

[WIA] WIDOM, H.: Asymptotic Expansions for Pseudodifferential Operators on Bounded Domains, Springer LN 1152, Berlin, Heidelberg, New York, Tokyo 1985

[WID] WIDOM, H.: A Complete Symbolic Calculus for Pseudodifferential Operators, Bull. Sc. Math. 2ième série, 104, 19 - 64 (1980)

MELLIN EXPANSIONS OF PSEUDO-DIFFERENTIAL OPERATORS AND CONORMAL ASYMPTOTICS OF SOLUTIONS

B.W. Schulze
Karl-Weierstraß-Institut für Mathematik
der AdW der DDR
1086 Berlin, Mohrenstrasse 39

Contents

1. Introduction

This exposition deals with the conormal asymptotics of solutions of equations, where the given operator admits an expansion into Mellin actions of conormal orders tending to $-\infty$. Mellin expansions exist for many 'singular' problems such as for pseudo-differential operators (ψDO's) on manifolds with conical singularities or mixed elliptic boundary problems (i.e. with jumping elliptic conditions, e.g. the Zaremba problem) that lead via reduction to the boundary to ψDO's without the transmission property. The Mellin expansions show a typical common structure of all these problems and give rise to parametrix constructions on a complete Mellin symbolic level and to the regularity of solutions in spaces with conormal asymptotics. It is our aim to illustrate this by typical examples.

Let Σ be the singular set of a problem (in concrete cases the vertex of a cone, an edge or a boundary of a manifold, a jump of the boundary conditions and so on). The method is then to perform a Mellin calculus along $\mathbb{R}_+ \ni t$, where t denotes the distance to $\sum$, and then to 'lift' the resulting algebra over $\mathbb{R}_+$ to higher dimensions by applying a ψDO calculus along the space of the remaining parameters. The idea can be modified (or iterated) for numerous other more complex situations such as boundary problems, operators on manifolds with higher singularities (where in general extra boundary and potential conditions with respect to $\sum$ are needed), furthermore for screen (crack) problems, and non-elliptic (e.g. parabolic) and non-linear problems.

Denote by M the Mellin transform, $Mu(z) = \int_0^\infty t^{z-1}u(t)dt$, $M : L^2(\mathbb{R}_+) \xrightarrow{\cong} L^2(\mathrm{Re}\, z = \frac{1}{2})$, and consider the 'Mellin ψDO' $op_M(h) = M^{-1}hM$, h being a multiplier in $L^2(\mathrm{Re}\, z = \frac{1}{2})$. Assume that h(z) is meromorphic with finitely many poles in every strip $S(l_1,l_2) = \{ z : \frac{1}{2} - l_1 < \mathrm{Re}\, z < \frac{1}{2} + l_2\}$, $l_1,l_2 \in \mathbb{R}_+$. Let $0 \leqq \gamma \leqq j$ and set $op_M^{-j}(h,\gamma) = t^{j-\gamma}op_M(T^\gamma h)t^\gamma$, $(T^\gamma h)(z) := h(z+\gamma)$. Then a Mellin expansion near $t = 0$ of an operator $A \in \mathcal{L}(L^2(\mathbb{R}_+))$ is of the form

$$A \sim \sum_{j=0}^{\infty} \omega(c_j t)\, op_M^{-j}(h_j,\gamma_j)\, \omega(c_j t), \tag{1.1}$$

$c_j > 0$ constants, ω a cut-off function (i.e. $\omega \in C_0^\infty(\mathbb{R}_+)$, $\omega \equiv 1$ near $t = 0$). The precise meaning of the expansion will be defined below and then we also admit operators between different spaces.

$$h_j(z) =: \sigma_M^{-j}(A)(z) \tag{1.2}$$

is called the Mellin symbol of A of conormal order -j.

Mellin expansions in the present form were introduced by Rempel, Schulze in [R5] and then generalized and further applied in [R6], [R7], [R8], [R9], [S1], [S2], [S3], [S4].

The discrete conormal asymptotics of solutions of concrete singular or degenerate problems in terms of the distance t to $\sum$ has the form

$$u(t) \sim \sum_{j=0}^{\infty} \sum_{k=0}^{m_j} \zeta_{jk}\, t^{-p_j} \log^k t \tag{1.3}$$

as $t \longrightarrow 0$, where $p_j \in \mathbb{C}$ is a sequence with $\mathrm{Re}\, p_j \longrightarrow -\infty$ as $j \longrightarrow \infty$, $m_j \in \mathbb{Z}_+$. The other parameters in u such as angle or tangent variables are dropped for the moment.

A behaviour of the form (1.3) for several classes of problems was established by many authors, cf. [K1], [K2], [K3], [K4], [M1], [M2], [M3], [M4], [P1], [E1], [J1]. The large scale of special investigations is motivated by applications in mechanics and other physical and technical disciplines, cf. e.g. [M7] and the references there. The exponents and coefficients in the asymptotics are useful for numerical computations, cf. [C4], [S5]. On the other hand there is an increasing interest in a systematic analysis of operators on manifolds with singularities. They may occur as factor spaces under actions of discrete groups that lead to conical points, edges, cusps, ..., cf. [M6]. It is then an interesting problem to establish an adequate analogue of the Atiyah-Singer index theorem on manifolds with singularities.

For each class of singular problems it is reasonable to look for an operator calculus with properties as they are known in analogous form for the classical calculus of ψDO's on closed compact manifolds (cf. [H1], [H2], [H3],...) or of boundary value problems for ψDO's with the transmission property (cf. [B1], [R2]). In other words we should have

(i) an algebra of operators (possibly a *-algebra) filtered by orders,
(ii) adequate scales of distribution spaces (Sobolev and C^∞ spaces with asymptotics),
(iii) a principal and a complete symbolic level with a composition rule on symbolic level ,
(iv) an ideal of 'smoothing' operators, such that an operator is determined by the complete symbol modulo a smoothing one,
(v) a concept of ellipticity, a parametrix construction in the algebra and the elliptic regularity with asymptotics, the index theory.

It turns out that this is a rather complex program, even for conical singularities. The analysis of concrete cases shows that the asymptotics of solutions depends on the given operator. For higher-dimensional singular sets $\sum$ this leads to the variable asymptotics, i.e. the p_j, m_j, ξ_{jk} depend on parameters and may have branchings and jumps with varying parameters. Then an adequate choice and the functional analysis of natural spaces with asymptotics as well as the structure of the smoothing operators is a delicate point (cf. [R9], [S2], [S3]). Moreover, we have to expect symbol hierarchies. In a simple form they already occur in Boutet de Monvel's algebra [B1], where there is an interior and a boundary symbolic level. The ellipticity is a condition to all components of the principal symbol. In theories with Mellin expansions we have in addition the ellipticity of the Mellin symbol of highest conormal order. This Mellin ellipticity is automatically satisfied for elliptic operators in the algebra of [B1].

In Section 2 we describe an algebra of ψDO's on manifolds with conical singularities, using results of [R7]. A novelty here is that we give an alternative concise description of the Green and flat operators in the cone algebra (Definitions 2.3, 2.4) which is essential for the generalizations to higher singularities, mentioned in Section 4. Theorem 2.11 also is new. The material of Section 2 is necessary to formulate the homotopy classification of the ellipticity for conical singularities as well as the examples and the arguments for the proof, which are given in Section 4. These results were not published before. The author is grateful to S. Rempel who contributed in discussions a number of details. Section 3 deals with the Mellin expansions of pseudo-differential actions for boundary symbols in the case of violated transmission property. The Mellin symbols of lower conormal order (Definition 3.1) as well as the whole approach extend earlier results of Eskin [E1] who obtained a principal symbolic calculus. The precise lower order Mellin symbols are published here for the first time. They will be used in [R9] to establish the conormal asymptotics for general mixed elliptic boundary problems in higher dimensions. Special examples for the asymptotics were discussed in [E1], but with other methods.

2. Mellin Expansions for Conical Singularities

Let M be a 'manifold' with conical singularity v, and $M \setminus \{v\}$ locally near v identified with $\mathbb{R}_+ \times X$, X being a closed compact C^∞-manifold. Denote by $\mathrm{Diff}^\mu(\mathbb{R}_+ \times X)$ the space of all differential operators that are in local coordinates x on X of the form

$$A(t,x,\partial_t,D_x) = \sum_{k+|\alpha|\le\mu} a_{k\alpha}(t,x)\left(-t\frac{\partial}{\partial t}\right)^k D_x^\alpha \tag{2.1}$$

$a_k \in C^\infty(\overline{\mathbb{R}}_+ \times U)$, U a coordinate neighbourhood on X. Globally with respect to X we can write

$$A = \sum_{k=0}^{\mu} a_k(t,x,D_x)(-t\frac{\partial}{\partial t})^k, \tag{2.2}$$

where $a_k \in C^\infty(\overline{\mathbb{R}}_+, \mathrm{Diff}^{\mu-k}(X))$, $\mathrm{Diff}^{\mu-k}(X)$ being the class of all differential operators on X of order $\mu-k$ with smooth coefficients. Note that the Laplace-Beltrami operator with respect to a Riemannian metric of the form $dt^2 + t^2 g$, g a Riemannian metric on X, belongs to our class (up to a weight factor which is not essential).

For studying the local properties of differential operators on M we mainly have to consider $\mathbb{R}_+ \times X$, for outside a neighbourhood of v the underlying space is a C^∞ manifold, where the usual calculus applies.

Now we have $-t\frac{\partial}{\partial t} = M^{-1} z\, M$, M the Mellin transform. Thus the operator (2.2) can be expressed by the Mellin transform. Consider the Taylor expansion $a_k(t,x,D_x) \sim \sum_{j=0}^{\infty} t^j a_k^{[j]}(x,D_x)$ for $t \longrightarrow 0$, $a_k^{[j]} \in \mathrm{Diff}^{\mu-k}(X)$ and define

$$\sigma_M^{-j}(A)(z) = \sum_{k=0}^{\mu} a_k^{[j]}(x,D_x)\, z^k,$$

considered as an operator-valued function of z. Then we get the following

<u>2.1. Theorem</u>. Let A, B be operators on $\mathbb{K} := \mathbb{R}_+ \times X$ as described, then

$$\sigma_M^{-l}(AB)(z) = \sum_{j+k=l} \sigma_M^{-j}(A)(z-k)\ \sigma_M^{-k}(B)(z)$$

for all $l \in \mathbb{Z}_+$.
Let $A \in \mathrm{Diff}^\mu(\mathbb{K})$ and set

$$\tilde{\sigma}_\psi^\mu(A)(t,x,\tau,\xi) = \sum_{k+|\alpha|=\mu} a_{k\alpha}(t,x)(-i\tau)^k \xi^\alpha$$

Then $(\tilde{\sigma}_\psi^\mu(A), \sigma_M^0(A))$ is a couple of principal symbols in the sense of (iii) in Section 1.

<u>2.2. Definition</u>. A is called elliptic (with respect to the weight $\gamma \in \mathbb{R}$ at t = 0) if

(i) $\tilde{\sigma}_{\psi}^{\mu}(A)$ does not vanish for $(\tau, \xi) \neq 0$,

(ii) $\sigma_M^0(A)(z) : H^s(X) \longrightarrow H^{s-\mu}(X)$ is an isomorphism for all z, $\mathrm{Re}\, z = \frac{1}{2} + \gamma$, $s \in \mathbb{R}$.

$H^s(X)$ denotes the usual Sobolev space over X with index s. The choice of s in (ii) is not essential. (i) implies (ii) for all $\gamma \in \mathbb{R}$ except for a discrete subset $\{\gamma_\nu\}_{\nu \in \mathbb{Z}} \subset \mathbb{R}$ with $\gamma_\nu \longrightarrow \pm\infty$ for $\nu \longrightarrow \pm\infty$. Now let us consider an analogue of the Sobolev spaces over $\mathbb{K}$. Define

$$\mathcal{H}^s(\mathbb{K}) = \left\{ u \in L^2(\mathbb{K}) : (t\tfrac{\partial}{\partial t})^k D_x^\alpha u \in L^2(\mathbb{K}),\ k+|\alpha| \leq s \right\},$$

$s \in \mathbb{Z}_+$. For $s \in \mathbb{R}$ a definition of $\mathcal{H}^s(\mathbb{K})$ was given in [R7], [S1]. Let

$$P = \left\{ (p_j, m_j, L_j) \right\}_{j \in \mathbb{Z}} \tag{2.3}$$

be a sequence with $p_j \in \mathbb{C}$, $\mathrm{Re}\, p_j < \frac{1}{2}$ for $j \in \mathbb{Z}_+$, $\mathrm{Re}\, p_j > \frac{1}{2}$ for $-(j+1) \in \mathbb{Z}_+$, $\mathrm{Re}\, p_j \longrightarrow \pm\infty$ as $j \longrightarrow \mp\infty$, $m_j \in \mathbb{Z}_+$, $L_j \subset C^\infty(X)$ a finite-dimensional subspace. Denote by $\mathcal{H}_P^s(\mathbb{K})$, $s \in \mathbb{R}$, the subspace of all $u \in \mathcal{H}^s(\mathbb{K})$ with

$$u(t,x) \sim \begin{cases} \sum_{j=0}^{\infty} \sum_{k=0}^{m_j} \xi_{jk}(x) t^{-p_j} \log^k t \text{ as } t \longrightarrow 0, \\ \sum_{j=-1}^{-\infty} \sum_{k=0}^{m_j} \xi_{jk}(x) t^{-p_j} \log^k t \text{ as } t \longrightarrow \infty, \end{cases}$$

$\xi_{jk} \in L_j$, $0 \leq k \leq m_j$. The precise definition of the asymptotics was given in [R7]; the asymptotics for $t \longrightarrow \infty$ is considered here for notational convenience. (2.3) is called a singularity type. $\mathcal{H}_P^s(\mathbb{K})$ is a Fréchet space in a natural way and $\mathcal{H}_P^\infty(\mathbb{K}) = \bigcap_{s \in \mathbb{R}} \mathcal{H}_P^s(\mathbb{K})$ is a nuclear Fréchet space.

It can easily be proved that the Mellin image of $\mathcal{H}_P^s(\mathbb{K})$ consists of $H^s(X)$-valued meromorphic functions with poles at p_j of multiplicities m_j+1 and Laurent coefficients in L_j. Incidentally we write $\pi_{\mathbb{C}} P = \{p_j\}_{j \in \mathbb{Z}}$.

There is a 'Borel theorem' that asserts that for every sequence of coefficients ξ_{jk} there is a $u \in \mathcal{H}^s_P(\mathbb{K})$, where the asymptotics has the given coefficients.

2.3. Definition. $A \in \mathcal{L}(L^2(\mathbb{K}))$ is called a Green operator if it induces continuous operators

$$A : L^2(\mathbb{K}) \longrightarrow \mathcal{H}^\infty_P(\mathbb{K}),\ A^* : L^2(\mathbb{K}) \longrightarrow \mathcal{H}^\infty_Q(\mathbb{K})$$

for certain singularity types P,Q depending on A. Denote by $\mathcal{R}_G(\mathbb{K})$ the class of all Green operators.

It is a simple consequence of this definition that when $1+A : \mathcal{H}^s(\mathbb{K}) \longrightarrow \mathcal{H}^s(\mathbb{K})$ is invertible, $A \in \mathcal{R}_G(\mathbb{K})$, then $(1+A)^{-1} = 1+A'$ for some other $A' \in \mathcal{R}_G(\mathbb{K})$.

Let $op_G(g)u(t,x) = (g(t,x,t',x'), Ju(t',x'))_{L^2(\mathbb{K})}$, J being the operator of complex conjugation, $g \in \mathcal{H}^\infty_P(\mathbb{K}) \otimes_\pi J\ \mathcal{H}^\infty_Q(\mathbb{K})$. Then every $A \in \mathcal{R}_G(\mathbb{K})$ is of the form $A = op_G(g)$ for such a kernel g. From this follows that $A \in \mathcal{R}_G(\mathbb{K})$ also induces continuous operators

$$A : \mathcal{H}^s(\mathbb{K}) \longrightarrow \mathcal{H}^\infty_P(\mathbb{K}),\ A^* : \mathcal{H}^s(\mathbb{K}) \longrightarrow \mathcal{H}^\infty_Q(\mathbb{K})$$

for all $s \in \mathbb{R}$. Note that the transform $I : u(t,x) \longrightarrow t^{-1} u(t^{-1},x)$ defines isomorphisms

$$I : \mathcal{H}^s(\mathbb{K}) \longrightarrow \mathcal{H}^s(\mathbb{K}),\quad \mathcal{H}^s_P(\mathbb{K}) \longrightarrow \mathcal{H}^s_{\check{P}}(\mathbb{K}),$$

where $\check{P}$ is another singularity type that can easily be expressed in terms of P. It is obvious that $IL^\mu_{cl}(\mathbb{K})I^{-1} = L^\mu_{cl}(\mathbb{K})$, $L^\mu_{cl}(\mathbb{K})$ being the space of all classical ψDO's over $\mathbb{K} = \mathbb{R}_+ \times X$ of order μ.

2.4. Definition. A classical ψDO $A \in L^\mu_{cl}(\mathbb{K})$ is called strictly degenerate at t = 0 if $A = \tilde{A}|_{\mathbb{K}}$ for some $\tilde{A} \in L^\mu_{cl}(\mathbb{R} \times X)$ with supp $\tilde{A}u$, supp $\tilde{A}^*u \subseteq \overline{\mathbb{R}}_+ \times X$ for all $u \in \mathcal{E}'(\mathbb{R} \times X)$. Denote by $\mathring{\mathcal{R}}^\mu(\mathbb{K})$ the class of all $A \in L^\mu_{cl}(\mathbb{K})$ such that A and IAI^{-1} are strictly degenerate at t = 0.

The operators $A \in \mathring{\mathcal{R}}^\mu(\mathbb{K}) \subset L^\mu_{cl}(\mathbb{K})$ can be characterized by the property to induce continuous operators

$$A, A^* : \mathcal{H}^s(\mathbb{K}) \longrightarrow \mathcal{H}^{s-\mu}_0(\mathbb{K}) \text{ for all } s \in \mathbb{R},$$

where 0 indicates the trivial singularity type, i.e. when $\pi_{\mathbb{C}} P = \emptyset$. Thus $\mathring{\mathfrak{N}}^{\infty}(\mathbb{K}) \cap \mathfrak{N}_G(\mathbb{K}) = \mathring{\mathfrak{N}}^{-\infty}(\mathbb{K})$.

<u>2.5. Theorem</u>. The spaces of operators $\mathfrak{N}_G(\mathbb{K}) + \mathring{\mathfrak{N}}^{\infty}(K)$ and $1 + \mathfrak{N}_G(\mathbb{K}) + \mathring{\mathfrak{N}}^{0}(\mathbb{K})$ form * algebras filtered by the orders (of the strictly degenerate terms) with $\mathfrak{N}_G(\mathbb{K})$ as two-sided ideals.

<u>2.6. Remark</u>. If $\omega_1, \omega_2, \omega_3$ are cut-off functions, $\omega_1 \omega_2 = \omega_1$, $\omega_2 \omega_3 = \omega_2$, then for every $G \in \mathfrak{N}_G(\mathbb{K})$, $N \in \mathring{\mathfrak{N}}^{0}(\mathbb{K})$ there exist an $N' \in \mathring{\mathfrak{N}}^{0}(\mathbb{K})$ and a constant $c > 0$ with

$$\omega_1(ct)(1+G+N)\,\omega_2(ct)(1+N')\,\omega_3(ct) = \omega_1(ct)(1+G')\,\omega_1(ct)$$

with some $G' \in \mathfrak{N}_G(\mathbb{K})$.

Let $A \in \mathrm{Diff}^{\mu}(\mathbb{K})$ be as at the beginning. Then there is a sequence of constants c_j, increasing sufficiently fast, such that

$$\omega A \omega - \sum_{j=0}^{\infty} \omega(c_j t)\,\mathrm{op}_M^{-j}(h_j, \gamma_j)\,\omega(c_j t) \in \mathfrak{N}_G(\mathbb{K}) + \mathring{\mathfrak{N}}^{\mu}(\mathbb{K}),$$

$h_j(z) := \sigma_M^{-j}(A)(z)$, $0 \leqq \gamma_j \leqq j$, $\gamma_j \longrightarrow \infty$, $j - \gamma_j \longrightarrow \infty$ as $j \longrightarrow \infty$, ω a fixed cut-off function.

Let A be elliptic in the sense of Definition 2.2 (for $\gamma = 0$). Then Theorem 2.1 yields a procedure to determine a sequence of Mellin symbols of a parametrix B, namely

$$\sigma_M^{0}(B)(z) = \left[\sigma_M^{0}(A)(z)\right]^{-1} \tag{2.4}$$

$$\sigma_M^{-1}(B)(z) = -\ \sigma_M^{0}(B)(z-1)\ \sigma_M^{-1}(A)(z)\ \sigma_M^{0}(B)(z),\ldots \tag{2.5}$$

The $\sigma_M^{-k}(B)(z)$ are operator-valued functions of z and belong to $\mathfrak{M}_{\mathcal{R}}^{-\mu}(X)$ in the sense of the following

<u>2.7. Definition</u>. $\mathfrak{M}_{\mathcal{R}}^{\nu}(X)$ consists of all operator-valued functions h(z) with

($\mathfrak{m}$.1) h(z) is meromorphic with values in $L_{cl}^{\nu}(X)$ and represents a parameter-depending ψDO of order ν with parameter $z \in \Gamma_{\rho}$, where $\Gamma_{\rho} := \{z \in \Gamma : |z| \geqq \rho\}$ and Γ is a closed cone in $\mathbb{C}$ with $i\mathbb{R} \subset \Gamma$, $h(z+\gamma)$ has the same property with $\rho = \rho(\gamma)$ for all $\gamma \in \mathbb{R}$,

($\mathcal{M}$.2) every strip $S(l_1,l_2)$ only contains a finite number of poles, $l_1,l_2 \in \mathbb{R}$, and the Laurent coefficients at every pole are in $L^{-\infty}(X)$ and finite-dimensional.

Denote by $\mathfrak{N}^{\nu}_{(o)}(\mathbb{K})$ the set of all operators $A \in \bigcap_{s \in \mathbb{R}} \mathcal{L}(\mathcal{H}^s(\mathbb{K}), \mathcal{H}^{s-\nu}(\mathbb{K}))$ for which there is a sequence $\sigma_M^{-k}(A) := h_k \in \mathfrak{M}^{\nu}_{\mathcal{R}}(X)$ and constants $c_k > 0$, $0 \leqq \gamma_k \leqq k$, $\gamma_k \longrightarrow \infty$, $k - \gamma_k \longrightarrow \infty$, $c_k \longrightarrow \infty$ as $k \longrightarrow \infty$, such that

$$\omega_1 A \omega_2 - \omega_1 \left\{ \sum_{k=0}^{\infty} \omega(c_k t) op_M^{-k}(h_k, \gamma_k) \omega(c_k t) \right\} \omega_2 \in \mathfrak{N}_G(\mathbb{K}) + \mathring{\mathfrak{N}}^{\nu}(\mathbb{K})$$

for all cut-off functions ω, ω_1, ω_2. Let us drop here the precise meaning of the convergence on the left. It can be proved that another choice of ω, ω_1, ω_2 and of $c_k, \gamma_k, k \in \mathbb{Z}_+$, only changes $\omega_1 \{ \dots \} \omega_2$ by some operator in $\mathfrak{N}_G(\mathbb{K}) + \mathring{\mathfrak{N}}^{\nu}(\mathbb{K})$. Moreover every $A \in \mathfrak{N}^{\nu}_{(o)}(\mathbb{K})$ induces a continuous operator

$$\omega_1 A \omega_2 : \mathcal{H}^s_P(\mathbb{K}) \longrightarrow \mathcal{H}^{s-\nu}_Q(\mathbb{K}), \ s \in \mathbb{R}$$

for every singularity type P with some singularity type Q depending on P and A, ω_1, ω_2 arbitrary cut-off functions.

<u>2.8. Theorem.</u> $\mathfrak{N}^{\infty}_{(o)}(\mathbb{K})$ is a *-algebra filtered by the orders and $\mathfrak{N}_G(\mathbb{K}) + \mathring{\mathfrak{N}}^{\infty}(\mathbb{K})$, $\mathfrak{N}_G(\mathbb{K})$ are two-sided ideals. Theorem 2.1. is valid for two operators $A,B \in \mathfrak{N}^{\infty}_{(o)}(\mathbb{K})$.

Now an operator of the form

$$C := \sum_{k=0}^{\infty} \omega(c_k t) \ op_M^{-k}(b_k, \delta_k) \omega(c_k t)$$

with $b_k(z) = \sigma_M^{-k}(B)(z)$, $c_k > 0$, $c_k \longrightarrow \infty$ sufficiently fast, $0 \leqq \delta_k \leqq k$, $\delta_k \longrightarrow \infty$, $k - \delta_k \longrightarrow \infty$ as $k \longrightarrow \infty$, has the property

$$\omega_1 A \omega C \omega_2 = \omega_1 (1+G+N) \omega_2$$

with certain $G \in \mathfrak{N}_G(\mathbb{K})$, $N \in \mathring{\mathfrak{N}}^{o}(K)$, ω, ω_1, ω_2 cut-off functions with $\omega_1 \omega = \omega_1$, $\omega \omega_2 = \omega$. Applying Remark 2.6. we can choose the cut-off functions ω, ω_1, ω_2, ω_3 in such a way that for some $N' \in \mathring{\mathfrak{N}}^{o}(\mathbb{K})$

$$(\omega_1 A \omega) \{ \omega_2 C \omega_2 (1+N') \omega_3 \} = \omega_1 (1+G') \omega_1,$$

$G' \in \mathcal{M}_G(\mathbb{K})$. Thus we get a parametrix of A near t = 0 modulo a Green operator. This method can be applied for arbitrary $A \in \mathcal{M}^{\mu}_{(o)}(\mathbb{K})$ satisfying the analogous ellipticity condition (with an adequate definition of $\tilde{\sigma}^{\mu}_{\psi}(A)$). In view of Theorem 2.8. we thus obtain

2.9. Proposition. Let $A \in \mathcal{M}^{\mu}_{(o)}(\mathbb{K})$ be elliptic (with respect to the weight $\gamma = 0$). Then there is a parametrix $B \in \mathcal{M}^{-\mu}_{(o)}(\mathbb{K})$ in the sense $\omega_1 A\ \omega\ B\omega_2 - \omega_1^2 \in \mathcal{M}_G(\mathbb{K})$, ω_i cut-off functions, $\omega_1\omega = \omega_1$, $\omega\omega_2 = \dot{\omega}$. $Au = f$ and $u \in \mathcal{H}^{-\infty}(\mathbb{K})$, $\omega f \in \mathcal{H}^s_Q(\mathbb{K})$ implies $\omega u \in \mathcal{H}^{s+\mu}_P(\mathbb{K})$ for any singularity type Q with some P depending on Q,A.

Now let M be a manifold with conical singularity v and $\check{M}$ the stretched manifold obtained from M by replacing a neighbourhood of v by $[0,1) \times X$. Then we can do the same over M, since the specific aspects are related to a neighbourhood of t = 0. In particular we have the spaces $\mathcal{H}^s(\check{M})$, $\mathcal{H}^s_P(\check{M})$, $s \in \mathbb{R}$, and the spaces $\mathcal{M}_G(\check{M})$, $\mathring{\mathcal{M}}^{\mu}(\check{M})$, $\mathcal{M}^{\mu}(\check{M})$ of operators. Then we get the

2.10. Theorem. $A \in \mathcal{M}^{\mu}(\check{M})$ is elliptic (with respect to the weight $\gamma = 0$) iff it defines a Fredholm operator $A : \mathcal{H}^s(\check{M}) \longrightarrow \mathcal{H}^{s-\mu}(\check{M})$ for fixed $s \in \mathbb{R}$. An elliptic operator has a parametrix $B \in \mathcal{M}^{-\mu}(\check{M})$. Moreover, $Au = f \in \mathcal{H}^s_Q(\check{M})$ and $u \in \mathcal{H}^{-\infty}(\check{M})$ imply $u \in \mathcal{H}^{s+\mu}_P(\check{M})$ with some singularity type P depending on Q and A.

It is useful to construct examples of elliptic operators in the class $\mathcal{M}^{\mu}(\check{M})$. Let us formulate a result of global reduction of orders. In Section 4 we derive a general class of canonical elliptic operators when the base of the cone is a sphere.

2.11. Theorem. For every $\mu \in \mathbb{R}$ there exists an elliptic operator $R^{\mu} \in \mathcal{M}^{\mu}(\check{M})$ which induces isomorphisms $R^{\mu} : \mathcal{H}^s(\check{M}) \longrightarrow \mathcal{H}^{s-\mu}(\check{M})$ for every $s \in \mathbb{R}$.

The technique of the proof is based on a parameter-depending version of $\mathcal{M}^{\mu}(\check{M})$ classes where the parameter is involved in the homogeneity of symbols. A parameter-depending elliptic operator can easily be found and it then becomes bijective for fixed large values of the parameter.

3. Mellin Expansions for Boundary Value Problems

As noted in Section 1 the conormal asymptotics may be expected also for solutions of mixed elliptic boundary value problems, i.e. boundary value problems for elliptic differential operators in an open smooth domain $\Omega \subset \mathbb{R}^{n+1}$, where $\partial\Omega$ is the union of manifolds $X_{\pm}$ with smooth boundary Y of codimension 1 in $\partial\Omega$, and on $X_{\pm}$ are given elliptic boundary conditions that may have a jump along Y. An example is the Zaremba problem with the Laplacian in Ω and Dirichlet conditions over X_-, Neumann conditions over X_+. Problems of this sort can be treated by reduction to the boundary and solving an associated transmission problem for a ψDO on $\partial\Omega$ with a jump along Y. By a local reflection argument this can be reduced to a pseudo-differential boundary problem on one side, say X_+, with extra boundary conditions along Y, cf. [R4], [R9]. All interesting mixed problems for differential operators lead to ψDO's on X_+ without the transmission property with respect to Y. For the Zaremba problem we get an operator with symbol $|\xi|$, ξ being the covariable on X_+. As for ψDO's with the transmission property (cf. [B1]) the symbolic calculus contains boundary symbols (cf. [R3] for the principal symbols). So we have to establish a boundary symbolic calculus that reflects the operators normal to the boundary modulo smoothing ones ('Green operators'). The operators are (up to finite-dimensional and Green operators) families of ψDO's on $\mathbb{R}_+$ parametrized by points in $T^*Y \setminus 0$. In addition we have extra Mellin operators.

The aspect of Mellin expansions already occurs on $\mathbb{R}_+$. Other specific aspects come from the dependence on $y \in Y$. This was systematically studied in [R9]. Here we want to discuss some typical points for ψDO's on $\mathbb{R}_+$. We restrict ourselves to the case of order zero and fix the parameters. Higher order operators require more preparations on the functional analysis of spaces on $\mathbb{R}_+$. This cannot be given in this short note.

For the singularity types we will use analogous notations as in the previous section. Here we have dim X = 0. Assume in addition $\mathrm{Re}\, p_j < \frac{1}{2}$ for all exponents in (1.3). Thus the singularity types here are of the form

$$P = \{(p_j, m_j)\}_{j \in \mathbb{Z}_+} \tag{3.1}$$

such that the functions in $\mathcal{S}_P(\mathbb{R}_+) := \bigcap_{s \in \mathbb{R}} \mathcal{H}^s_P(\mathbb{R}_+)$ are flat of infinite order at $t = \infty$. Set $\mathcal{S}_o(\mathbb{R}_+) := \mathcal{S}_\emptyset(\mathbb{R}_+)$ and $\mathcal{S}(\overline{\mathbb{R}}_+) = \mathcal{S}_{P_o}(\mathbb{R}_+)$ for $P_o = \{(-j,o)\}_{j\in\mathbb{Z}_+}$.

Now let $a(t,\tau) \in S^o_{cl}(\mathbb{R})$ be given. Assume throughout this section that a is independent of t for large $|t|$. Denote by F the Fourier transform on $\mathbb{R}$, $Fu(\tau) = \int e^{-it\tau}u(t)dt$, and set $op(a) = F^{-1}aF : L^2(\mathbb{R}) \longrightarrow L^2(\mathbb{R})$. Let $e^+ : L^2(\mathbb{R}_+) \longrightarrow L^2(\mathbb{R})$ be the operator of extension by zero and $r^+ : L^2(\mathbb{R}) \longrightarrow L^2(\mathbb{R}_+)$ the restriction operator. Define

$$op_\psi(a) := r^+ \, op(a)e^+ : L^2(\mathbb{R}_+) \longrightarrow L^2(\mathbb{R}_+) . \tag{3.2}$$

Since we are interested in Mellin expansions with decreasing conormal orders, we may assume for the moment that a is independent of t. For general $a(t,\tau)$ we then apply the Taylor expansion

$$a(t,\tau) \sim \sum_{k=0}^{\infty} t^k a^{[k]}(\tau) \text{ as } t \longrightarrow 0,\ a^{[k]}(\tau) \in S^o_{cl}(\mathbb{R}) .$$

Let $a(\tau) \in S^o_{cl}(\mathbb{R})$ and consider the asymptotic expansion $a(\tau) \sim \sum_{j=0}^{\infty} a^{\pm}_j (i\tau)^{-j}$ as $\tau \longrightarrow \pm\infty$. Set

$$g^+(z) = (1-e^{-2\pi iz})^{-1},\ g^-(z) = (1-e^{2\pi iz})^{-1} .$$

<u>3.1. Definition</u>. The meromorphic functions

$$\sigma^{-j}_M(a)(z) := (a^+_j g^+(z) + a^-_j g^-(z))\, f_j(z),\ j \in \mathbb{Z}_+$$

with $f_o(z) = 1$, $f_j(z) = \prod_{k=1}^{j} (k-z)^{-1}$, $j \geq 1$, are called the Mellin symbols of $a(\tau) \in S^o_{cl}(\mathbb{R})$ of conormal order $-j$. For $a(t,\tau) \in S^o_{cl}(\mathbb{R})$ we define

$$\sigma^{-j}_M(a)(z) = \sum_{l+k=j} \sigma^{-l}_M(a^{[k]})(z),$$

$$\sigma_M(a)(z) = \{ \sigma^{-j}_M(a)(z)\}_{j\in\mathbb{Z}_+} .$$

$\sigma^o_M(a)(z)$ was already defined by Eskin in [E1].

<u>3.2. Definition</u>. An $A \in \mathcal{L}(L^2(\mathbb{R}_+))$ is called a Green operator of the

class $\mathcal{L}_G(\mathbb{R}_+)$ if it induces continuous operators $A : L^2(\mathbb{R}_+) \longrightarrow \mathcal{S}_P(\mathbb{R}_+)$, $A^* : L^2(\mathbb{R}_+) \longrightarrow \mathcal{S}_Q(\mathbb{R}_+)$ with singularity types P,Q depending on A.

Any $A \in \mathcal{L}_G(\mathbb{R}_+)$ is of the form $Au(t) = \int_0^\infty g(t,s)u(s)ds$ with some kernel $g \in \mathcal{S}_P(\mathbb{R}_+) \hat{\otimes}_\pi J\, \mathcal{S}_Q(\mathbb{R}_+)$. We also write $Au = op_G(u)$.

<u>3.3. Theorem</u>. For every $a(t,\tau) \in S^0_{cl}(\mathbb{R})$ and arbitrary cut-off functions $\omega, \omega_1, \omega_2$ there is a sequence of constants $c_j > 0$ such that for some $G \in \mathcal{L}_G(\mathbb{R}_+)$

$$\omega_1 \Big\{ op_\psi(a) - \sum_{j=0}^{\infty} \omega(c_j t) op_M^{-j}(h_j, \tfrac{j}{2})\, \omega(c_j t) \Big\} \omega_2 - G \in \mathring{\mathcal{R}}^0(\mathbb{R}_+) .$$

The strictly degenerate ψDO's play no major role in this section. But Theorem 3.3. expresses a Mellin expansion of $op_\psi(a)$, and from this we can conclude a number of useful mapping properties.

<u>3.4. Theorem</u>. $op_\psi(a)$, $a \in S^0_{cl}(\mathbb{R})$, induces continuous operators

$$op_\psi(a) : \mathcal{S}_P(\mathbb{R}_+) \longrightarrow \mathcal{S}_Q(\mathbb{R}_+)$$

for every singularity type P with some Q depending on a and P. A symbol $a(t,\tau)$ has the transmission property with respect to $t = 0$ if $a_j^{[k],+} = a_j^{[k],-}$ for all $j,k \in \mathbb{Z}_+$. Then if P is a singularity type with $P_0 \subseteq P$ we have continuous operators $op_\psi(a) : \mathcal{S}_P(\mathbb{R}_+) \longrightarrow \mathcal{S}_P(\mathbb{R}_+)$.

<u>3.5. Definition</u>. Denote by $\mathcal{R}$ the set of all sequences $R = \{(q_j, n_j)\}_{j \in \mathbb{Z}}$, $q_j \in \mathbb{C}$, $n_j \in \mathbb{Z}_+$, where $\operatorname{Re} q_j \longrightarrow \pm\infty$ as $j \longrightarrow \mp\infty$ and set $\mathfrak{M}^{-\infty}_{\mathcal{R}} = \varinjlim_{R \in \mathcal{R}} \mathfrak{M}^{-\infty}_R$ with $\mathfrak{M}^{-\infty}_R$ as the space of all meromorphic functions

$h(z)$ with poles at $z = q_j$ of multiplicities $n_j + 1$ and

$$\sup_{\operatorname{Re} z = \gamma} |(1+|z|)^N \chi(z) h(z)| < \infty \quad \text{for all } N \in \mathbb{Z}_+,\ \gamma \in \mathbb{R}$$

and for every $\chi \in C^\infty(\mathbb{C})$, $\chi \equiv 0$ in a neighbourhood of $\pi_{\mathbb{C}} R = \bigcup \{q_j\}$, $\chi \equiv 1$ outside another neighbourhood of $\pi_{\mathbb{C}} R$. By $(a \#_\psi b)(t,\tau)$, $a,b \in S^0_{cl}(\mathbb{R})$, we denote any symbol in $S^0_{cl}(\mathbb{R})$ with asymptotic expansion $\sum \frac{1}{k!} D^k_\tau a\, \partial^k_t b(t,\tau)$. Moreover if $h(z) = \{h_j(z)\}_{j \in \mathbb{Z}_+}$, $l(z) =$

$\{l_j(z)\}_{j\in\mathbb{Z}_+}$ are two sequences of meromorphic functions we set

$$(h \#_M l)(z) = \Big\{ \sum_{\nu=l+k} h_j(z-k)\, l_k(z) \Big\}_{\nu\in\mathbb{Z}_+}$$

<u>3.6. Theorem</u>. For a,b $\in S^0_{cl}(\mathbb{R})$ we have

$$\sigma_M(a \#_\psi b) - \sigma_M(a) \#_M \sigma_M(b) = \{h_j\}_{j\in\mathbb{Z}_+}$$

with $h_j \in \mathfrak{M}^{-\infty}_{\mathcal{R}}$.

Denote by $\mathcal{L}_{M+G}(\mathbb{R}_+)$ the class of all operators

$$C = \sum_{j=0}^{\infty} \omega(c_j t)\, op_M^{-j}(h_j,\gamma_j)\, \omega(c_j t) + G$$

$h_j \in \mathfrak{M}^{-\infty}_{\mathcal{R}}$, ω a cut-off function, c_j, γ_j constants, $0 \leq \gamma_j \leq j$, $\gamma_j \longrightarrow \infty$ $j - \gamma_j \longrightarrow \infty$ as $j \longrightarrow \infty$ and $c_j \longrightarrow \infty$ sufficiently fast (the precise condition on the c_j will be dropped here). Set $\sigma_M^{-j}(C)(z) = h_j(z)$. Moreover let $\mathcal{L}(\mathbb{R}_+)$ be the class of all operators A = $op_\psi(a) + C$, $a \in S^0_{cl}(\mathbb{R}_+)$, $C \in \mathcal{L}_{M+G}(\mathbb{R}_+)$, and $\sigma_M^{-j}(A)(z) = \sigma_M^{-j}(a)(z) + \sigma_M^{-j}(C)(z)$, $\sigma_M(A)(z) :=$ $\{\sigma_M^{-j}(A)(z)\}_{j\in\mathbb{Z}_+}$, $\sigma_\psi(A)(t,\tau) := a(t,\tau)$.

It is obvious from Theorem 3.4. that every operator $A \in \mathcal{L}(\mathbb{R}_+)$ induces continuous operators

$$A : L^2(\mathbb{R}_+) \longrightarrow L^2(\mathbb{R}_+),\ A : \mathcal{S}_P(\mathbb{R}_+) \longrightarrow \mathcal{S}_Q(\mathbb{R}_+)$$

for every P with some resulting singularity type Q that depends on P and A.

<u>3.7. Theorem</u>. $\mathcal{L}(\mathbb{R}_+)$ is a *-algebra with $\mathcal{L}_{M+G}(\mathbb{R}_+)$ and $\mathcal{L}_G(\mathbb{R}_+)$ as two-sided ideals. $A \in \mathcal{L}(\mathbb{R}_+)$, $\sigma_\psi(A) \in S^{-\infty}(\mathbb{R})$, $\sigma_M(A) = 0$ imply $A \in \mathcal{L}_G(\mathbb{R}_+)$. For A, B $\in \mathcal{L}(\mathbb{R}_+)$ we have

$$\sigma_\psi(AB)(t,\tau) = \sigma_\psi(A)(t,\tau) \#_\psi \sigma_\psi(B)(t,\tau) \quad \text{mod } S^{-\infty}(\mathbb{R}),$$

$$\sigma_M(AB)(z) = \sigma_M(A)(z) \#_M \sigma_M(B)(z) .$$

<u>3.8. Definition</u>. An operator $A \in \mathcal{L}(\mathbb{R}_+)$ is called elliptic (with respect to the weight $\gamma \in \mathbb{R}$ at t = 0) if

(i) $\sigma_\psi(A)(t,\tau) \neq 0$ for $|\tau| \geq c_t$ with a constant $c_t \geq 0$, $c_\infty = 0$,
(ii) $\sigma_M^0(A)(z) \neq 0$ for $\mathrm{Re}\, z = \frac{1}{2} + \gamma$.

3.9. Definition. An operator B is called a parametrix of $A \in \mathcal{A}(\mathbb{R}_+)$ if $AB - 1, BA - 1 \in \mathcal{A}_G(\mathbb{R}_+)$.

3.10. Theorem. $A \in \mathcal{A}(\mathbb{R}_+)$ is elliptic (with respect to the weight $\gamma = 0$) iff it defines a Fredholm operator $A : L^2(\mathbb{R}_+) \longrightarrow L^2(\mathbb{R}_+)$. An elliptic operator $A \in \mathcal{A}(\mathbb{R}_+)$ has a parametrix $B \in \mathcal{A}(\mathbb{R}_+)$. Moreover $Au = f \in \mathcal{S}_Q(\mathbb{R}_+)$ and $u \in L^2(\mathbb{R}_+)$ imply $u \in \mathcal{S}_P(\mathbb{R}_+)$ with some singularity type P depending on Q and A.

An algebra of operators of the form $op_\psi(a) + \omega op_M(h) + r$ in $L^2(\mathbb{R}_+)$, r compact, was studied by Eskin in [E1], § 15. There was established the principal symbolic calculus and the Fredholm property in $L^2(\mathbb{R}_+)$ under the ellipticity condition. Our results show that the compact remainders r admit Mellin expansions and reflect the asymptotics.

Remark that Theorem 3.10. can be extended to a natural substitute of Sobolev spaces such that it becomes completely analogous to Theorem 2.10. This requires more information on adequate distribution spaces on $\mathbb{R}_+$ that we have dropped here, for details cf. [R9]. Other extensions concern

(a) higher order operators,
(b) operators on manifolds (of arbitrary dimension) with boundary.

For dealing with (a) one has to modify the notion of Green operators in the sense of 'positive class' operators (cf. the analogous notion in the special case of [B1]), and there has to be given a natural rule to define an action of ψDO's on spaces on $\mathbb{R}_+$. (b) needs extra trace and potential conditions with respect to the boundary and a systematic technique to the branching of the p_j with varying parameter on the boundary, cf. [R9]. Details on the functional analysis for spaces with branching asymptotics also may be found in [S3] (in a variant for manifolds with edges). The application then to parametrix constructions and the conormal asymptotics for mixed elliptic boundary problems require the boundary reduction in the reverse direction and an expression of the asymptotics of potential operators in Boutet de Monvel's algebra applied to distributions with asymptotics along the jump of the conditions. This also was done in [R9].

4. The Index of Operators

Let us conclude this survey with some remarks on the index of operators that admit Mellin expansions. First consider the case of conical singularities.

An operator $A \in \mathfrak{M}^{\mu}(\check{M})$ considered as continuous map $A : \mathcal{H}^{s}(\check{M}) \longrightarrow \mathcal{H}^{s-\mu}(\check{M})$ equals the closure of A with the domain $C_0^\infty(\text{int } \check{M})$. If A is elliptic we can pass to the elliptic operators A*A and AA* in $\mathfrak{M}^{2\mu}(\check{M})$ (the * refers to the $L^2(\check{M})$ scalar product) and then

$$\text{ind } A = \dim \ker A^*A - \dim \ker AA^* \tag{4.1}$$

where A*A (AA*) may be considered as the unique self-adjoint extensions of the operators over $C_0^\infty(\text{int } \check{M})$ in $L^2(\check{M})$. In this sense the L^2-index (4.1) coincides with the index of A in the sense of Theorem 2.10. The same is true of elliptic operators acting between distributional sections of vector bundles. Since the index of A only depends of the couple of principal symbols $(\tilde{\sigma}_\psi^\mu(A), \sigma_M^0(A))$, we may ask for a description of the set of all stable homotopy classes in a similar sense as in the classical calculus over smooth compact manifolds, cf. Atiyah, Singer [A1]. The answer gives the following

4.1. Theorem. The stable homotopy classes of elliptic principal symbols may be described by $K(\tilde{T}^*\check{M} \setminus 0) \oplus \mathbb{Z}^N$, where N denotes the number of connection components of the base X of the cone near the vertex, $K(\tilde{T}^*\check{M} \setminus 0)$ the K-group over the compressed cotangent bundle of $\check{M}$ minus the zero-section.

Explicit index expressions for special elliptic operators in a geometric context have been derived by several authors, cf. e.g. [M6] and the references there, and [B3]. The L^2-index can also be studied for elliptic complexes of operators in $\mathfrak{M}^\infty$. A Hodge decomposition theorem was derived in [R9] and there was established the asymptotics of the harmonic forms near the vertex as an immediate consequence of Theorem 2.10. A similar result could be derived for complexes of boundary value problems, with conical singularities on the boundary, using an analogue of the calculus in [P2] with the results of [R8]. A more particular result was obtained by Mei-Chi-Shaw [M5].

Now let us come to some details on the proof of Theorem 4.1. First of all it is clear that the assertion refers to operators acting between distributional sections of vector bundles E,F over M. Denote by Vect(.) the set of vector bundles on the space in the brackets, where isomorphic bundles are identified. Let $\mathfrak{M}^{\mu}(\check{M};E,F)$ be the class of cone operators of order μ associated with $E,F \in \mathrm{Vect}(\check{M})$.

Every $A \in \mathfrak{M}^{\mu}(\check{M};E,F)$ has a couple of principal symbols

$$\tilde{\sigma}^{\mu}_{\psi}(A) : \tilde{\pi}^*E \longrightarrow \tilde{\pi}^*F, \qquad (4.2)$$

$\tilde{\pi} : \tilde{T}^*\check{M} \setminus 0 \longrightarrow \check{M}$ being the canonical projection,

$$\sigma^{0}_{M}(A) : \Gamma_{\frac{1}{2}} \times H^{s}(X,E') \longrightarrow \Gamma_{\frac{1}{2}} \times H^{s-\mu}(X,F'),$$

$\Gamma_{\eta} = \{\mathrm{Re}\, z = \eta\}$, the prime indicates restrictions of bundles to X. Then the ellipticity is defined in an analogous way as in Definition 2.2. Moreover we have an obvious notion of homotopy $A_0 \simeq A$, of operators in $\mathfrak{M}^{\mu}(\check{M};E,F)$ through elliptic operators. Then ind A_0 = ind A_1. Two operators $A_j \in \mathfrak{M}^{\mu}(\check{M};E_j,F_j)$, $j = 0,1$, are called stable homotopic (or stable equivalent) if there exist vector bundles $G_j \in \mathrm{Vect}(\check{M})$ and isomorphisms $h_0 : E_0 \oplus G_0 \longrightarrow E_1 \oplus G_1$, $h_1 : F_0 \oplus G_0 \longrightarrow F_1 \oplus G_1$ so that

$$A_0 \oplus \mathrm{id}_{G_0} - I_{h_0}^{-1}(A_1 \oplus \mathrm{id}_{G_1})I_{h_1}$$

through elliptic operators, I_{h_j} being the operator in $\mathfrak{M}^0$ induced by h_j.

Let $\mathcal{E}$ denote the set of stable equivalence classes of elliptic operators over M. Then the composition of operators induces in $\mathcal{E}$ the structure of an Abelian group. Denote by $\sigma(\mathcal{E})$ the set of stable equivalence classes of couples of elliptic symbols $\sigma(A) = (\tilde{\sigma}^{\mu}_{\psi}(A), \sigma^{0}_{M}(A))$, equipped with the analogous group structure. Then we have an isomorphism $\sigma : \mathcal{E} \longrightarrow \sigma(\mathcal{E})$. Applying the analogue of Theorem 2.11. to the classes $\mathfrak{M}^{\mu}(\check{M};E,E)$ we can prove that $\mathcal{E}$ and $\sigma(\mathcal{E})$ are generated by the stable homotopy classes of zero order operators.

Let $\sigma_{\psi}(\mathcal{E})$ denote the group of stable equivalence classes of isomorphisms $\tilde{\pi}^*E \longrightarrow \tilde{\pi}^*F$, cf. (4.2), which are homogeneous of order zero. Then we have a canonical surjective homomorphism

$$p : \sigma(\mathcal{E}) \longrightarrow \sigma_{\psi}(\mathcal{E}). \qquad (4.3)$$

The surjectivity follows by the same arguments that are applied for proving the existence of an operator in the cone class for a given homogeneous principal symbol. As in the standard index theory (cf. [A1], [R2]) we have

$$\sigma_\psi(\mathcal{E}) = K(\tilde{T}^*\check{M} \setminus 0) \cong K(T^*\check{M} \setminus 0)$$

with canonical identifications. In order to describe the kernel of (4.3) we set

$$\mathcal{K}^{(1)}(E) = \{A \in \mathfrak{N}^0(M;E,E) : \tilde{\sigma}_\psi^0(A) = id_{\tilde{\pi}^*E}\} .$$

The above notion of stable equivalance of operators induces a corresponding notion in the subclass $\mathcal{K}^{(1)} = \bigcup \mathcal{K}^{(1)}(E)$ (union over E) where we impose the extra condition that homotopies go through $\mathcal{K}^{(1)}$.

Let $\mathcal{R}^{(1)}$ be the Abelian group of corresponding stable equivalence classes and $\sigma(\mathcal{R}^{(1)})$ the corresponding group of couples of symbols. It is clear that there is a canonical homomorphism $\varkappa : \sigma(\mathcal{R}^{(1)}) \longrightarrow$ ker p. A crucial point in the proof of Theorem 4.1 is

$$\ker p = \operatorname{im} \varkappa \ , \ \ker \varkappa = 1 \ , \tag{4.4}$$

in other words,

4.2. Proposition. The sequence

$$1 \longrightarrow \sigma(\mathcal{R}^{(1)}) \xrightarrow{\ \varkappa\ } \sigma(\mathcal{E}) \xrightarrow{\ p\ } \sigma_\psi(\mathcal{E}) \longrightarrow 1$$

is exact.

Moreover we have

4.3. Proposition. There is a canonical isomorphism $\sigma(\mathcal{R}^{(1)}) \longrightarrow \mathbb{Z}^N$, where N is the number of connection components of X.

Then we get Theorem 4.1 as a corollary. Let us sketch the details of the proof. For $A \in \mathcal{R}^{(1)}(E)$ we know that

$$\sigma_M^0(A)(z) : L^2(X,E') \longrightarrow L^2(X,E'),\ z \in \Gamma_{\frac{1}{2}}$$

is a ψDO over X depending on the parameter z, and the homogeneous principal symbol is the identity in π^*E, $\pi : T^*X \setminus 0 \longrightarrow X$. Thus

$\sigma^{0}_{M}(A)(z) - 1 = K(z)$ is a ψDO of order -1 with parameter. It is easily seen that it tends to zero in $\mathcal{L}(L^2(X,E'))$ as $|\mathrm{Im}\, z| \longrightarrow \infty$. In addition $K(z)$ is compact in $L^2(X,E')$. Thus $\sigma(\tilde{\mathcal{R}}^{(1)})$ is isomorphic to the group of stable equivalance classes of families of isomorphisms

$$S^1 \ni z \longrightarrow (1+K(z)) : L^2(X,E') \cong L^2(X,E')$$

where $K(z)$ is compact for every z and S^1 is identified with $\Gamma_{\frac{1}{2}}$. Then Proposition 4.3 is an obvious consequence. The equation $(4.4)_{\overline{2}}$ is a consequence of the following simple statements.

4.4. Lemma. Let $\{A(\lambda)\}_{0\leq\lambda=1}$ be a homotopy through elliptic operators in $\mathcal{M}^{0}(\check{M};E,F)$. Then there exists a homotopy $\{B(\lambda)\}_{0\leq\lambda\leq 1}$ through elliptic operators in $\mathcal{M}^{0}(\check{M};E,F)$ so that $B(\lambda)$ is a parametrix of $A(\lambda)$ for $0\leq\lambda\leq 1$.

4.5. Lemma. Let $\{\sigma_\psi(\lambda)\}_{0\leq\lambda\leq 1}$ be a homotopy through isomorphisms $\tilde{\pi}^*E \longrightarrow \tilde{\pi}^*F$, homogeneous of order zero. Then there exists a homotopy through elliptic operators $\{A(\lambda)\}_{0\leq\lambda\leq 1}$ in $\mathcal{M}^{0}(\check{M};E,F)$ such that $\sigma_\psi(\lambda) = \tilde{\sigma}^{0}_{\psi}(A(\lambda))$, $0\leq\lambda\leq 1$.

The proofs are formally analogous as the corresponding results for ψDO's on closed compact C^∞ manifolds. Here we have to use the symbolic structures of Section 2.

Now let $\{A'(\lambda)\}$, $\{A''(\lambda)\}$, $0\leq\lambda\leq 1$, be two homotopies through elliptic operators in $\mathcal{M}^{0}$ with $\sigma_\psi(\lambda) = \tilde{\sigma}^{0}_{\psi}(A(\lambda)) = \tilde{\sigma}^{0}_{\psi}(A''(\lambda))$, $0\leq\lambda\leq 1$. Then we find $K(\lambda)A'(\lambda) = A''(\lambda)$ for a certain homotopy $K(\lambda)$ of elliptic operators, $K(0)$, $K(1) \in \mathcal{K}^{(1)}$. We have used Lemma 4.4., i.e. $K(\lambda) = A''(\lambda) B'(\lambda)$ for a homotopy of parametrices $B'(\lambda)$ of $A'(\lambda)$. The injectivity of $\varkappa$ follows from

4.6. Lemma. Let $\{A^{0}(\lambda)\}_{0\leq\lambda\leq 1}$ be a homotopy through $\mathcal{M}^{0}(\check{M};E,E)$ with $A^{0}(0)$, $A^{0}(1) \in \mathcal{K}^{(1)}(E)$. Then there is another homotopy

$$\{A^{1}(\lambda,\eta)\}_{0\leq\lambda\leq 1,\, 0\leq\eta\leq 1}$$

through elliptic operators in $\mathcal{M}^{0}(\check{M};F,F)$ for another $F \in \mathrm{Vect}(\check{M})$ such that for certain $G_0, G_1 \in \mathrm{Vect}(\check{M})$ and isomorphisms $h_0, h_1 : E \oplus G_0 \longrightarrow F \oplus G_1$, $A^{0}(\lambda,0) \oplus id_{G_0} = I^{-1}_{h_0}(A^{1}(\lambda,0) \oplus id_{G_1})I_{h_1}$, $0\leq\lambda\leq 1$.

For proving this we have to use again systematically the basic calculus of Section 2.

Let us return now to the construction of examples of elliptic operators over manifolds with conical singularities. At the beginning of Section 2 we have defined the class of differential operators that are totally characteristic neat $t = 0$. For $\check{M}$ we have similarly the class $\mathrm{Diff}^{\mu}(\check{M})$. Now let M be a closed compact C^{∞} manifold and fix a point $m_0 \in M$. By introducing polar coordinates (t,x) locally near m_0, where m_0 corresponds to $t = 0$, x varies on S^n, $n+1 = \dim M$, we can pass to a manifold $\check{M}$ with a conical singularity. Let $A \in \mathrm{Diff}^{\mu}(M)$ (= class of differential operators on M of order μ). Then A transforms to some $A^{\vee}$ by substituting (t,x) locally near m_0 which is up to the factor $t^{-\mu}$ of the form (2.1). Denote by g^{μ} a strictly positive function on M which equals t^{μ} close to $t = 0$. Then our procedure can be written as a linear mapping $b : \mathrm{Diff}^{\mu}(M) \longrightarrow \mathrm{Diff}^{\mu}(\check{M})g^{-\mu}$, $bA := A^{\vee}$.

4.7 Theorem. Let $A \in \mathrm{Diff}^{\mu}(M)$ be elliptic in the standard sense. Then $g^{-n/2}bAg^{n/2} \in \mathrm{Diff}^{\mu}(\check{M})$ is elliptic in $\mathfrak{N}^{\mu}(\check{M})$.

The proof follows by using the fact that the ellipticity of A implies the Fredholm property of $A:H^{s}(M) \longrightarrow H^{s-\mu}(M)$ between the standard Sobolev spaces over M. The space $H^{\mu}(M)$ may be canonically identified with $g^{\mu-n/2}\ \mathcal{H}^{\mu}(\check{M})$ and $L^2(M)$ with $g^{-n/2}L^2(\check{M})$. Then $g^{-n/2}\ bA\ g^{n/2}$ corresponds to A on M and hence it is a Fredholm operator $\mathcal{H}^{\mu}(\check{M}) \longrightarrow \mathcal{H}^{0}(\check{M})$. From Theorem 2.10 then follows the ellipticity.

Remark that the ellipticity in the cone algebra contains a'global spectral condition' along the base which is an analogue of Agmons condition in the spectral theory for parameter depending elliptic operators. In concrete cases it may be difficult to verify this for a special weight. The same is true of the positions of the other poles in the complex Mellin plane. So it is an important task to study classes of examples under this point of view. This is not automatically solved by the constructions of Section 2. Theorem 4.7 shows a method to give general answers in terms of operators that are obtained from elliptic ones over a simpler space. The same arguments apply for boundary value problems. One may hope that higher singularities such as edges and corners may be treated in an analogous way. The index theory for corners in the sense of Theorem 4.1 also seems to be analogous though much more technical.

The informations on the case with edges are not so complete, but one may expect certain topological phenomena that are known in an analogous form for elliptic boundary value problems of the class of [B1], such as a topological obstruction for the existence of extra boundary and potential conditions with respect to the edge. The idea is to consider ψDO's along the edge Y with symbols that are operator-valued with values in $\mathfrak{N}^{\infty}(\check{M})$, where the edge is locally identified with $Y \times \check{M}$. Let S^*Y be the unit cosphere bundle induced by T^*Y (with a Riemannian metric). Then an operator-valued symbol $A(y,\eta)$ which is point-wise elliptic induces an index element $\mathrm{ind}_{S^*Y} A \in K(S^*Y)$ (cf. [B1], [R2]). The obstruction would be that $\mathrm{ind}_{S^*Y} A \in \pi^*K(Y)$, $\pi : S^*Y \longrightarrow Y$. Analogous obstructions may be expected in higher iterations, i.e. when the manifold with singularities is inductively identified with $\mathbb{R}^{k_m} \times (\mathbb{R}_+)^{l_m} \times \ldots \times \mathbb{R}^{k_1} \times (\mathbb{R}_+)^{l_1} \times X$. This would lead to symbol hierarchies and hierarchies of index elements.

For the elliptic boundary value problems for ψDO's without the transmission property there is an index theorem which is analogous to that of [B1]. This was proved in [R3]. The principal Mellin symbol, first given on $Y \times \{\mathrm{Re}\ z = \frac{1}{2}\}$ (Y the boundary) is transformed to a function on $Y \times [-1,1] \subset \{$conormal bundle of Y$\}$. The ellipticity of the ψDO symbol together with the Mellin symbol on $Y \times [-1,1]$ then leads to a K-theoretic interpretation in terms of an index element over S^*Y. The stable homotopy classes of elliptic boundary value problems can be represented by problems for ψDO's with the transmission property. An index theorem for elliptic transmission problems was given in [R1].

References

[A1] M.F. Atiyah, I.M. Singer: The index of elliptic operators. Ann. of Math. 87 (1968), 484-530

[B1] L. Boutet de Monvel: Boundary problems for pseudo-differential operators. Acta Math. 126 (1971), 11-51

[B2] P. Bolley, M. Dauge, J. Camus: Regularité Gevrey pour le problème de Dirichlet dans des domaines à singularités coniques. Comm. Part. Diff. Equ. 10, 4(1985), 391-432

[B3] J. Brüning, R. Seeley: The resolvent expansion for second order regular singular operators. Preprint No. 80, Univ. Augsburg 1985

[C1] J. Cheeger: Spectral geometry of spaces with cone-like singularities. Proc. Nat. Acad. Sci. USA 76 (1979), 2103-2106

[C2] H.O. Cordes, E.A. Herman: Pseudo-differential operators on a half line. J. Math. Mech. 18 (1969), 893-908

[C3] M. Costabel: Starke Elliptizität von Randintegraloperatoren erster Art. TH Darmstadt, Preprint 868, 1984

[C4] M. Costabel, E. Stephan: A direct boundary integral equation method for transmission problems. J. Math. Anal. Appl. 106 (1985) 367-413

[E1] G.I. Eskin: Boundary problems for elliptic pseudo-differential equations. Nauka, Moskva 1973 (Transl. of Mathem. Monographs 52, Amer. Math. Soc., Providence, Rhode Island 1980)

[H1] L. Hörmander: Pseudo-differential operators and non-elliptic boundary problems. Ann. of Math. 83, 1 (1966), 129-209

[H2] L. Hörmander: Pseudo-differential operators and hypoelliptic equations. Amer. Math. Soc., Symp. on Singular Integral Operators (1966), 138-183

[H3] L. Hörmander: The analysis of linear partial differential operators. I, II, III, IV Grundlehren der Mathem. Wiss., Springer Berlin, Heidelberg, New York, Tokyo 1983-1985

[J1] P. Jeanquartier: Transformation de Mellin et développements asymptotiques. L'Enseignement Mathématique 25, 3-4 (1980), 285-308

[K1] V.A. Kondrat'ev: Boundary value problems for elliptic equations in domains with conical points. Trudy Mosk. Mat. Obšč. 16 (1967) 209-292

[K2] A.I. Komeč: Elliptic boundary problems for pseudo-differential operators on manifolds with conical points. Matem. sb. 86, 2 (1971), 268-298

[K3] A.I. Komeč: Elliptic boundary problems on manifolds with piecewise smooth boundary. Matem. sb. 92, 1 (1973), 89-134

[K4] V.A. Kondrat'ev, O.A. Oleynik: Boundary problems for partial differential equations in non-smooth domains. Uspechi Mat. Nauk 38, 2 (1983), 3-76

[L1] J.E. Lewis, C. Parenti: Pseudodifferential operators of Mellin type. Comm. Part. Diff. Equ. 8, 5 (1983), 477-544

[M1] V.G. Maz'ja, B.A. Plamenevskij: On the coefficients in the asymptotics of solutions of elliptic boundary value problems in domains with conical points. Math. Nachr. 76 (1977), 29-60 (Amer. Math. Soc. Transl. (2) 123, 1984, 57-88)

[M2] V.G. Maz'ja, B.A. Plamenevskij: Schauder estimates of solutions of elliptic boundary value problems in domains with edges on the boundary. Part. Diff. Equ., Proc. Sem. S.L. Sobolev 1988, 2, Institut Math. Sibirsk. Otdel. Akad. Nauk SSSR, Novosibirsk 1978, 69-102 (Amer. Math. Soc. Transl. (2) 123, 1984, 141-170)

[M3] R. Melrose, G.A. Mendoza: Elliptic boundary problems in spaces with conical points. Proc. Journees 'Equ. Dériv. Part.' St. Jean-de-Monts 1981, Conf. 4

[M4] R. Melrose, G.A. Mendoza: Elliptic operators of totally characteristic type. MSRI 047-83, Berkeley 1983

[M5] Mei-Chi-Shaw: Hodge theory on domains with conic singularities. Comm. in Part. Diff. Equ. 8, 1 (1983), 65-88

[M6] W. Müller: L^2-index of elliptic operators on manifolds with cusps of rank one. Report d.Karl-Weierstraß-Instituts für Mathematik, R-Math-06/85, Berlin 1985

[M7] E. Meister: Einige gelöste und ungelöste kanonische Probleme der Beugungstheorie. Preprint No. 918, Techn.Hochschule Darmstadt 1985

[P1] B.A.Plamenevskij: On algebras generated by pseudo-differential operators with isolated singularities of symbols. Dokl. Akad. Nauk SSSR 248, 2(1979) (Soviet Math. Dokl. 20, 5 (1979), 1013-1017)

[P2] U. Pillat, B.-W. Schulze: Elliptische Randwert-Probleme für Komplexe von Pseudodifferentialoperatoren. Math. Nachr. 94 (1980), 173-210

[R1] S. Rempel: The index theorem for elliptic transmission problems. Seminar Analysis 1983/84, 141-162, Karl-Weierstraß-Institut für Mathematik, Berlin 1984

[R2] S. Rempel , B.-W. Schulze: Index theory of elliptic boundary problems. Akademie Verlag Berlin 1982

[R3] S. Rempel, B.-W. Schulze
Parametrices and boundary symbolic calculus for elliptic boundary problems without the transmission property. Math. Nachr. 105 (1982), 43-149

[R4] S. Rempel, B.-W. Schulze: A theory of pseudo-differential boundary value problems with discontinuous conditions I, II. Ann. Glob. Analysis and Geometry 2, 2 (1984), 163-251, 2, 3 (1984), 289-384

[R5] S. Rempel, B.-W. Schulze: Pseudo-differential and Mellin operators in spaces with conormal singularity (boundary symbols). Report Institute of Mathematics, Acad. Sc. I1/84, Berlin 1984

[R6] S. Rempel, B.-W. Schulze: Complete Mellin symbols and the conormal asymptotics in boundary value problems. Proc. Journees 'Equations aux Dériv. Part! St. Jean-de-Monts 1984, Conf. No. V

[R7] S. Rempel, B.-W. Schulze: Complete Mellin and Green symbolic calculus in spaces with conormal asymptotics. Ann. Glob. Analysis and Geometry 4, 2 (1986), 137-224

[R8] S. Rempel, B.-W. Schulze: Mellin symbolic calculus and asymptotics for boundary value problems. Seminar Analysis d. Karl-Weierstraß-Instituts für Mathematik 1984/85, Berlin 1985

[R9] S. Rempel, B.-W. Schulze: Asymptotics for elliptic mixed boundary problems (pseudo-differential and Mellin operators in spaces with conormal singularity). Math. Research Series, Akademie Verlag Berlin (to appear 1989)

[S1] B.-W. Schulze: Ellipticity and continuous conormal asymptotics on manifolds with conical singularities. Prepublications Université de Paris-Sud, Orsay 1986; Math. Nachr. (to appear)

[S2] B.-W. Schulze: Opérateurs pseudo différentiels et asymptotiques sur des variétés à singularités. Séminaire Equ. aux Dériv. Part., Ecole Polytechnique 1985-1986

[S3] B.-W. Schulze: Regularity with continuous and branching asymptotics on manifolds with edges. Preprint des SFB 72, Bonn 1986

[S4] B.-W. Schulze: Pseudo-differential operators on manifolds with corners (in preparation)

[S5] E. Stephan: Boundary integral equations for mixed boundary value problems, screen and transmission problems in $\mathbb{R}^3$. Preprint No. 848, Techn.Hochschule Darmstadt 1984

[T1] N. Teleman: The index theorem for topological manifolds. Acta Mathematica 153, 1-2 (1984), 117-152

Semiclassical resonances generated by non-degenerate critical points.

by

Johannes Sjöstrand ,

Dept. of Mathematics , University of Lund ,

Box 118 , S-22100 Lund , Sweden.

0 . Introduction.

In this paper , we continue our study (started in [5] , [3]) of resonances in situations , where the Hamilton flow has a particularly simple structure .

Let P be a semiclassical differential operator on $\mathbb{R}^n$ with analytic coefficients , satisfying all the general assumptions of [5] , section 8 , that permit to define resonances in a small neighborhood of $0 \in \mathbb{C}$. (The case when the coefficients are only C^∞ on a compact set and analytic outside , could certainly also be treated to the price of certain technical complications .) Let $p(x,\xi)$ be the principal symbol (defined as in [5]) . The most interesting case that we have in mind is of course when

$$(0.1) \qquad P = - h^2 \Delta + V(x) \quad ,$$

and in this case

$$(0.2) \qquad p(x,\xi) = \xi^2 + V(x) \ .$$

In the appendix of [3] , we defined in $p^{-1}([-\varepsilon,\varepsilon])$ for $\varepsilon > 0$ small enough , the outgoing tail ,

$$\Gamma_+ = \{\rho \in p^{-1}([-\varepsilon,\varepsilon]) \ ; \ \exp(tH_p)(\rho) \not\to \infty \ , \text{ when } \ t \to -T_-(\rho)\}.$$

Here $]-T_-(\rho),T_+(\rho)[$ is the maximal interval of definition for the flow of the Hamilton field , $H_p = \Sigma p'_{\xi_j}\partial_{x_j} - p'_{x_j}\partial_{\xi_j}$. Similarly , we defined the incoming tail Γ_- , and we put $\Gamma^0_\pm = \Gamma_\pm \cap p^{-1}(0)$. If $\Omega \subset\subset \mathbb{R}^{2n}$, then $\Gamma^0_\pm \cap \Omega$ is the limit in the sense of sets , of $\Gamma_\pm \cap \Omega$, when $\varepsilon \to 0$. The sets $K = \Gamma_+ \cap \Gamma_-$ and $K^0 = \Gamma^0_+ \cap \Gamma^0_-$ are compact . They are the unions of completely trapped trajectories in $p^{-1}([-\varepsilon,\varepsilon])$ and $p^{-1}(0)$ respectively . Again $K^0 = \lim K$, as $\varepsilon \to 0$.

When $K^0 = \emptyset$, it is easy to see,(using for instance the arguments of the appendix of [3] ,) that there is an escape function $G(x,\xi)$,(defined as in [5] ,) such that $H_p(G) > 0$ everywhere on $p^{-1}([-\varepsilon,\varepsilon])$, provided that $\varepsilon > 0$ is small enough . It is then an easy consequence of the general theory of [5] that there are no resonances in a fixed h-independent neighborhood of 0 , when $h > 0$ is small enough .

In [3] , we studied the case when K^0 is a closed trajectory of hyperbolic type . In this paper we shall study the case when K^0 is reduced to a point , verifying certain conditions below . The treatment of this case is similar to the one of [3] , and even slightly simpler . This means that we can leave out some details at places where the proofs are almost identical .

Our first assumption is

(0.3) K^0 is reduced to a point $\{(x_0,\xi_0)\}$.

Clearly , p and dp vanish at (x_0,ξ_0) .

When discussing further assumptions , we shall start with the Schrödinger operator case . Then necessarily $\xi_0 = 0$, and after a translation , we may assume that also $x_0 = 0$. Then 0 is a critical point of V with critical value 0 . We assume

(0.4) 0 is a non-degenerate critical point .

Let $(n-d,d)$ be the signature of $V''(0)$. The case $d = 0$ has been thouroughly discussed by Helffer-Sjöstrand in [5] . (See also Combes-Duclos-Klein-Seiler [2]) In this case the result of this paper will be less refined , since it will not take into account possible exponentially small corrections , due to tunneling .

After a linear change of coordinates , we obtain

$$(0.5) \qquad p(x,\xi) = \sum_{1}^{n-d} (\lambda_j/2)(\xi_j^2+x_j^2) + \sum_{n-d+1}^{n} (\lambda_j/2)(\xi_j^2-x_j^2) + O(|(x,\xi)|^3) ,$$

where $\lambda_j > 0$. Then the fundamental matrix F_p of p at $(0,0)$ (i.e. the matrix of the linearized Hamilton field) has the eigen-values $\pm i\lambda_j$, $1\leq j\leq n-d$, and $\pm\lambda_j$, $n-d+1\leq j\leq n$. After a real linear symplectic change of the (x'',ξ'')-coordinates , we may write

(0.6) $$p(x,\xi) = p'(x',\xi') + Ax''\cdot\xi'' + O(|(x,\xi)|^3) \quad ,$$

where p' is a positive definite quadratic form , whose fundamental matrix $F_{p'}$ has the eigen values $\pm i\lambda_j$, $1\le j\le n-d$, and A is a real (diagonalizable) matrix with the positive eigenvalues μ_j , $n-d+1\le j\le n$.

We now return to the general case , and we shall make assumptions , which in the Schrödinger case are implied by (0.4) . We assume ,

(0.7) $$p''((x_0,\xi_0)) \text{ is non-degenerate .}$$

Then F_p is also non-degenerate . Since F_p is real and antisymmetric with respect to the symplectic form , we have : $\lambda \in \sigma(F_p) \Rightarrow \bar{\lambda}\ , -\lambda\ , -\bar{\lambda} \in \sigma(F_p)$. Let d be the number of eigenvalues with real part > 0 . Then the eigenvalues of F_p are of the form

(0.8) $$\pm i\lambda_j \ , \ 1\le j\le n-d \ , \text{ and } \ \pm\mu_j \ , \ 1\le j\le d \quad ,$$

where $\lambda_j > 0$ and $\mathrm{Re}\,\mu_j > 0$. It is then easy to find real symplectic coordinates centered at (x_0,ξ_0) , such that (0.6) holds , where now p' is a non-degenerate quadratic form with $\sigma(F_{p'}) = \pm i\lambda_j$, while A is a real matrix with eigenvalues $\mu_1,\ldots,\mu_d$. Our last assumption is then that

(0.9) $$p' \text{ is positive definite .}$$

If $\theta > 0$ is small enough , then by the stable manifold theorem , we have a complex n-dimensional manifold Λ_+ , passing through the origin , invariant under the H_p-flow , and such that $e^{-i\theta}H_p$ is "expansive" on Λ_+ . It is easy to see that Λ_+ is a complex Lagrangian manifold whose tangent space at $(0,0)$ is the sum of the eigenspaces of F_p, associated to the eigenvalues $\mu_1,..\mu_d,i\lambda_1,..,i\lambda_{n-d}$. Similarly, we can define the stable incoming manifold, Λ_- for $e^{-i\theta}H_p$, with tangent space at the origin associated to $-\mu_1,..,-\mu_d,-i\lambda_1,..,-i\lambda_{n-d}$. Λ_+ and Λ_- intersect transversally at $(0,0)$, and we may introduce complex symplectic coordinates (x,ξ), such that $\Lambda_+ : \xi=0$, $\Lambda_- : x=0$. Then,

$$p(x,\xi) = B(x,\xi)x\cdot\xi \quad , \tag{0.10}$$

where $B(0,0)$ has the eigenvalues $\mu_1,..,\mu_d,i\lambda_1,..,i\lambda_{n-d}$.

At least formally, we can represent P in the new coordinates, as an h-pseudodifferential operator with symbol,

$$B(x,\xi)x\cdot\xi + h\, p_{-1}(x,\xi) + O(h^2) \quad , \tag{0.11}$$

where $B_0 = B(0,0)$ has the eigenvalues μ_j, $i\lambda_j$.

If S_P is the subprincipal symbol (, invariantly defined from the original symbol of P ; $p+h\tilde{p}_{-1}+h^2\tilde{p}_{-2}+..$, as $(\tilde{p}_{-1}+\frac{1}{2}i\sum p''_{\xi_j x_j})(0,0)$), then

$$p_{-1}(0,0) = S_P-\tfrac{1}{2}i\ \mathrm{tr}(B) \quad . \tag{0.12}$$

Consider,

$$P_0 = B_0 x\cdot\tilde{D}_x + h\, p_{-1}(0,0) \quad , \tag{0.13}$$

where we use the notation from our earlier works ; $\tilde{D} = hD = (h/i)\partial$. If $\mathcal{P}^N$ denotes the space of polynomials of degree $< N$, and we write $\mathcal{P}^N = E_0\oplus..\oplus E_{N-1}$, where E_j denotes the space of j-homogeneous polynomials, then $P_0 : \mathcal{T}^N \longrightarrow \mathcal{T}^N$ can be represented as a block diagonal matrix $((P_0)_{j,k})$, with $(P_0)_{j,k} = 0$ for $j \neq k$, and where $(P_0)_{j,j}$ has the eigenvalues,

$$h(p_{-1}(0,0) + i^{-1}\ \alpha\cdot z\ ) \quad , \tag{0.14}$$

for $\alpha \in \mathbb{N}^n$, $|\alpha| = j$. Here $z = (z_1,..,z_n) = (i\lambda_1,..,i\lambda_{n-d}, \mu_1,..,\mu_d)$. Fix $C_0 > 0$ such that none of the values (0.14) for $\alpha \in \mathbb{N}^n$ falls on the boundary of the disc $D(0,C_0h)$. Let $\Gamma^0(h)$ be the set of values (0.14) in $D(0,C_0h)$. We count the elements of $\Gamma^0(h)$ with their natural multiplicity. Since all the z_j belong to a proper cone in $\mathbb{C}$, the set $\Gamma^0(h)$ has a fixed finite number of elements.

Let us first state a weak version of our main result, which has the advantage of being easier to formulate :

<u>Theorem 0.1.</u> Fix $C_0 > 0$ as above. Then for $h > 0$ small enough, there is a bijection $b = b(h)$ from $\Gamma^0(h)$ onto the set of resonances of P in $D(0,C_0h)$ (also counted with their natural multiplicity), such that $b(h)(\mu) - \mu = o(h)$, $h \longrightarrow 0$, uniformly for $\mu \in \Gamma^0(h)$.

To formulate the complete result, we fix N_0 so large that no elements (0.14) are inside $D(0,C_0h)$, for $|\alpha| \geq N_0$. Then we have,

<u>Theorem 0.2.</u> There exists a matrix $F^{\infty}_{-+}(z,h) : \mathcal{T}^{N_0} \longrightarrow \mathcal{T}^{N_0}$, depending holomorphically on $z \in D(0,C_0)$, which is a symbol in h with an asymptotic expansion, $F^{\infty}_{-+}(z,h) \sim \sum_0^{\infty} A_j(z)\, h^{\frac{1}{2}j}$, where $A_0 = h^{-1}P_0 - z$, such that the following holds :

Let $\Gamma^{\infty}(h)$ be the set of roots in $D(0,C_0h)$ of $\det F_{-+}^{\infty}(E/h,h) = 0$, counted with their natural multiplicity . Then for sufficiently small h , $\Gamma^{\infty}(h)$ is equal to the set of resonances of P inside $D(0,C_0h)$.

From this result , we deduce that all resonances in a disc $D(0,C_0h)$ are given by asymptotic expansions in certain fractiona powers of h . In principle , it should not be too difficult to give sufficient conditions , which guarantee that only integer powers of h appear in these expansions . We content ourselves with stating a simple result in this direction :

Proposition 0.3. Fix α , and assume that the corresponding element $z_0(h)$ of $\Gamma^0(h)$ is simple . For $h > 0$ small enough , let $z(h)$ be the corresponding simple resonance , given by Theorem 0.1 . Then $z(h) \sim h(E_0+E_1h+E_2h^2+..)$.

Once the main theorems have been proved , it is easy to establish this result by simple WKB-constructions , for instance as in [4] , and we shall not go through the proof .

In section 4 , we shall also give an example of a Schrödinger operator with a resonance of multiplicity 2 , such that the corresponding 2-dimensional Jordan block has a non-vanishing nilpotent part . (Equivalently , the resolvent has a pole of second order at this resonance .)

In a recent preprint , Briet , Combes and Duclos [1] , give some estimates on resonance free domains , and they discuss as an application , the case when the potential has a strict non-degenerate maximum (, which corresponds to the case ; $d=n$ in our theorems). They also announce a forthcoming result about actual resonances , which intersects with the case $d=n$ of our results .

1. The geometry .

Choose real symplectic coordinates , such that (0.6) holds . Let $(\ |\ )$ and $(\ |\ )^*$ denote scalar products on $\mathbb{R}^d$ and $(\mathbb{R}^d)^*$ such that $(Ax''|x'')$ and $({}^tA\xi''|\xi'')^*$ are >0 , for all $x''\neq 0$, $\xi''\neq 0$. As a local "escape function" , we put

$$(1.1)\qquad G=\frac{1}{2}((x''|x'')-(\xi''|\xi'')^*)\quad .$$

Then,

$$(1.2)\qquad H_pG=(Ax''|x'')+({}^tA\xi''|\xi'')^*+O(|(x,\xi)|^3)$$
$$\sim|(x'',\xi'')|^2+O(|(x,\xi)|^3)\ .$$

Since $p\geq C^{-1}|(x',\xi')|^2-C|(x'',\xi'')|^2$, we have $|(x',\xi')|\leq$ const.$\times|(x'',\xi'')|$ on $p^{-1}(0)$, so

$$(1.3)\qquad H_pG\sim|(x'',\xi'')|^2\sim|(x,\xi)|^2\ \text{ on }\ p^{-1}(0)\ .$$

(Our discussion is of course local in a neighborhood of (0,0).)

In the appendix of [3] , we constructed (in general) an escape function $\tilde{G}$ (in the sense of [5]),vanishing in a neighborhood of K , such that $H_p\tilde{G}\geq 0$ on $p^{-1}([-\varepsilon,\varepsilon])$, (when $\varepsilon>0$ is small enough) and such that we have strict inequality , outside an arbitrarily small neighborhood of K. Using a partition of unity , we can glue a large multiple of $\tilde{G}$ to G , so that we obtain an escape function $\hat{G}$, such that $\hat{G}=G$ in a neighborhood of (x_0,ξ_0) and such that $H_p\hat{G}>0$ outside this neighborhood in $p^{-1}([-\varepsilon,\varepsilon])$ (, after decreasing $\varepsilon>0$ if necessary). We denote this function by G .

Remark 1.1. By the stable manifold theorem , there exist locally unique H_p - invariant manifolds $L_{\pm}$ of dimension d , passing through (x_0,ξ_0) , such that $T_{(x_0,\xi_0)}(L_{\pm})$ are the sums of the eigenspaces of F_p , associated to the eigenvalues $\mu_1,..,\mu_d$, and $-\mu_1,..,-\mu_d$ respectively . We extend L_+ (and L_-) by integrating the Hamilton field for all positive (negative) times . Then $L_{\pm}$ become globally defined closed analytic manifolds , contained in $\Gamma^0_{\pm}$. We claim that $L_{\pm} = \Gamma^0_{\pm}$. To prove this , let $]-\infty,0] \ni t \longmapsto \gamma(t) \in \Gamma^0_+$ be a trajectory . For large negative times , $\gamma(t)$ is in a small neighborhood of (x_0,ξ_0) , and must therefore stay all the time in the region ,

$$(1.4) \qquad G \geq \delta|(x'',\xi'')|^2 \quad ,$$

for some sufficiently small $\delta > 0$. In fact, if $\gamma(t_0)$ is outside this region, it is easy to see that $\gamma(t_1)$ belongs to the region $G < -\delta|(x'',\xi'')|^2$, for some smaller time $t_1 < t_0$ because of (1.3) . But then $G(\gamma(t_1)) < 0$, and since G decreases when t decreases , $\gamma(t)$ cannot reach (x_0,ξ_0) as $t \longrightarrow -\infty$, which is a contradiction . Since we are then in the region (1.4) , it follows from (1.3) , that $|\gamma(t)|$ tends to 0 exponentially fast , when $t \longrightarrow 0$. This property however characterizes L_+ .

As in [5] , we define Λ_{tG} for small $t > 0$, as the image of the map

$$(1.5) \qquad \mathbb{R}^{2n} \ni (x,\xi) \longmapsto (x,\xi)+itH_G(x,\xi) \in \mathbb{C}^{2n}$$

We shall next study $p|_{\Lambda_{tG}}$ close to (x_0,ξ_0) . Using (1.5) as a parametrization for Λ_{tG} , we get ,

(1.6) $$p|_{\Lambda_{tG}} = p(x,\xi)-itH_pG+O(t^2|(x",\xi")|^2) \quad .$$

Thus for $t > 0$ small enough :

(1.7) $$\text{Re } p|_{\Lambda_{tG}} \geq C^{-1}|(x',\xi')|^2 - C|(x",\xi")|^2 \quad ,$$

and using (1.2) , we also have ,

(1.8) $$- \text{Im } p|_{\Lambda_{tG}} \geq (t/C)|(x",\xi")|^2 - Ct|(x',\xi')|^3 \ .$$

restricting to $|(x,\xi)| \leq \varepsilon$, we get,

(1.9) $$\begin{aligned} - \text{Im } p|_{\Lambda_{tG}} &\geq (t/C)|(x",\xi")|^2 - C\varepsilon t|(x',\xi')|^2 \\ &\geq (t/\tilde{C})|(x",\xi")|^2 - \tilde{C}\varepsilon t \text{ Re } p|_{\Lambda_{tG}} \quad . \end{aligned}$$

This is uniformly true for $t > 0$ small and for ε in any fixed compact interval in $]0,1]$, sufficiently close to 0. Chosing a smaller value for ε when $\text{Re } p \geq 0$, we see that $p|_{\Lambda_{tG}}$ takes its values in a proper cone :

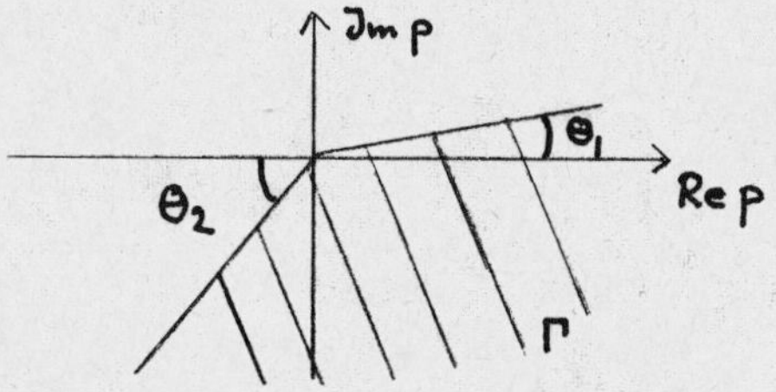

Here $0 < \theta_1 < \theta_2$ are proportional to t , but if we fix $t > 0$ and decrease ε (which amounts to decreasing the neighborhood of (0,0) where we study $p|_{\Lambda_{tG}}$), we can choose θ_1 arbitrarily small without changing θ_2 .

Combining (1.7),(1.8) , we also get the "transversal ellipticity" property,

(1.9) $\quad |p|_{\Lambda_{tG}}| \geq (t/C)|(x,\xi)|^2 \quad .$

We recall from the introduction , that $T_{(0,0)}(\Lambda_+)$ is the sum of the eigenspaces of F_p associated to the eigenvalues $\mu_1,..,\mu_d,i\lambda_1,..,i\lambda_{n-d}$. These eigenvalues are precisely the ones which belong to the rotated cone $i\Gamma$. The linearized situation we now have , is wellknown from the study [7] of hypoelliptic operators with double characteristics , if we think of $T_{(0,0)}\Lambda_{tG}$ as the new real space . We then know that Λ_+ is strictly positive with respect to Λ_{tG} (in the terminology of [6] , [8]). Similarly Λ_- is strictly negative with respect to Λ_{tG} .

2. An additional transformation.

We first make a standard FBI-transform T as in [8] , defined microlocally near (x_0,ξ_0) and such that the associated canonical tranformation , κ_T maps (x_0,ξ_0) to $(0,0)$, and $\mathbb{R}^{2n}$ to $\Lambda_{\Phi_0} := (2/i)\partial_x\Phi_0$. Here Φ_0 is a real analytic strictly pluri-subharmonic function , defined in a neighborhood of 0 in $\mathbb{C}^n$. We then know that the fiber $\{x=0\}$ is a Lagrangian manifold , which is strictly negative with respect to Λ_{Φ_0} . If we now denote $\kappa_T(\Lambda_\pm)$ simply by $\Lambda_\pm$, we can choose T so that $\{x=0\}$ and Λ_- are transversal . Then $\Lambda_\pm$ are given by $\xi = \partial_x\phi_\pm$, where $\phi_\pm$ are holomorphic, and $-\operatorname{Im}\phi_+ - \Phi_0(x) \leq 0$. The image of Λ_{tG} is of the form Λ_{Φ_t} , where $\Phi_t = \Phi_0+tG(x)+O(t^2|x|^2)$. Here , we consider

G also as a function on $\Lambda_{\Phi_0} \simeq \mathbb{C}^n_x$. The strict positivity of Λ_+ with respect to Λ_{tG} , then implies that

(2.1) $$\operatorname{Im} \phi_+ + \Phi_t \sim |x|^2 \quad .$$

Using the strict negativity of Λ_- with respect to Λ_{tG} , we shall prove :

Lemma 2.1. There is a totally real linear subspace , $L \subset \mathbb{C}^n$, of real dimension n , such that ,

(2.2) $$\Phi_t + \operatorname{Im} \phi_- \sim -|x|^2 \quad \text{on } L \, .$$

Proof. Consider the quadratic form $q = \Phi''_t(0) + \operatorname{Im} \phi''_-(0)$, which is strictly plurisubharmonic and non-degenerate . If Λ_- were very close to $\{x=0\}$, then the last term in the expression for q would dominate and the signature would be (n,n) . Now the set of strictly negative Lagrangian planes, transversal to $\{x=0\}$ and $T_{(0,0)}(\Lambda_{\Phi_t})$, is connected, so the signature is always (n,n) . The existence of a suitable L then follows from [8] , chapter 3 .

Considering still the situation after application of κ_T , we now introduce a new canonical transformation , κ , which maps Λ_+ to $\{\xi=0\}$ and Λ_- to $\{x=0\}$. Then it is easy to see that κ is given by $(y,-\phi'_y(x,y)) \longrightarrow (x,\phi'_x(x,y))$, where the generating function $\phi(x,y)$ verifies : $\det \phi''_{x,y} \neq 0$, and,

(2.3) $$\phi(0,y) = - \phi_-(y) \quad .$$

Let $f(x,y,h)$ be a classical analytic symbol defined near $(x,y)=(0,0)$. Using the lemma , we see that if $u \in H_{\Phi_t,0}$, then we can define $Fu \in H_{\tilde{\Phi}_t,0}$, by chosing a nice contour for the integral expression ,

$$(2.4) \qquad Fu(x,h) = \int e^{i\phi(x,y)/h} f(x,y,h)\, u(y)\, dy \ .$$

Here , we use the terminology of [8] . $\tilde{\Phi}_t$ is a new strictly pl.s.h. function , determined up to a constant by the relation $\Lambda_{\tilde{\Phi}_t} = \kappa(\Lambda_{\Phi_t})$. Notice that the strict positivity of $\kappa(\Lambda_+) : \xi=0$ (that we shall from now on denote simply by "Λ_+",) means that ,

$$(2.5) \qquad \tilde{\Phi}_t \sim |x|^2 \ .$$

Up to exponentially small errors , we can invert F by an operator ,

$$(2.6) \qquad Gv(y,h) = \int e^{-i\phi(x,y)/h} g(x,y,h)\, v(x)\, dx \ .$$

(See [8] , chapter 4 .)

As in [3] , we can describe the norms on the space $H(\Lambda_{tG},1)$ (defined in [5]) in such a way that the crucial microlocal control of u in a neighborhood of (x_0,ξ_0) is given by ,

$$||FTu||^2_{H(\Omega)} = \int_\Omega |FTu(x,h)|^2 e^{-2\Phi(x)/h} L(dx) \ ,$$

where Ω is a small neighborhood of 0 , and $\Phi = \tilde{\Phi}_t$.

Let P also denote the classical analytic pseudo-differential operator , obtained formally from the original P by conjugation with FT . Then we get the new principal symbol ,

$$(2.7)\qquad p(x,\xi) = B(x,\xi)x\cdot\xi \quad ,$$

where $B(0,0)$ has the eigenvalues $\mu_1,\ldots,\mu_d,i\lambda_1,\ldots,i\lambda_{n-d}$.

3. Proof of Theorem 0.2.

We shall follow the arguments of section 4,5 of [3] , but only explicit the parts which are not quite identical . We consider the additionally transformed situation , introduced in the preceding section . Let Ω be a small neighborhood of $0 \in \mathbb{C}^n$ and write Φ instead of $\tilde{\Phi}_t$. Put

$$(3.1)\qquad ||u||^2_{H(\Omega)} = \int_\Omega |u(x)|^2 e^{-2\Phi(x)/h} L(dx) \quad ,$$

where $L(dx)$ is the Lebesgue measure . Let $H(\Omega)$ be the space of holomorphic functions on Ω with finite $H(\Omega)$-norm .We also introduce the weighted spaces $H^m(\Omega)$, which coincide with $H(\Omega)$ as linear spaces , but which are equipped with a different norm ,

$$(3.2)\qquad ||u||^2_{H^m(\Omega)} = \int_\Omega (1+h^{-\frac{1}{2}}|x|)^{2m}|u(x)|^2 e^{-2\Phi(x)/h} L(dx) .$$

If $\Omega_1 \subset\subset \Omega_2$, then we can realize $h^{-1}P$ as a bounded operator $H^m(\Omega_2) \longrightarrow H^{m-1}(\Omega_1)$. More generally , if A is a classical analytic symbol $= O(1)(1+h^{-\frac{1}{2}}(|x|+|\xi|))^{m_0}$, then we have a realization , which is $O(1) : H^m \longrightarrow H^{m-m_0}$. This applies in particular to $(h^{-\frac{1}{2}}x)^\alpha(h^{-\frac{1}{2}}\tilde{D}_x)^\beta$. We also introduce,

$$(3.3)\qquad \tau_N u(x) = \sum_{|\alpha|<N} (\alpha!)^{-1}(\partial_x^\alpha u)(0)x^\alpha =$$

$$= \sum_{|\alpha|<N} (\alpha!)^{-1}(h^{\frac{1}{2}}\partial_x)^\alpha u(0)(h^{-\frac{1}{2}}x)^\alpha \quad .$$

Then for all m_1 , $m_2 \in \mathbb{R}$, the operator τ_N is $O(1)$: $H^{m_1}(\Omega) \longrightarrow H^{m_2}(\Omega)$.

Using the sector property (1.9) and the transversal ellipticity property (1.10) , we see that if $\theta > 0$ is small enough , then

$$- \operatorname{Im} e^{-i\theta} p\Big|_{\Lambda_\Phi} \sim |x|^2 \quad ,$$

in Ω , provided that Ω is small enough . We then obtain as in [3] , Proposition 4.5 ,

<u>Proposition 3.1.</u> Let $0 \in \Omega_1 \subset\subset \Omega_2 \subset\subset \Omega_3 \subset\subset \Omega$. If N is sufficiently large , then for all $u \in H(\Omega_3) \cap O^N$:

(3.4) $$h\|u\|_{H^2(\Omega_3)} \leq C(\|Pu\|_{H^0(\Omega_2)} + h\|u\|_{H^2(\Omega_3 \setminus \Omega_1)}) .$$

Here we use a realization of P , which goes from $H(\Omega_3)$ to $H(\Omega_2)$, and we let O^N denote the space of holomorphic functions which vanish to the order N at 0 .

Let $\mathcal{T}^N$ be the space of all polynomials in n variables, of degree < N . We consider $\mathcal{T}^N$ as a finite dimensional vector space equipped with the basis ; $(h^{-\frac{1}{2}}x)^\alpha$, $|\alpha| < N$. We denote by ,

$$\mathcal{T}^N = \bigoplus_0^{N-1} E_j \quad ,$$

the decomposition of $\mathcal{T}^N$ into spaces of j-homogeneous polynomials . The operator $P^{(N)} = \tau_N P \tau_N$, acting on $\mathcal{T}^N$, can then be represented by a matrix ; $(P^{(N)}_{j,k})_{0 \leq j,k \leq N-1}$.

We also put ,

(3.5) $$P_0 = \Sigma\Sigma\, b_{j,k}(0,0) x_k \tilde{D}_{x_j} + hp_{-1}(0,0) \quad .$$

Then P_0 acts like h times a constant matrix in the spaces E_k , so it contributes to $P_{j,k}^{(N)}$ with a diagonal part:

$$P_0^{(N)} = ((P_0)_{j,k}) \quad ,$$

where $(P_0)_{j,k}=0$, if $j\neq k$, and $(P_0)_{j,j} = h \times$ constant matrix. The action of $(P_0)_{j,j}$ is easily analyzed by using the Jordan decomposition of $B(0,0) = (b_{j,k}(0,0))$, and we see that $(P_0)_{j,j}$ has the eigenvalues ,

(3.6) $\quad (h/i)\, \alpha\cdot z + h\, p_{-1}(0,0) \quad ,$

where $\alpha \in \mathbb{N}^n$, $|\alpha|=j$, $z = (z_1,\ldots,z_n)$.

The problem of analyzing $P^{(N)}$ is then reduced to that of analyzing $\tau_N(P-P_0)\tau_N$. As in [3] , we get ,

<u>Proposition 3.2.</u> $\tau_N(P-P_0)\tau_N$ can be represented by a matrix of classical symbols $(m_{j,k}(h))$, where $m_{j,k}$ is of order $-(1+|j-k|/2)$ for $j \neq k$, and of order -2 for $j = k$.

<u>Proposition 3.3.</u> If $0 < M \leq N$, then $\tau_M P(1-\tau_N) = \tau_M(P-P_0)(1-\tau_N)$ and $(1-\tau_N)P\tau_M = (1-\tau_N)(P-P_0)\tau_M$ are $O(h^{(3+N-M)/2})$ as bounded operators $H^{m_2}(\Omega_2) \longrightarrow H^{m_1}(\Omega_1)$. Here m_1 , m_2 are arbitrary , and $0 \in \Omega_1 \subset\subset \Omega_2$.

Put $U^{(N)} = h^{-1} P^{(N)}$, $U_0^{(N)} = h^{-1} P_0^{(N)}$. Let $K = \overline{D(0,C_0)}$ be the closed disc in $\mathbb{C}$ with center 0 and radius C_0 , such that no values $i^{-1}\alpha\cdot z + p_{-1}(0,0)$, $\alpha \in \mathbb{N}^n$, fall on the boundary . Choose N_0 so large that $(U_0)_{j,j}$ has no eigenvalues in K , when $j \geq N_0$. Take $N \geq N_0$. Then for $z \in K$, we consider the Grushin problem :

$$(3.7)\quad \begin{cases} (U^{(N)}-z)u + R_-^{(N)}u_- = v \\ R_+^{(N)}u = v_+ \end{cases},$$

where $u\,,\, v \in \mathfrak{J}^N$, $u_-\,,\, u_+ \in \mathbb{C}^M$, $M = \dim \mathfrak{J}^N$,

$$(3.8)\quad \begin{cases} R_+^{(N)}u = \{(h^{\frac{1}{2}}\partial_x)^\alpha u(0)\}_{|\alpha|<N} \\ R_-^{(N)}u_- = \sum (\alpha!)^{-1}\, u_-^\alpha\, (h^{-\frac{1}{2}}x)^\alpha \end{cases}.$$

Using Proposition 3.2 , we see that this problem has a unique bounded solution ,

$$(3.9)\quad \begin{pmatrix} u \\ u_- \end{pmatrix} = \begin{pmatrix} F^{(N)} & F_+^{(N)} \\ F_-^{(N)} & F_{-+}^{(N)} \end{pmatrix} \begin{pmatrix} v \\ v_+ \end{pmatrix},$$

if we use the natural h-dependent norm on $\mathfrak{J}^N$, which makes $R_+^{(N)}$ and $R_-^{(N)}$ trivially bounded . Notice that this norm is uniformly equivalent to $h^{-n/2}||u||_{H^j(\Omega)}$, for any j , if Ω is one of the neighborhoods of 0 above .

Let Ω_1 , Ω_2 , Ω_3 be as in Proposition 3.1 , and consider the Grushin problem

$$(3.10)\quad \begin{cases} (h^{-1}P-z)u + R_-^{(N)}u_- = v \quad , \text{ in } \Omega_2 \,, \\ R_+^{(N)}u = v_+ \end{cases},$$

$u_-\,,\, v_+ \in \mathbb{C}^M$, $u \in H^2(\Omega_3)$, $v \in H(\Omega_2)$. Then as in [3] , we obtain for N sufficiently large the a priori estimate ,

$$(3.11)\quad (||u||_{H^2(\Omega_1)} + h^{n/2}||u_-||) \leq C(||v||_{H^0(\Omega_2)} + h^{\frac{1}{2}n}||v_+|| + O(h^\infty)||u||_{H^2(\Omega_3)})$$

The operator FT can be realized as an operator : $H(\Lambda_{tG},1) \longrightarrow H(\Omega)$, and we shall now denote it simply by T . As in [3] we take a suitable microlocal inverse S and a suitable cutoff-operator $\chi(x,D_x,h)$. We then consider the Grushin problem for the original operator P :

$$(3.12) \qquad \begin{cases} (h^{-1}P-z)u + R_-u_- = v \\ R_+u = v_+ \end{cases} ,$$

where u_- , $v_+ \in \mathbb{C}^M$, $v \in H(\Lambda_{tG},1)$, $u \in H(\Lambda_{tG},m)$, $R_+u = R_+^{(N)}T\chi u$, $R_-u_- = SR_-^{(N)}u_-$, and where m is the order function associated to P .(See [5] .) As in [3] , we then show that this problem is well-posed for $z \in K$, even for $N = N_0$, with an inverse ,

$$\begin{pmatrix} F & F_+ \\ F_- & F_{-+}^{\infty} \end{pmatrix} ,$$

and that we have the a priori estimate ,

$$(3.13) \qquad (||u||+h^{\frac{1}{2}n}||u_-||) \leq C(||v||+h^{\frac{1}{2}n}||v_+||) ,$$

where the norms are taken in the obvious spaces . Finally , as in [3] , we determine the asymptotic structure of F_{-+}^{∞} by direct WKB-constructions (easier than in [3]) , which are quite obvious from [4] and [3] , and we show by trace computations , that the resonances of P in $D(0,C_0h)$ are precisely the elements of the set $\Gamma^{\infty}(h)$.

4. Examples of resonances which are double poles for the resolvent.

It is rather clear that the constructions of this section could be applied also to the closed trajectory situation of [3]. It should also be possible to generalize them, and obtain poles of any finite order. We hope to make these extensions in a future paper.

Our examples will be produced by a perturbation argument. As the unperturbed operator, we take,

$$P_0 = -h^2\Delta + V_0(x) \quad , \tag{4.1}$$

on $\mathbb{R}^2$, where $V_0(x) = -|x|^2$. Then the general theory of [5] applies with the escape function $G(x,\xi) = x\cdot\xi$ (suitably modified in the region where $|\xi| >> |x|$) and with $R(x) = r(x) = (1+x^2)^{\frac{1}{2}}$. The only trapped trajectory is the stationary one at (0,0), so the results of this paper apply. (We could eaily make other choices of V_0. For instance, we could let V_0 be a real and analytic function, depending only on $|x|$, which is $= -1+o(1)$ as $x \longrightarrow \infty$ in a complex region : $|\mathrm{Im}\, x| \leq C^{-1}|\mathrm{Re}\, x|$, satisfying : $V_0(0) = 0$, $V_0''(0) < 0$, and $V_0(x) < 0$, $\partial_{|x|}V_0(x) < 0$, for $x \neq 0$.)

The resonances near 0 of P_0 are then of the form,

$$-ih(2+2|\alpha|) \quad , \tag{4.2}$$

for $\alpha = (\alpha_1,\alpha_2) \in \mathbb{N}^2$, $|\alpha|=\alpha_1+\alpha_2$. In particular, $\lambda_0(h) = -i4h$ is a resonance of multiplicity 2, and the corresponding eigenspace is generated by

$$\phi_j(x) = h^{-1}\, x_j\, e^{ix^2/2h} \quad , \; j = 1,2 \quad . \tag{4.3}$$

(Thanks to the factor h^{-1} , ϕ_1 , ϕ_2 are approximately normalized in $H(\Lambda_{tG},1)$, $t > 0$.)

The general theory of [5] , now tells us that $H(\Lambda_{-tG},1)$ is the L^2-dual space of $H(\Lambda_{tG},1)$ and that we can also define "incoming" resonances for P_0 , associated to $H(\Lambda_{-tG},1)$. (The usual ones will be called "outgoing".) The incoming resonances are then precisely the complex conjugates of the outgoing ones . If $F_\lambda \subset H(\Lambda_{tG},1)$ is the spectral space corresponding to the outgoing resonance λ , and $F^*_\mu \subset H(\Lambda_{-tG},1)$ the spectral space corresponding to the incoming resonance μ , then $F_\lambda \perp F^*_\mu$, if $\lambda \neq \overline{\mu}$, and F_λ and F^*_μ are duals to each other when $\lambda = \overline{\mu}$. In the latter case , we also have $\pi_{F^*_\mu} = (\pi_{F_\lambda})^*$ for the corresponding spectral projections .

Returning to our specific operator (4.1) , we have the incoming resonance $\overline{\lambda}_0 = 4ih$ of multiplicity 2 , and for a suitable value $C_0 \neq 0$, we have the dual basis of incoming eigenfunctions ,

$$\text{(4.4)} \qquad \phi^*_j = C_0\, h^{-1}\, x_j\, e^{-ix^2/2} \quad , \; , \; j = 1,2 \; .$$

Here we recall that the L^2-duality between $H(\Lambda_{tG},1)$ and $H(\Lambda_{-tG},1)$ is defined by the natural density argument. In the present case , we approach ϕ_j and ϕ^*_j in $H(\Lambda_{\pm tG},1)$ by $e^{-\varepsilon x^2}\phi_j$ and $e^{-\varepsilon x^2}\phi^*_j$, and for the modified functions , in the computation of their L^2-inner product we can shift the integration contour to $e^{i\theta}\mathbb{R}^n$, $0 < \theta \leq \pi/4$. The shifted integrals converge as $\varepsilon \longrightarrow 0$, and we get ,

$$(4.5) \qquad (\phi_j|\phi_k^*) = \overline{C_0}\, h^{-2} \int_{e^{i\theta}\mathbb{R}^n} x_j\, x_k\, e^{ix^2/h}\, dx \quad ,$$

which is then a convergent integral .

We shall now perturb P_0 , by changing V_0 into $V = V_0 + \Delta V$. Here $\Delta V(x) = h^l\, q_{2m}(x)\, e^{-x^2/2}$, and q_{2m} , is a real 2m-homogeneous polynomial . We also require that $l+m \geq 2$, $m \geq 1$. When $l = 0$, we shall assume that q_{2m} is small , so that the geometry of the bicharacteristic flow of $p = \xi^2+V(x)$ is essentially unchanged and so that the results of this paper apply with the same escape function . (When $l \geq 1$, we consider P as a pseudodifferential operator with leading symbol $p = \xi^2+V_0(x)$.) Let $\varepsilon_0 > 0$ be small and let $D = D(\lambda_0,\varepsilon_0 h)$ be the open disc of center λ_0 and radius $\varepsilon_0 h$. It follows easily from the earlier results and proofs of this paper that D contains precisely 2 resonances .

For $z \in D$, let us construct $(P-z)^{-1} : H \longrightarrow \tilde{H}$, where $H = H(\Lambda_{tG},1)$, $\tilde{H} = H(\Lambda_{tG},\tilde{r}^2)$, $\tilde{r}(x,\xi) = (\xi^2+r(x)^2)^{1/2}$. We take the standard norm on H , but on $\tilde{H}$ we relax the $H(\Lambda_{tG},\tilde{r}^2)$-norm microlocally near $(0,0)$, to

$$||(h+|x|^2)FTu||_{L^2(\Omega,e^{-2\Phi/h}L(dx))} \quad ,$$

with T , F , Ω , Φ as in section 3 .(The detailed definition of the $\tilde{H}$-norm is rather clear from the general theory of [5] , and we omit it .) Then ,

$$C^{-1}h||u||_{H(\Lambda_{tG},\tilde{r}^2)} \leq ||u||_{\tilde{H}} \leq C||u||_{H(\Lambda_{tG},\tilde{r}^2)} \quad ,$$

so we have $||u||_H \leq (C/h)||u||_{\tilde{H}}$.

The analysis of section 3 , combined with the generalities of [5] shows easily that $(P_0-z)^{-1}$ is $O(1) : H \longrightarrow \tilde{H}$, and that $P-P_0$ is $O(h^l) : \tilde{H} \longrightarrow H$ (respectively $O(\varepsilon)$, when $l=0$ and $q_{2m}=O(\varepsilon)$). Thus $(P-P_0)(P_0-z)^{-1}$ is $O(h^l)$ (respectively $O(\varepsilon)$) as a bounded operator in H . Then we get the convergent Neumann series ,

$$(4.6) \qquad (P-z)^{-1} = \sum_0^\infty (P_0-z)^{-1}((P-P_0)(P_0-z)^{-1})^k \quad ,$$

so

$$(4.7) \qquad (P-z)^{-1} = O(1) : H \longrightarrow \tilde{H} \quad ,$$

$$(4.8) \qquad (P-z)^{-1}-(P_0-z)^{-1} = O(h^l) \text{ (resp. } O(\varepsilon) \text{) } : H \longrightarrow \tilde{H} \ .$$

This implies ,

$$(4.9) \qquad (P-z)^{-1} = O(h^{-1}) : H \longrightarrow H \quad ,$$

$$(4.10) \quad (P-z)^{-1}-(P_0-z)^{-1} = O(h^{l-1}) \text{ (resp. } O(\varepsilon h^{-1}) \text{) } : H \to H \ .$$

From this we conclude that the spectral projection

$$(4.11) \qquad \pi = (2\pi i)^{-1} \int_{\partial D(\lambda_0,\varepsilon_0 h)} (z-P)^{-1}\, dz \quad ,$$

corresponding to the resonances of P in $D(\lambda_0,\varepsilon_0 h)$, is close in $\mathcal{L}(H,H)$ to the corresponding projection for P_0 . Thus (as we already mentionned) there are precisely two resonances for P in the disc , and if $F = \pi(H)$ is the corresponding 2-dimensional space , then a basis for F is given by $v_j = \pi(\phi_j)$, $j = 1,2$.

Starting from $(P-z)\phi_j = (\lambda_0-z)\phi_j + \Delta V\phi_j$, we get

$$(P-z)^{-1}\phi_j = (\lambda_0-z)^{-1}\phi_j - (P-z)^{-1}(\lambda_0-z)^{-1}\Delta V\phi_j \quad ,$$

and integration over the boundary of the disc gives ,

$$\text{(4.12)} \qquad v_j - \phi_j = (2\pi i)^{-1} \int (P-z)^{-1}(\lambda_0-z)^{-1}\Delta V\phi_j \, dz \quad .$$

From the explicit form of ϕ_j and ΔV , we see that

$$\text{(4.13)} \qquad \Delta V\phi_j = O(h^{1+m}) \quad \text{in} \quad H \quad ,$$

so by (4.12) and (4.9) ,

$$\text{(4.14)} \qquad v_j-\phi_j = O(h^{1+m-1}) \quad \text{in} \quad H \ .$$

Let $H^* = H(\Lambda_{-tG},1)$ and let $F^* = \pi^*(H^*)$ be the spectral subspace corresponding to the incoming resonances in D . Let v_1^* , $v_2^* \in F^*$ be the basis dual to v_1 , v_2 . The matrix of $P|_F$ with respect to the basis v_1,v_2 is then $((Pv_k|v_j^*))$. We have $(Pv_k|v_j^*) = (P\phi_k|v_j^*)$, since $P(v_k-\phi_k)$ is in the kernel of π , which is orthogonal to F^* . On the other hand ,

$$(P\phi_k|v_j^*) = \lambda_0(\phi_k|v_j^*) + (\Delta V\phi_k|v_j^*) \quad ,$$

and $(\phi_k|v_j^*) = (v_k|v_j^*) = \delta_{j,k}$, again since $v_k-\phi_k$ is orthogonal to F^* . Hence the matrix of $P|_F$ is of the form:

$$\text{(4.15)} \qquad \lambda_0 I + ((\Delta V\phi_k|v_j^*)) \quad .$$

Here we want to replace v_j^* by ϕ_j^* , so we shall estimate ,

$$v_j^* - \phi_j^* = (v_j^*-\pi^*\phi_j^*)+(\pi^*\phi_j^*-\phi_j^*) \ .$$

The proof of (4.14) gives also

$$\text{(4.16)} \qquad \pi^*\phi_j^* - \phi_j^* = O(h^{1+m-1}) \quad \text{in} \quad H^*.$$

To estimate $v_j^* - \pi^* \phi_j^*$, we shall see to what extent $\pi^* \phi_j^*$ fails to be a dual basis for v_j :

$$(4.17) \qquad (v_j | \pi^* \phi_k^*) = (v_j | \phi_k^*) = \delta_{j,k} + (v_j - \phi_j | \phi_k^*) \ .$$

Since $v_j - \phi_j$ is orthogonal to F^* , we get

$$(v_j - \phi_j | \phi_k^*) = (v_j - \phi_j | \phi_k^* - \pi^* \phi_k^*) \ .$$

Estimating this with (4.14) and (4.16) , we get from (4.17):

$$(4.18) \qquad (v_j | \pi^* \phi_k^*) = \delta_{j,k} + O(h^{2(l+m-1)}) \ ,$$

so

$$(4.19) \qquad \pi^* \phi_k^* - v_k^* = O(h^{2(l+m-1)}) \ .$$

Combining this with (4.16) , we get

$$(4.20) \qquad v_j^* - \phi_j^* = O(h^{l+m-1}) \quad \text{in} \quad H \ ,$$

since $l+m \geq 2$. It follows from this and (4.13) that

$$(4.21) \qquad (\Delta V \phi_k | v_j^*) = (\Delta V \phi_k | \phi_j^*) + O(h^{2(l+m)-1}) \ ,$$

so in view of (4.15) , we have proved :

<u>Theorem 4.1.</u> The matrix of $P|_F$ with respect to the basis v_1, v_2 is given by

$$(4.22) \qquad \lambda_0 I + ((\Delta V \phi_k | \phi_j^*)) + O(h^{2(l+m)-1}) \ .$$

Notice that the middle term in (4.22) is $O(h^{l+m})$. When $l+m = 2$, it is then $O(h^2)$, while the last term is $O(h^3)$. The theorem remains of course valid , when ΔV is a finite sum of terms $h^l q_{2m} e^{-x^2/2}$, provided that $m \geq 1$, and that $l+m$ takes the same constant value ≥ 2 for the different terms .

We shall now examine the middle terms in (4.22) in the cases $l=m=1$ and $l=0, m=2$. In both cases, we get by stationary phase,

$$(4.23)\qquad \overline{C}_0^{-1}\, h^{-2}\, (\Delta V\phi_k|\phi_j^*) = O(h) + \int_{e^{i\pi/4}\mathbb{R}^2} q_{2m}(x)x_jx_k e^{ix^2}\, dx =$$

$$= O(h) - i^m \int_{\mathbb{R}^2} q_{2m}(y)y_jy_k\, e^{-y^2}\, dy \underset{\text{def.}}{=} O(h) - i^m a_{j,k}\,.$$

<u>Lemma 4.2.</u> Both when $m=1$ and when $m=2$, $(a_{j,k})$ may be any real symmetric matrix.

<u>Proof.</u> Apart from a universal constant factor, we have

$$a_{j,k} \sim (-1)^{m+1} q_{2m}(\partial_y)\partial_{y_j}\partial_{y_k} e^{-y^2/4}\Big|_{y=0}\,.$$

Using this, we get for $m=1$: When $q_2(y) = y_1^2$, then $a_{j,k} = 0$ for $j \neq k$ and $0 < a_{2,2} < a_{1,1}$. Using linear combinations of y_1^2 and y_2^2 we can therefore produce any diagonal matrix. Since $a_{1,2} = a_{2,1} \neq 0$ when $q_2(y) = y_1y_2$, we get any symmetric matrix in this case.

For $m = 2$ the same argument works: $q_4 = y_1^4$ gives $a_{1,2}=a_{2,1}=0$, $0<a_{2,2}<a_{1,1}$ and $q_4 = y_1^3\, y_2$ gives $a_{1,2}=a_{2,1} \neq 0$.

Using that i^m is imaginary for $m = 1$ and real for $m = 2$, we get from (4.23), Lemma 4.2 and Theorem 4.1:

<u>Theorem 4.3.</u> Consider perturbations of the form $V = (hq_2+q_4)e^{-x^2/2}$ and write the expression (4.22) for the matrix of $P|_F$ in the form:

$$(4.24)\qquad \lambda_0 I + h^2(\mathcal{M}_0(q_2,q_4)+O(h)) = \lambda_0 I + h^2\mathcal{M}(q_2,q_4,h)\,.$$

Then the map $(q_2,q_4) \longmapsto \mathcal{M}_0(q_2,q_4)$ is surjective and real-linear from $E_2^{\mathbb{R}} \oplus E_4^{\mathbb{R}}$ to the space of complex symmetric 2×2 matrices . Here $E_j^{\mathbb{R}}$ is the space of real j-homogeneous polynomials.

Actually , all the arguments developped so far in this section , are also valid for complex valued perturbations (, provided that we pass to the formal adjoint P^* , when we discuss the incoming resonances). Then the description (4.24) of the matrix of $P|_F$ is still valid , when (q_2,q_4) belongs to a neighborhood of $(0,0)$ in $E_2 \oplus E_4$, and $\mathcal{M}_0$ is then a complex-linear function of q_2,q_4 , while $\mathcal{M}$ is holomorphic . From Cauchy's inequalities we get

$$(4.25) \qquad d\mathcal{M} = d\mathcal{M}_0 + O(h) \quad ,$$

where the differentials are taken with respect to (q_2,q_4) .

We now restrict (q_2,q_4) to a neighborhood U of $(0,0)$ in $E_2^{\mathbb{R}} \oplus E_4^{\mathbb{R}}$. Assume for simplicity , that U is large enough , so that $\mathcal{M}_0(U)$ contains the complex symmetric matrix ,

$$M_0 = \begin{pmatrix} 1 & i \\ i & -1 \end{pmatrix} \quad .$$

(Otherwise , replace M_0 by a small multiple .) In a neighborhood of M_0 in the space of all complex matrices ,

$$Q = \begin{pmatrix} a & b \\ c & d \end{pmatrix} \quad ,$$

the ones with double eigenvalues form a complex hypersurface H , given by

$$D(Q) = bc + (a-d)^2/4 = 0 .$$

On H the double eigenvalue $\lambda(Q) = \frac{1}{2}\,\mathrm{trace}(Q)$ is a holomorphic function of Q , and after shrinking the neighborhood W of M_0 if necessary , $Q - \lambda(Q)I$ is nilpotent and $\neq 0$ for all Q in W . Thus , H is also the set of matrices in W with a non-trivial nilpotent part in their Jordan decomposition . Let $\Gamma_0 \subset W$ be the complex curve of matrices

$$M_a = \begin{pmatrix} 1+a & i \\ i & -1 \end{pmatrix} ,$$

for $a \in \mathbb{C}$, a small . Then $D(M_a) = a + a^2/4$ has a simple root , $a = 0$, so Γ_0 intersects H transversally at M_0 . Let $\tilde{\Gamma}_0$ be the inverse image of Γ_0 for $\mathcal{M}_0$, and let Γ_h be the image of $\tilde{\Gamma}_0$ under $\mathcal{M}$. Then Γ_h is a smooth curve of real dimension 2 which differs from Γ_0 by $O(h)$ in the C^∞ topology . Thus Γ_h intersects H transversally , when h is small enough . If $a(h)$ is the corresponding a-value , then $a(h) = O(h)$.

Summing up , we have proved ,

Theorem 4.4. We can find a real-linear map $\mathbb{C} \longrightarrow E_2^{\mathbb{R}} \oplus E_4^{\mathbb{R}}$, $a \longmapsto (q_{2,a}, q_{4,a})$ and a neighborhood of $0 \in \mathbb{C}$, such that if $V_a = V_0 + V_a$ is the corresponding perturbed potential and P_a the corresponding Schrödinger operator , then for h small enough , there is exactly one value $a = a(h)$ in the neighborhood of 0 for which the Jordan form of $P_a\big|_F$ has a non-trivial nilpotent part . Moreover , $a(h) = O(h)$. Here $F = F_a$ denotes the two-dimensional spectral subspace , associated to the two resonances of P_a , close to $\lambda_0(h$

Bibliography.

(We only give a few references , particularly important for the special problem under consideration , and we refer to [1] , [2] ,[5] for more references to the very abundant litterature on resonances .)

1. P.Briet,J.M.Combes,P.Duclos, On the location of resonances for Schrödinger operators in the semiclassical limit I . Preprint.

2. J.M.Combes,P.Duclos,M.Klein,R.Seiler, The shape resonance. Preprint.

3. C.Gérard,J.Sjöstrand, Semiclassical resonances generated by a closed trajectory of hyperbolic type. Preprint.

4. B.Helffer,J.Sjöstrand, Multiple wells in the semi-classical limit I. Comm.P.D.E.,9(4)(1984),337-408.

5 B.Helffer,J.Sjöstrand, Resonances en limite semiclassique, Bull. de la S.M.F.,to appear.

6. A.Melin,J.Sjöstrand, Fourier integral operators with complex valued phase functions. Springer L.N. in Math.,n°459, 120-223.

7. J.Sjöstrand, Parametrices for pseudodifferential operators with multiple characteristics. Ark.f.Mat.,12, n°1(1974),85-130.

8. J.Sjöstrand, Singularités analytiques microlocales. Astérisque, n° 95 (1982).

REMARKS ON AN INVERSE BOUNDARY VALUE PROBLEM

John Sylvester*
Duke University

and

Gunther Uhlmann**
University of Washington

The problem of imaging the interior of the earth arises naturally in geophysics [Cl]. The problem of finding a method for electrical prospection based on just measurements at the boundary in order to determine the impedance (conductivity) of the earth's interior was proposed in the case that the impedance depend only on depth by Slichter [S] and it was first considered and formulated in generality by Calderón [(C]). Electrical impedance methods have also been used to measure certain cardiac parameters [H-W] and, in general, arises as a natural way of determining conductivity contrasts in the human body. We formulate now more precisely the mathematical problem.

Let Ω be a bounded smooth domain in $\mathbb{R}^n$, $n \geq 2$ and let γ be a strictly positive function in $L^\infty(\overline{\Omega})$. (We shall assume throughout $\gamma \in C^\infty(\overline{\Omega})$.) We consider the differential operator defined by

$$L_\gamma u = \text{div}\,(\gamma \nabla u) = \gamma \Delta u + \nabla\gamma \cdot \nabla u.$$

We solve the Dirichlet problem, given $\varphi \in H^{1/2}(\partial\Omega)$, find u solving

$$\begin{aligned} L_\gamma u &= 0 \quad \text{in} \quad \Omega \\ u\,\big|_{\partial\Omega} &= \varphi. \end{aligned} \tag{1}$$

We associate to u as in (1) its Dirichlet integral

$$Q_\gamma(\varphi) = \int_\Omega \gamma |\nabla u|^2. \tag{2}$$

γ is called here the *conductivity* of Ω ($\frac{1}{\gamma}$ measures the resistivity of Ω) and $Q_\gamma(\varphi)$ measures the power needed to maintain a potential φ on the boundary. The problem, proposed by Calderón, is

* Supported by NSF grant DMS-8600797
** Supported by NSF grant DMS-8601118 and an Alfred P. Sloan Research Fellowship

whether the measurement of $Q_\gamma(\varphi)$ for all $\varphi \in H^{1/2}(\partial\Omega)$ determines γ in Ω, i.e., is the map

$$\gamma \xrightarrow{Q} Q_\gamma \quad \text{injective?} \tag{3}$$

By polarizing the quadratic form (2), knowing $Q_\gamma(\varphi)\forall\varphi \in H^{1/2}(\partial\Omega)$, determines

$$Q_\gamma(\varphi,\psi) = \int_\Omega \gamma\nabla u \cdot \nabla v \tag{4}$$

with u, v solutions of $L_\gamma w = 0$ in Ω with $u\,|_{\partial\Omega} = \varphi$, $v\,|_{\partial\Omega} = \psi$.

Using Green's formula in (4) we have

$$Q_\gamma(\varphi,\psi) = \int_{\partial\Omega} \gamma u \frac{\partial v}{\partial\nu} \tag{5}$$

where ν denotes outer unit normal at $\partial\Omega$. Therefore Q_γ determines a unique self adjoint map

$$\begin{aligned} \Lambda_\gamma : H^1(\partial\Omega) &\longrightarrow L^2(\partial\Omega) \\ u\,|_{\partial\Omega} &\longrightarrow \left(\gamma\frac{\partial u}{\partial\nu}\right)\Big|_{\partial\Omega}. \end{aligned} \tag{6}$$

$\left(\gamma\frac{\partial u}{\partial\nu}\right)\Big|_{\partial\Omega}$ measures the electrical flux density entering or leaving the boundary. Λ_γ is the Neumann map called also here the *voltage* to *current* map. Calderón's question can then be rephrased: is the map

$$\gamma \xrightarrow{\Lambda} \Lambda_\gamma \quad \text{injective?} \tag{7}$$

¿From the practical point of view Λ_γ involves only measurements at the boundary. For every potential on the boundary we measure the induced current.

Calderón [C] proved that the linearized map dQ is injective at the constants.

Theorem 1. *$dQ\,|_{\gamma=1}$ is injective.*

Proof. Let $\delta \in C_0^\infty(\Omega)$, and

$$\gamma(t,x) = 1 + t\delta(x). \tag{8}$$

Let $Q_{\gamma(t,\cdot)}$ be the curve of quadratic forms associated to $\gamma(t,\cdot)$ as in (2), parametrized by t. We have, as in (4),

$$Q_{\gamma(t,\cdot)}(\varphi,\psi) = \int_\Omega \gamma\nabla u \cdot \nabla v \tag{9}$$

where

$$\begin{aligned} L_\gamma u &= 0 \quad \text{in} \quad \Omega \\ u\,|_{\partial\Omega} &= \varphi \end{aligned} \quad , \quad \begin{aligned} L_\gamma v &= 0 \quad \text{in} \quad \Omega \\ v\,|_{\partial\Omega} &= \psi. \end{aligned}$$

We differentiate (9) with respect to t (we denote $\frac{\partial}{\partial t}$ by $\cdot$).

$$\dot{Q}_\gamma(\varphi,\psi) = \int_\Omega \dot{\gamma}\nabla u\cdot\nabla v + \int_\Omega \gamma(\nabla\dot{u}\cdot\nabla v + \nabla\dot{v}\cdot\nabla u). \tag{10}$$

We integrate the second term in (10) by parts and we obtain

$$\dot{Q}_\gamma(\varphi,\psi) = \int_\Omega \dot{\gamma}\nabla u\cdot\nabla v + \int_{\partial\Omega} \gamma\Big(\dot{\varphi}\frac{\partial\psi}{\partial\nu} + \dot{\psi}\frac{\partial\varphi}{\partial\nu}\Big). \tag{11}$$

If we choose φ and ψ independent of t, we have

$$\dot{Q}_\gamma(\varphi,\psi) = \int_\Omega \delta\nabla u\cdot\nabla v. \tag{12}$$

Making use of the identity

$$L_\gamma(uv) = uL_\gamma v + vL_\gamma u + 2\gamma\nabla u\cdot\nabla v$$

and integrating by parts in (12), we get

$$\dot{Q}_\gamma(\varphi,\psi) = \int_\Omega L_\gamma\Big(\frac{\delta}{\gamma}\Big)uv. \tag{13}$$

In the case $t=0$, (13) becomes

$$\dot{Q}_1(\varphi,\psi) = \int_\Omega \Delta\delta uv \tag{14}$$

where

$$\begin{aligned} \Delta u &= 0 \quad \text{in} \quad \Omega \\ u\,\big|_{\partial\Omega} &= \varphi \end{aligned} \quad , \quad \begin{aligned} \Delta v &= 0 \quad \text{in} \quad \Omega \\ v\,\big|_{\partial\Omega} &= \psi. \end{aligned} \tag{15}$$

Following Calderón [C] we choose complex plane wave solutions to (15)

$$\begin{aligned} u &= e^{x\cdot\xi_1} \qquad (\varphi = e^{x\cdot\xi_1}\,\big|_{\partial\Omega}) \\ v &= e^{x\cdot\xi_2} \qquad (\psi = e^{x\cdot\xi_2}\,\big|_{\partial\Omega}) \end{aligned} \tag{16}$$

with $\xi_j \in \mathbb{C}^n$, $j=1,2$, satisfying

$$\xi_j\cdot\xi_j = 0 \quad \text{for} \quad j=1,2, \quad \text{or if}$$

$$\xi_j = \eta_j + ik_j \tag{17}$$

$$|\eta_j| = |k_j|, \quad \eta_j\cdot k_j = 0, \quad j=1,2.$$

Calderón chose

$$\xi_1 = \frac{\eta}{2} - i\frac{k}{2}$$
$$\xi_2 = -\frac{\eta}{2} - i\frac{k}{2}$$

with $\eta, k \in \mathbf{R}^n$ satisfying (17). We get

$$\dot{Q}_1(e^{x\cdot\xi_1}\big|_{\partial\Omega}, e^{x\cdot\xi_2}\big|_{\partial\Omega}) = \int_\Omega \Delta\delta e^{-ix\cdot k} = \widehat{\Delta\delta}(k) \tag{18}$$

where $\widehat{\ }$ denotes the Fourier transform and we have extended δ to be zero outside Ω. This shows that if $\dot{Q}_1 = 0$ then $\widehat{\Delta\delta}(k) = 0\ \forall k$ and therefore $\Delta\delta = 0$ in Ω by the Fourier inversion formula and then $\delta = 0$ in Ω since $\delta\big|_{\partial\Omega} = 0$. □

Calderón's proof gives actually a left inverse. From (18) we have

$$\delta = (\Delta_D)^{-1}[\dot{Q}_1(e^{x\cdot\xi_1}\big|_{\partial\Omega}, e^{x\cdot\xi_2}\big|_{\partial\Omega})]^\vee$$

where $\vee$ denotes the inverse Fourier transform and $(\Delta_D)^{-1}$ is the solution operator to the Dirichlet problem. However the linearized map dQ is not onto and therefore the implicit function theorem cannot be applied to construct a local left inverse for Q. This difficulty was overcome in [S-U, I] to obtain a local uniqueness result in dimension 2 (Theorem 3). Furthermore in [S-U, II] it was obtained a global uniqueness result (Theorem 4) for $n \geq 3$. We shall describe briefly the main ideas in the proof below.

We use the following result at the boundary proved by Kohn and Vogelius ([K-V, I]), namely that knowledge of Q_γ (or Λ_γ) determines the Taylor series of γ at the boundary of Ω.

Theorem 2. *Let γ_i $(i = 0, 1)$ be $C^\infty(\overline{\Omega})$ with a positive lower bound. Let $x_0 \in \partial\Omega$ and let U be a neighborhood of x_0 relative to $\overline{\Omega}$. Suppose that*

$$Q_{\gamma_0}(\varphi) = Q_{\gamma_1}(\varphi)\ \forall\varphi \in H^{1/2}(\partial\Omega) \text{ with supp } \varphi \subseteq B \cap \partial\Omega,$$

then $\partial^\alpha\gamma_0(x_0) = \partial^\alpha\gamma_1(x_0)\ \forall\alpha$.

As a corollary of the theorem we see that a real analytic γ is a priori determined by Λ_γ. Kohn and Vogelius ([K-V ,II]) have extended this result to cover piecewise analytic γ.

Sketch of proof of Theorem 2.

A different sketch of the proof than that of Kohn and Vogelius who used elliptic regularity follows. It is well known that Λ_γ, the voltage to current map, is a classical pseudodifferential operator of order 1 on $\partial\Omega$. Its principal symbol $\sigma_{\Lambda_\gamma}(x,\xi) = \gamma(x)|\xi|$, and therefore knowing Λ_γ we can determine γ at the boundary and all of its tangential derivatives. Now, the ***full symbol*** of Λ_γ can be written asymptotically as an infinite sum of functions λ_k homogeneous of degree $1-k$. $\lambda_k(x,\xi)$ involves the normal derivative of γ of order k at x with a non-zero coefficient plus terms involving normal derivatives of order strictly less than k at x and tangential derivatives of order at most k. Then an inductive argument proves that we can determine all the derivatives of γ at the boundary from the full symbol of Λ_γ. □

Theorem 3. *[S-U, II]. Let $n \geq 3$, $\gamma_0, \gamma_1 \in C^\infty(\overline{\Omega})$ with a positive lower bound so that*

$$Q_{\gamma_0} = Q_{\gamma_1}.$$

Then

$$\gamma_0 = \gamma_1 \quad in \quad \overline{\Omega}.$$

Sketch of proof. Let γ be any smooth strictly positive function in $\overline{\Omega}$. One of the main ideas is to construct solutions of $L_\gamma u = 0$ in Ω of the form

$$u = e^{x\cdot\xi}\gamma^{-1/2}(1+\psi(x,\xi)) \tag{19}$$

where $\xi \in \mathbb{C}^n$ with $\xi\cdot\xi = 0$ as in Calderón's computation (Theorem 1). Using ideas from geometrical optics we would like that the solutions (19) behave like the complex plane waves $e^{x\cdot\xi}$ for $|\xi|$ large.

We want, then,

$$\begin{aligned} \psi(x,\xi) &\longrightarrow \quad 0 \\ \text{as} \quad |\xi| &\to \infty \end{aligned} \tag{20}$$

uniformly in $\overline{\Omega}$.

The "transport equation" for ψ is the singular perturbation problem

$$\Delta\psi + \xi\cdot\nabla\psi - q\psi = q \text{ in } \Omega \tag{21}$$

where

$$q = \frac{\Delta\gamma^{1/2}}{\gamma^{1/2}}.$$

However, if we give boundary conditions for (21) at $\partial\Omega$, ψ will not satisfy, in general, the decay condition (20). Actually, we would expect that the dominant term in (21) for large $|\xi|$ is $\xi\cdot\nabla\psi$. In dimension 3 if $\xi = \eta + ik$ with $\xi\cdot\xi = 0$, $\xi\cdot\nabla\psi$ is the Cauchy Riemann equation in the planes perpendicular to η and certainly we cannot then impose general boundary conditions on ψ. Unable to characterize the boundary values of ψ satisfying (21) and (20) (this remains an interesting open question for reconstruction) we extended γ suitable and we look for solutions of (21) in the whole space and with growth conditions at infinity in the x-variable. We proved in [S-U, II].

Lemma 1. *Let $\gamma \in C^\infty(\mathbb{R}^n)$, $n \geq 3$ with γ strictly positive and $\gamma = 1$ outside a large ball containing Ω. Then there is a unique solution $\psi \in L^2_\delta$, $-1 < \delta < 0$ of*

$$\Delta\psi + \xi \cdot \nabla\psi - q\psi = q \text{ in } \mathbb{R}^n$$

satisfying

$$||\psi||_{H^s_\delta} \leq \frac{C}{|\xi|}, \quad s > \frac{n}{2}$$

with C depending on s, Ω, δ, q and H^s_δ is the weighted Sobolev space built over the weighted L^2_δ space with

$$||\psi||^2_{L^2_\delta} = \int_{\mathbb{R}^n} (1 + |x|^2)^\delta |\psi(x)|^2 dx.$$

Now we proceed with our sketch of proof of Theorem 3. Let

$$\gamma(t, x) = (1 - t)\gamma_0 + t\gamma_1 \text{ in } \Omega, \ 0 \leq t \leq 1 \tag{22}$$

$$\tilde{\gamma}(t, x) = \tilde{\gamma}_0 \quad \text{in} \quad \complement\Omega$$

where $\tilde{\gamma}_0$ is a smooth extension of both γ_0 and γ_1 (this is possible by Theorem 2) with $\tilde{\gamma}_0 = 1$ outside a ball that contains Ω. We consider solutions $L_\gamma u = 0$, $L_\gamma v = 0$ in $\mathbb{R}^n$ with γ as in (22) of the form (using Lemma 1)

$$\begin{aligned} u(x, \xi_1, t) &= e^{x\cdot\xi_1}\gamma^{-1/2}(1 + \psi(x, \xi_1, t)) \\ v(x, \xi_2, t) &= e^{x\cdot\xi_2}\gamma^{-1/2}(1 + \psi(x, \xi_2;, t)) \end{aligned} \tag{23}$$

with $\xi_k \cdot \xi_k = 0$, $k = 1, 2$, and Re $(\xi_1 + \xi_2) = 0$. We have (this is completely analogous to the computation made before with γ as in (8)).

$$\dot{Q}_\gamma(u\,|_{\partial\Omega}\,,\ v\,|_{\partial\Omega}) = \int_\Omega \dot{\gamma}\nabla u \cdot \nabla v + \int_{\partial\Omega} \gamma\Big(\dot{u}\frac{\partial v}{\partial\nu} + \dot{v}\frac{\partial u}{\partial\nu}\Big) \tag{24}$$

The difference is now that u and v depend on t at the boundary. However we have (see [S-U, II]).

Lemma 2. $u(x,\xi_k,0) = u(x,\xi_k,1) \forall x \in \complement\Omega,\ k=1,2.$ *The proof uses the fact that* $\gamma(\cdot,t)$ *is independent of* t *in* $\complement\Omega$ *and the fact that the Neumann map for* $\gamma(\cdot,0)$ *is equal to the Neumann map for* $\gamma(\cdot,1)$.

Using Lemma 2 and integration by parts we can write the boundary integral in (24) as an integral over a large ball. The growth condition on ψ at infinity (see [S-U, II] for more details) gives:

Lemma 3. *Let* u, v *be as in (23). Then*

$$\int_{\partial\Omega} \gamma\Big(\dot{u}\frac{\partial v}{\partial \nu} + v\frac{\partial \dot{u}}{\partial \nu}\Big) = 0.$$

Integrating (24) in t and using the fact that the boundary values of $u(x,\xi_k,0)$ and $u(x,\xi_k,1)$, $k=1,2$ are the same we obtain and "average linearization"

$$\int_0^1\int_\Omega \dot{\gamma}\nabla u\cdot\nabla v = 0. \tag{25}$$

Proceeding as in step (12) to (13) we get

$$\int_0^1\int_\Omega L_\gamma\Big(\frac{\dot{\gamma}}{\gamma}\Big)uv = 0. \tag{26}$$

Now we make special choices of $\xi_1,\ \xi_2$ in (23), namely

$$\begin{aligned}\xi_1 &= \varsigma + i\Big(\frac{k}{2} + r\eta\Big)\\ \xi_2 &= -\varsigma + i\Big(\frac{k}{2} - r\eta\Big)\end{aligned} \tag{27}$$

where $k\in\mathbb{R}^n,\ r\in\mathbb{R}$ and $\eta,\varsigma\in\mathbb{R}^n$ satisfying

$$\langle k,\eta\rangle = \langle k,\varsigma\rangle = \langle \eta,\varsigma\rangle = 0,$$

$|\eta| = 1$, $|\varsigma|^2 = \frac{|k|^2}{4} + r^2$, so that $\xi_k\cdot\xi_k = 0,\ k=1,2.$

The idea is that

$$e^{x\cdot(\xi_1+\xi_2)} = e^{ix\cdot k} \tag{28}$$

with ξ_1,ξ_2 as in (27). The right hand side of (28) is the exponential in the Fourier transform. However, for fixed k, $\psi(x,\xi_1,t), \psi(x,\xi_2,t)$ approach zero, uniformly in $\overline{\Omega}$, as r approaches infinity. The choice (27) is only possible in dimension three or larger.

Now (26) becomes

$$0 = \int_0^1 dt \int_{\mathbf{R}^n} \frac{1}{\gamma} L_\gamma\left(\frac{\dot{\gamma}}{\gamma}\right) e^{ix\cdot k}(1+\psi(x,\xi_1,t))(1+\psi(x,\xi_2,t)).$$

Letting r approach infinity and applying Lemma 1 we obtain

$$0 = \int_{\mathbf{R}^n} \left[\int_0^1 dt \frac{1}{\gamma} L_\gamma\left(\frac{\dot{\gamma}}{\gamma}\right)\right] e^{ix\cdot k} \quad \forall k \in \mathbf{R}^n.$$

Therefore

$$0 = \int_0^1 dt[\Delta \log \gamma + \tfrac{1}{2}|\nabla \log \gamma|^2]^\bullet$$

and using the fundamental theorem of calculus

$$0 = \Delta(\log \gamma_1 - \log \gamma_0) + \tfrac{1}{2}[|\nabla(\log \gamma_1)|^2 - |\nabla(\log \gamma_0)|^2]$$

or

$$0 = \Delta(\log \gamma_1 - \log \gamma_0) + \tfrac{1}{2}\nabla(\log \gamma_1 + \log \gamma_0) \cdot \nabla(\log \gamma_1 - \log \gamma_0),$$

a linear equation for $\log \gamma_1 - \log \gamma_0$ which vanishes on $\partial\Omega$ by the Kohn-Vogelius result. The maximum principle applies to give

$$\log \gamma_1 - \log \gamma_0 = 0 \quad \text{in} \quad \Omega$$

or

$$\gamma_1 = \gamma_0 \quad \text{in} \quad \Omega$$

proving the theorem. □

The global uniqueness problem in the two dimensional case remains open at present. The difficulty arises since in this case the inverse problem is formally determined. For $n \geq 2$, the kernel of the Neumann map is a function in $\partial\Omega \times \partial\Omega$ depending on $2(n-1)$ variables. The function γ depends on n variables and $2n-2 = n$ for $n = 2$ and $2n-2 > n$ for $n \geq 3$. This freedom for $n \geq 3$ was explained in the choices of ξ_1, ξ_2 as in (27). For $n = 2$, we can construct solutions of the form (21). Lemma 1 is valid although the proof is different because the term $\xi \cdot \nabla\psi$ is actually a Cauchy-Riemann equation (see [S-U, I]). The other ingredients, Lemma 2 and the average linearization are also true. However in this case we also need a low frequency estimate which was proven essentially by Calderón ([C]) in the case γ close to a constant. We have (see [S-U, I]):

Theorem 4. *Let $\gamma_0, \gamma_1 \in C^\infty(\overline{\Omega})$ with a positive lower bound and*

$$Q_{\gamma_0} = Q_{\gamma_1}.$$

Then $\exists\ \varepsilon(\Omega)$ such that if

$$\|\gamma_i - 1\|_{C^8(\overline{\Omega})} < \varepsilon(\Omega),\ i = 0, 1,\ \text{then } \gamma_0 = \gamma_1 .$$

We shall study further the transport equation (21) and derive new results (Theorem 5). From this point on we assume that $n = 2$.

Theorem 5. *Let $n = 2$, $\gamma_0, \gamma_1 \in C^\infty(\overline{\Omega})$ with a positive lower bound and*

$$Q_{\gamma_0} = Q_{\gamma_1}.$$

Then

$$\int_\Omega (q_0 - q_1)(w) w^m dw \wedge d\overline{w} = \int_\Omega (q_0 - q_1)(w) \overline{w}^m dw \wedge d\overline{w} = 0$$

for all m integers $m \geq 0$, where $q_i = \frac{\Delta \gamma_i^{1/2}}{\gamma_i^{1/2}}$ and $w = x_1 + i x_2 \in \mathbb{R}^2$.

In other words $(q_0 - q_1)$ is orthogonal in $L^2(\Omega)$ to the set of analytic and anti-analytic functions in $\mathbb{C}$. (Therefore, since $q_0 - q_1$ is real, orthogonal to the set of harmonic functions in Ω). We easily obtain

Corollary 1. *Suppose γ_0, γ_1 satisfy the conditions of Theorem 5, then*

$$\int_\Omega |\nabla \log \gamma_0|^2 = \int_\Omega |\nabla \log \gamma_1|^2.$$

Proof. Take $m = 0$ in Theorem 4. Then use

$$q_0 - q_1 = \Delta \log(\gamma_0 - \gamma_1) + |\nabla \log \gamma_0|^2 - |\nabla \log \gamma_1|^2 \quad \text{and integrate.}$$ □

Corollary 2. *Let γ_0, γ_1 be as in Theorem 5 with $\gamma_0 = C$, then $\gamma_1 = C$.*

Proof. Using Corollary 1, we have

$$\int_\Omega |\nabla \log \gamma_1|^2 = 0.$$

Therefore $\gamma_1 =$ constant, and since γ_0 coincides with γ_0 in the boundary, $\gamma_1 = \gamma_0 = C$. □

This means that we can distinguish constants from their Neumann maps.

Before going into the proof of Theorem 5 we point out that the transport equation (21) for $n = 2$ can be factorized and we want to solve with $\psi \in L^2_\delta$, $-1 < \delta < 0$.

$$\overline{\partial}(\partial + (k_2 + ik_1))\psi - q\psi = q \text{ in } \mathbf{R}^n \tag{29}$$

where $q = \frac{\Delta\gamma^{1/2}}{\gamma^{1/2}}$ and $\xi = \eta + ik$, $\eta, k \in \mathbf{R}^2$.

For $|k|$ large, the dominant term in (29) is the Cauchy-Riemann operator $\overline{\partial}$ and of course, we cannot give general boundary condition in $\partial\Omega$. However, as was proved by Nirenberg and Walker [N-W], given $f \in L^2_{\delta+1}$, $\exists!\ u \in L^2_\delta$ solving $Lu = f$ in $\mathbf{R}^2$, where L represents ∂ or $\overline{\partial}$.

This is one of the reasons why looking for solutions of (29) in the whole space works.

We proved in [S-U, I]:

Lemma 4. *Given $-1 < \delta < 0$, there exists a constant $C(\delta)$ such that, if $q \in L^2_{\delta+1}$ and $|k| \geq C||q(1+|x|^2)^{1/2}||_{L^\infty}$ then there exists a unique solution to*

$$\Delta\psi + (k_1 + ik_2)\overline{\partial}\psi - q\psi = q$$

such that $\psi, \nabla\psi \in L^2_\delta$. Moreover, ψ may be written in the form

$$\psi(x,k) = a(x,k) + e^{-ix\cdot k}c(x,k)$$

where

$$a(x,k) = \frac{a_1(x)}{k_2 + ik_1} + \sum_{j=2}^{\infty} \frac{a_j(x,k)}{(k_2 + ik_1)^j}$$

and

$$c(x,k) = \sum_{j=1}^{\infty} \frac{c_j(x,k)}{(k_2 + ik_1)^j}$$

with

$$||c_j||_{H^1_\delta}, ||a_j||_{H^1_\delta} \leq c||q(1+|x|^2)^{1/2}||^{j-1}_{L^\infty}||q||_{L^2_{\delta+1}} \tag{30}$$

Now we are in a position to prove the theorem.

Proof of Theorem 5.

We can rearrange the series for ψ as in Lemma 4 in the following way:

$$\psi = \frac{a_1(x)}{k_2 + ik_1} + \frac{-\partial a_1}{(k_2 + ik_1)^2} + \sum_{j=2}^{\infty} \frac{a_j}{(k_2 + ik_1)^j} \tag{31}$$

$$+e^{-ix\cdot k}\Big(\frac{h}{(k_2+ik_1)^2}+\sum_{j=2}^{\infty}\frac{c_j}{(k_2+ik_1)^j}$$

where

(32) $$c_1 = -\frac{e^{ix\cdot k}\partial a_1}{k_2+ik_1}+\frac{h}{k_2+ik_1}$$

with

$$\partial h = e^{ix\cdot k}\partial^2 a_1$$

since the right hand side of (32) satisfies the same equation as c_1, namely

$$\partial c_1 = e^{-ix\cdot k}\partial a_1.$$

Now a_1 is determined by solving

(32*b*) $$\partial a_1 = q$$

Now from ((31) and the property (30) of the c_j, a_j's, we deduce that

(33) $$\psi(x,k) = \frac{a_1(x)}{k_2+ik_1}+O\Big(\frac{1}{|k|^2}\Big) \text{ for } |k| \text{ large}$$

where the lower order term in (33) is uniformly bounded for compact subsets of $\mathbf{R}^2$.

Let us denote ψ^0 and ψ^1 respectively as the ψ's associated with γ_0 and γ_1. We also denote by a_1^0, a_1^1 the first term in (33). Using now Lemma 2 (which is also valid for $n=2$) we get:

$$\psi^0 = \psi^1 \quad \text{in} \quad \complement\Omega$$

and therefore by (33)

(34) $$a_1^0 = a_1^1 \quad \text{in} \quad \complement\Omega.$$

Now

$$\overline{\partial}(a_1^0 - a_1^1) = q_0 - q_1$$

and $q_0 - q_1$ has compact support. Therefore

(35) $$(a_1^0 - a_1^1)(z) = \int \frac{(q_0-q_1)(w)}{z-w}dw\wedge d\overline{w}$$

with $z = x_1 + ix_2$.

By (34)

$$(a_1^0 - a_1^1)(z) = \frac{1}{z}\sum_{n=0}^{\infty} z^n \int (q_0 - q_1)(w) w^n dw \wedge d\overline{w}$$

for $|z|$ large enough. Therefore we conclude

$$\int (q_0 - q_1)(w) w^n dw \wedge d\overline{w} = 0 \; \forall n.$$

Changing ξ to $\overline{\xi}$ in the transport equation (21) changes (29) to

$$\partial(\overline{\partial} + (k_2 - ik_1))\psi - q\psi = q$$

and the equation for the analog of a_1 in Lemma 4 is

$$\partial a_1 = q.$$

Repeating the argument above, one gets

$$(a_1^0 - a_1^1)(z) = \int \frac{1}{z - w}(q_0 - q_1)(w) dw \wedge d\overline{w}$$

and therefore

$$\int \overline{w}^n (q_0 - q_1)(w) dw \wedge d\overline{w} = 0$$

for all n, thus proving the theorem. □

References.

[C] Calderón, A. P., "On an inverse boundary value problem," Seminar on Numerical Analysis and its Applications to Continuum Physics, *Soc. Brasileira de Matemática*, Río de Janeiro, 1980, 65–73.

[C] Claerbout, Jon I., *Imaging the Earth's Interior*. Blackwell Scientific Publications, 1985.

[H-W] Henderson, R. and Webster, J., "An impedance camera for spatially specific measurements of the thorax," *IEEE Trans. Bio. Engl.*, Dec. 1977.

[K-V,I] Kohn, R., and Vogelius, M., "Determining conductivity by boundary measurements," *Comm. Pure Appl. Math.* **37**(1984), 289–298.

[K-V,II] ——————, *Comm. Pure Appl. Math.* **38**(1985), 643–667.

[N-W] Nirenberg, L., and Walker, H., "Null spaces of elliptic partial differential operators in $\mathbb{R}^n$," *J. Math. Anal. Appl.* **42**(1973), 271–301.

[S] Slichter, L. B., *Physics* **4**, Sept. 1933.

[S-U,I] Sylvester, J., and Uhlmann, G., "A uniqueness theorem for an inverse boundary value problem in electrical prospection," *Comm. Pure Appl. Math.* **39**(1986), 91–112.

[S-U,II] ——————, "A global uniqueness theorem for an inverse boundary value problem," to appear in *Annals of Math*.

PROPAGATION OF QUASI-HOMOGENEOUS MICROLOCAL SINGULARITIES OF SOLUTIONS TO NONLINEAR PARTIAL DIFFERENTIAL EQUATIONS

Masao YAMAZAKI

Department of Mathematics,
Faculty of Science,
University of Tokyo
Hongo, Bunkyo-ku, Tokyo 113 JAPAN

§0. Introduction.

Among a number of studies on microlocal analysis for nonlinear partial differential equations, Bony [1] was the first where general nonlinear equations were treated. He showed the microlocal hypoellipticity at non-characteristic points and the propagation of singularities along simple bicharacteristic strips, by making use of paradifferential operators as the main tool. Along this line, Meyer [8] improved Bony's microlocal hypoellipticity theorem.

On the other hand, Lascar [7] introduced the notion of quasi-homogeneous wave front set, and obtained the propagation of singularities of solutions to the equations of Schrödinger type.

The main purpose of this article is to generalize the result of Bony [1] on the propagation of singularities to the quasi-homogeneous case. That is, we give a weight m_l to each coordinate variable x_l , regard the differential operator $\partial/\partial x_l$ as an operator of order m_l , and consider the function spaces, principal symbols and bicharacteristic strips suitable to this setting. This treatment is natural for a number of important nonlinear equations; for example, nonlinear parabolic equations, nonlinear equations of Schrödinger type, and the KdV equation. In this direction, Yamazaki [11], [14] showed the microlocal hypoellipticity in general function spaces of Besov type. Also, Godin [3] and Sakurai [9] showed the propagation of singularties for a class of semilinear equations. Then Sakurai [10] obtained the propagation theorem along simple bicharacteristic strips for general semilinear equations. Here we aim to generalize his result to general nonlinear equations. Also, we do not assume that the principal symbol is real, as in Hörmander [4].

The secondary purpose is to relax the condition posed *a priori* on the smoothness of the solutions, and to obtain sharper microlocal regularity. Recently, Kobayashi-Nakamura [5] showed that the microlocal hypoellipticity theorem is not applicable for solutions with strong singularity, by constructing solutions of semilinear hyperbolic equations with singular spectra on the non-characteristic points. Hence it is of interest to relax the condition on the smoothness of solutions for which microlocal analysis is applicable. The condition obtained in this article depends on the non-linearity of the equations.

The main tool employed here is the technique of paradifferential operators as in Bony [1] and in Meyer [8]. However, for the purposes above, we need a quasi-homogeneous version of paradifferential operators and sharp estimates for them, given in Yamazaki [13, I and II]. Besides, for the second purpose, we use a sharp estimate of the error term of the linearization by means of paradifferential operators. This estimate is given in Yamazaki [14]. Hence we use some of the results of these papers.

The outline of this article is as follows. In Section 1 we list our notations and make some preliminary remarks and definitions. Then we state our main theorem in Section 2. In Section 3 we recall some properties of paradifferential operators given in [13], and prove a new property. Section 4 is devoted to the proof of a sharp Gårding inequality, improving a similar one given in Bony [1]. Finally, in Section 5, we prove our main theorem.

§1. Notations and preliminary remarks.

We start with some notations. For a multi-index $\alpha = (\alpha_1, \cdots, \alpha_n) \in \mathbb{N}^n$ and two vectors $x = (x_1, \cdots, x_n)$, $\xi = (\xi_1, \cdots, \xi_n) \in \mathbb{R}^n$, we put $|\alpha| = \alpha_1 + \cdots + \alpha_n$, $\alpha! = \alpha_1! \cdots \alpha_n!$, $x^\alpha = x_1^{\alpha_1} \cdots x_n^{\alpha_n}$ and $x \cdot \xi = x_1\xi_1 + \cdots + x_n\xi_n$. Here $\mathbb{N}$ denotes the set of natural numbers (= nonnegative integers).

Next, for the coordinate variable $x = (x_1, \cdots, x_n) \in \mathbb{R}^n$, let dx denote the Lebesgue measure on $\mathbb{R}^n$, and put $\bar{d}x = (2\pi)^{-n}dx$. We omit the domain of integration if it is the whole space $\mathbb{R}^n$. Next we put $\partial_{x_\ell} = \partial/\partial x_\ell$, $\partial_x^\alpha = (\partial_{x_1})^{\alpha_1} \cdots (\partial_{x_n})^{\alpha_n}$, $i = \sqrt{-1}$ and $D_x^\alpha = (-i)^{|\alpha|}\partial_x^\alpha$.

Further, let $\mathcal{S}$ and $\mathcal{S}'$ denote the space of rapidly decreasing functions and that of tempered distributions on $\mathbb{R}^n$ respectively. For $u(x), v(\xi) \in \mathcal{S}'$, let $\hat{u}(\xi) = \mathcal{F}[u](\xi)$ denote the Fourier transform of $u(x)$ and $\mathcal{F}^{-1}[v](x)$ the inverse Fourier transform of $v(\xi)$; that is, $\hat{u}(\xi) = \int \exp(-ix\cdot\xi)u(x)dx$ and $\mathcal{F}^{-1}[v](x) = \int \exp(ix\cdot\xi)v(\xi)\bar{d}\xi$. Next, for $u(x), v(x) \in \mathcal{S}'$, we put $(u|v) = \int u(x)\overline{v(x)}dx$ if this integral is well-defined. For a symbol $P(x,\xi)$ on $\mathbb{R}^n$, put $\hat{P}(\eta,\xi) = \mathcal{F}[P(\cdot,\xi)](\eta)$. Finally, for an open set Ω in $\mathbb{R}^n$, let $\mathcal{D}'(\Omega)$ denote the space of all distributions on Ω.

Now we state our general assumptions. We give a weight $M = (m_1, \cdots, m_n) \in (\mathbb{R}^+)^n$ satisfying $\min_{\ell=1,\cdots,n} m_\ell = 1$ to the coordinate variable $x \in \mathbb{R}^n$, where $\mathbb{R}^+$ denotes the set of nonnegative real numbers. Then we put $|M| = m_1 + \cdots + m_n$ and $\Theta = \{ \ell = 1, \cdots, n;\ m_\ell = 1 \}$. The dimension n and the weight M are fixed throughout this article.

Next, we define the action of $t \in \mathbb{R}^+$ to $x \in \mathbb{R}^n$ by $t^M x = (t^{m_1}x_1, \cdots, t^{m_n}x_n)$. For $t > 0$ and $s \in \mathbb{R}$, we put $t^{sM}x = (t^s)^M x$. In particular, we write $t^{-M}x = (t^{-1})^M x$. Now, for $\xi \neq 0$, let $[\xi]$ [resp. $\langle\xi\rangle$] denote the unique positive root of $|t^{-M}\xi| = 1$ [resp. $t^{-2} + |t^{-M}\xi|^2 = 1$], regarded as an equation with respect to t. For $\xi = 0$, we put $[0] = 0$ and $\langle 0 \rangle = 1$. Then it is easily seen that the function $\langle\xi\rangle$ satisfies the conditions of the basic weight function in Kumano-go [6,Chap. 7]. (For detailed proof, see [13,I,Section 1].) For an open set $\Omega \subset \mathbb{R}^n$, a subset $V \subset T^*\Omega\backslash 0 = \Omega\times(\mathbb{R}^n\backslash\{0\})$ is called *M-conic* if $(x,\xi) \in V$ and $t > 1$ imply $(x, t^M\xi) \in V$.

Now we give the definition of our function spaces.

Definition. For $s \in \mathbb{R}$, we define the *anisotropic Sobolev space* $H^{M,s}$ by

$$H^{M,s} = \{ u(x) \in \mathcal{S}';\ \|u(x)\|_{M,s} = \left(\int \langle\xi\rangle^{2s}|\hat{u}(\xi)|^2\bar{d}\xi\right)^{1/2} < \infty \}.$$

At the end of this section we microlocalize this space in the same manner as in Lascar [7].

Definition. Let Ω be an open set in $\mathbb{R}^n$, and let $u(x) \in \mathcal{D}'(\Omega)$ and $(\mathring{x},\mathring{\xi}) \in T^*\Omega\backslash 0$.

Then $u(x)$ is said to be *microlocally in* $H^{M,s}$ *at* $(\mathring{x},\mathring{\xi})$ if there exist a function $\phi(x) \in C_0^\infty(\Omega)$ and an M-conic neighborhood V of $(\mathring{x},\mathring{\xi})$ satisfying $\phi(\mathring{x}) \neq 0$ and $\int_U \langle\xi\rangle^{2s}|\mathcal{F}[\phi u](\xi)|^2 d\xi < \infty$, where $U = \{ \xi \in \mathbb{R}^n;\ (x,\xi) \in V \text{ for some } x \in \Omega \}$.

And $u(x)$ is said to be *locally in* $H^{M,s}$ *at* $\mathring{x}$ if there exists a function $\phi(x) \in C_0^\infty(\Omega)$ satisfying $\phi(\mathring{x}) \neq 0$ and $\phi(x)u(x) \in H^{M,s}$.

§2. Statement of the main theorem.

We consider nonlinear partial differential equations on Ω of the following form:

$$(2.1)\quad \sum_{k=1}^{K} \partial_x^{\beta(k)}\Big(F_k(x;\ u(x),\cdots,\partial_x^\alpha u(x),\cdots)\Big) = f(x),$$

where Ω is an open set of $\mathbb{R}^n$, $K \in \mathbb{N}$, $\beta(k) \in \mathbb{N}^n$ ($k = 1,\cdots,K$), $u(x),f(x) \in \mathcal{D}'(\Omega)$ and the function $F_k(x;\ X,\cdots,X_\alpha,\cdots)$ is of the form $X_{\alpha(k,1)}\cdots X_{\alpha(k,L_k)}G_k(x;\ X,\cdots,X_\alpha,\cdots)$. (We put $L_k = 0$ and $G_k = F_k$ if F_k is not divisible by any X_α.)

Here the number L_k is determined and the multi-indices $\alpha(k,1),\cdots,\alpha(k,L_k)$ are arranged in such a way that the relation $\lambda_{1k} \geq \lambda_{2k} \geq \cdots \geq \lambda_{0k}$ should hold, where

$$\lambda_{0k} = \max\{ M\cdot\alpha;\ G_k \text{ depends on } X_\alpha \}$$

$$(\lambda_{0k} = -\infty \text{ if } G_k \text{ does not depend on any } X_\alpha)$$

and

$$\lambda_{jk} = \begin{cases} M\cdot\alpha(k,j) & (\text{if } 1 \leq j \leq L_k) \\ \lambda_{0k} & (\text{if } j > L_k). \end{cases}$$

Next we suppose that $u(x)$ is a solution of (2.1) such that one of the following two conditions holds.

(HR) *The functions* $G_k(x; X,\cdots,X_\alpha,\cdots)$ *are* C^∞ *with respect to* $x \in \Omega$ *and every* $X_\alpha \in \mathbb{R}$, *and* $u(x)$ *is a real-valued distribution.*

(HC) *The functions* $G_k(x; X,\cdots,X_\alpha,\cdots)$ *are* C^∞ *with respect to* $x \in \Omega$, *and entire with respect to each* X_α, *and* $u(x)$ *is a complex-valued distribution.*

We write the formal development of (2.1) as

(2.2) $A(x; u(x),\cdots,\partial_x^\alpha u(x),\cdots) = f(x)$,

and we put $a_\beta(x) = \dfrac{\partial A}{\partial(\partial_x^\beta u)}\Big(x; u(x),\cdots,\partial_x^\alpha u(x),\cdots\Big)$ for $\beta \in \mathbb{Z}^n$.

Now we define the *weighted order* of the equation (2.1) by

$$m = \max \{ M\cdot\alpha;\ \partial_x^\alpha u(x) \text{ appears in } (2.2) \},$$

and the *weighted principal symbol* of (2.1) by $P_m(x,\xi) = \sum_{M\cdot\beta=m} a_\beta(x)(i\xi)^\beta$, and we say that a point $(x,\xi) \in T^*\Omega\backslash 0$ is *non-characteristic with respect to* M if $P_m(x,\xi) \neq 0$.

Remark 2.1. If the equation (2.1) is not semilinear, the weighted principal symbol may depend on the choice of the solution $u(x)$, and it may not be smooth. We shall return to this problem in Remark 2.4.

Further, in order to state our main theorem, we introduce some notions following Lascar [7].

Definition. We call the vector field

$$X(x,\xi) = \sum_{l\in\Theta} \left(\frac{\partial}{\partial\xi_l}(Re\ P_m(x,\xi))\cdot\frac{\partial}{\partial x_l} - \frac{\partial}{\partial x_l}(Re\ P_m(x,\xi))\cdot\frac{\partial}{\partial\xi_l}\right)$$

the quasi-homogeneous Hamiltonian vector field associated to the solution $u(x)$. Next, if $X(\mathring{x},\mathring{\xi}) \neq 0$ at a point $(\mathring{x},\mathring{\xi}) \in T^*\Omega\backslash 0$ such that $P_m(x,\xi) = 0$, then we say that the equation (2.1) is *of principal type at* $(\mathring{x},\mathring{\xi})$ *with respect to the weight* M *and the solution* $u(x)$. An integral curve of X contained in the set $V_0 = \{ (x,\xi) \in T^*\Omega\backslash 0;\ Re\ P_m(x,\xi) = 0 \}$ is called a *null bicharacteristic strip associated to the weight* M *and the solution* $u(x)$.

Further, we introduce some numbers. First we put

$$\rho = \max_{1\le k\le K}\left(\lambda_{0k}+|M|/2,\ \max_{h\ge 2}\left\{|M|/2+(\textstyle\sum_{j=1}^{h}\lambda_{jk}-|M|)/h\right\}\right),$$

and

$$\sigma = \max\left\{\rho\ ,\ \max_{1\le k\le K,\ h\ge 2}\{|M|/2+(M\cdot\beta(k)+\textstyle\sum_{j=1}^{h}\lambda_{jk}+1-m)/(h-1)\}\right\}.$$

Next, for every $k = 1,\cdots,K$ and $s > \sigma$, put

$$(2.3)\quad \mu_k(s) = \min_{h\ge 2}\left\{\textstyle\sum_{j=1}^{h}(s-\lambda_{jk}-|M|/2)+|M|/2\right\} - M\cdot\beta(k).$$

Now let $\mu(s)$ be a real number such that $\mu(s) \le \mu_k(s)$ holds for every $k = 1,\cdots,N$ and that $\mu(s) < \mu_k(s)$ holds for every k satisfying $\lambda_{3k} \ge s-|M|/2$ and

(2.4) There exists an integer $j \in \{1,\cdots,L_k\}$ satisfying $\lambda_{jk} = s-|M|/2$ such that $\partial_x^{\alpha(k,j)}u(x)$ is not essentially bounded.

Example 2.3. If the equation (2.1) is fully nonlinear of weighted order m, we have $\sigma = m+1+|M|/2$ and $\mu(s) = 2s-2m-|M|/2$. If the equation has better linearity, then σ is smaller and $\mu(s)$ is greater. Especially, if (2.1) is linear, we can take σ arbitrarily small and $\mu(s)$ arbitrarily large. Some other examples are given in Yamazaki [14].

Then our main result is the following.

Theorem. *Let $\Gamma = \{\ \gamma(t);\ \ a \le t \le b\ \} \subset V_0$ be a null bicharacteristic strip associated to M and $u(x)$, and suppose that the following six conditions are satisfied.*

(2.5) *$u(x)$ is locally in $H^{M,s}$ at every point of Ω, where $s > \sigma$.*

(2.6) *The equation (2.1) is of principal type with respect to the solution $u(x)$.*

(2.7) *For every $\ell,\ell' \in \Theta$, the derivative $\partial_{x_\ell} \mathrm{Re}\, P_m(x,\xi)$ is Lipshitz continuous with respect to $x_{\ell'}$.*

(2.8) *We can take an M-conic neighborhood V of Γ in $T^*\Omega\backslash 0$ and a disjoint partition $I_1,\cdots,I_N$ of the set $\{\beta \in \mathbb{N}^n;\ m-1 < M\cdot\beta \le m\}$ such that the conditions $\mathrm{Im}\sum_{\beta\in I_\nu} a_\beta(x)(i\xi)^\beta \ge 0$ and*

$$|\mathrm{Im}\ i^\beta\left(a_\beta(x+x')-2a_\beta(x)+a_\beta(x-x')\right)| \le C\sum_{\ell=1}^{n}|x'_\ell|^{(M\cdot\beta-m+1)\tau(\nu)/m_\ell} \qquad (\beta \in I_\nu)$$

hold for every $(x,\xi) \in V$ *and every* $\nu = 1,\cdots,N$ *with some constant* C, *where* $\tau(\nu) = 2/(m+1-\max_{\beta\in I_\nu} M\cdot\beta)$.

(2.9) $f(x)$ *is microlocally in* $H^{M,t}$ *at every point of* Γ, *where* $t \leq \mu(s)$.

(2.10) $u(x)$ *is microlocally in* $H^{M,t+m-1}$ *at* $\gamma(b)$.

Then $u(x)$ *is microlocally in* $H^{M,t+m-1}$ *at every point of* Γ.

Remark 2.4. The condition $s > \rho$ guarantees that the nonlinear terms are well-defined and can be linearized by means of the paradifferential operators. This condition depends, at least formally, on the expression (2.1).

On the other hand, the condition

$$s > \max_{1\leq k\leq K,\ h\geq 2} \{|M|/2+(M\cdot\beta(k)+ \textstyle\sum_{j=1}^{h} \lambda_{jk}-m+1)/(h-1)\}$$

implies that we can take

(2.11) $\mu(s)+m-1 > s$,

which is necessary for the main theorem to be meaningful. Besides, (2.11) implies that $P_b(x,\xi)$ is continuous for every $b > m-1$, and that $P_m(x,\xi)$ is differentiable with respect to x_l for every $l \in \Theta$. However, this is not strong enough to guarantee (2.7) and the latter inequality in (2.8) (The Hölder-type condition). If $\mu(s)+m-2 > s$, then these conditions will be automatically satisfied. These facts will be verified in Section 5.

The condition (2.11) depends only on the formal development (2.2), not on the expression (2.1).

Remark 2.5. The microlocal ellipticity theorem holds under a somewhat weaker condition

$$s > \max\left\{\rho,\ \max_{1\leq k\leq K,\ h\geq 2} \{|M|/2+(M\cdot\beta(k)+ \textstyle\sum_{j=1}^{h} \lambda_{jk}-m)/(h-1)\}\right\}.$$

For the precise statement of the theorem, and the meaning and the optimality of this condition, see Yamazaki [14].

Remark 2.6. Our main theorem generalizes Theorem 6.1 of Bony [1], where the case $M = (1,\cdots,1)$ is treated. Moreover, our theorem improves the values of σ and $\mu(s)$.

§3. Some properties of paradifferential operators.

In this section we recall the definition of quasi-homogeneous paradifferential operators introduced in [13, I and II], recall some properties shown there, and prove a new one.

First we introduce a partition of unity, which is an anisotropic version of the Littlewood-Paley decomposition. Let $\Psi(t)$ be a C^∞ function on $\mathbb{R}^+$ such that $0 \leq \Psi(t) < 1$, $\operatorname{supp}\Psi \subset [0,13/10)$ and $\Psi(t) \equiv 1$ on $[0,11/10]$. Next, for $\xi \in \mathbb{R}^n$, we put $\Psi_j(\xi) = \Psi(2^j[\xi])$ ($j \in \mathbb{N}$) and $\Psi_j(\xi) \equiv 0$ ($j \in \mathbb{Z}\setminus\mathbb{N}$) and $\Phi_j(\xi) = \Psi_j(\xi)-\Psi_{j-1}(\xi)$. Then we have $\sum_{j=0}^\infty \Phi_j(\xi) \equiv 1$ and $\Phi_j(\xi) = \Phi_1(2^{1-j}\xi)$ for $j \geq 1$.

Then the space $H^{M,s}$ can be characterized by means of this family of functions. In fact, we have the following proposition, which coincides with Proposition 3.11 of [13, I].

Proposition 3.1. *For every* $s \in \mathbb{R}$, *there exists a constant* C *such that the inequality*

$$C^{-1}\|u\|_{M,s} \leq \left(\sum_{j=0}^\infty \|\mathcal{F}^{-1}[\Phi_j(\xi)\hat{u}(\xi)](x)\|^2_{M,0}\right)^{1/2} \leq C\|u\|_{M,s}$$

holds for every $u \in H^{M,s}$.

Next we introduce two symbol classes.

Definition. For real numbers m and σ, let $S^{M,m,\sigma}$ denote the set of symbols $P(x,\xi)$ on $\mathbb{R}^n$ such that, for every $\alpha \in \mathbb{N}^n$, there exists a constant C_α such that the following inequality holds:

$$\sup_{x\in\mathbb{R}^n} |\mathcal{F}_\eta^{-1}[\Phi_j(\eta)\partial_\xi^\alpha \hat{P}(\eta,\xi)](x)| \leq C_\alpha 2^{-\sigma j}\langle\xi\rangle^{m-M\cdot\alpha}.$$

Further, a subset $A \subset S^{M,m,\sigma}$ is called *bounded* if, for every $\alpha \in \mathbb{N}^n$, such that the above inequality holds for every $P(x,\xi) \in A$.

Next, let $S^{M,m}$ denote the set $\bigcap_{\sigma\in\mathbb{R}} S^{M,m,\sigma}$.

Remark 3.2. The class $S^{M,m,\sigma}$ coincides with the class $S(B^{M,\sigma}_{\infty,\infty})^m$ in [13], and the class $S^{M,m}$ coincides with the class

$S^m_{\langle\xi\rangle,1,0}$ of Kumano-go [6,Chap. 7].

Now we define the *paradifferential operator associated to the symbol* $P(x,\xi)$ as follows.

Definition. For a symbol $P(x,\xi)$ and $u \in \mathcal{S}'$, we put

$$\pi_1(P(X,D),u)(x) = \sum_{j=2}^{\infty} \mathcal{F}^{-1}[\int \Psi_{j-2}(\xi-\eta)\hat{P}(\xi-\eta,\eta)\Phi_j(\eta)\hat{u}(\eta)\bar{d}\eta](x).$$

The difference between pseudodifferential operators and paradifferential operators is described by the following proposition, which is a consequence of Theorems A, B, C of [13,I] and Remark 3.2.

Proposition 3.3. *Let* $P(x,\xi)$ *be a symbol in* $S^{M,m,\sigma}$, *and let* s *be a real number greater than* $-\sigma$. *If* t *satisfies* $t \le \min\{ s,0 \}$ *and* $t < \max\{ s,0 \}$, *there exists a constant* C *such that*

$$\|P(x,D)u(x)-\pi_1(P(x,D),u)(x)\|_{M,t+\sigma} \le C\|u\|_{M,s+m}$$

holds for every $u \in H^{M,s+m}$.

On the other hand, the paradifferential operators have boundedness properties and satisfy the symbol calculus with error terms. Namely, we have the following two propositions. Remark 3.2 implies that the former is a special case of Theorems A and B of [13,I] and that the latter is a special case of Theorem A of [13,II].

Proposition 3.4. *Suppose that* $m,s \in \mathbb{R}$ *and that* $P(x,\xi)$ *belongs to* $S^{M,m,\sigma}$. *If* t *satisfies* $t \le \min\{ \sigma,0 \}$ *and* $t < \max\{ \sigma,0 \}$, *there exists a constant* C *such that the inequality*

$$\|\pi_1(P(x,D),u)\|_{M,s+t} \le C\|u\|_{M,s+m}$$

holds for every $u \in H^{M,s+m}$.

Proposition 3.5. *Suppose that* $l,m,s \in \mathbb{R}$, $\rho = \min\{\sigma,\tau\} \ge 0$ *and* $\max\{\sigma,\tau\} > |M|/2$. *Then, for symbols* $P(x,\xi) \in S^{M,m,\sigma}$, $Q(x,\xi) \in S^{M,l,\tau}$ *and every* $\alpha \in \mathbb{N}^n$ *satisfying* $M\cdot\alpha \le \rho$, *the symbol* $R_\alpha(x,\xi) = \frac{1}{\alpha!}\partial_\xi^\alpha P(x,\xi)D_x^\alpha Q(x,\xi)$ *belongs to* $S^{M,m+l-M\cdot\alpha,\rho-M\cdot\alpha}$. *Moreover, there exists a constant* C *such that the inequality*

$$\|\pi_1\Big(P(x,D),\pi_1(Q(x,D),u)\Big)(x)-\sum_{M\cdot\alpha\le\rho}\pi_1(R_\alpha(x,D),u)(x)\|_{M,s+\rho}\le C\|u(x)\|_{M,s+l+m}$$

holds for every $u(x) \in H^{M,s+l+m}$.

Remark 3.6. If A and B are bounded sets of $S^{M,m,\sigma}$ and $S^{M,\ell,\tau}$ respectively, there exists a constant C such that the inequalities in the conclusions of Propositions 3.3, 3.4 and 3.5 hold for every $P(x,\xi) \in A$ and $Q(x,\xi) \in B$.

At the end of this section we prove a new property concerning the dual of paradifferential operators.

Proposition 3.7. *Let* $\sigma \geq 0$, $m, s \in \mathbb{R}$, *and suppose that* $P(x,\xi)$ *is a symbol in* $S^{M,m,\sigma}$. *Then, for every* $\alpha \in \mathbb{N}^n$ *satisfying* $M\cdot\alpha \leq \sigma$, *the symbol* $Q_\alpha(x,\xi) = \frac{1}{\alpha!}\partial_\xi^\alpha D_x^\alpha \overline{P(x,\xi)}$ *belongs to* $S^{M,m-M\cdot\alpha,\sigma-M\cdot\alpha}$. *Moreover, there exists a constant* C *such that the inequality*

$$|(\pi_1(P(x,D),u)|v) - \sum_{M\cdot\alpha\leq\sigma} (u|\pi_1(Q_\alpha(x,D),v))| \leq C\|u\|_{M,s+m}\|v\|_{M,s+\sigma}$$

holds for every $u \in H^{M,s+m}$ *and* $v \in H^{M,s+\sigma}$.

Proof. Since $Q_\alpha(x,\xi) \in S^{M,m-M\cdot\alpha,\sigma-M\cdot\alpha}$ follows from Remark 4.7 of [13,I], it suffices to prove the inequality. First we put

$$u_j(x) = \mathcal{F}^{-1}[\Phi_j(\xi)\hat{u}(\xi)](x), \quad v_j(x) = \mathcal{F}^{-1}[\Phi_j(\xi)\hat{v}(\xi)](x),$$

$$I_{j,k,h} = \iint \Phi_{j-k}(\xi-\eta)\hat{P}(\xi-\eta,\eta)\hat{u}_j(\eta)\bar{d}\eta\cdot\overline{v_{j-h}(\xi)}\cdot\bar{d}\xi$$

and

$$J_{j,h} = \iint \Psi_{j-h-3}(\xi-\eta)\hat{P}(\xi-\eta,\eta)\hat{u}_j(\eta)\bar{d}\eta\cdot\overline{v_{j-h}(\xi)}\cdot\bar{d}\xi.$$

Then the condition on $\operatorname{supp}\Phi_j$ implies

$$(3.1)\qquad \iint \Psi_{j-2}(\xi-\eta)\hat{P}(\xi-\eta,\eta)\hat{u}_j(\eta)\bar{d}\eta\cdot\overline{v(\xi)}\cdot\bar{d}\xi$$

$$= \sum_{h=-1}^{1} \iint \Psi_{j-2}(\xi-\eta)\hat{P}(\xi-\eta,\eta)\hat{u}_j(\eta)\bar{d}\eta\cdot\overline{v_{j-h}(\xi)}\cdot\bar{d}\xi \; +$$

$$+ \iint \Phi_{j-2}(\xi-\eta)\hat{P}(\xi-\eta,\eta)\hat{u}_j(\eta)\bar{d}\eta\cdot\overline{v_{j-2}(\xi)}\cdot\bar{d}\xi$$

$$= \sum_{h=-1}^{1} J_{j,h} + I_{j,3,1} + I_{j,2,0} + I_{j,2,1} + I_{j,2,2}\,.$$

On the other hand, putting

$$P_{j,k}(x,\xi) = \mathcal{F}_\eta^{-1}[\Phi_{j-k}(\eta)\hat{P}(\eta,\xi)\Big(\Psi_{j+2}(\eta)-\Psi_{j-3}(\eta)\Big)](x),$$

we can write

$$I_{j,k,h} = \iiint \exp(i(\eta-\xi)\cdot x)P_{j,k}(x,\eta)dx\cdot\int\exp(-i\eta\cdot y)u_j(y)dy\cdot\overline{\hat{v}_{j-h}(\xi)}\bar{d}\xi\bar{d}\eta$$

$$= \iiint \exp(i\eta\cdot(x-y))P_{j,k}(x,\eta)u_j(y)\overline{v_{j-h}(x)}dydx\bar{d}\eta.$$

Introducing a differential operator L on $\mathbb{R}^n_\eta$ by the formula $Lw = w+ \sum_{\ell=1}^n \partial_{\eta_\ell}^2\Big(\langle\eta\rangle^{2m_\ell}w\Big)$ and integrating by parts, we obtain

$$I_{j,k,h} = \iiint\Big(1+ \sum_{\ell=1}^n \langle\eta\rangle^{2m_\ell}|x_\ell-y_\ell|^2\Big)^{-n}\exp(i\eta\cdot(x-y))\cdot$$

$$(L^nP_{j,k})(x,\eta)u_j(y)\overline{v_{j-h}(x)}\cdot dydx\bar{d}\eta.$$

This formula, together with Hölder's inequality and the fact that $L^nP(x,\eta) \in S^{M,m,\sigma}$ for every n, implies

$$|I_{j,k,h}|$$

$$\le \int \sup_{x\in\mathbb{R}^n}|(L^nP_{j,k})(x,\eta)|\cdot\int\Big(1+ \sum_{\ell=1}^n \langle\eta\rangle^{2m_\ell}|z_\ell|^2\Big)^{-n}dz\bar{d}\eta\cdot\|u_j\|_{M,0}\|v_{j-h}\|_{M,0}$$

$$\le C\int_{E_j}\langle\eta\rangle^{m-|M|}d\eta\cdot 2^{-(j-k)\sigma}\|u_j\|_{M,0}\|v_{j-h}\|_{M,0},$$

where $E_j = \{\ \eta \in \mathbb{R}^n;\ \ \langle\eta\rangle < 2^{j+4}/3\ \}$ if $j \le 2$ and $E_j = \{\ \eta \in \mathbb{R}^n;\ 2^{j-3} < \langle\eta\rangle < 2^{j+4}/3\ \}$ if $j \ge 3$.

This implies

$$(3.2)\quad |I_{j,k,h}| \le C\cdot 2^{k\sigma+j(m-\sigma)}\|u_j\|_{M,0}\|v_{j-h}\|_{M,0}.$$

Next, choose a natural number N greater than σ, and put

$$Q_{\alpha,j,h}(x,\xi) = \frac{1}{\alpha!}\cdot\mathcal{F}_\zeta^{-1}[\Psi_{j-h-3}(\zeta)\zeta^\alpha\overline{\partial_\xi^\alpha\hat{P}(-\zeta,\xi)}\Big(\Psi_{j+1}(\xi)-\Psi_{j-2}(\xi)\Big)](x)$$

for $\alpha \in \mathbb{N}^n$ satisfying $|\alpha| \le N$.

Then the condition on $\operatorname{supp} \Phi_j$ and Taylor's formula yield

$$\text{(3.3)}\quad \overline{J_{j,k}} = \iint \Psi_{j-k-3}(\xi-\eta)\overline{\hat{P}(\xi-\eta,\eta)\cdot\hat{u}_j(\eta)}\Big(\Psi_{j+1}(\eta)-\Psi_{j-2}(\eta)\Big)\bar{d}\eta\cdot v_{j-h}(\xi)\cdot\bar{d}\xi$$

$$= \textstyle\sum_{|\alpha|\leq N} K_{\alpha,j,h,0} + \sum_{|\alpha|=N}\displaystyle\int_0^1 N(1-\theta)^{N-1}K_{\alpha,j,h,\theta}\, d\theta,$$

where

$$K_{\alpha,j,h,\theta} = \iint \hat{Q}_{\alpha,j,h}(\eta-\xi,\xi+\theta(\eta-\xi))\hat{v}_{j-h}(\xi)\overline{\hat{u}_j(\eta)}\bar{d}\xi\bar{d}\eta$$

$$= \iint v_{j-h}(x)\overline{u_j(y)}\iint \exp(-i\xi\cdot x+i\eta\cdot y)\hat{Q}_{\alpha,j,h}(\eta-\xi,\xi+\theta(\eta-\xi))\bar{d}\xi\bar{d}\eta dxdy$$

$$= \iint v_{j-h}(x)\overline{u_j(y)}\iint \exp\Big(i\xi'(y-x)+i\eta'(\theta x+(1-\theta)y)\Big)\hat{Q}_{\alpha,j,h}(\eta',\xi')\bar{d}\xi'\bar{d}\eta' dxdy$$

$$= \iint v_{j-h}(x)\overline{u_j(y)}\iint\Big(1+\textstyle\sum_{\ell=1}^n \langle\xi\rangle^{2m_\ell}|x_\ell-y_\ell|^2\Big)^{-n}\exp(i\xi'(y-x))$$

$$(L^n Q_{\alpha,j,h})(\theta x+(1-\theta)y,\xi')\bar{d}\xi' dxdy.$$

If $M\cdot\alpha > \sigma$, then

$$\text{(3.4)}\quad |K_{\alpha,j,h,\theta}| \leq \iint\Big\{\sup_{x,y\in\mathbb{R}^n}\Big((L^n Q_{\alpha,j,h})(\theta x+(1-\theta)y,\xi)\Big)\|v_{j-h}\|_{M,0}\|u_j\|_{M,0}\times$$

$$\int\Big(1+\textstyle\sum_{\ell=1}^n \langle\xi'\rangle^{2m_\ell}|z_\ell|^2\Big)^{-n}dz\Big\}\bar{d}\xi$$

$$\leq C\int_E \langle\xi\rangle^{m-M\cdot\alpha-|M|}d\xi\cdot\textstyle\sum_{k=0}^{j-h-3} 2^{k(M\cdot\alpha-\sigma)}\|u_j\|_{M,0}\|v_{j-h}\|_{M,0}$$

$$\leq C\cdot 2^{h(\sigma-M\cdot\alpha)+j(m-\sigma)}\|u_j\|_{M,0}\|v_{j-h}\|_{M,0}.$$

On the other hand, for $\alpha\in\mathbb{N}^n$ satisfying $M\cdot\alpha\leq\sigma$, we have

$$\text{(3.5)}\quad \textstyle\sum_{h=-1}^1 K_{\alpha,j,h,0} = \sum_{h=-1}^1 \displaystyle\iint \hat{v}_{j-h}(\xi)\Psi_{j-h-3}(\eta-\xi)\hat{Q}_\alpha(\eta-\xi,\xi)\hat{u}_j(\eta)\bar{d}\eta\bar{d}\xi$$

$$= (\pi_1(Q_\alpha(x,D),v)(x)|u_j(x)) + \textstyle\sum_{k=-1}^2 I'_{\alpha,j,k},$$

where

$$I'_{\alpha,j,k} = \iint \Phi_{j+k-2}(\eta-\xi)\hat{Q}_\alpha(\eta-\xi,\xi)\hat{v}_{j+k}(\xi)\hat{u}_j(\eta)\bar{d}\eta\bar{d}\xi.$$

Then we can show

$$(3.6)\quad |I'_{\alpha,j,k}| \leq C\cdot 2^{k(\sigma-M\cdot\alpha)+j(m-\sigma)}\|u_j\|_{M,0}\|v_{j+k}\|_{M,0}$$

in the same way as (3.1).

Now the formulae (3.1) - (3.6), together with Schwarz's inequality and Proposition 3.1, yield

$$|(\pi_1(P(x,D),u)(x)|v(x)) - \sum_{M\cdot\alpha\leq\sigma}(u(x)|\pi_1(Q_\alpha(x,D),v)(x))|$$

$$\leq \sum_{j=0}^{\infty}\Big\{|I_{j,3,1}| + \sum_{k=0}^{2}|I_{j,2,k}| + \sum_{k=-1}^{2}\sum_{M\cdot\alpha\leq\sigma}|I'_{\alpha,j,k}| +$$

$$+ \sum_{h=-1}^{1}\Big(\sum_{M\cdot\alpha>\sigma,\,|\alpha|<N}|K_{\alpha,j,h,0}| + \sum_{|\alpha|=N}\int_0^1 N(1-\theta)^{N-1}|K_{\alpha,j,h,\theta}|d\theta\Big)\Big\}$$

$$\leq C\sum_{j=0}^{\infty}2^{j(m-\sigma)}\|u_j\|_{M,0}\sum_{k=-2}^{2}\|v_{j+k}\|_{M,0}$$

$$\leq C\Big(\sum_{j=0}^{\infty}2^{2j(s+m)}\|u_j\|_{M,0}^2\Big)^{1/2}\cdot\Big(\sum_{j=0}^{\infty}2^{2j(-s-\sigma)}\|v_j\|_{M,0}^2\Big)^{1/2}$$

$$\leq C\|u\|_{M,s+m}\|v\|_{M,-s-\sigma}.$$

This completes the proof.

§4. A sharp Gårding inequality for non-smooth symbols.

In this section we shall prove a sharp Gårding inequality for quasi-homogeneous paradiffential operators. An inequality of this type for pseudodifferential operators with smooth symbols is given in Sakurai [10, Proposition 2.5], but we modify part of the proof in [10] in order to obtain a sharper result for non-smooth symbols. Even in the homogeneous case, our theorem improves similar inequality given in Bony [1]. As well as the results mentioned here, we use the wave packet transform introduced by Córdoba-Fefferman [2] in the proof.

Proposition 4.1. *Let* μ,δ *be real numbers, and assume that* $0 < \lambda(1) \leq \cdots \leq \lambda(K) = 2-2\delta \leq 1$. *Then, if the symbols* $P_1(x,\xi)$, $\cdots$, $P_K(x,\xi)$ *satisfy the conditions*

(4.1) $|\partial_\xi^\alpha P_k(x,\xi)| \le C_\alpha \langle\xi\rangle^{2\mu+\lambda(k)-M\cdot\alpha}$ *for every* k *and* $\alpha \in \mathbb{N}^n$,

(4.2) $|\partial_\xi^\alpha P_k(x+x',\xi)-2\partial_\xi^\alpha P_k(x,\xi)+\partial_\xi^\alpha P_k(x-x',\xi)|$

$$\le C_\alpha \sum_{l=1}^n |x_l'|^{\lambda(k)/\delta m_l}\langle\xi\rangle^{2\mu+\lambda(k)-M\cdot\alpha} \quad \textit{for every } k \textit{ and } \alpha \in \mathbb{N}^n$$

and

(4.3) $\sum_{k=1}^K Re\ P_k(x,\xi) \ge -C_0\langle\xi\rangle^{2\mu}$ *for every* k,

there exists a constant C *such that the inequality*

$$\sum_{k=1}^K Re\ (\pi_1(P_k(x,D),u)(x)|u(x)) \ge -C\|u\|_{M,\mu}^2$$

holds for every $u \in C_0^\infty(\mathbb{R}^n)$.

Proof. First, putting $v = \langle D\rangle^\mu u$, we obtain

$$Re\Big(\pi_1(P_k(x,D)\langle D\rangle^{-2\mu},v)|v\Big)$$

$$= Re\Big(\pi_1\Big(\langle D\rangle^\mu,\pi_1\Big(P_k(x,D)\langle D\rangle^{-2\mu},\pi_1(\langle D\rangle^\mu u)\Big)\Big)\Big|u\Big) + R_1$$

$$= Re\Big(\pi_1(P_k(x,D),u)|u\Big) + R_1 + R_2$$

from Propositions 3.3, 3.4, 3.5 and the condition $\lambda(k) \le 1$, where $|R_1|,|R_2| \le C\|u\|_{M,\mu}^2$ with some constant C. On the other hand, the symbols $P_k(x,\xi)\langle\xi\rangle^{-2\mu}$ ($k = 1,\cdots,K$) satisfy assumptions (4.1), (4.2) and (4.3) with μ replaced by 0. This implies $P_k(x,\xi)\langle\xi\rangle^{-2\mu} \in S^{M,\lambda(k),\lambda(k)/\delta}$, and it follows from the fact $\delta < 1$ and Proposition 3.3 that the inequality

$$\|P_k(x,D)\langle D\rangle^{-2\mu}v-\pi_1\Big(P_k(x,D)\langle D\rangle^{-2\mu},v\Big)\|_{M,0} \le C\|v\|_{M,0}$$

holds with some constant C. Hence we may assume $\mu = 0$ and show

(4.4) $\sum_{k=1}^K Re\ (P_k(x,D)u(x)|u(x)) \ge -C\|u\|_{M,0}^2$.

First, for $u(x) \in C_0^\infty(\mathbb{R}^n)$, we define the wave packet transform of $u(x)$ by the formula

$$(Wu)(y,\xi) = \frac{\langle\xi\rangle^{\delta|M|/2}}{2^{n/2}\pi^{3n/4}}\int \exp\Big(i(y-x)\cdot\xi - \sum_{l=1}^n \langle\xi\rangle^{2\delta m_l}|y_l-x_l|^2/2\Big)u(x)dx,$$

and put $\mathscr{A}_k = \int P_k(x,\xi)(Wu)(x,\xi)\overline{(Wu)(x,\xi)}dy\bar{d}\xi$ for every $k = 1,\cdots,K$.

Then assumption (4.3) yields

(4.5) $\sum_{k=1}^{K} Re\, \mathscr{A}_k + C_0\int |(Wu)(x,\xi)|^2 dx\bar{d}\xi \geq 0.$

On the other hand, putting

$$Q_k(x,\xi,z) = \frac{\langle\xi\rangle^{\delta|M|}}{\pi^{n/2}}\int \exp\left(-\sum_{\ell=1}^{n}\langle\xi\rangle^{2\delta m_\ell}(|x_\ell-y_\ell|^2+|y_\ell-z_\ell|^2)/2\right)P(y,\xi)dy,$$

we can write

(4.6) $\mathscr{A}_k = \left(\int \exp(i(x-z)\cdot\xi)Q_k(x,\xi,z)u(z)dz \middle| u(x)\right).$

We now calculate each $\mathscr{A}_k$. First, in view of assumption (4.1), we have the estimate

$$|\partial_\xi^\alpha \partial_x^\beta \partial_z^\gamma Q_k(x,\xi,z)| \leq C_{\alpha\beta\gamma}\langle\xi\rangle^{\lambda(k)-M\cdot\alpha+\delta M(\beta+\gamma)}$$

for every $\alpha,\beta,\gamma \in \mathbb{N}^n$.

This estimate and Theorem 2.5 of Kumano-go [6,Chap.7] yield the equality

(4.7) $\int \exp(i(x-z)\cdot\xi)Q_k(x,\xi,z)u(z)dz$

$$= P_k(x,D)u(x)+R_k(x,D)u(x)+\sum_{h=1}^{n} S_{kh}(x,D)u(x)+T_k(x,D)u(x),$$

where

$$R_k(x,\xi) = Q_k(x,\xi,x)-P_k(x,\xi),$$

$$S_{kh}(x,\xi) = \partial_{\xi_h} D_{x_h} Q_k(y,\xi,x)\Big|_{y=x}$$

and $T_k(x,\xi)$ belongs to the class $S^{\lambda(k)-2+2\delta}_{\langle\xi\rangle,1,\delta}$ of Kumano-go [6,Chap.7]. Since $\lambda(k)-2+2\delta = \lambda(k)-\lambda(K) \geq 0$ and $\delta < 1$, Theorem 1.6 of Kumano-go [6,Chap.7] yields the estimate

(4.8) $\left(T_k(x,D)u(x)|u(x)\right) \leq C\|u(x)\|^2_{M,0}.$

We turn to $S_{kh}(x,D)u(x)$. Put $\gamma = \lambda(k)/2\delta$. Then, since $\gamma < \lambda(k)/\delta$ and $\gamma \leq \lambda(K)/2\delta = \delta^{-1}-1 \leq 1$, we can deduce the inequality

(4.9) $|\partial_\xi^\alpha P_k(x-z,\xi)-\partial_\xi^\alpha P_k(x,\xi)| \leq C_\alpha \sum_{\ell=1}^{n}|z_\ell|^{\gamma/m_\ell}\langle\xi\rangle^{2\mu+\lambda(k)-M\cdot\alpha}$

for every $\alpha \in \mathbb{N}^n$ from assumption (4.2) and the property of non-smooth symbols (See [13,II,Theorem 4.1,Assertion 1)]).

On the other hand, we have the equalities

$$(4.10)\quad \partial_{x_h} Q_k(y,\xi,x)\Big|_{y=x}$$

$$= \pi^{-n/2}\langle\xi\rangle^{\delta|M|}\int\langle\xi\rangle^{2\delta m_h}|x_h-y_h|\exp\Big(-\textstyle\sum_{l=1}^{n}\langle\xi\rangle^{2\delta m_l}|x_l-y_l|^2\Big)P_k(y,\xi)dy$$

$$= \pi^{-n/2}\langle\xi\rangle^{\delta|M|}\int\langle\xi\rangle^{2\delta m_h}|z_h|\exp\Big(-\textstyle\sum_{l=1}^{n}\langle\xi\rangle^{2\delta m_l}|z_l|^2\Big)\cdot\Big(P_k(x-z,\xi)-P_k(x,\xi)\Big)dz$$

and

$$(4.11)\quad \partial_{x_h} Q_k(y,\xi,x)\Big|_{y=x}-\partial_{x_h} Q_k(y,\xi,x')\Big|_{y=x'}$$

$$= \frac{\langle\xi\rangle^{\delta|M|}}{\pi^{n/2}}\int\langle\xi\rangle^{2\delta m_h}|z_h|\exp\Big(-\textstyle\sum_{l=1}^{n}\langle\xi\rangle^{2\delta m_l}|z_l|^2\Big)\cdot\Big(P_k(x-z,\xi)-P_k(x'-z,\xi)\Big)dz.$$

From (4.9), (4.10) and (4.11) we obtain

$$(4.12)\quad |\partial_\xi^\alpha S_{kh}(x,\xi)| \le C\langle\xi\rangle^{\lambda(k)-M\cdot\alpha-m_h+\delta(m_h-\gamma)}$$

and

$$(4.13)\quad |\partial_\xi^\alpha S_{kh}(x,\xi)-\partial_\xi^\alpha S_{kh}(x',\xi)| \le C\langle\xi\rangle^{\lambda(k)-M\cdot\alpha-m_h+\delta m_h}\textstyle\sum_{l=1}^{n}|x_l-x_l'|^{\gamma/m_l}.$$

On the other hand, we have the inequality

$$(4.14)\quad \lambda(k)-m_h+\delta m_h-\delta\gamma \le \lambda(k)-1+\delta-\delta\gamma = \lambda(k)/2-\delta\gamma = 0.$$

In view of (4.12), (4.13), (4.14) and the fact $\delta < 1$, we can apply the main theorem of [12] to obtain

$$(4.15)\quad \Big(S_{kh}(x,D)u(x)\,|\,u(x)\Big) \le C\|u(x)\|_{M,0}^2 \qquad \text{for every } h = 1,\cdots,n.$$

Finally we consider $R_k(x,D)u(x)$. In view of assumption (4.2) and the equalities

$$2R_k(x,\xi) = \frac{\langle\xi\rangle^{\delta|M|}}{\pi^{n/2}}\int\exp\Big(-\textstyle\sum_{l=1}^{n}\langle\xi\rangle^{2\delta m_l}|z_l|^2\Big)\Big(P(x+z,\xi)-2P(x,\xi)+P(x-z,\xi)\Big)dz$$

and

$$R_k(x+x',\xi)-2R_k(x,\xi)+R_k(x-x',\xi) =$$

$$\frac{\langle\xi\rangle^{\delta|M|}}{\pi^{n/2}}\int\exp\Big(-\sum_{l=1}^{n}\langle\xi\rangle^{2\delta m_l}|z_l|^2\Big)\Big(P_k(x+x'-z,\xi)-2P_k(x-z,\xi)+P_k(x-x'-z,\xi)\Big)dz,$$

we obtain $|\partial_\xi^\alpha R_k(x,\xi)| \le C\langle\xi\rangle^{-M\cdot\alpha}$ and

$$|\partial_\xi^\alpha R_k(x+x',\xi)-2\partial_\xi^\alpha R_k(x,\xi)+\partial_\xi^\alpha R_k(x-x',\xi)| \le C\langle\xi\rangle^{\lambda(k)-M\cdot\alpha}\sum_{l=1}^n |x'_l|^{\lambda(k)/\delta m_l}.$$

In view of these two inequalities and the fact $\delta < 1$, we can apply the main theorem of [12] again to obtain

$$(4.16)\quad \Big(R_k(x,D)u(x)|u(x)\Big) \le C\|u(x)\|^2_{M,0}$$

Combining (4.6) and (4.7) and making use of (4.8), (4.15) and (4.16), we obtain

$$|\mathcal{A}_k-(P_k(x,D)u(x)|u(x))| \le C\|u(x)\|^2_{M,0}$$

for some constant C. Taking the real part of both sides, we conclude

$$Re\ (P_k(x,D)u(x)|u(x))-Re\ \mathcal{A}_k \ge -C\|u(x)\|^2_{M,0}\ .$$

In the same way as above, we can deduce

$$-\int|(Wu)(x,\xi)|^2dx \ge -C\|u(x)\|^2_{M,0}\ .$$

Adding up these two formulae and substituting into (4.5), we obtain (4.4). Now the proof is complete.

§5. Proof of the main theorem.

In this section we shall prove our main theorem. First we remark that we can divide Γ into finite parts and consider them separately in the proof.

Next we may assume that the quasi-homogeneous Hamiltonian vector field $X(x,\xi)$ does not vanish on Γ. In fact, the conclusion is not trivial at $(\mathring{x},\mathring{\xi})$ only if $Im\ P_m(\mathring{x},\mathring{\xi}) = 0$, and in this case X does not vanish around $(\mathring{x},\mathring{\xi})$ in view of (2.6). We may further assume that $X(x,\xi)$ is never proportional to $X_0(\xi) = \sum_{l=1}^n m_l\xi_l(\partial/\partial\xi_l)$ on Γ. In fact, if $X(x,\xi) = CX_0(\xi)$ holds for some $(x,\xi) \in \Gamma$, then $\Gamma_0 = \{(x,t^M\xi);\ t > 0\}$ is another null bicharacteristic strip associated to M and $u(x)$ passing through (x,ξ), since the fact $P_m(x,t^M\xi) = t^m P_m(x,\xi)$ implies that $X_0(t^M\xi) = tX_0(\xi)$ is proportional to $X(x,t^M\xi)$. On the other hand, assumption (2.7) guarantees the uniqueness of such null bicharacteristic strips. Hence we have $\Gamma \subset \Gamma_0$, in which case the conclusion is trivial. By shrinking V if necessary, we may assume that $X(x,\xi)$ is never proportional to $X_0(\xi)$ on V.

Further, we introduce a notation. For two symbols $A(x,\xi)$ and $B(x,\xi)$, let $\{A,B\}(x,\xi)$ denote the *partial Poisson bracket* defined by

$$\{A,B\}(x,\xi) = \sum_{l\in\Theta} \left(\partial_{\xi_l} A(x,\xi)\cdot\partial_{x_l} B(x,\xi) - \partial_{x_l} A(x,\xi)\cdot\partial_{\xi_l} B(x,\xi)\right).$$

Then we have $X(A)(x,\xi) = \{Re\ P_m, A\}(x,\xi)$ for every symbol $A(x,\xi)$.

Now we start the proof. First we recall the linearization employed in [14], and verify Remark 2.4. For every $z = (x,\xi) \in \Gamma$, choose an open M-conic neighborhood of z of the form $U_z \times W_z$, and functions $\phi_z(x) \in C_0^\infty(\Omega)$, $\psi_z(\xi) \in C_0^\infty(\mathbb{R}^n)$ satisfying the conditions

$$\begin{cases} 0 \leq \phi_z(x) \leq 1,\ 0 \leq \psi_z(\xi) \leq 1, \\ \psi(t^M\xi) = \psi(\xi) \quad \text{if} \quad t \geq 1 \quad \text{and} \quad |\xi| \geq 1, \\ \phi_z(x) \equiv 1 \quad \text{on} \quad U_z,\ \psi_z(\xi) \equiv 1 \quad \text{on} \quad W_z, \\ \operatorname{supp} \phi(x) \times \operatorname{supp} \psi(\xi) \subset V, \\ \phi_z(x)(\psi_z(D)f)(x) \in H^{M,s}. \end{cases}$$

Then the compact set Γ is covered by a finite number of open sets of the form $U_z \times W_z$. Dividing Γ into finite parts if necessary, we may assume $\Gamma \subset U_z \times W_z$. Now we put $\phi(x) = \phi_z(x)$, $\psi(\xi) = \psi_z(\xi)$, and choose functions $\chi(x) \in C_0^\infty(\Omega)$ and $\omega(x) \in C_0^\infty(U_z)$ satisfying $0 \leq \chi(x) \leq 1$, $0 \leq \omega(x) \leq 1$, $\Gamma \subset V_1 = \{\, x;\ \omega(x) = 1 \,\} \times W_z$ and $\chi(x) \equiv 1$ on some neighborhood of $\operatorname{supp} \phi(x)$. Further, we put $v(x) = \chi(x)u(x)$.

Then, since $v(x) \in H^{M,s}$ follows from assumption (2.5), the situation is the same as in [14, Section 5]. Hence (5.13), (5.14) and (5.19) of [14] imply the existence of a symbol $R(x,\xi)$ such that

$$(5.1)\quad \pi_1(R(x,D),v)(x) \in H^{M,t}.$$

Next, corresponding to (5.15) and (5.16) of [14], we have

$$(5.2)\quad R(x,\xi) = \sum_{b\in\Lambda} R_b(x,\xi)$$

and

$$(5.3)\quad \Lambda = \{\ M\cdot\alpha;\ \alpha \in \mathbb{N}^n,\ s-\mu(s)-|M|/2 \le M\cdot\alpha \le m\ \}.$$

Moreover, (5.17) of [14], Remark 4.7 of [13,I] and Remark 3.2 imply

$$(5.4)\quad R_b(x,\xi) \in S(B^{M,\mu(s)-s+b}_{\infty,\infty})^b = S^{M,b,\mu(s)-s+b}. \qquad \text{for every } b \in \Lambda.$$

Hence, applying Proposition 3.4 to $R_b(x,\xi)$ with $b < s-\mu(s)$ and observing the inequality $t \le \mu(s)$, we obtain

$$(5.5)\quad \textstyle\sum_{b\in\Lambda'} \pi_1(R_b(x,D),v)(x) = g(x) \in H^{M,t}$$

from (5.1), (5.2), (5.3) and (5.4), where

$$(5.6)\quad \Lambda' = \{\ M\cdot\alpha;\ \alpha \in \mathbb{N}^n,\ s-\mu(s) \le M\cdot\alpha \le m\ \}.$$

Furthermore, as we have verified in [14], we have

$$(5.7)\quad R_m(x,\xi) = \omega(x)\psi(\xi)P_m(x,\xi).$$

More generally, we obtain

$$(5.8)\quad R_b(x,\xi) = \omega(x)\psi(\xi)\textstyle\sum_{M\cdot\beta=b} a_\beta(x)(i\xi)^\beta$$

for every $m-1 < b \le m$ in the same way. Now Remark 2.4 follows immediately from this fact and (5.4).

Next, choose a function $B(x,\xi) \in C^\infty(T^*\Omega\backslash 0)$ satisfying $X(B) \ge \langle\xi\rangle^{m-1}$ on V_0 and $B(x,t^M\xi) = B(x,\xi)$ for $t \ge 1$ and $|\xi| \ge 1$, and put

$$\varkappa = (1/4)\min\Big\{1,\ \mu(s)+m-1-s,\ \min\{m_l-1;\ l\notin\Theta\},\ \min\ \{m-b;\ b\in\Lambda',\ b\ne m\}\Big\}.$$

Then (2.11) implies $\varkappa > 0$. Further, in view of (2.10), we can take an M-conic neighborhood V_∞ of $\gamma(b) \in \Gamma$ such that $v(x)$ is microlocally in $H^{M,t+m-1}$ at every point of V_∞.

Then, as usual, our main theorem is obtained by repeated application of the following

Proposition 5.1. *In addition to the above assumption, assume that* $v(x)$ *is microlocally in* $H^{M,r}$ *at every point of* V_2, *where* V_2 *is an open* M*-conic neighborhood of* Γ *contained in* V_1. *Then there exists another open* M*-conic neighborhood* V_3 *of* Γ *contained in* V_2 *such that* $v(x)$ *is microlocally in* $H^{M,r'}$ *at every point of* V_3, *where* $r' = \min\ \{\ r+\varkappa,\ t+m-1\ \}$.

In the proof we use the following lemma, which is a modification of Lemma 6.4 of Bony [1].

Lemma 5.2. *For every open M-conic neighborhood V' of γ contained in V_1, we can take a function $A(x,\xi) \in C^\infty(V')$ satisfying*

(5.9) $A(x,t^M\xi) = A(x,\xi)$ *for* $t \geq 1$ *and* $|\xi| \geq 1$,

(5.10) $A(x,\xi) > 0$ *on a conic neighborhood of* Γ,

(5.11) $XA \geq 0$ *on a conic neighborhoof of* $V'\backslash V_\infty$,

and

(5.12) *The set* $\{ (x,\xi) \in V', |\xi| = 1, A(x,\xi) > 0 \}$ *is compact.*

Proof. By dividing Γ into finite parts, taking $V'' \subset V'$ sufficiently small, renumbering the coordinate variables and changing the orientation if necessary, we may assume $V'' \subset D_n = \{ (x,\xi); \xi_n > 0 \}$. Now put $P_n = \{ (x,\xi); \xi_n = 1 \}$ and define a mapping τ from D_n onto P_n by the formula $\tau(x,\xi) = (x, \left(\xi_n^{-1/m_n}\right)^M \xi)$. Then, since X is not proportional to X_0 on $V'' \subset V'$, the vector field $Y = \tau_* X$ does not vanish on $\tau(V'')$.

For a positive number ε, we say that a piecewise C^1-curve $Z(t)$ in P_n is an *ε-integral curve of* Y if $|Z'(t)-Y(Z(t))| < \varepsilon$ holds for almost every t. For $z \in P_n$, let $T_\varepsilon(z)$ denote the supremum of such numbers T that there exists an ε-integral curve $\{Z(t); 0 \leq t \leq T\}$ of Y satisfying $|Z(0)-\tau(\gamma(a))| < \varepsilon$ and $Z(T) = z$. If such an ε-integral curve does not exist, put $T_\varepsilon(z) = 0$.

Now identify P_n with $\mathbb{R}_z^{2n-1}$, and introduce a function $\chi_0(z) \in C_0^\infty(\mathbb{R}^{2n-1})$ satisfying $\chi_0(z) \geq 0$ and $\int_{\mathbb{R}^{2n-1}} \chi_0(z)dz = 1$. Then, shrinking V'' and taking ε and δ sufficiently if necessary, we can verify in the same way as in the proof of Lemma 6.4 of Bony [1] that the function $A_0(z) = \chi_1(z)\int_{\mathbb{R}^{2n-1}} T_\varepsilon(z-\delta\zeta)\chi_0(\zeta)d\zeta$ satisfies

$$\begin{cases} A_0(z) > 0 \text{ on a neighborhood of } \tau(\Gamma), \\ YA_0 \geq 0 \text{ on } \tau(V''), \\ K = \operatorname{supp} A_0(z) \cap \left(\tau(V''\backslash V_\infty)\right) \text{ is a compact subset of } \tau(V''). \end{cases}$$

Now let $\chi_1(z) \in C_0^\infty(\tau(V''))$ be a function satisfying $\chi_1(z) \equiv 1$ on a neighborhood of K, and let $\chi_2(t)$ be a smooth function on $\mathbb{R}$ such that $\chi_2(t) \equiv 0$ on $]-\infty,4/3]$ and that $\chi_2(t) \equiv 1$ on $[3/2,\infty[$. Then we put $A(x,\xi) = \chi_2(\langle\xi\rangle)\chi_1(\tau(x,\xi))A_0(\tau(x,\xi))$. Then, for $(x,\xi) \in V''\backslash V_\infty$ satisfying $\langle\xi\rangle \geq 2$, we have

$$X(A(x,\xi)) = X(\tau^* A_0(x,\xi)) = \Big((\tau_* X)A_0\Big)(\tau(x,\xi)) = (YA_0)(\tau(x,\xi)) > 0.$$

It is easy to see that this function $A(x,\xi)$ satisfies other conditions. The proof of Lemma 5.2 is now complete.

Proof of Proposition 5.1. We follow the idea of Hörmander [4]. Replacing r by $t+m-1-\mu$ if necessary, we may assume

(5.13) $r' = r+\varkappa \leq t+m-1 \leq \mu(s)+m-1.$

First, fix a symbol $Q(x,\xi) \in S^{M,0}$ with support contained in V_1 such that $Q(x,\xi) \equiv 1$ on some open conic neighborhood V_3 of Γ, and put $Q_\varepsilon(x,\xi) = Q(x,\xi)\langle\varepsilon^M\xi\rangle^{-\varkappa} \in S^{M,-\varkappa}$ for a positive parameter ε. Next, apply Lemma 5.2 to obtain a symbol $A(x,\xi)$ satisfying (5.9), (5.10), (5.11) and (5.12) with $V' = V_3$. Now we put

$$S_\lambda(x,\xi) = A(x,\xi)\exp\Big(\lambda B(x,\xi)\Big)\langle\xi\rangle^{r+\varkappa+(1-m)/2} \in S^{M,r+\varkappa+(1-m)/2}$$

for a positive parameter λ. Further, we put $w_\varepsilon = \pi_1(Q_\varepsilon(x,D),v)$ and $w_{\lambda,\varepsilon} = \pi_1(S_\lambda(x,D),w_\varepsilon)$. Then Proposition 3.4 yields $w_\varepsilon \in H^{M,r+\varkappa}$ and $w_{\lambda,\varepsilon} \in H^{M,(m-1)/2}$. We also remark the following:

(5.14) $\{Q_\varepsilon(x,\xi);\ 0 < \varepsilon \leq 1\}$ is bounded in $S^{M,0,\sigma}$ for every $\sigma \in \mathbb{R}$.

From this fact, Proposition 3.4 and Remark 3.6 we obtain

(5.15) $\{\ \pi_1\Big(Q_\varepsilon(x,D),\cdot\Big);\ 0 < \varepsilon \leq 1\ \}$ is bounded with respect to the operator norm on $H^{M,\tau}$ for every $\tau \in \mathbb{R}$. In particular, the set $\{\ w_\varepsilon;\ 0 < \varepsilon \leq 1\ \}$ is bounded in $H^{M,r}$.

In the following each C_j denotes a constant independent of λ and ε, and each $C'_{j,\lambda}$ denotes a constant depending on λ but not on ε.

Now we put $h(x) = \sum_{b>m-1} \pi_1(R_b(x,D),w_{\lambda,\varepsilon})(x)$. Then we have

$$h(x)+i\pi_1\Big(\{R_m,S_\lambda\}(x,D),w_\varepsilon\Big)(x) = \sum_{j=1}^{5} I_j(x),$$

where

$$I_1 = \sum_{b\in\Lambda', b\leq m-\varkappa} \left\{\pi_1(R_b(x,D), w_{\lambda,\varepsilon}) - \pi_1\left(S_\lambda(x,D), \pi_1(R_b(x,D), w_\varepsilon)\right)\right\},$$

$$I_2 = \pi_1(R_m(x,D), w_{\lambda,\varepsilon}) - \pi_1\left(S_\lambda(x,D), \pi_1(R_m(x,D), w_\varepsilon)\right) + i\pi_1\left(\{R_m, S_\lambda\}(x,D), w_\varepsilon\right),$$

$$I_3 = \sum_{b\in\Lambda'} \pi_1\left(S_\lambda(x,D), \{\pi_1(R_b(x,D), w_\varepsilon) - \pi_1(Q_\varepsilon(x,D), \pi_1(R_b(x,D), v))\}\right),$$

$$I_4 = \pi_1\left(S_\lambda(x,D), \pi_1(Q_\varepsilon(x,D), g)\right)$$

and

$$I_5 = -\sum_{b\leq m-1, b\in\Lambda'} \pi_1(R_b(x,D), w_{\lambda,\varepsilon}).$$

We now estimate $\|I_j(x)\|_{M,(1-m)/2}$ for every $j = 1,2,3,4,5$.

First, the inequality $\|I_5\|_{M,(1-m)/2} \leq C_1\|w_{\lambda,\varepsilon}\|_{M,(m-1)/2}$ follows from Proposition 3.4 and (5.4).

Next, Proposition 3.4 and (5.15) yield the estimate $\|I_4\|_{M,(1-m)/2} \leq C'_{2,\lambda}\|g\|_{M,r+\varkappa+1-m}$, and (5.5) and (5.13) imply that $\|g\|_{M,r+\varkappa+1-m}$ is dominated by a constant independent of ε.

On the other hand, since $Q(x,\xi) \equiv 1$ on supp $S(x,\xi)$, it follows from (5.4), (5.14), Proposition 3.5 and Remark 3.6 that we have $\|I_3\|_{M,(1-m)/2} \leq C'_{3,\lambda}\|v\|_{M,1-m+r+\varkappa-\mu(s)+s}$. It follows from the condition (5.13) and the fact $v \in H^{M,s}$ that the right-hand side is dominated by a constant independent of ε.

Also, by virtue of Proposition 3.5 and Proposition 3.4, we see that the estimate $\|I_j\|_{M,(1-m)/2} \leq C'_{4,\lambda}\|w_\varepsilon\|_{M,r}$ holds for $j = 1,2$.

From these facts and (5.15) we conclude

$$(5.16)\quad \|h(x) + i\pi_1\left(\{R_m, S_\lambda\}(x,D), w_\varepsilon\right)(x)\|_{M,(1-m)/2} \leq C_1\|w_{\lambda,\varepsilon}\|_{M,(m-1)/2} + C'_{5,\lambda}.$$

We now turn to the calculation of $Im(h|w_{\lambda,\varepsilon})$. First, Proposition 3.7 yields

$$2Im(h|w_{\lambda,\varepsilon}) = \sum_{b>m-1}\left\{\left(\pi_1(R_b(x,D), w_{\lambda,\varepsilon})\middle| w_{\lambda,\varepsilon}\right) - \left(w_{\lambda,\varepsilon}\middle|\pi_1(R_b(x,D), w_{\lambda,\varepsilon})\right)\right\}$$

$$= \sum_{b>m-1}\left(\pi_1(2Im\, R_b(x,D), w_{\lambda,\varepsilon})\middle| w_{\lambda,\varepsilon}\right) + I_6\ ,$$

where $|I_6| \leq C_6\|w_{\lambda,\varepsilon}\|^2_{M,(m-1)/2}$.

On the other hand, in view of (2.8) and (5.8), we can write

$$\sum_{b>m-1} Im\, R_b(x,\xi) = \sum_{\nu=1}^N \tilde{R}_\nu(x,\xi) = \sum_{\nu=1}^N \omega(x)\psi(\xi)\cdot Im \sum_{\beta\in I_\nu}\left(a_\beta(x)(i\xi)^\beta\right),$$

and $\tilde{R}_\nu(x,\xi) \geq 0$ follows from the choice of $\omega(x)$ and $\psi(\xi)$ and the positivity condition in (2.8). In view of this fact and the Hölder-type condition in (2.8), application of Proposition 4.1 to each $\tilde{R}_\nu(x,\xi)$ yields the estimate

$$(5.17)\quad Im(h|w_{\lambda,\varepsilon}) \geq -C_7\|w_{\lambda,\varepsilon}\|^2_{M,(m-1)/2}\,.$$

From (5.16) and (5.17) we conclude

$$(5.18)\quad -Re\Big(\pi_1\Big(\{R_m,S_\lambda\}(x,D),w_\varepsilon\Big)\Big|w_{\lambda,\varepsilon}\Big) = Im\Big(-i\pi_1\Big(\{R_m,S_\lambda\}(x,D),w_\varepsilon\Big)\Big|w_{\lambda,\varepsilon}\Big)$$

$$\geq -C_8\|w_{\lambda,\varepsilon}\|^2_{M,(m-1)/2} - C'_{9,\lambda}\|w_{\lambda,\varepsilon}\|_{M,(m-1)/2}\,.$$

We now calculate $Re\Big(\pi_1\Big(-\{R_m,S_\lambda\}(x,D)+\lambda' S_\lambda(x,D)\langle D\rangle^{m-1},w_\varepsilon\Big)\Big|w_{\lambda,\varepsilon}\Big)$, where λ' is a constant determined later. Applying Propositions 3.4, 3.5 and 3.7, we obtain

$$(5.19)\quad Re\Big(\pi_1\Big(-\{R_m,S_\lambda\}(x,D)+\lambda' S_\lambda(x,D)\langle D\rangle^{m-1},w_\varepsilon\Big)\Big|w_{\lambda,\varepsilon}\Big)$$

$$= Re\Big\{\Big(\pi_1\Big(S_\lambda(x,D),\pi_1(-\{R_m,S_\lambda\}(x,D)+\lambda' S_\lambda(x,D)\langle D\rangle^{m-1},w_\varepsilon)\Big)\Big|w_\varepsilon\Big)+I_7\Big\}$$

$$= Re\Big\{\Big(\pi_1\Big((-1/2)\{R_m,S_\lambda{}^2\}(x,D)+\lambda' S_\lambda{}^2(x,D)\langle D\rangle^{m-1},w_\varepsilon\Big)\Big|w_\varepsilon\Big)+I_7+I_8\Big\}$$

$$= \Big(\pi_1\Big((-1/2)Re\ \{R_m,S_\lambda{}^2\})(x,D)+\lambda' S_\lambda{}^2(x,D)\langle D\rangle^{m-1},w_\varepsilon\Big)\Big|w_\varepsilon\Big)+Re\ I_7+I_8+I_9,$$

where the estimate $|I_7|,|I_8|,|I_9| \leq C'_{10,\lambda}\|w_\varepsilon\|^2_{M,r}$ holds with a constant $C'_{10,\lambda}$ independent of ε.

We turn to the calculation of the partial Poisson bracket. First, we have the equality

$$-Re\ \{R_m,S_\lambda{}^2\}(x,\xi) = -X\Big(A(x,\xi)^2\exp(2\lambda B(x,\xi))\langle\xi\rangle^{2r+2\varkappa+1-m}\Big)$$

$$= -A(x,\xi)\exp(2\lambda B(x,\xi))\langle\xi\rangle^{2r+2\varkappa+1-m}\times$$

$$\Big(2X(A)(x,\xi)+2\lambda X(B)(x,\xi)A(x,\xi)+(2r+2\varkappa+1-m)\langle\xi\rangle^{-1}X(\langle\xi\rangle)A(x,\xi)\Big).$$

Here $X(A)(x,\xi) \geq 0$ holds on a conic neighborhood of $V'\backslash V_\infty$. On the other hand, $X(B)(x,\xi) \geq \langle\xi\rangle^{m-1}$ and $X(\langle\xi\rangle) \leq C_{11}\langle\xi\rangle^m$ hold on a conic neighborhood of supp $A(x,\xi)$. Hence there exists a symbol $T_\lambda(x,\xi) \in S^{M,r+\varkappa+(m-1)/2,\mu(s)-s+m-1} \subset S^{M,r+\varkappa+(m-1)/2,4\varkappa}$ with support contained in V_∞ such that the symbol $U_\lambda(x,\xi)$ defined by

(5.20)
$$U_\lambda(x,\xi) = (-1/2)Re\ \{R_m, S_\lambda{}^2\}(x,\xi) + S_\lambda(x,\xi)T_\lambda(x,\xi) + (\lambda - C_{12})S_\lambda{}^2(x,\xi)\langle\xi\rangle^{m-1}$$

satisfies $U_\lambda(x,\xi) \le 0$ and $U_\lambda(x,\xi) \in S^{M,2r+2\varkappa,4\varkappa}$. Hence, in view of the inequality $\varkappa \le 1/4$, we can apply Proposition 4.1 again to obtain $\left(\pi_1(U_\lambda(x,D),w_\varepsilon)|w_\varepsilon\right) \le C'_{13,\lambda}\|w_\varepsilon\|^2_{M,r}$. Putting $\lambda' = \lambda - C_{12}$, combining this formula with (5.18), (5.19) and (5.20) and observing (5.15), we conclude

$$(5.21)\quad (\lambda - C_{12})Re\left(\pi_1(S_\lambda(x,D)\langle D\rangle^{m-1}, w_\varepsilon)\middle| w_{\lambda,\varepsilon}\right) - \left|\left(\pi_1((S_\lambda T_\lambda)(x,D),w_\varepsilon)\middle| w_\varepsilon\right)\right|$$
$$\le C'_{14,\lambda} + C_8\|w_{\lambda,\varepsilon}\|^2_{M,(m-1)/2} + C'_{9,\lambda}\|w_{\lambda,\varepsilon}\|_{M,(m-1)/2}\ .$$

Applying Propositions 3.3, 3.4 and 3.5, we have

$$(5.22)\quad \|\pi_1(S_\lambda(x,D)\langle D\rangle^{m-1},w_\varepsilon) - \langle D\rangle^{m-1}w_{\lambda,\varepsilon}\|_{M,(1-m)/2} \le C'_{15,\lambda}\|w_\varepsilon\|_{M,r}.$$

On the other hand, Propositions 3.4 and 3.7 yield

$$(5.23)\quad \left|\left(\pi_1((S_\lambda T_\lambda)(x,D),w_\varepsilon)\middle| w_\varepsilon\right) - \left(\pi_1(T_\lambda(x,D),w_\varepsilon)\middle| w_{\lambda,\varepsilon}\right)\right| \le C'_{16,\lambda}\|w_\varepsilon\|^2_{M,r}.$$

Further, the condition (5.13) and Propositions 3.4 and 3.5 imply

(5.24)
$$\|\pi_1(T_\lambda(x,D),w_\varepsilon) - \sum_{M\cdot\alpha\le\mu(s)-s+m-1}\frac{1}{\alpha!}\,\pi_1\left(((\partial_\xi^\alpha T_\lambda)\cdot(D_x^\alpha Q_\varepsilon))(x,D),v)\right)\|_{M,(1-m)/2}$$
$$\le C'_{17,\lambda}\|v\|_{M,r+\varkappa-\mu(s)+s-m+1} \le C'_{17,\lambda}\|v\|_{M,s}\ .$$

Finally, the estimate

$$(5.25)\quad \|\pi_1\left(((\partial_\xi^\alpha T_\lambda)\cdot(D_x^\alpha Q_\varepsilon))(x,D),v)\right)\|_{M,r'-r+\varkappa+(1-m)/2} \le C'_{18,\lambda}$$

follows from the choice of V_∞ and the support condition of T_λ.

Substituting (5.22)-(5.25) into (5.21) and observing (5.15), we conclude

$$(\lambda - C_{12})\|w_{\lambda,\varepsilon}\|^2_{M,(m-1)/2} \le C_8\|w_{\lambda,\varepsilon}\|^2_{M,(m-1)/2} + C'_{19,\lambda}\|w_{\lambda,\varepsilon}\|_{M,(m-1)/2} + C'_{20,\lambda}.$$

Taking $\lambda > C_8 + C_{12}$, we see that $\|w_{\lambda,\varepsilon}\|^2_{M,(m-1)/2} \le C'$ holds with a constant C' independent of ε. However, $w_{\lambda,\varepsilon}$ converges to $\pi_1\left(S_\lambda(x,D),\pi_1(Q(x,D),v)\right)$ in $\mathcal{S}'$. Hence this limit belongs to $H^{M,(m-1)/2}$. Now the conclusion follows from Propositions 3.3, 3.4 and the construction of the symbols $S_\lambda(x,\xi)$ and $Q(x,\xi)$.

References

[1] J. M. Bony, Calcul symbolique et propagation des singularités pour les équations aux dérivées partielles non linéaires, Ann. Scient. Ec. Norm. Sup., 4ème série, 14 (1981), 209-246.

[2] A. Córdoba and C. Fefferman, Wave packets and Fourier integral operators, Comm. Partial Diff. Eq., 3 (1978), 976-1005.

[3] P. Godin, Lectures at the symposium "Microlocal Analysis and Partial Differential Equations" held at Mathematische Forschungsinstitut Oberwolfach, 1983.

[4] L. Hörmander, On the existence and the regularity of solutions of linear pseudo-differential equations, L'Enseignement Math., 17 (1971), 99-163.

[5] T. Kobayashi and G. Nakamura, Singular solutions for semilinear hyperbolic equations I, preprint.

[6] H. Kumano-go, Pseudo-Differential Operators, MIT Press, Cambridge and London, 1981.

[7] R. Lascar, Propagation des singularités des solutions d'équations pseudo-différentielles quasi homogènes, Ann. Inst. Fourier, Grenoble, 27 (1977), 79-123.

[8] Y. Meyer, Remarques sur un théorème de J. M. Bony, Supplemento ai rendiconti del Circolo matematico di Palermo, serie II, 1 (1981), 1-20.

[9] T. Sakurai, Propagation of singularities of solutions to semilinear Schrödinger equations, Proc. Japan Acad., 61 (1985), Ser. A, 31-34.

[10] ———, Propagation of regularities of solutions to semilinear partial differential equations of quasi-homogeneous type, to appear in J. Fac. Sci. Univ. Tokyo, IA.

[11] M. Yamazaki, Régularité microlocale quasi homogène des solutions d'équations aux dérivées partielles non linéaires, C. R. Acad. Sc. Paris, 298 (1984), Série I, 225-228.

[12] ———, The L^p-boundedness of pseudodifferential operators satisfying estimates of parabolic type and product type. II, Proc. Japan Acad. 61 (1985), 95-98.

[13] ———, A quasi-homogeneous version of paradifferential operators. I, Boundedness on spaces of Besov type, J. Fac. Sci. Univ. Tokyo IA, 33 (1986), 131-174; II, A symbol calculus, to appear in the same journal.

[14] ———, A Quasi-homogeneous version of the microlocal analysis for nonlinear partial differential equations, preprint.

SELBERG TRACE FORMULAE, ΨDO'S AND EQUIDISTRIBUTION THEOREMS FOR CLOSED GEODESICS AND LAPLACE EIGENFUNCTIONS

Steve Zelditch

0. Introduction.

The upper half-plane $H = \{z \in \mathbb{C} : \text{Im } z > 0\}$ has a metric of constant negative curvature: $ds^2 = y^{-2}(dx^2 + dy^2)$. The group $G = PSL_2(\mathbb{R}) = SL_2(\mathbb{R})/\pm I$ acts transitively on H by the isometries: $gz = \frac{az+b}{cz+d}$, $g = \begin{pmatrix} a & b \\ c & d \end{pmatrix}$. Thus H may be identified with the rank one symmetric space, G/K where $K = SO(2)$ is the isotropy subgroup of i $(k \cdot i = i$ for $k \in K)$. Associated to the above (Poincare) metric on H is the Laplacian $\Delta = y^2(\partial^2/\partial x^2 + \partial^2/\partial y^2)$. If we view $C^\infty(H)$ as the right K-invariant elements of $C^\infty(G)$ then Δ may be identified with the Casimir operator Ω acting in this subspace (Ω is the Laplacian for the (Lorentz) metric on G coming from the Cartan-Killing form).

Now let Γ be a discrete subgroup of G. Γ may be viewed as a non-Euclidean lattice. Since Γ acts isometrically, the Poincare metric and Laplacian descend to the quotient $X_\Gamma = \Gamma\backslash H = \Gamma\backslash G|K$. One says Γ is co-compact if X_Γ is compact or co-finite if X_Γ has finite area. We will assume here that Γ is co-compact although Γ co-finite would really suffice. X_Γ is then a compact smooth surface of constant negative curvature, the higher genus analogue of the flat torus $\mathbb{C}/\Lambda$ where Λ is a lattice $Z \oplus Z\tau$ in $\mathbb{C}$. The assumption on Γ is equivalent to requiring that all $\gamma \in \Gamma$ are hyperbolic, i.e. diagonalizable over $\mathbb{R}$. It is not so simple to write down explicit examples of such Γ. One class of examples (the arithmetic ones) are the quaternion groups (cf. [I])

$$\Gamma(n,p) = \left\{ \begin{bmatrix} A+B\sqrt{n} & (C+D\sqrt{n})\sqrt{p} \\ (C-D\sqrt{n})\sqrt{p} & A-B\sqrt{n} \end{bmatrix} ; A,B,C,D \in Z,\ A^2-nB^2-pC^2+npD^2 = \hat{1} \right\}$$

which are hyperbolic if $p \equiv 1 \pmod 4$.

However, every metric on a surface X of genus $g \geq 2$ is conformally equivalent to a metric of constant curvature, and this metric arises from the Poincare metric when X is identified with $\Gamma\backslash H$ $(\Gamma = \pi_1(X))$.

Good references for background material are Hejhal's books [He] and Iwaniec's survey articles (see [I]). The references [H] and [L] contain introductory material on harmonic analysis and representation theory for $SL_2(\mathbb{R})$.

We would now like to concentrate on the following data associated to X_Γ:

(A) The eigenvalues $-\lambda_k$ and orthonormal eigenfunctions u_k of Δ. Set Δ-sp $= \{-\lambda_k\}$.

(B) The closed geodesics γ and their lengths L_γ. Set L-sp $= \{L_\gamma\}$.

Here, γ denotes a curve in $S^*X_\Gamma = \Gamma|G$, namely a periodic orbit of the geodesic flow G^t (right translation by $a_t = \begin{pmatrix} e^{t/2} & 0 \\ 0 & e^{-t/2} \end{pmatrix}$). There is a one-one correspondence between closed geodesics γ and conjugacy classes of hyperbolic elements in Γ: namely a hyperbolic element fixes (not pointwise) a geodesic $\bar{\gamma}$ in H (its axis), and conjugate γ's yield $\bar{\gamma}$'s which project to the same geodesic in X_Γ. A hyperbolic element diagonalizes to $\begin{bmatrix} e^{L_\gamma/2} & 0 \\ 0 & e^{-L_\gamma/2} \end{bmatrix}$, L_γ being the length of the associated closed geodesic, so L_γ may be considered a conjugacy class invariant.

The problems we consider in this paper and in the references ([Z1]-[Z4]) involve the asymptotic behaviour of the above data (A) and (B) as eigenvalues or lengths tend to ∞. The theme we will outline below is that the eigenfunctions and closed geodesics become uniformly distributed relative to Liouville measure on S^*X_Γ (i.e. Haar measure on $(\Gamma\backslash G)$ as the spectral parameter tends to ∞. This theme is not new. In the case of closed geodesics, the equidistribution theorem was proved by Bowen ([B]) in 1972. In the case of eigenfunctions, the theorem was at least announced by A.I. Snirelman [Sn] in 1974; he elsewhere sketched an incomplete proof. What will be new here is the technique of applying certain generalizations of Selberg's trace formula (STF) to these problems. This will give much more precise versions of Bowen's theorem (rates of equidistribution, equidistribution theorems for closed geodesics satisfying constraints). The technique also leads to a fairly easy proof of the equidistribution theorem for eigenfunctions on finite area surfaces. Actually, in the compact case the detailed trace formulae here are not really necessary for the equidistribution theory of eigenfunctions, and the proof extends readily to compact manifolds with ergodic geodesic flow (cf. [C de V]). As yet, we do not know how to improve this result (as in the geodesic case) to get a rate of equidistribution. This may however be expected to be a hard problem, since in special cases, some well known conjectures on the behaviour of so-called Rankin-Selberg zeta functions are essentially equivalent to giving such decay rates.

The trace formula we will generalize below is the following (STF):

$$\sum_n h(r_n) = (\mathrm{id}) + (\mathrm{hyp})$$

where

$$(\mathrm{id}) = \frac{A}{4\pi}\int_{-\infty}^{\infty} h(r)\, r\, \mathrm{th}\, \pi r \, dr, \qquad (\ A = \text{area}\ )$$

$$(\text{hyp}) = \sum_{\{\gamma\}} \frac{L_{\gamma 0}}{\text{sh}L_\gamma/2}\,\hat{h}(L_\gamma),$$

$$h \in C_0^\infty(R),\ h \text{ even},\ \hat{h} = \text{fourier transform of } h$$

$$\lambda_n = \frac{1}{4} + r_n^2$$

γo = primitive closed geodesic corresponding to γ (once around).

Two standard applications of (STF) are:

(A) Weyl law with remainder: $N(\lambda) = \#\{\sqrt{\lambda_k} \le \lambda\} = \frac{A}{4\pi}\lambda^2 + R(\lambda)$ where A = area of X_Γ and $R(\lambda) << \lambda/\log\lambda$.

(b) Prime geodesic theorem: $\Psi_\Gamma(T) = \sum_{L_\gamma \le T} L_{\gamma o} = \sum_{j=0}^{M} \frac{e^{(1/2+t_j)T}}{1/2 + t_j} + O(Te^{3/4T})$.

Here $r_j = it_j$ for $0 \le \lambda_j < 1/4$.

A basic principle is that the behaviour of spectral functions as $r_k \to \infty$ is controlled by small lengths (principally the (id) term) while the behaviour of geodesic functions is controlled by small eigenvalues ($r_j \in iR$).

In this (mainly expository) paper we will be interested in the functions:

$$N(\lambda,\sigma) = \sum_{\sqrt{\lambda_k} \le \lambda} (O\rho\ (\sigma)u_k, u_k)$$

$$S(\lambda, X,\sigma) = \sum_{\sqrt{\lambda_k} \le \lambda} (O\rho(\sigma)u_k, u_k)X^{i\sqrt{\lambda_k}}$$

$$|N|^\kappa(\lambda,\sigma) = \sum_{\sqrt{\lambda_k} \le \lambda} |(O\rho(\sigma)u_k, u_k)|^k$$

(B)

$$\Psi_\Gamma(T,\chi,\sigma) = \sum_{L\gamma \le T} \chi(\gamma) \int_{\gamma o} \sigma \quad (\chi : \Gamma \to U(1))$$

σ is assumed to be an automorphic form in the precise sense that $\sigma \in C^\infty(S^*X_\Gamma) = C^\infty(\Gamma\backslash G)$,

$$\begin{cases} \Omega\sigma = (s-1)(s+1)\sigma \\ \frac{1}{i}W\sigma = m\sigma \end{cases}$$

Here Ω is the Casimir on $\Gamma\backslash G$ and $W \sim \begin{bmatrix} 0 & 1 \\ -1 & 0 \end{bmatrix}$ generates K. Such a σ is a form of weight m lying in an irreducible $H(s)$ for G. We will always assume $\sigma \not\equiv 1$, so that $\langle \sigma, 1 \rangle = \int_{\Gamma\backslash G} \sigma = 0$. In fact we will assume further that σ is one of: $\{u_k, X_+u_k, \Psi_m\}$, $X_+ \sim \begin{bmatrix} 0 & 1 \\ 0 & 0 \end{bmatrix}$ and Ψ_m is a holomorphic form. These generate $L^2(\Gamma\backslash G)$

over G^t.

THEOREM A.

(i) $N(\lambda,\sigma) \ll \lambda/\log \lambda$

$$(ii)\ S(\lambda,X,\sigma) = \begin{cases} \left[\dfrac{\lambda}{1-e^{-L_\gamma}}\right]\int_{\gamma_o}\sigma + O_X(\lambda/\log \lambda) & \text{if } X = e^{L_\gamma} \\ O_X(\lambda\ /\log \lambda) & \text{if } X \neq e^{L_\gamma}) \end{cases}$$

(iii) $\frac{1}{\lambda^2}|N|^k(\lambda,\sigma) = o(1)$

THEOREM B.

(i) There exists constants $\gamma^j_{s,m,\chi}$ so that

$$\Psi_\Gamma(T,\chi,\sigma) = \sum_{j=1}^{M} (Op(\sigma)u_j^\chi, u_j^\chi)\gamma^j_{s,m,\chi}\ \frac{e^{(1/2+t_j^\chi)T}}{1/2+t_j^\chi} + O(Te^{3/4T})$$

where u_k^χ are the eigenfunctions of Δ on $L^2_\chi(\Gamma\backslash G/K) = \{u:u(\gamma z) = \chi(\gamma)u(z)\}$, and the sum over j runs over the small eigenvalues $r_j^\chi = it_j^\chi$ of Δ on L^2_χ.

(ii) $\frac{1}{\Psi_\Gamma(T)}\,\Psi_\Gamma(T,1,\sigma) \sim e^{(t_1-1/2)T}$

(iii) $\frac{1}{\Psi_\Gamma(T)}\,\Sigma_{L_\gamma}\leq T\ |\int_{\gamma_o}\sigma| = o(1)$

We remark that all of Theorem A is probably valid on compact negatively curved manifolds. However Theorem B has a more subtle generalization (due to Bowen and Margulis): geodesics become equidistributed relative to a measure (the measure of maximal entropy) which is typically singular relative to Liouville measure. It is unknown to what extent the trace formula methods used here could be generalized to the variably curved case.

This paper is organized as follows. In §1, we outline the derivation of the trace formulae for co-compact Γ. Full details on this will appear in [Z2]. In §2 we sketch the applications to Theorems A and B for the case $\sigma = u_k$. This is of course the simplest case, but the method of proof is more or less the same for all σ and there is no space for a more complete treatment here. Full details on Theorem A appear in [Z1] and [Z5], on Theorem B in [Z2], [Z3].

Finally, we would like to thank P. Sarnak for very helpful conversations on trace formulae, prime geodesic theorems, etc. In particular, the proof of Theorem

(B) is based on Sarnak's proof of the prime geodesic theorem ([Sa]). Moreover, Theorems A(i),(ii) are analogues of well known theorems in analytic number theory ([T],3.9), and of the case $\sigma \equiv 1$ which Sarnak pointed out to us. However, responsibility for the accuracy of the results rests with solely the author; since they are all very recent, this responsibility is heavier than usual.

1. Trace Formulae.

In this section we will state the specialized trace formulae we will be using, and give some details on their derivation. Roughly speaking, the formulae read:

(TF) $$\sum_{r_k} (\mathrm{Op}(\sigma)u_k, u_k)h(r_k) = \sum_{\{\gamma\}} \int_{\gamma_0} \sigma \; Mh(L_\gamma)$$

where σ is an automorphic form, $\mathrm{Op}(\sigma)$ is the ΨDO with complete symbol σ in the sense of [Z6], and M is an explicit transform depending only on the weight and Casimir eigenvalue of σ. The special shape of the formula resides in the dependence of the right side only on the geodesic integrals and on the eigenvalue parameters of σ. (Strictly speaking, this requires an assumption since another term involving $\int_{\gamma_0} X_+\sigma$ could appear. But it is correct for the generating cases $\sigma = \{u_k, X_+u_k, \Psi_m\}$.)

We begin by recalling the following definitions, facts, and notation:

(a) $G = PSL_2(R) = ANK; A = \{a_t = \exp tH\}$, $N = \{n_u = \exp uX_+\}$, $K = \{k_\theta = \exp \theta W\}$,

$$H = \begin{bmatrix} 0 & 1 \\ -1 & 0 \end{bmatrix}, \quad X_+ = \begin{bmatrix} 0 & 1 \\ 0 & 0 \end{bmatrix}, \quad W = \begin{bmatrix} 0 & 1 \\ -1 & 0 \end{bmatrix}.$$

(b) R_g = right translation by g on $\Gamma\backslash G$, or on $L^2(\Gamma\backslash G)$. $G^t = R_{a_t}$. For $\phi \in C_0^\infty(G), R_\phi = \int_G \phi(g)R_g dg$.

(c) $S_{m,n} = \{\phi \in C(G), \phi(k_{\theta_1} g\, k_{\theta_2}) = e^{im\theta_1}\phi(g)e^{in\theta_2}\}$.

(d) $L^2(\Gamma\backslash G) = \oplus_{s_k} H(s_k) = C \oplus$ [continuous series] + [discrete series]; if $H(s_k)$ belongs in the continuous series, it has a unique K-invariant vector u_{s_k} (or u_k for short). If the Casimir $\Omega = (s_k-1)(s_k+1)$ on $H(s_k)$ then $\Delta u_k = (s_k-1)(s_k+1)u_k$ (this is four times the usual Δ). If $H(s_m)$ is in the discrete series it has a lowest (or highest) weight vector $\Psi_m, \frac{1}{i}W\,\Psi_m = m\Psi_m$ $(s_m = m-1)$.

(e) $\sigma_{s,m}$ will denote a vector in $H(s)$ of weight m (weight = eigenvalue of $\frac{1}{i}W$).

(f) $G \approx D \times B$ where D is the unit disc and B is the boundary. Using the Helgason-Fourier transform on D, one can define the ΨDO $\mathrm{Op}(\sigma)$ with symbol $\sigma \in C^\infty(D\times B)$ to be a properly supported operator with kernel:

$$\int_0^\infty \int_B \sigma(z,b)e^{(i\lambda+1)<z,b>} e^{(-i\lambda+1)<w,b>} \lambda \,\mathrm{th}\, \frac{\pi\lambda}{2} d\lambda \; db$$

(see [H] for notation). $\mathrm{Op}(\sigma)$ commutes with left translation by $\gamma \iff \sigma(\gamma z, \gamma b) = \sigma(z,b)$, i.e. $\sigma \in C^\infty(\Gamma\backslash D \times B) = C^\infty(\Gamma\backslash G) = C^\infty(S^*X_\Gamma)$.

A basic observation [Z3] is that operators of the form $\mathrm{Op}(\sigma)f(\Delta)$ (or more generally $Op(\chi\sigma)f(\Delta)$ where $\chi(\delta(z,w))$ is a properly supported cutoff, $\delta(\cdot,\cdot)$ = distance) are exactly the operators of the form σR_ϕ, where σ has weight m and $\phi \in S_{m,0}$. It is somewhat more convenient to work with the σR_ϕ so henceforth we will emphasize the point of view that ΨDO's on $L^2(X_\Gamma)$ are compositions of a multiplication and a convolution on $L^2(\Gamma\backslash G)$, preserving weight 0.

Our first step may now be stated: generalize the trace formula from the traces $Tr h(R)$ (or $Tr R_\phi$) to the traces $Tr Op(\chi\sigma)h(R)$ (or $Tr\, \sigma\, R_\phi$), where $-\Delta = I + R^2$.

Following the standard proof,

$$\begin{aligned} Tr\, \sigma\, R_\phi &= \sum_{\gamma\in\Gamma} \int_{\Gamma\backslash G} \sigma(x)\phi(x^{-1}\gamma x)dx \\ &= \sum_{\{\gamma\}} \int_{\Gamma_\gamma\backslash G} \sigma(x)\phi(x^{-1}\gamma x)dx \end{aligned}$$

where $\{\gamma\}$ runs over conjugacy classes

$$\begin{aligned} &= \sum_{\{\gamma\}} \int_{G_\gamma\backslash G} \phi(x^{-1}\gamma x) \left[\int_{\Gamma_\gamma\backslash G_\gamma} \sigma(g_1 x)\ dg_1\right] dx \\ &= \sum_{\{\gamma\}} I_{\{\gamma\}}(\phi,\sigma) \\ &= (id) + (hyp), \end{aligned}$$

according to the type of conjugacy class. For γ = id, (id) = $\int_{\Gamma\backslash G} \sigma(x)dx = 0$ if $\sigma = \sigma_{s,m} \perp 1$.

A simple change of variables gives:

$$I_{\{\gamma\}}(\sigma_{s,m},\phi) = \int_{A\backslash G} \phi(x^{-1}\delta_\gamma x) \left[\int_{<\delta_{\gamma o}>\backslash A} \sigma(\alpha_j^{-1}ax)da\right] dx$$

where $\delta_\gamma \in A$ is conjugate to γ, $<\delta_{\gamma o}>$ is the cyclic group generated by the diagonal conjugate of the primitive element, and $\gamma = \alpha_\gamma^{-1}\delta_\gamma\alpha_\gamma$. When σ has weight m and $\phi \in S_{m,0}$ we have:

$$I_{\{\gamma\}}(\sigma_{s,m},\phi) = \int_N \phi(n_u^{-1}\ \delta_\gamma\ n_u) \left[\int_{<\delta_{\gamma o}>\backslash A} \sigma(\alpha_\gamma^{-1}an_u)da\right] dn_u.$$

Consider $I_{\{\gamma\}}(\sigma,u) = \int_{<\delta_{\gamma o}>\backslash A} \sigma(\alpha_j^{-1}an_u)\ da$.

Everything hinges on the fact that the equations $\Omega\,\sigma_{s,m} = \lambda_s\sigma_{s,m}$ (continuous series) or $E^-\sigma_{s,m} = 0$ (discrete series, E^- = lowering operator [L]) imply that $I_{\{\gamma\}}(\sigma,u)$ satisfies an ODE in u. In the continuous series case, $I_{\{\gamma\}}(\sigma_{s,m},u)$ satisfies a hypergeometric ODE, so $I_{\{\gamma\}}(\sigma_{s,m})(u) = (\int_{\gamma o}\sigma_{s,m})\,F_{s,m}(u) + (\int_{\gamma o} X_+\sigma_{s,m})\,G_{s,m}(u)$. In the discrete series case, one gets a first order ODE, and finds that $I_{\{\gamma\}}(\sigma_{s,m},u) = (\int_{\gamma o}\sigma_{s,m})\,(1-iu)^{-m/2}$.

One further has:

(a) if m = 0, then $\phi(n_u^{-1}\delta_\gamma n_u)$ is even in u, $F_{s,m}(u)$ is even, $G_{s,m}(u)$ is odd, so

$$I_{\{\gamma\}}(\sigma_{s,0},\phi) = \left[\iint_{\gamma o}\sigma_{s,0}\right] H_s\phi(\delta_\gamma)$$

where

$$H_s\phi(\delta_\gamma) = \int_{-\infty}^{\infty}\phi(n_u^{-1}\delta_\gamma n_u)\ F_{s,m}(u)du.$$

(b) when m = 2, we can define a kind of conjugate $\sigma^*_{s,m}$ and ϕ^*, so that

$$I_{\{\gamma\}}(\sigma_{s,2},\phi) - I_{\{\gamma\}}(\sigma^*_{s,2},\phi^*) = \left[\iint_{\gamma o} X_+\sigma_{s,0}\right]\int_{-\infty}^{\infty}\phi(n_u^{-1}\delta_\gamma n_u)F^*_{s,2}(u)du.$$

(c) $I_{\{\gamma\}}(\sigma_{m,m},\phi) = \left[\int_{\gamma o}\sigma_{m,m}\right]\int_{-\infty}^{\infty}\phi(n_u^{-1}\delta_\gamma n_u)\ (1 - iu)^{-m/2}du.$

It turns out that the vectors $\sigma_{s,0}$ and $X_+\sigma_{s,0}$ and their A-translates span H(s), while $\sigma_{m,m}$ is cyclic for A in $H(s_m)$. Since we are only interested in geodesic integrals, it will suffice to consider the cases (a)-(c) (see [Z1],[Z2]).

We thus have:

$$\mathrm{Tr}\ \sigma_{s,m}R_\phi = \sum_{\{\gamma\}}\left[\iint_{\gamma o}\sigma\right]H_{s,m}\ \phi(\delta_\gamma)$$

where $H_{s,m}$ is a kind of generalized Harish transform, and where Tr should be interpreted as $\mathrm{Tr}(\sigma R_\phi - \sigma^* R^*_\phi)$ in case (b). We also note that σR_ϕ acts on $L^2_\chi(\Gamma\backslash G)$ = $\{u^\chi \in L^2$ with $u^\chi(\gamma g) = \chi(\gamma)u^\chi(g)$, $\chi : \Gamma \to U(1)\}$. So we can take the trace $\mathrm{Tr}_\chi\sigma R_\phi$ on that space. This is important because $\int_{\gamma o}\sigma + \int_{\gamma_o^{-1}}\sigma = 0$ if σ is odd under $(x,\xi) \to (x,-\xi)$ (up to terms with vanishing integral), while $H_{s,2}\phi(\delta_\gamma) = H_{s,2}\phi(\delta_\gamma^{-1})$. So putting in a character saves some formulae from triviality. It also opens up the possibility of averaging over the character variety, giving equidistribution theorems or closed geodesics in a fixed homology class (this will appear in a forthcoming paper; the idea is taken from a recent work of Phillips-Sarnak on counting closed geodesics in a fixed homology class).

Thus:

$$\mathrm{Tr}_\chi\ \sigma_{s,m}\ R_\phi = \sum_{\{\gamma\}}\chi(\gamma)\int_{\gamma_o}\sigma_{s,m}\ H_{s,m}\phi(\delta_\gamma).$$

On the spectral side, we note that when m = 0, $\mathrm{Tr}_\chi u_k R_\phi = \Sigma_j(u_j u_k^\chi, u_k^\chi)S\phi(s_k)$ where S is the spherical transform. It is well-known ([L]) that S = MH on $S_{0,0} = C_0^\infty(G\backslash\backslash K)$ where M is the Mellin transform and H is the Harish transform, $H\phi(a) = \int_N \phi(n^{-1}an)dn$. We conclude:

$$\sum_k (u_j u_k^\chi, u_k^\chi)\ MH\phi(s_k) = \sum_{\{\gamma\}}\chi(\gamma)\left[\iint_{\gamma o}u_j\right]H_j\phi(\delta_\gamma).$$

One can work out the spectral side in a similar way for cases (b)-(c). We get:

$$\sum_k (Op(X_+ u_j)u_k^X, u_k^X)\ MH_2\phi(s_k) = \sum_{\{\gamma\}} \chi(\gamma) \left[\int_{\gamma o} X_+ u_j\right] H_{j,2}\phi(\delta_\gamma)$$

and

$$\sum_k (Op(\Psi_m)u_k, u_k)\ MH_m\phi(s_u) = \sum_{\{\gamma\}} \left[\int_{\gamma o} \Psi_m\right] H_{m,m}\phi(\delta_\gamma)$$

where $H_k\phi(a) = \int_N \phi(n^{-1}an)dn$ on $S_{k,0}$.

These are the trace formulae. The next step is to choose ϕ appropriately for the applications.

2. Applications.

In this section we will outline proofs of Theorems A(i), (ii) and Theorem B for the case $\sigma = u_k$. The proofs for other automorphic σ involve similar techniques but with the trace formulae of (b) and (c) instead of (a). We will moreover sketch different proofs of Theorem A(i) and Theorem A(ii): the first using a smoothing technique of Berard [Be] (and of [S], see also [E]), the second using a well-known method of analytic number theory.

To begin with, let us state more precisely the relevant trace formulae (for the other σ see [Z2], [Z3]). (TF): For h even, $h \in C_0^\infty$:

$$\sum_{r_n} (u_k u_n, u_n)\ h(r_n) = \sum_{\{\gamma\}} \frac{\int_{\gamma o} u_k}{shL_\gamma/2} M_k h(L_\gamma)$$

where

$$cM_k h(L_\gamma) = \hat{h}(L_\gamma) + \int_{L_\gamma}^\infty \hat{h}(t) \frac{\partial}{\partial t} F(1/4+s_k/4, 1/4-s_k/4, 1, -\left[\frac{ch\ t - ch\ L_\gamma}{(shL_\gamma/2)^2}\right] dt,$$

c being an inessential constant. ([Z2], Theorem 4.1).

This version puts the transform on the right side. This is optimal for applications to spectral sums. For geodesic sums, we must invert M_k. The result is: $(TF)_2$ ([Z3], Proposition 2.10): For $\Psi \in C_0^\infty(R)$,

$$\sum_{n=0}^\infty (u_k u_n, u_n)\ M_k^{-1}\Psi(s_n) = \sum_{\{\gamma\}} \left[\int_{\gamma o} u_k\right] \Psi(v_\gamma)$$

where: $v_\gamma = 2(shL_\gamma/2)^2$, and

$$M_k^{-1}\Psi(s) = \int_{Rez=\sigma} M\Psi(s)\{\Gamma_k^+(z,s)+\Gamma_k^-(z,s)\} \frac{dz}{i}\ (1 > \sigma > 1/2 + t_1)$$

where:

$$\sum_\pm \Gamma^\pm(z,s) = 2^{z-3/2} \frac{\Gamma(z)\Gamma(z-1/2)\Gamma(-2z+2)}{\Gamma(z+s_k/4-1/4)\Gamma(z-s_k/4-1/4)} \left[\sum_\pm \frac{\Gamma(\pm s+z-1/2)}{\Gamma(\pm s-z+3/2)}\right]$$

and M is the Mellin transform.

We now have:

THEOREM (Ai). $N(\lambda, u_k) \ll \lambda/\log\lambda$

PROOF: Fix $\Psi \in S(\mathbb{R})$, $\int \Psi = 1$, Ψ even, supp $\Psi \subset (-1,1)$. Let $\Psi_\epsilon(x) = 1/\epsilon\ \Psi(x/\epsilon)$.

Also, set $h_T(r) = \chi_T((1+r^2)^{1/2})$ where χ_λ is the characteristic function of $[0,\lambda]$. Set $h_{\lambda,\epsilon} = h_\lambda * \varphi_\epsilon$. Then set $\epsilon = 1/\log\lambda$.

$$(1) \qquad \sum_j (u_k u_j, u_j) h_\lambda(r_j) = \sum_j (u_k u_j, u_j) h_{\lambda,\epsilon}(r_j) + O(\lambda/\log\lambda).$$

The argument is identical to that in ([Sa],[E]). Idea: break up $\sum_{r_j} (u_k u_j, u_j)[h_\lambda(r_j) - h_{\lambda,\epsilon}(r_j)]$ into the sums $\sum_{r_j \le \lambda-\epsilon} + \sum_{r_j \in [\lambda-\epsilon,\lambda+\epsilon]} + \sum_{r_j \ge \lambda+\epsilon}$. Use a decomposition of the terms of the outer sums into intervals of size ϵ, and the estimate $|h_{\lambda,\epsilon}(r) - h_\lambda(r)| \ll \left[\frac{|\lambda-r|}{\epsilon}\right]^{-N}$ for the r occurring in these sums. Then use $N(\lambda+\epsilon) - N(\lambda) \ll \epsilon\lambda$ for all these terms, with $\epsilon = 1/\log\lambda$. The fact that $N(\lambda+1/\log\lambda) - N(\lambda) \ll \lambda/\log\lambda$ follows from (STF).

We can now apply $(TF)_1$:

$$(2) \qquad \sum_j (u_k u_j, u_j) h_{\lambda,\epsilon}(r_j) = \sum_{\{\gamma\}} \left[\int_{\gamma_0} u_k\right] \frac{1}{\operatorname{sh} L_\gamma/2} M_k h_{\lambda,\epsilon}(L_\gamma),$$

$$= \int_{\ell_0}^{1/\epsilon} (\operatorname{sh}\ell/2)^{-1} M_k h_{\lambda,\epsilon}(\ell)\, d\varphi_\Gamma(\ell, u_k)$$

where

$$M_k h_{\lambda,\epsilon}(\ell) = \hat{h}_\lambda(\ell)\hat{\varphi}(\epsilon\ell) + \int_\ell^\infty \hat{h}_\lambda(t)\hat{\varphi}(\epsilon,t) \frac{\partial}{\partial t} F_k(\ell,t)dt$$

(in an obvious notation) and φ_Γ is from §0.

We can use routine arguments to get:

$$\int_{\ell_0}^{1/\epsilon} (\operatorname{sh}\ell/2)^{-1} \hat{h}_\lambda(\ell)\, \hat{\varphi}(\epsilon\ell)\, d\varphi_\Gamma(\ell, u_k)$$

$$\ll \int_{\ell_0}^{1/\epsilon} e^{-\ell/2}\, 1/\ell\, d\varphi_\Gamma(\ell)$$

$$\ll \epsilon e^{1/2\epsilon} = \lambda^{1/2} / \log\lambda$$

where we use $(|d\Psi_\Gamma|)(\ell,u_k) \ll d\varphi_\Gamma(\ell) \sim e^\ell$. For the second term, we integrate by parts to get

$$\int_{\ell_0}^{1/\epsilon} \varphi_\Gamma(\ell, u_k) \frac{\partial}{\partial\ell}\left[(\operatorname{sh}\ell/2)^{-1} \int_\ell^\infty \hat{h}_\lambda(t)\, \hat{\varphi}(\epsilon t)\, \frac{\partial}{\partial t} F_k(\ell,t))dt\right] d\ell.$$

The only term not handled as above is the one involving $\frac{\partial^2}{\partial t \partial t} F_k(\ell,t)$. However, using:

$$\frac{d}{dz} F(a,b,c,z) = (\text{const})\, F(a+1,\ b+1,\ c+1,\ z)$$

$$F(a,b,c,z) = A(-z)^{-a}F(a,1+a-c,1+a-b,z^{-1}) + B(-z)^{-b}F(b,1+b-c,1+b-a,z^{-1}),$$

we see that $\frac{\partial^2}{\partial t\partial t} F(\ell,t) << 1$ for $t > \ell$.

So

$$\int_{\ell}^{\infty} \hat{h}_{\lambda}(t)\, \hat{\Psi}(\epsilon t)\, \frac{\partial^2}{\partial \ell \partial t} F_k(\ell,t)\, dt$$

$$<< \int_{\ell}^{\infty} 1/t\, \hat{\Psi}(\epsilon t)\, dt \quad << \quad \log \epsilon .$$

So this term behaves like $\int_{\ell_o}^{1/\epsilon} e^{-\ell/2}\ |\Psi_{\Gamma}(\ell,u_k)|\,d\ell << \log\epsilon \quad e^{1/2\epsilon} << (\log\log T)T^{1/2}$.

This concludes the proof. Q.E.D.

For the proof of Theorem A(ii) we will introduce the analogue of $\frac{1}{2s-1}\,\frac{Z'(s)}{Z(s)}$ where $Z(s)$ is the Selberg zeta function. Although not indispensible for the proof, it may be of independent interest, especially in view of the major error committed in making the approximations in (i).

To this end, let $h(r) = \frac{1}{r^2+(s-1/2)^2} - \frac{1}{r^2+(\beta-1/2)^2}$. The geometric side becomes $L(u_k,s) - L(u_k,\beta)$ where for Res > 1,

$$L(u_k,s) = c\sum_{\{\gamma\}} \left[\int_{\gamma o} u_k\right] \frac{1}{shL_{\gamma}/2} \int_{L_{\gamma}}^{\infty} e^{-(s-1/2)t} F_k(t)\, dt$$

and where β is a parameter, Re $\beta > 1$, introduced to ensure convergence on the spectral side ([Hej], §1). (TF) gives the relation

$$L(u_k,s) - L(u_k,\beta) = \sum (u_k u_n, u_n) \left[\frac{1}{r_k^2+(s-1/2)^2} - \frac{1}{r_k^2+(\beta-1/2)^2}\right]$$

from which it is evident that $L(u_k,1-s) = L(u_k,s)$ and that $L(u_k,s)$ has a meromorphic continuation to $\mathbb{C}$. Unlike $Z'(s)/Z(s)$, $L(u_k,s)$ is not a logarithmic derivative.

It is clear that

$$\sum_{r_n \text{real},\leq T} (u_k u_n, u_n)\, X^{ir_n} = \int_{R[T,\epsilon,A]} L(u_k,s)(2s-1)X^s\, ds$$

where $R_{[T,\epsilon,A]}$ is the rectangle with top along Im s = T, bottom along Im s = $\epsilon > 0$ and sides along Res = 2, Res = -A. The top side is assumed not to run across a pole. ϵ is only significant in that the contour avoids the exceptional eigenvalues.

Following [T, 3.9] or [H, §11], the various sides can be estimated as follows:

(i) $I_1(T,\epsilon,A) = \int_{-A+i\epsilon}^{-A+iT} L(u_k,s)(2s-1)X^s ds \to 0$ as $A \to \infty$ (use the functional equation to move the contour to Res > 1, and the result is obvious from the geometric formula).

This leaves

(ii)

$$I_2(\epsilon,\infty) = \int_{-\infty+i\epsilon}^{2+i\epsilon} +$$

$$+ I_3(T) = \int_{-\infty+iT}^{2+iT} +$$

$$+ I_4(\epsilon,T) = \int_{2+i\epsilon}^{2+iT}$$

$$= I_3(T) + I_4(\epsilon,T) + O(1)$$

(iii) $I_3(T) = \int_{-\infty+iT}^{-1+iT} + \int_{-1+iT}^{2+iT}$

It is routine that $\int_{-\infty+iT}^{-1+iT} (2s-1)X^s\, L(u_k,s)\, ds << \frac{X^{-1}}{\ln X} = O(1)$. So the significant terms are $I_4(\epsilon,T)$ and $I_5(T) = \int_{-1+iT}^{2+iT}$. $I_5(T)$ is handled in essentially the same way as in ([Hej], §11) or in [T, 3.9]. The starting point is that

$$L(u_k,s) = \frac{1}{2s-1} \sum_{|r_n-T|\le 1/\ln T} \frac{(u_k u_n, u_n)}{s-1/2-ir_n} + O(1)$$

for $s = \sigma+iT$, $-1 \le \sigma \le 2$, as $T \to \infty$. This is easily proved using (Ai). It then follows that $I_5(\epsilon,T) <<_X T/\ln T$. The argument is identical to the case $u_k \equiv 1$ and will be omitted.

The term $I_4(\epsilon,T) = \int_{2+i\epsilon}^{2+iT}$ remains. Here the trace formula gives: $I_4(\epsilon,T) = \sum_{\{\gamma\}} \frac{\int_{\gamma o} u_k}{shL_\gamma/2}\{I + II\}$ where $I = \int_{2+i\epsilon}^{2+iT} e^{-(s-1/2)L_\gamma} X^s\, ds$ and $II = \int_{L_\gamma}^{\infty} \left\{\int_{2+i\epsilon}^{2+iT} e^{-(s-1/2)t} X^s\, ds\right\} \frac{\partial}{\partial t} F_k(t)\, dt$. When $X = e^{L_\gamma}$, $I = e^{1/2\, L_\gamma} T + O(1)$, else $I = O_X(1)$. On the other hand $II = O_X(1)$. These estimates can be summed, and we get

THEOREM (Aii). $\sum_{\substack{r_j \text{ real} \\ r_j \le T}} (u_k u_j, u_j) X^{ir_j} = T \frac{e^{1/2\, L_\gamma}}{shL_\gamma/2} \int_{\gamma o} u_k + O_X(T/\ln T)$ if $X = e^{L_\gamma}$ or $= O_X(T/\ln T)$ if $X \ne e^{L_\gamma}$. Q.E.D.

REMARK: We expect $\sum_{\sqrt{\lambda_j}\le T} (Au_j, u_j) e^{it\sqrt{\lambda_j}} = T \frac{e^{1/2\, L_\gamma}}{|\det(I-P_\gamma)|^{1/2}} \int_{\gamma o} \sigma_A + o_t(T)$ if $t = L_\gamma$, respectively $o(T)$ if $t \ne L_\gamma$, only under the assumption that the closed geodesics form a set of measure 0 (P_γ = Poincare map). The tool for this is the Duistermaat-Guillemin trace formula (only the leading order term at L_γ should contribute; in progress).

For the Theorem A(iii) we refer to [Z1] of [CdeV]. Unlike A(i) and A(ii), A(iii) ues some microlocal machinery: Egorov theorem, Garding inequality, Friedrichs symmetrization, as well as various ergodic theorems for G^t. As yet, we do not know a proof proceeding from the trace formula by means of classical analysis.

We now turn to Theorem B.

THEOREM B. $\sum_{L_\gamma \le T} \int_{\gamma o} u_k = \sum_{j=1}^{M} \gamma_{k,j}(u_k u_j, u_j) e^{T(1/2+1/2t_j)} + O(Te^{3/4T})$.

PROOF (SKETCH): From $(TF)_2$ we have:

$$\sum_{r_j} (u_k u_j, u_j) M_k^{-1}\varphi(s_j) = \sum_{\{\gamma\}} (\int_{\gamma o} u_k)\varphi(v_\gamma),$$

$M_k^{-1}\varphi$ given above.

Let $\chi_T(v)$ be the characteristic function of $[1, v_T]$, $v_T = 1/2(\text{sh}T/2)^2$. Let φ_T be the odd extension of χ_T to $[0,\infty]$ in the multiplicative sense, so $M\varphi_T(0) = 0$. Let $\varphi \in C_0^\infty(e^{-1}, e^1)$, $\varphi_\epsilon(x) = \epsilon^{-1}\varphi(x^{1/\epsilon})$, and let $\varphi_{T,\epsilon} = \varphi_T * \varphi_\epsilon$. Then $\varphi_{T,\epsilon} \in C_0^\infty(R^+)$, and $M\varphi_{T,\epsilon} = M\varphi_T(s)M\varphi(\epsilon s)$ is entire. Plugging into the trace formula, we get

$$\sum_{L_\gamma \le T} \int_{\gamma o} u_k = \sum_{\{\gamma\}} \left[\int_{\gamma o} u_k\right]\varphi_{T.\epsilon}(v_\gamma) + O(\epsilon e^T)$$

$$= \sum_{r_j \text{imag}} (u_k u_j, u_j) \int_{\text{Re}z=\sigma} M\varphi_T(z) M\phi(\epsilon z)\{\Gamma_k^+(z, s_j) + \Gamma_k^-(z, s_j)\}\, dz$$

$$+ \sum_{r_j \text{real}} (u_k u_j, u_j)[\text{Same}]$$

For $\sum_{r_j \text{imag}}$, we shift the line Rez = σ leftwards, first encountering poles at $z = 1/2 + t_j$. We pick up residues $\sum_{r_j} (u_k u_j, u_j)\gamma_{jk} e^{(1/2+1/2t_j)T} + O(1)$, with $\gamma_{jk} = 2^{1/2t_j - 1} \frac{\Gamma(1/2+1/2t_j)\Gamma(1/2t_j)\Gamma(-t_j+1)}{\Gamma(1/4+1/2t_j+s_k/4)\Gamma(1/4+1/2t_j-s_k/4)}$. The line can then be shifted further left. The residues encountered on the way to Rez = $-\delta$, $1 >> \delta > 0$ give lower exponents of T. Since $M\varphi_T(0) = 0$ we pick up no residue at $z = 0$, and once Rez < 0, the integrals are absolutely convergent after letting $\epsilon \to 0$. Consequently the exceptional sum over r_j imaginary produces the desired asymtote.

The next step is to check that the infinite sum over r_j is $O(Te^{3/4T})$. Shifting left, we first pick up poles on Rez = 1/2. The apparent pole at $z = 1/2$ is spurious, so we only pick poles at $z = 1/2 + ir_j$.

Using Stirling's formula and Theorem (Ai), the sum over r_j of the residues can be bounded by $\epsilon^{-1}\, Te^{1/2T}$.

We then shift left further. No pole is picked up at $z = 0$, while on Rez = $-\delta$ the infinite sum becomes absolutely convergent and bounded by $e^{T/2+2\delta T}$.

It follows that

$$\sum_{L_\gamma \leq T} \int_{\gamma_0} u_k = \sum_{j=1}^{M} (u_k u_j, u_j)\gamma_{jk} e^{(1/2+1/2t_j)T} + O(\epsilon\, e^T) + O(\epsilon^{-1} T e^{T/2}).$$

The error is optimized by setting $\epsilon = e^{-T/4}$. Q.E.D.

REFERENCES

[Bé] P. Bérard, On the wave equation on a compact Riemannian manifold without conjugate points, Math. Z. 155 (1977), 249-276.

[B] R. Bowen, The equidistribution theory of closed geodesics, Am. J. Math. 94 (1972), 413-423.

[CdeV] Y. Colin de Verdière, Ergodicité et fonctions properes du Laplacien, Seminaire Bony-Sjostrand-Meyer, 1984-85, Expose No. XI (1985).

[D-G] H. Duistermant and V. Guillemin, The spectrum of positive elliptic operators and periodic bicharacteristics, Invent. Math. 29 (1975), 39-79.

[E] I. Efrat, Cusp forms in higher rank, to appear.

[He] D. Hejhal, "The Selberg Trace Formula for $PSL_2(\mathbb{R})$," Vols. 1 and 2, Lecture Notes in Math, No. 548 and 1001, Springer-Verlag, 1976 and 1983.

[H] S. Helgason, "Topics in Harmonic Analysis on Homogeneous Spaces," Birkhauser, Boston, 1979.

[I] H. Iwaniec, Non-holomorphic modular forms and their applications (Preprint)

[L] S. Lang, "$SL_2(\mathbb{R})$," Addison-Wesley, 1975.

[P-S] R. Phillips and P. Sarnak, Geodesics in homology classes (Preprint, 1986).

[S] P. Sarnak, Thesis (Stanford, 1980).

[Sn] A.I. Snirelman, Ergodic Properties of Eigenfunctions, Usp. Math. Nauk. 29 (1974), 181-182.

[T] E.C. Titchmarsh, "The zeta function of Riemann," Hafner Publishing Co., NY (1972).

[Z1] S. Zelditch, Uniform distribution of eigenfunctions on compact hyperbolic surfaces, to appear in Duke Math. J.

[Z2] S. Zelditch, Selberg Trace Formulae, Pseudodifferential operators and the geodesic integrals of automorphic forms (Preprint, 1985).

[Z3] S. Zelditch, Trace Formulae for compact $\Gamma\backslash PSL_2(\mathbb{R})$ and the equidistribution theory of closed geodesics (Preprint, 1986).

[Z4] S. Zelditch, Trace Formulae for finite area $\Gamma\backslash PSL_2(\mathbb{R})$ and equidistribution theorems for closed geodesics and Laplace eigenfunctions (in preparation).

[Z5] S. Zelditch, On some oscillatory spectral sums (in preparation).

[Z6] S. Zelditch, Pseudo-differential analysis on hyperbolic surfaces, to appear in J. Fun. Anal.

LECTURE NOTES IN MATHEMATICS
Edited by A. Dold and B. Eckmann

Some general remarks on the publication of proceedings of congresses and symposia

Lecture Notes aim to report new developments - quickly, informally and at a high level. The following describes criteria and procedures which apply to proceedings volumes.

1. One (or more) expert participant(s) of the meeting should act as the responsible editor(s) of the proceedings. They select the papers which are suitable (cf. points 2, 3) for inclusion in the proceedings, and have them individually refereed (as for a journal). It should not be assumed that the published proceedings must reflect conference events faithfully and in their entirety. Contributions to the meeting which are not included in the proceedings can be listed by title. The series editors will normally not interfere with the editing of a particular proceedings volume - except in fairly obvious cases, or on technical matters, such as described in points 2, 3. The names of the responsible editors appear on the title page of the volume.

2. The proceedings should be reasonably homogeneous (concerned with a limited area). For instance, the proceedings of a congress on "Analysis" or "Mathematics in Wonderland" would normally not be sufficiently homogeneous.

   One or two longer survey articles on recent developments in the field are often very useful additions to such proceedings - even if they do not correspond to actual lectures at the congress. An extensive introduction on the subject of the congress would be desirable.

3. The contributions should be of a high mathematical standard and of current interest. Research articles should present new material and not duplicate other papers already published or due to be published. They should contain sufficient information and motivation and they should present proofs, or at least outlines of such, in sufficient detail to enable an expert to complete them. Thus resumes and mere announcements of papers appearing elsewhere cannot be included, although more detailed versions of a contribution may well be published in other places later.

   Surveys, if included, should cover a sufficiently broad topic, and should in general not simply review the author's own recent research. In the case of surveys, exceptionally, proofs of results may not be necessary.

   The editors of a volume are strongly advised to inform contributors about these points at an early stage.

.../...

4. Proceedings should appear soon after the meeeting. The publisher should, therefore, receive the complete manuscript within nine months of the date of the meeting at the latest.

5. Plans or proposals for proceedings volumes should be sent to one of the editors of the series or to Springer-Verlag Heidelberg. They should give sufficient information on the conference or symposium, and on the proposed proceedings. In particular, they should contain a list of the expected contributions with their prospective length. Abstracts or early versions (drafts) of some of the contributions are very helpful.

6. Lecture Notes are printed by photo-offset from camera-ready typed copy provided by the editors. For this purpose Springer-Verlag provides editors with technical instructions for the preparation of manuscripts and these should be distributed to all contributing authors. Springer-Verlag can also, on request, supply stationery on which the prescribed typing area is outlined. Some homogeneity in the presentation of the contributions is desirable.

   Careful preparation of manuscripts will help keep production time short and ensure a satisfactory appearance of the finished book. The actual production of a Lecture Notes volume normally takes 6 -8 weeks.

   Manuscripts should be at least 100 pages long. The final version should include a table of contents.

7. Editors receive a total of 50 free copies of their volume for distribution to the contributing authors, but no royalties. (Unfortunately, no reprints of individual contributions can be supplied.) They are entitled to purchase further copies of their book for their personal use at a discount of 33 1/3%, other Springer mathematics books at a discount of 20% directly from Springer-Verlag.

   Commitment to publish is made by letter of intent rather than by signing a formal contract. Springer-Verlag secures the copyright for each volume.

LECTURE NOTES

ESSENTIALS FOR THE PREPARATION
OF CAMERA-READY MANUSCRIPTS

Springer
Springer-Verlag
Berlin Heidelberg New York
London Paris Tokyo

The preparation of manuscripts which are to be reproduced by photo-offset requires special care. Manuscripts which are submitted in technically unsuitable form will be returned to the author for retyping. There is normally no possibility of carrying out further corrections after a manuscript is given to production. Hence it is crucial that the following instructions be adhered to closely. If in doubt, please send us 1 - 2 sample pages for examination.

Typing area. On request, Springer-Verlag will supply special paper with the typing area outlined.

The CORRECT TYPING AREA is 18 x 26 1/2 cm (7,5 x 11 inches).

Make sure the TYPING AREA IS COMPLETELY FILLED. Set the margins so that they precisely match the outline and type right from the top to the bottom line. (Note that the page-number will lie outside this area). Lines of text should not end more than three spaces inside or outside the right margin (see example on page 4).

Type on one side of the paper only.

Type. Use an electric typewriter if at all possible. CLEAN THE TYPE before use and always use a BLACK ribbon (a carbon ribbon is best).

Choose a type size large enough to stand reduction to 75%.

Word Processors. Authors using word-processing or computer-typesetting facilities should follow these instructions with obvious modifications. Please note with respect to your printout that
i) the characters should be sharp and sufficiently black;
ii) if the size of your characters is significantly larger or smaller than normal typescript characters, you should adapt the length and breadth of the text area proportionally keeping the proportions 1:0.68.
iii) it is not necessary to use Springer's special typing paper. Any white paper of reasonable quality is acceptable.
IF IN DOUBT, PLEASE SEND US 1-2 SAMPLE PAGES FOR EXAMINATION. We will be glad to give advice.

Spacing and Headings (Monographs). Use ONE-AND-A-HALF line spacing in the text. Please leave sufficient space for the title to stand out clearly and do NOT use a new page for the beginning of subdivisions of chapters. Leave THREE LINES blank above and TWO below headings of such subdivisions.

Spacing and Headings (Proceedings). Use ONE-AND-A-HALF line spacing in the text. Start each paper on a NEW PAGE and leave sufficient space for the title to stand out clearly. However, do NOT use a new page for the beginning of subdivisions of a paper. Leave THREE LINES blank above and TWO below headings of such subdivisions. Make sure headings of equal importance are in the same form.

The first page of each contribution should be prepared in the same way. Therefore, we recommend that the editor prepares a sample page and passes it on to the authors together with these ESSENTIALS. Please take

.../...

the following as an example.

MATHEMATICAL STRUCTURE IN QUANTUM FIELD THEORY

John E. Robert
Fachbereich Physik, Universität Osnabrück
Postfach 44 69, D-4500 Osnabrück

Please leave THREE LINES blank below heading and address of the author. THEN START THE ACTUAL TEXT OF YOUR CONTRIBUTION.

Footnotes. These should be avoided. If they cannot be avoided, place them at the foot of the page, separated from the text by a line 4 cm long, and type them in SINGLE LINE SPACING to finish exactly on the outline.

Symbols. Anything which cannot be typed may be entered by hand in BLACK AND ONLY BLACK ink. (A fine-tipped rapidograph is suitable for this purpose; a good black ball-point will do, but a pencil will not). Do not draw straight lines by hand without a ruler (not even in fractions).

Equations and Computer Programs. Equations and computer programs should begin four spaces inside the left margin. Should the equations be numbered, then each number should be in brackets at the right-hand edge of the typing area.

Pagination. Number pages in the upper right-hand corner in LIGHT BLUE OR GREEN PENCIL ONLY. The final page numbers will be inserted by the printer.

There should normally be NO BLANK PAGES in the manuscript (between chapters or between contributions) unless the book is divided into Part A, Part B for example, which should then begin on a right-hand page.

It is much safer to number pages AFTER the text has been typed and corrected. Page 1 (Arabic) should be THE FIRST PAGE OF THE ACTUAL TEXT. The Roman pagination (table of contents, preface, abstract, acknowledgements, brief introductions, etc.) will be done by Springer-Verlag.

Corrections. When corrections have to be made, cut the new text to fit and PASTE it over the old. White correction fluid may also be used.

Never make corrections or insertions in the text by hand.

If the typescript has to be marked for any reason, e.g. for TEMPORARY page numbers or to mark corrections for the typist, this can be done VERY FAINTLY with BLUE or GREEN PENCIL but NO OTHER COLOR: these colors do not appear after reproduction.

Table of Contents. It is advisable to type the table of contents later, copying the titles from the text and inserting page numbers.

Literature References. These should be placed at the end of each paper or chapter, or at the end of the work, as desired. Type them with single line spacing and start each reference on a new line.
Please ensure that all references are COMPLETE and PRECISE.

Vol. 1090: Differential Geometry of Submanifolds. Proceedings, 1984. Edited by K. Kenmotsu. VI, 132 pages. 1984.

Vol. 1091: Multifunctions and Integrands. Proceedings, 1983. Edited by G. Salinetti. V, 234 pages. 1984.

Vol. 1092: Complete Intersections. Seminar, 1983. Edited by S. Greco and R. Strano. VII, 299 pages. 1984.

Vol. 1093: A. Prestel, Lectures on Formally Real Fields. XI, 125 pages. 1984.

Vol. 1094: Analyse Complexe. Proceedings, 1983. Edité par E. Amar, R. Gay et Nguyen Thanh Van. IX, 184 pages. 1984.

Vol. 1095: Stochastic Analysis and Applications. Proceedings, 1983. Edited by A. Truman and D. Williams. V, 199 pages. 1984.

Vol. 1096: Théorie du Potentiel. Proceedings, 1983. Edité par G. Mokobodzki et D. Pinchon. IX, 601 pages. 1984.

Vol. 1097: R.M. Dudley, H. Kunita, F. Ledrappier, École d'Éte de Probabilités de Saint-Flour XII – 1982. Edité par P.L. Hennequin. X, 396 pages. 1984.

Vol. 1098: Groups – Korea 1983. Proceedings. Edited by A.C. Kim and B.H. Neumann. VII, 183 pages. 1984.

Vol. 1099: C.M. Ringel, Tame Algebras and Integral Quadratic Forms. XIII, 376 pages. 1984.

Vol. 1100: V. Ivrii, Precise Spectral Asymptotics for Elliptic Operators Acting in Fiberings over Manifolds with Boundary. V, 237 pages. 1984.

Vol. 1101: V. Cossart, J. Giraud, U. Orbanz, Resolution of Surface Singularities. Seminar. VII, 132 pages. 1984.

Vol. 1102: A. Verona, Stratified Mappings – Structure and Triangulability. IX, 160 pages. 1984.

Vol. 1103: Models and Sets. Proceedings, Logic Colloquium, 1983, Part I. Edited by G.H. Müller and M.M. Richter. VIII, 484 pages. 1984.

Vol. 1104: Computation and Proof Theory. Proceedings, Logic Colloquium, 1983, Part II. Edited by M.M. Richter, E. Börger, W. Oberschelp, B. Schinzel and W. Thomas. VIII, 475 pages. 1984.

Vol. 1105: Rational Approximation and Interpolation. Proceedings, 1983. Edited by P.R. Graves-Morris, E.B. Saff and R.S. Varga. XII, 528 pages. 1984.

Vol. 1106: C.T. Chong, Techniques of Admissible Recursion Theory. X, 214 pages. 1984.

Vol. 1107: Nonlinear Analysis and Optimization. Proceedings, 1982. Edited by C. Vinti. V, 224 pages. 1984.

Vol. 1108: Global Analysis – Studies and Applications I. Edited by Yu.G. Borisovich and Yu.E. Gliklikh. V, 301 pages. 1984.

Vol. 1109: Stochastic Aspects of Classical and Quantum Systems. Proceedings, 1983. Edited by S. Albeverio, P. Combe and M. Sirugue-Collin. IX, 227 pages. 1985.

Vol. 1110: R. Jajte, Strong Limit Theorems in Non-Commutative Probability. VI, 152 pages. 1985.

Vol. 1111: Arbeitstagung Bonn 1984. Proceedings. Edited by F. Hirzebruch, J. Schwermer and S. Suter. V, 481 pages. 1985.

Vol. 1112: Products of Conjugacy Classes in Groups. Edited by Z. Arad and M. Herzog. V, 244 pages. 1985.

Vol. 1113: P. Antosik, C. Swartz, Matrix Methods in Analysis. IV, 114 pages. 1985.

Vol. 1114: Zahlentheoretische Analysis. Seminar. Herausgegeben von E. Hlawka. V, 157 Seiten. 1985.

Vol. 1115: J. Moulin Ollagnier, Ergodic Theory and Statistical Mechanics. VI, 147 pages. 1985.

Vol. 1116: S. Stolz, Hochzusammenhängende Mannigfaltigkeiten und ihre Ränder. XXIII, 134 Seiten. 1985.

Vol. 1117: D.J. Aldous, J.A. Ibragimov, J. Jacod, Ecole d'Été de Probabilités de Saint-Flour XIII – 1983. Édité par P.L. Hennequin. IX, 409 pages. 1985.

Vol. 1118: Grossissements de filtrations: exemples et applications. Seminaire, 1982/83. Edité par Th. Jeulin et M. Yor. V, 315 pages. 1985.

Vol. 1119: Recent Mathematical Methods in Dynamic Programming. Proceedings, 1984. Edited by I. Capuzzo Dolcetta, W.H. Fleming and T. Zolezzi. VI, 202 pages. 1985.

Vol. 1120: K. Jarosz, Perturbations of Banach Algebras. V, 118 pages. 1985.

Vol. 1121: Singularities and Constructive Methods for Their Treatment. Proceedings, 1983. Edited by P. Grisvard, W. Wendland and J.R. Whiteman. IX, 346 pages. 1985.

Vol. 1122: Number Theory. Proceedings, 1984. Edited by K. Alladi. VII, 217 pages. 1985.

Vol. 1123: Séminaire de Probabilités XIX 1983/84. Proceedings. Edité par J. Azéma et M. Yor. IV, 504 pages. 1985.

Vol. 1124: Algebraic Geometry, Sitges (Barcelona) 1983. Proceedings. Edited by E. Casas-Alvero, G.E. Welters and S. Xambó-Descamps. XI, 416 pages. 1985.

Vol. 1125: Dynamical Systems and Bifurcations. Proceedings, 1984. Edited by B.L.J. Braaksma, H.W. Broer and F. Takens. V, 129 pages. 1985.

Vol. 1126: Algebraic and Geometric Topology. Proceedings, 1983. Edited by A. Ranicki, N. Levitt and F. Quinn. V, 423 pages. 1985.

Vol. 1127: Numerical Methods in Fluid Dynamics. Seminar. Edited by F. Brezzi, VII, 333 pages. 1985.

Vol. 1128: J. Elschner, Singular Ordinary Differential Operators and Pseudodifferential Equations. 200 pages. 1985.

Vol. 1129: Numerical Analysis, Lancaster 1984. Proceedings. Edited by P.R. Turner. XIV, 179 pages. 1985.

Vol. 1130: Methods in Mathematical Logic. Proceedings, 1983. Edited by C.A. Di Prisco. VII, 407 pages. 1985.

Vol. 1131: K. Sundaresan, S. Swaminathan, Geometry and Nonlinear Analysis in Banach Spaces. III, 116 pages. 1985.

Vol. 1132: Operator Algebras and their Connections with Topology and Ergodic Theory. Proceedings, 1983. Edited by H. Araki, C.C. Moore, Ş. Strătilă and C. Voiculescu. VI, 594 pages. 1985.

Vol. 1133: K.C. Kiwiel, Methods of Descent for Nondifferentiable Optimization. VI, 362 pages. 1985.

Vol. 1134: G.P. Galdi, S. Rionero, Weighted Energy Methods in Fluid Dynamics and Elasticity. VII, 126 pages. 1985.

Vol. 1135: Number Theory, New York 1983–84. Seminar. Edited by D.V. Chudnovsky, G.V. Chudnovsky, H. Cohn and M.B. Nathanson. V, 283 pages. 1985.

Vol. 1136: Quantum Probability and Applications II. Proceedings, 1984. Edited by L. Accardi and W. von Waldenfels. VI, 534 pages. 1985.

Vol. 1137: Xiao G., Surfaces fibrées en courbes de genre deux. IX, 103 pages. 1985.

Vol. 1138: A. Ocneanu, Actions of Discrete Amenable Groups on von Neumann Algebras. V, 115 pages. 1985.

Vol. 1139: Differential Geometric Methods in Mathematical Physics. Proceedings, 1983. Edited by H. D. Doebner and J. D. Hennig. VI, 337 pages. 1985.

Vol. 1140: S. Donkin, Rational Representations of Algebraic Groups. VII, 254 pages. 1985.

Vol. 1141: Recursion Theory Week. Proceedings, 1984. Edited by H.-D. Ebbinghaus, G.H. Müller and G.E. Sacks. IX, 418 pages. 1985.

Vol. 1142: Orders and their Applications. Proceedings, 1984. Edited by I. Reiner and K. W. Roggenkamp. X, 306 pages. 1985.

Vol. 1143: A. Krieg, Modular Forms on Half-Spaces of Quaternions. XIII, 203 pages. 1985.

Vol. 1144: Knot Theory and Manifolds. Proceedings, 1983. Edited by D. Rolfsen. V, 163 pages. 1985.

Vol. 1145: G. Winkler, Choquet Order and Simplices. VI, 143 pages. 1985.

Vol. 1146: Séminaire d'Algèbre Paul Dubreil et Marie-Paule Malliavin. Proceedings, 1983–1984. Edité par M.-P. Malliavin. IV, 420 pages. 1985.

Vol. 1147: M. Wschebor, Surfaces Aléatoires. VII, 111 pages. 1985.

Vol. 1148: Mark A. Kon, Probability Distributions in Quantum Statistical Mechanics. V, 121 pages. 1985.

Vol. 1149: Universal Algebra and Lattice Theory. Proceedings, 1984. Edited by S. D. Comer. VI, 282 pages. 1985.

Vol. 1150: B. Kawohl, Rearrangements and Convexity of Level Sets in PDE. V, 136 pages. 1985.

Vol 1151: Ordinary and Partial Differential Equations. Proceedings, 1984. Edited by B.D. Sleeman and R.J. Jarvis. XIV, 357 pages. 1985.

Vol. 1152: H. Widom, Asymptotic Expansions for Pseudodifferential Operators on Bounded Domains. V, 150 pages. 1985.

Vol. 1153: Probability in Banach Spaces V. Proceedings, 1984. Edited by A. Beck, R. Dudley, M. Hahn, J. Kuelbs and M. Marcus. VI, 457 pages. 1985.

Vol. 1154: D.S. Naidu, A.K. Rao, Singular Pertubation Analysis of Discrete Control Systems. IX, 195 pages. 1985.

Vol. 1155: Stability Problems for Stochastic Models. Proceedings, 1984. Edited by V.V. Kalashnikov and V.M. Zolotarev. VI, 447 pages. 1985.

Vol. 1156: Global Differential Geometry and Global Analysis 1984. Proceedings, 1984. Edited by D. Ferus, R.B. Gardner, S. Helgason and U. Simon. V, 339 pages. 1985.

Vol. 1157: H. Levine, Classifying Immersions into $\mathbb{R}^4$ over Stable Maps of 3-Manifolds into $\mathbb{R}^2$. V, 163 pages. 1985.

Vol. 1158: Stochastic Processes – Mathematics and Physics. Proceedings, 1984. Edited by S. Albeverio, Ph. Blanchard and L. Streit. VI, 230 pages. 1986.

Vol. 1159: Schrödinger Operators, Como 1984. Seminar. Edited by S. Graffi. VIII, 272 pages. 1986.

Vol. 1160: J.-C. van der Meer, The Hamiltonian Hopf Bifurcation. VI, 115 pages. 1985.

Vol. 1161: Harmonic Mappings and Minimal Immersions, Montecatini 1984. Seminar. Edited by E. Giusti. VII, 285 pages. 1985.

Vol. 1162: S.J.L. van Eijndhoven, J. de Graaf, Trajectory Spaces, Generalized Functions and Unbounded Operators. IV, 272 pages. 1985.

Vol. 1163: Iteration Theory and its Functional Equations. Proceedings, 1984. Edited by R. Liedl, L. Reich and Gy. Targonski. VIII, 231 pages. 1985.

Vol. 1164: M. Meschiari, J.H. Rawnsley, S. Salamon, Geometry Seminar "Luigi Bianchi" II – 1984. Edited by E. Vesentini. VI, 224 pages. 1985.

Vol. 1165: Seminar on Deformations. Proceedings, 1982/84. Edited by J. Ławrynowicz. IX, 331 pages. 1985.

Vol. 1166: Banach Spaces. Proceedings, 1984. Edited by N. Kalton and E. Saab. VI, 199 pages. 1985.

Vol. 1167: Geometry and Topology. Proceedings, 1983–84. Edited by J. Alexander and J. Harer. VI, 292 pages. 1985.

Vol. 1168: S.S. Agaian, Hadamard Matrices and their Applications. III, 227 pages. 1985.

Vol. 1169: W.A. Light, E.W. Cheney, Approximation Theory in Tensor Product Spaces. VII, 157 pages. 1985.

Vol. 1170: B.S. Thomson, Real Functions. VII, 229 pages. 1985.

Vol. 1171: Polynômes Orthogonaux et Applications. Proceedings, 1984. Edité par C. Brezinski, A. Draux, A.P. Magnus, P. Maroni et A. Ronveaux. XXXVII, 584 pages. 1985.

Vol. 1172: Algebraic Topology, Göttingen 1984. Proceedings. Edited by L. Smith. VI, 209 pages. 1985.

Vol. 1173: H. Delfs, M. Knebusch, Locally Semialgebraic Spaces. XVI, 329 pages. 1985.

Vol. 1174: Categories in Continuum Physics, Buffalo 1982. Seminar. Edited by F.W. Lawvere and S.H. Schanuel. V, 126 pages. 1986.

Vol. 1175: K. Mathiak, Valuations of Skew Fields and Projective Hjelmslev Spaces. VII, 116pages. 1986.

Vol. 1176: R.R. Bruner, J.P. May, J.E. McClure, M. Steinberger, H_∞ Ring Spectra and their Applications. VII, 388 pages. 1986.

Vol. 1177: Representation Theory I. Finite Dimensional Algebras. Proceedings, 1984. Edited by V. Dlab, P. Gabriel and G. Michler. XV, 340 pages. 1986.

Vol. 1178: Representation Theory II. Groups and Orders. Proceedings, 1984. Edited by V. Dlab, P. Gabriel and G. Michler. XV, 370 pages. 1986.

Vol. 1179: Shi J.-Y. The Kazhdan-Lusztig Cells in Certain Affine Weyl Groups. X, 307 pages. 1986.

Vol. 1180: R. Carmona, H. Kesten, J.B. Walsh, École d'Été de Probabilités de Saint-Flour XIV – 1984. Édité par P.L. Hennequin. X, 438 pages. 1986.

Vol. 1181: Buildings and the Geometry of Diagrams, Como 1984. Seminar. Edited by L. Rosati. VII, 277 pages. 1986.

Vol. 1182: S. Shelah, Around Classification Theory of Models. VII, 279 pages. 1986.

Vol. 1183: Algebra, Algebraic Topology and their Interactions. Proceedings, 1983. Edited by J.-E. Roos. XI, 396 pages. 1986.

Vol. 1184: W. Arendt, A. Grabosch, G. Greiner, U. Groh, H.P. Lotz, U. Moustakas, R. Nagel, F. Neubrander, U. Schlotterbeck, One-parameter Semigroups of Positive Operators. Edited by R. Nagel. X, 460 pages. 1986.

Vol. 1185: Group Theory, Beijing 1984. Proceedings. Edited by Tuan H.F. V, 403 pages. 1986.

Vol. 1186: Lyapunov Exponents. Proceedings, 1984. Edited by L. Arnold and V. Wihstutz. VI, 374 pages. 1986.

Vol. 1187: Y. Diers, Categories of Boolean Sheaves of Simple Algebras. VI, 168 pages. 1986.

Vol. 1188: Fonctions de Plusieurs Variables Complexes V. Séminaire, 1979–85. Edité par François Norguet. VI, 306 pages. 1986.

Vol. 1189: J. Lukeš, J. Malý, L. Zajíček, Fine Topology Methods in Real Analysis and Potential Theory. X, 472 pages. 1986.

Vol. 1190: Optimization and Related Fields. Proceedings, 1984. Edited by R. Conti, E. De Giorgi and F. Giannessi. VIII, 419 pages. 1986.

Vol. 1191: A.R. Its, V.Yu. Novokshenov, The Isomonodromic Deformation Method in the Theory of Painlevé Equations. IV, 313 pages. 1986.

Vol. 1192: Equadiff 6. Proceedings, 1985. Edited by J. Vosmansky and M. Zlámal. XXIII, 404 pages. 1986.

Vol. 1193: Geometrical and Statistical Aspects of Probability in Banach Spaces. Proceedings, 1985. Edited by X. Femique, B. Heinkel, M.B. Marcus and P.A. Meyer. IV, 128 pages. 1986.

Vol. 1194: Complex Analysis and Algebraic Geometry. Proceedings, 1985. Edited by H. Grauert. VI, 235 pages. 1986.

Vol.1195: J.M. Barbosa, A.G. Colares, Minimal Surfaces in $\mathbb{R}^3$. X, 124 pages. 1986.

Vol. 1196: E. Casas-Alvero, S. Xambó-Descamps, The Enumerative Theory of Conics after Halphen. IX, 130 pages. 1986.

Vol. 1197: Ring Theory. Proceedings, 1985. Edited by F.M.J. van Oystaeyen. V, 231 pages. 1986.

Vol. 1198: Séminaire d'Analyse, P. Lelong – P. Dolbeault – H. Skoda. Seminar 1983/84. X, 260 pages. 1986.

Vol. 1199: Analytic Theory of Continued Fractions II. Proceedings, 1985. Edited by W.J. Thron. VI, 299 pages. 1986.

Vol. 1200: V.D. Milman, G. Schechtman, Asymptotic Theory of Finite Dimensional Normed Spaces. With an Appendix by M. Gromov. VIII, 156 pages. 1986.